# Engineering Materials and Processes

**Series Editor**
Brian Derby, Materials Science Center, University of Manchester, Manchester, UK

The Engineering Materials and Processes series focuses on all forms of materials and the processes used to synthesise and formulate them as they relate to the various engineering disciplines.The series deals with a diverse range of materials: ceramics; metals (ferrous and non-ferrous); semiconductors; composites, polymers, biomimetics etc. Each monograph in the series is written by a specialist and demonstrates how enhancements in materials and the processes associated with them can improve performance in the field of engineering in which they are used.

*Professor Derby and Springer welcome new book ideas from potential authors. If you are interested in writing for the series please contact: Anthony Doyle (Executive Editor - Engineering, Springer) at anthony.doyle@springer.com.*

More information about this series at https://link.springer.com/bookseries/4604

Shameem Hasan · Veera M. Boddu ·
Dabir S. Viswanath · Tushar K. Ghosh

# Chitin and Chitosan

## Science and Engineering

Shameem Hasan
Advanced Isotope Technologies, LLC
Mobile, AL, USA

Dabir S. Viswanath
Emeritus Professor
Austin, TX, USA

Veera M. Boddu ⓘD
Wide Area and Infrastructure
Decontamination Branch
Center for Environmental Solutions
and Emergency Response
Office of Research and Development
U.S. Environmental Protection Agency
Research Triangle Park, NC, USA

Tushar K. Ghosh
Department of Chemical Engineering
University of Missouri
Columbia, MO, USA

ISSN 1619-0181         ISSN 2365-0761  (electronic)
Engineering Materials and Processes
ISBN 978-3-031-01231-0    ISBN 978-3-031-01229-7  (eBook)
https://doi.org/10.1007/978-3-031-01229-7

© The Editor(s) (if applicable) and The Author(s), under exclusive license to Springer Nature Switzerland AG 2022
This work is subject to copyright. All rights are solely and exclusively licensed by the Publisher, whether the whole or part of the material is concerned, specifically the rights of translation, reprinting, reuse of illustrations, recitation, broadcasting, reproduction on microfilms or in any other physical way, and transmission or information storage and retrieval, electronic adaptation, computer software, or by similar or dissimilar methodology now known or hereafter developed.
The use of general descriptive names, registered names, trademarks, service marks, etc. in this publication does not imply, even in the absence of a specific statement, that such names are exempt from the relevant protective laws and regulations and therefore free for general use.
The publisher, the authors, and the editors are safe to assume that the advice and information in this book are believed to be true and accurate at the date of publication. Neither the publisher nor the authors or the editors give a warranty, expressed or implied, with respect to the material contained herein or for any errors or omissions that may have been made. The publisher remains neutral with regard to jurisdictional claims in published maps and institutional affiliations.

This work is not a product of the United States Government or the U.S. Environmental Protection Agency. The author/editor is not doing this work in any governmental capacity. The views expressed are his/her own and do not necessarily represent those of the United States or U.S. EPA. Any mention of trade names, manufacturers or products does not imply an endorsement by the United States Government or the U.S. Environmental Protection Agency. EPA and its employees do not endorse any commercial products, services, or enterprises.

This Springer imprint is published by the registered company Springer Nature Switzerland AG
The registered company address is: Gewerbestrasse 11, 6330 Cham, Switzerland

# Preface

The current demands of polysaccharides in various industrial applications have increased the interest in chitin and chitosan due to their biocompatibility, biodegradability, and lack of toxicity and allergenicity. Chitin, a structural polysaccharide, can be extracted from marine sources. It is the second most abundant natural raw material, next only to cellulose. Chitosan is a partially deacetylated derivative of chitin. Chitosan is a polysaccharide of considerable interest because of its abundance and unusual combination of properties, which include toughness, biodegradability, and relative inertness and ease of converting into gels and films.

Chitin, chitosan, and their derivatives are used in a plethora of applications from wastewater treatment to prosthetics, and the literature is growing at a very rapid pace. There is an enormous amount of information being generated and published in the open literature on different areas pertaining to synthesis, characterization, and application of chitin, chitosan, and their derivatives. One of the main objectives of this monograph is to provide the reader insights into the chitin formation, molecular structure of chitin, chitin extraction process, and the properties of chitin, chitosan, and their different preparation processes.

This monograph has 12 chapters, and each chapter is devoted to one particular application of chitosan and its derivatives with the exception of Chaps. 1–5. Chapter 1 deals with the natural sources and types of polysaccharides in general. The development, processing, and life cycle analysis of chitin and chitosan materials pose many challenges. Chapter 2 provides information on chitin formation chemistry, extraction of chitin, and chitosan preparation processes. It outlines the relationship between structural and biological properties of chitosan, different chitin and chitosan preparation processes, and their limitations. Chapters 3 and 4 provide a comprehensive source of technical information related to the science and engineering aspects of chitin and chitosan and their characterization. Chapter 5 summarizes the recent progress of applied research related to the synthesis and preparation of chitosan derivatives, including water-soluble chitosan, hydrogel, nanoparticles, nanofibers, and chitosan-coated inorganic particles. The remaining seven chapters are devoted to the application of chitosan and its derivatives in different fields. This book covers published work from 2000 to 2020, mainly in the areas of wastewater treatment, sensors, fuel cells, textiles, food, pharmaceuticals, and biomedical applications. The topics explored in this book may be of interest to academic and industrial personnel involved in research related to food, pharmaceuticals, bioprocessing, biomedical, chemical and environmental engineering-related industries. The information related to the application of chitin and chitosan that are provided in this book can be applied to other closely related fields such as biology, bioprocessing, and bio-engineering.

In closing, it is a pleasure to thank V. K. Palsetia and three other reviewers for reading the manuscript, who in spite of their busy schedules cooperated in getting the manuscript completed in a short time. Their contribution significantly improved the quality of the manuscript. We also wish to express our gratitude to the University of Missouri-Columbia and **Springer Nature Switzerland AG** for their help and cooperation. Special thanks to Mr. Oliver Jackson for his help and coordination with the administrative processes associated with the publication process.

| | |
|---|---|
| Mobile, AL, USA | Shameem Hasan |
| Research Triangle Park, NC, USA | Veera M. Boddu |
| Austin, TX, USA | Dabir S. Viswanath |
| Columbia, MO, USA | Tushar K. Ghosh |

# Contents

| | | | |
|---|---|---|---|
| **1** | **Polysaccharides: Chitin and Chitosan** | | 1 |
| | 1.1 | Introduction | 1 |
| | | 1.1.1 Polysaccharides in General | 1 |
| | | 1.1.2 Storage Polysaccharides | 2 |
| | | 1.1.3 Structural Polysaccharides | 5 |
| | | 1.1.4 Polysaccharides from Marine Source | 8 |
| | | 1.1.5 Mucosubstances | 8 |
| | | 1.1.6 Polysaccharides as a Human Nutrition Source | 9 |
| | | 1.1.7 Chitosan Market | 10 |
| | | 1.1.8 Summary | 10 |
| | References | | 14 |
| **2** | **Preparation of Chitin and Chitosan** | | 17 |
| | 2.1 | Chitin and Chitosan | 17 |
| | 2.2 | Chitosan Preparation Methods | 21 |
| | | 2.2.1 Introduction | 21 |
| | | 2.2.2 Extraction of Chitin | 25 |
| | 2.3 | Preparation of Chitosan | 30 |
| | 2.4 | Chitin and Chitosan Properties | 32 |
| | 2.5 | Summary | 43 |
| | References | | 43 |
| **3** | **Chitosan Characterization** | | 51 |
| | 3.1 | Introduction | 51 |
| | | 3.1.1 Molecular Weight | 52 |
| | | 3.1.2 Degree of Deacetylation (DD) and Dissociation Constant of Chitosan | 55 |
| | | 3.1.3 Swelling Behavior and Solubility | 58 |
| | | 3.1.4 Percentage of Moisture (Loss on Drying) | 59 |
| | | 3.1.5 Percentage of Ash | 60 |
| | 3.2 | Morphology of the Chitosan Using SEM, EDS X-Ray Analysis, and TEM | 61 |

|  |  | 3.2.1 | Scanning Electron Microscopy (SEM) Analysis | 61 |
|---|---|---|---|---|

- 3.2.1 Scanning Electron Microscopy (SEM) Analysis .......  61
- 3.2.2 Energy-Dispersive Spectroscopic (EDS) X-Ray Microanalysis .......  62
- 3.2.3 Transmission Electron Micrograph (TEM) ............  64
- 3.2.4 Thermogravimetric Analysis (TGA) of Chitosan ......  64
- 3.2.5 Determination of Surface Area and Surface Charge of Chitosan .......  66
- 3.2.6 Moisture Sorption onto Chitosan ....................  69
- 3.2.7 Mechanical Properties of Chitin and Chitosan .........  71
- 3.2.8 Diffuse Reflective UV-Vis (DRUV) Spectroscopy .....  71
- 3.2.9 X-Ray Photoelectron Spectroscopy (XPS) Analysis .......  72
- 3.2.10 Electron Spin Resonance (ESR) Spectroscopy ........  73
- 3.3 Summary .......  75
- References .......  76

**4 The Structural Difference Between Chitin and Chitosan** .......... 79
- 4.1 Molecular Structure of Chitin ............................. 79
- 4.2 Chitin and Chitosan: Structural Differences from Spectroscopic Analysis ............................... 84
  - 4.2.1 X-Ray Diffraction (XRD) Pattern of Chitin and Chitosan .......  84
  - 4.2.2 Differential Scanning Calorimetry (DSC) ............ 87
  - 4.2.3 Fourier Transform Infrared (FTIR) Spectroscopy ...... 90
  - 4.2.4 Nuclear Magnetic Resonance (NMR) Spectroscopy .......  93
  - 4.2.5 Raman Spectroscopy ............................. 97
- 4.3 Summary .......  97
- References .......  99

**5 Preparation and Application of Chitosan Derivatives** ............. 103
- 5.1 Physicochemical Crosslinking of Chitosan ................... 103
- 5.2 Radiation-Induced Crosslinking ............................ 107
- 5.3 Water-Soluble Chitosan Derivatives ........................ 107
- 5.4 Overall Synthesis Process for Water-Soluble Chitosan Derivatives .......  109
  - 5.4.1 Alkylation of Chitosan .......................... 109
  - 5.4.2 Sulfated Chitosan ............................... 114
  - 5.4.3 Phosphorylation of Chitosan ...................... 115
  - 5.4.4 Quaternized Chitosan ............................ 116
- 5.5 Preparation of Chitosan-Based Hydrogel .................... 117
  - 5.5.1 Introduction and Background ..................... 117
  - 5.5.2 Preparation of Chitosan Hydrogel ................. 120
  - 5.5.3 Chitosan-Based "Smart" Hydrogels ................ 123
- 5.6 Chitosan Nanoparticles ................................... 125

|   |   |   |   |
|---|---|---|---|
| 5.7 | Chitosan Nanofibers | | 131 |
| 5.8 | Chitosan Coated on to Inorganic Materials for the Preparation of Adsorbent Bead | | 139 |
| 5.9 | Summary | | 145 |
| References | | | 146 |

# 6 Adsorption—Heavy Metals Removal ... 157

| | | | |
|---|---|---|---|
| 6.1 | Introduction | | 157 |
| 6.2 | Adsorption Methods | | 167 |
| | 6.2.1 | Experimental Procedure for Equilibrium Uptake of Metal Ions on Adsorbent from Aqueous Solution | 168 |
| | 6.2.2 | Adsorption Controlling Parameters | 170 |
| | 6.2.3 | Effect of Contact Time and Sorbate Concentration in Solution on Adsorption Process | 172 |
| | 6.2.4 | Equilibrium Considerations | 172 |
| | 6.2.5 | Single Component Monolayer Models | 174 |
| | 6.2.6 | Single Component Multilayer Models | 189 |
| 6.3 | Error Analysis for Isotherm Studies | | 190 |
| 6.4 | Thermodynamic Parameters | | 193 |
| | 6.4.1 | Pseudo-First-Order Kinetics and Equilibrium Adsorption Isotherm | 193 |
| | 6.4.2 | Pseudo-Second-Order Kinetic Model | 194 |
| | 6.4.3 | The Elovich Equation | 194 |
| | 6.4.4 | Intraparticle Diffusion | 195 |
| 6.5 | Thermodynamic Parameters | | 198 |
| | 6.5.1 | Polanyi's Potential Theory | 201 |
| | 6.5.2 | Activation Energy | 204 |
| 6.6 | Dynamic Adsorption | | 205 |
| | 6.6.1 | Fundamentals of Dynamic Adsorption | 207 |
| | 6.6.2 | LUB Equilibrium Method | 208 |
| 6.7 | Sorption Mechanisms | | 211 |
| 6.8 | Case Study I | | 215 |
| | 6.8.1 | Sorption Mechanism of Arsenic on to Chitosan–Iron Bead | 215 |
| | 6.8.2 | Summary | 229 |
| 6.9 | Case Study 2 | | 230 |
| | 6.9.1 | Sorption Mechanism of Uranium on to Chitosan-Coated Perlite (CP) Beads | 230 |
| | 6.9.2 | Results and Discussion | 231 |
| | 6.9.3 | Summary | 240 |
| References | | | 240 |

## 7 Chitosan-Based Sensors ... 249
- 7.1 Introduction ... 249
- 7.2 Fabrication of Chitosan Membranes and Films ... 251
- 7.3 Chitosan-Based Sensors ... 253
  - 7.3.1 Glucose Biosensor ... 253
  - 7.3.2 Chitosan-Based Sensor by Layer-by-Layer (LBL) Method ... 261
  - 7.3.3 Chitosan Nanocomposite-Based Sensor ... 267
- 7.4 Summary ... 284
- References ... 285

## 8 Application of Chitosan in the Medical and Biomedical Field ... 291
- 8.1 Skin Wound ... 291
- 8.2 Application of Biopolymer in Wound Healing ... 293
- 8.3 Application of Chitosan in Medical and Biomedical Fields ... 294
  - 8.3.1 Wound Dressing and Healing ... 294
  - 8.3.2 Scaffolds for Bone Tissue Engineering ... 301
  - 8.3.3 Antimicrobial Application of Chitosan ... 303
  - 8.3.4 Anticancer and Antitumor Activity of Chitosan and Its Derivatives ... 307
- 8.4 Chitosan-Based Medical Isotope Separation ... 310
  - 8.4.1 Separation of $^{99m}$Tc from $^{99}$Mo Using Chitosan-Based Adsorbent ... 312
- 8.5 Summary ... 315
- References ... 316

## 9 Application of Chitosan in Textiles ... 323
- 9.1 Chitosan Use in Textiles ... 323
- 9.2 Summary ... 334
- References ... 334

## 10 Chitosan for the Agricultural Sector and Food Industry ... 339
- 10.1 Introduction ... 339
- 10.2 Application of Chitosan in Agriculture and Agro-based Industry ... 340
- 10.3 Summary ... 351
- References ... 352

## 11 Applications of Chitosan in Fuel Cells ... 357
- 11.1 Proton Exchange Membrane Fuel Cells ... 357
- 11.2 Proton Exchange Membrane (PEM) Fuel Cell ... 358
  - 11.2.1 Membrane ... 359
  - 11.2.2 Membrane/Electrode Assembly ... 360
  - 11.2.3 Efficiency, Power, and Energy of Polymer Electrolyte Membrane Fuel Cell ... 363
  - 11.2.4 The Polymer Electrolyte Membrane Fuel Cell Stack ... 365

|  |  | 11.2.5 | Water Management in a Fuel Cell | 366 |
|---|---|---|---|---|
|  | 11.3 | Fuel | | 366 |
|  | 11.4 | Direct Methanol Fuel Cell (DMFC) | | 366 |
|  | 11.5 | Application of Chitosan in Fuel Cells | | 369 |
|  | 11.6 | Summary | | 373 |
|  | References | | | 374 |
| **12** | **Chitosan Uses in Cosmetics** | | | **377** |
|  | 12.1 | Introduction | | 377 |
|  | 12.2 | Target Organs for Cosmetics Products | | 377 |
|  |  | 12.2.1 | Skin | 378 |
|  |  | 12.2.2 | Hair | 379 |
|  |  | 12.2.3 | Teeth | 380 |
|  | 12.3 | Chitin and Chitosan and Their Derivatives in the Cosmetics | | 381 |
|  |  | 12.3.1 | Synthesis of Chitosan Derivatives | 382 |
|  |  | 12.3.2 | Skincare Cosmetics | 382 |
|  |  | 12.3.3 | Hair Care Cosmetics | 388 |
|  |  | 12.3.4 | Teeth | 390 |
|  | 12.4 | Summary | | 398 |
|  | References | | | 399 |

**Index** .................................................................. 405

# Chapter 1
# Polysaccharides: Chitin and Chitosan

**Abstract** Polysaccharides are the most abundant natural form of macromolecules. This chapter deals with the natural sources and types of polysaccharides in general.

## 1.1 Introduction

Chitin and chitosan are becoming the most versatile polysaccharides-based materials of the twenty-first century. Chitin is essentially a homopolymer of 2-acetamido-2-deoxy-β-D-glucopyranose. When chitin is deacetylated to about 50% or more, it becomes soluble in dilute acids and is referred to as chitosan. The general discussion of polysaccharide materials and their differences from chitin is as follows.

### 1.1.1 Polysaccharides in General

Polysaccharides are an abundant form of macromolecules such as carbohydrate-based biomaterials in nature. It is usually linear to highly branched-chain carbohydrate biopolymers composed of a simple carbohydrate monomer called monosaccharides $(CH_2O)_n$. The monosaccharide units, such as glucose $(C_6H_{12}O_6)$, fructose $(C_6H_{12}O_6)$, glyceraldehyde $(C_3H_6O_3)$, are bound together by glycosidic linkage (covalent bond) in polysaccharide structure [1]. Polysaccharides are long polymers of monosaccharides, such as starch $(C_6H_{10}O_5)_n$, glycogen $(C_{24}H_{42}O_{21})$, cellulose $(C_6H_{10}O_5)_n$, and chitin $(C_8H_{13}O_5)_n$. In nature, polysaccharides play a crucial role in maintaining the life cycle of plants and animals [2]. Polysaccharides can be storage polysaccharides, structural polysaccharides, and mucosubstances, based on their function.

## 1.1.2 Storage Polysaccharides

Storage polysaccharides serve as reserve food. When needed, hydrolysis of storage polysaccharides occurs, and the released sugars become available to the living cells for energy and biosynthetic activity. Starch and glycogen are examples of two main storage polysaccharides (Fig. 1.1).

### 1.1.2.1 Starch

Starch is a long-chain glucose polymer and is considered one of plants' primary energy-related carbohydrate reserves (Fig. 1.1). Starch grains occur singly as simple grains or in groups as compound grains. Starch consists of linear polymer amylose and highly branched polymer amylopectin [3]. Amylose is more soluble in water than amylopectin. Figure 1.2 shows the schematic of amylose and amylopectin. Amylose is in a helically arranged continuous straight chain where each turn contains about six glucose units. The repeated glucose units in amylose are linked together by α (1 → 4) glycosidic bonds and a few branches of α (1 → 6) linkages [4]. Amylopectin has a structure similar to amylose. Amylopectin has additional α (1 → 6) glycosidic branching points, which occur every 10–60 glucose units. Besides, straight-chain amylopectin bears several side chains that may be branched further, as shown in Fig. 1.2.

Figure 1.3 shows a typical schematic illustration of the inner architecture of starch granules. In general, starch granules consist of 65–90% of amylopectin and the rest as amylose. The internal structure of starch granules consists of the crystalline and amorphous regions, which together form the crystalline and amorphous growth rings

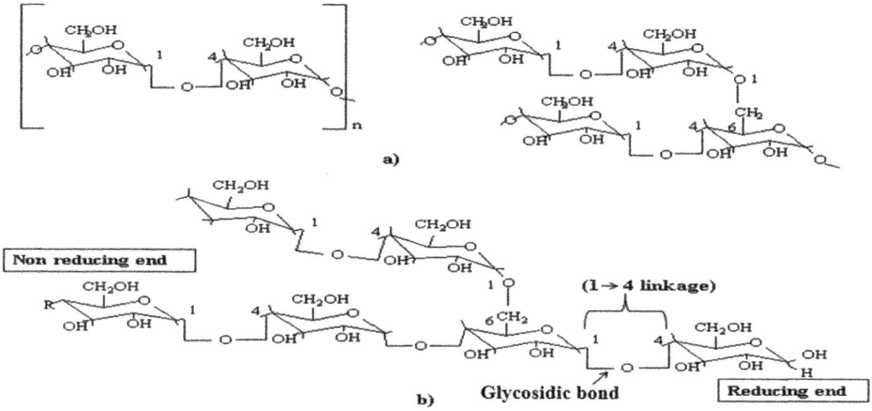

**Fig. 1.1** Storage polysaccharides, **a** starch components (linear amylose and segment of amylopectin molecule), and **b** glycogen (n, called the degree of polymerization, represents the number of glucose groups)

1.1 Introduction

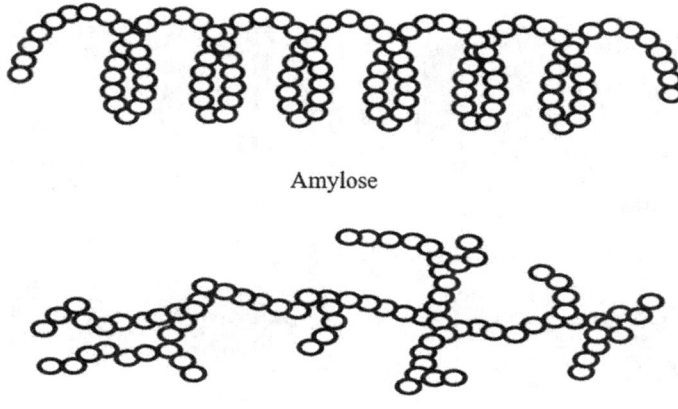

**Fig. 1.2** Schematic diagram of amylose and amylopectin. Copyright © 2019, Springer-Verlag GmbH Germany, part of Springer Nature, Willfahrt, A. et al., Applied Physics A: Materials Science & Processing, *Appl. Phys. A* **125**, 474 (2019). https://doi.org/10.1007/s00339-019-2767-6. Reproduced from Ref. [3] with permission

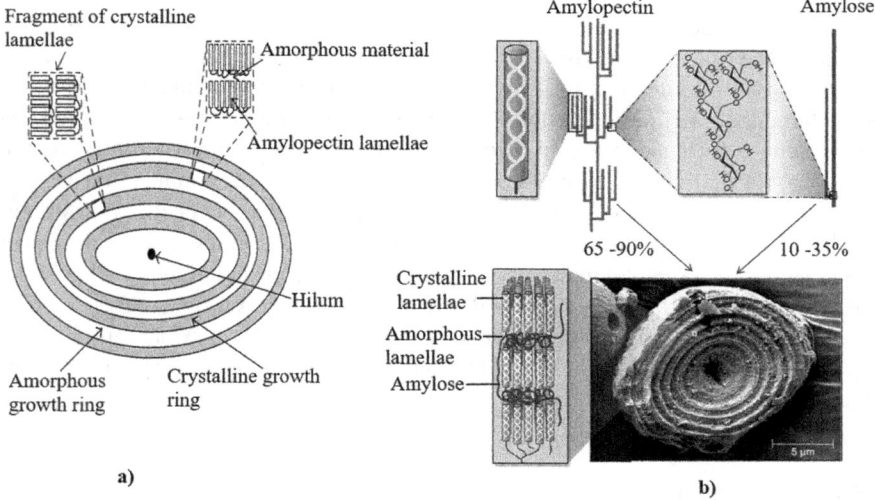

**Fig. 1.3 a** Schematic of starch granule structure [4a] and **b** molecular architecture of starch granules [4b]. Copyright © 2011 Ning Lin et al. Creative Commons Attribution License, Journal of Nanomaterials, vol 2011, Article ID 573687, 13 pages, 2011. https://doi.org/10.1155/2011/573687. Reproduced from Ref. [4a]; Copyright © 2020 Seung, D., New Phytologist © 2020 New Phytologist Trust, New Phytologist (2020) 228: 1490–1504, https://doi.org/10.1111/nph.16858, the Creative Commons Attribution License. Reproduced from Ref. [4b]

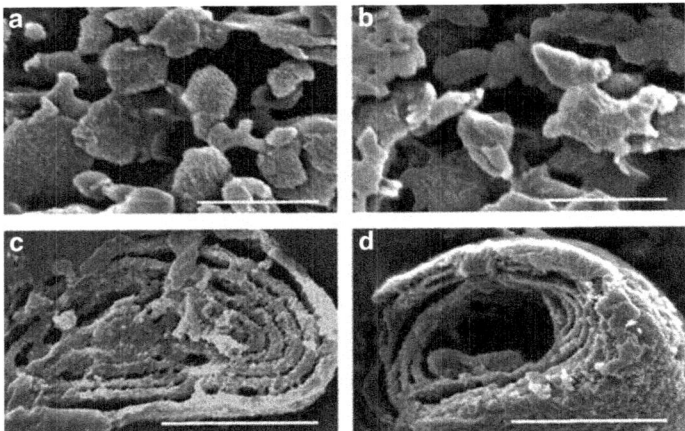

**Fig. 1.4** Scanning electron micrographs of partially digested starch granules from the wild type (**a, b**) and Arabidopsis mutants (**c, d**). Granules were ground in liquid nitrogen and partially digested with α-amylase to reveal internal growth ring structures. Copyright © 2002, Oxford University Press, *Plant Physiology*, June 2002, Vol. 129, pp. 516–529. Reproduced from Ref. [5] with permission

(Fig. 1.3). The dominant component in the crystalline region is the amylopectin lamellae. The branch structure of the adjacent amylopectin chain can pack together to form crystalline lamellae. As shown in Fig. 1.3b, the branch point resides in amorphous lamellae. Alternating crystalline and amorphous lamellae make starch granules matrix semi-crystalline in nature. Amylose mostly resides in the amorphous region and does not participate in semi-crystalline granules. Starch granules has several crystalline and semi-crystalline shells or layers arranged concentrically or eccentrically around a proteinaceous point (i.e., hilum). Some starch consists entirely of amylopectin, for example, maize and some cereals-based starch. On the other hand, starch sources from pea may have as much as 98% of amylose. Zeeman et al. characterize starch synthesis, composition, and granule structure in Arabidopsis leaves [5]. Figure 1.4 shows the scanning electron micrograph of typical starch granules obtained from Arabidopsis leaves.

### 1.1.2.2 Glycogen

Glycogen is a long-chain polymer of glucose. It is a principal energy-related polysaccharide food reserve for animals, bacteria, and fungi. Figure 1.5 shows the schematic structure of storage polysaccharide glycogen. It is popularly called animal starch and is mainly stored inside the liver and muscles.

The polysaccharide glycogen has an extensive branched chain than the chemical structure of starch. However, like amylopectin, the chemical structure of the glycogen also has $\alpha$ (1 → 6) branch points at every 8–4 glucose residues [7].

## 1.1 Introduction

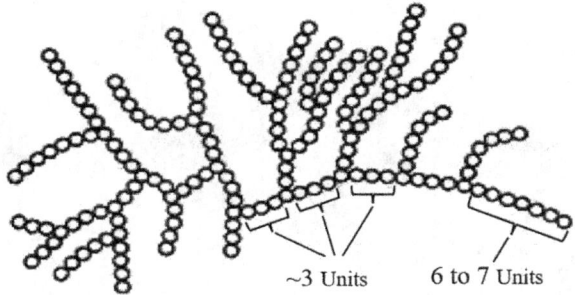

**Fig. 1.5** Schematic diagram of the glycogen structure. Copyright © 1991 Published by Elsevier Ltd., Carbohydrate Polymers, Volume 16, Issue 1, 1991, 37–82. Reproduced from Ref. [6] with permission

### 1.1.3 Structural Polysaccharides

Structural polysaccharides play a significant role in forming the cell walls in plants and the skeleton of animals. Cellulose and chitin are examples of two main types of structural polysaccharides.

#### 1.1.3.1 Cellulose

Cellulose is a structural polysaccharide. It is the principal load-bearing component of plant cell walls. Cellulose is formed by the D-glucopyranosyl units, where the repeated monomers are linked by $\beta\ (1 \rightarrow 4)$ glycoside bond [8, 9]. Cellulose provides

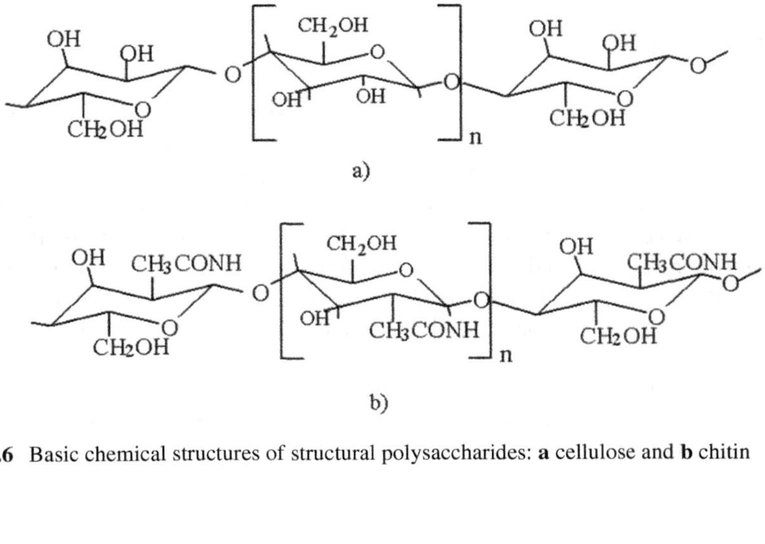

**Fig. 1.6** Basic chemical structures of structural polysaccharides: **a** cellulose and **b** chitin

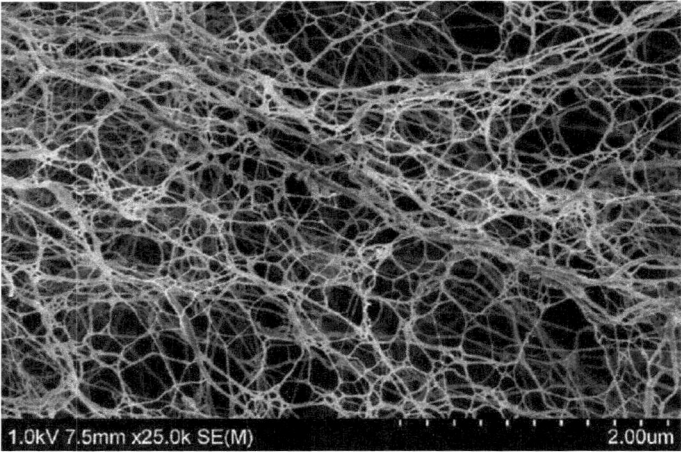

**Fig. 1.7** Cellulose microfibrils composing the surface structure. Copyright © 2019 Núria B., Felicia L., and Vincent B., Creative Commons CC BY license. Cellulose (2019) 26:3083–3094. https://doi.org/10.1007/s10570-019-02285-4. Reproduced from Ref. [10]

rigidity to the plant cells. Figure 1.6 shows the chemical structure of structural polysaccharides.

The bonds between each cellulose molecule are powerful. The intermolecular strengthening of the chain occurs by forming a hydrogen bond between the hydroxyl group at position 3 and the oxygen atom of the next residue (Fig. 1.6). Figure 1.7 shows a typical field-emission scanning electron micrograph of cellulose microfibrils. A microfibril, visible under the electron microscope, is formed when about 2000 cellulose chains or molecules are packed together.

### 1.1.3.2 Chitin

Chitin is a structural polysaccharide like cellulose. The structure of chitin is similar to cellulose's chemical structure except for replacing the hydroxyl group at the carbon with the acetylamino group ($-NHCOCH_3$) [11]. Chitin is the second most abundant biopolymer after cellulose, but it does not occur in cellulose-producing organisms. Chitin is a complex carbohydrate of heteropolysaccharides. It is obtained from skeletons of crustaceans and cartilaginous fish tissues. Fungus cellulose is the chitin present in the fungal walls. It is soft and leathery, therefore, provides both strength and elasticity. Chitin becomes hard when impregnated with specific proteins and calcium carbonate. In chitin, the basic unit is not glucose but a nitrogen-containing glucose derivative known as N-acetyl glucosamine. The monomers are joined together by 1-4β-linkages (Fig. 1.6b). It has an unbranched configuration, and the molecules that occur in parallel are held together by hydrogen bonds. Table 1.1 shows a typical example of polysaccharides from natural sources.

## 1.1 Introduction

**Table 1.1** Typical example of polysaccharides from natural sources [8–14]

| Polysaccharide | Composition | Sugar content/function | Link type | Source |
|---|---|---|---|---|
| Starch | Amylose: linear D-glucose units that contribute 20–30% of the structure of starch macromolecules. Amylopectin: highly branched glucose components that made up 70–80% of the starch macromolecules | Glucose/carbohydrate energy storage: plant cell | Glucose units mainly bond together in amylose through $\alpha(1 \to 4)$ glycoside linkage. In amylopectin, branching occurs with $\alpha(1 \to 6)$ linkage, and branch points occur every 24–30 glucose residues | Starches, especially corn, potatoes, rice |
| Glycogen | Like starch, it is composed of repeating units of glucose monomer held by a glycosidic linkage to form the polymer | Glucose/carbohydrate energy storage: animal | $\alpha(1 \to 4)$ and $\alpha(1 \to 6)$ branched glycoside linkage | Liver and muscle cells of all animals |
| Cellulose | The D-glucopyranosyl units mainly form a linear polymer. Cellulose contains 44.44% carbon, 6.17% hydrogen, and 49.39% oxygen. The cellulose is $(C_6H_{10}O_5)_n$; where n represents the number of glucose groups, ranging from hundreds to tens of thousands | Glucose/structure: stability of plant cell walls | $\beta(1 \to 4)$ glycosidic bond (linear) | Plant kingdom |
| Xylan | Xylans are polysaccharides formed from xylose residues. It contains five carbon atoms and an aldehyde functional group | Xylose/structural | $\beta(1 \to 4)$ glycosidic bond | All land plants |
| Inulin | Inulin consists of chain-terminating glucose moieties and a repetitive fructose moiety | Fructose/natural storage polysaccharides | $\beta(2 \to 1)$ link | Artichoke, chicory |

(continued)

**Table 1.1** (continued)

| Polysaccharide | Composition | Sugar content/function | Link type | Source |
|---|---|---|---|---|
| Dextran | Dextran is a group of glucose polymers made by certain bacteria | Glucose/function is unknown | $\alpha$ (1 → 4) and $\alpha$ (1 → 6) | Primarily bacteria |
| Agar | Agar consists of two polysaccharides: agaropectin and linear polymer agarose | Galactose | $\alpha$ (1 → 3) | Seaweeds |
| Chitin | Chitin is a long-chain unbranched polysaccharide made of N-acetylglucosamine residues | Structure: crustacean animal cell | The repeated monomers in chitin are linked through $\beta$ (1 → 4) glycosidic (covalent) bonds | Insects and crustacean skeleton |

## *1.1.4 Polysaccharides from Marine Source*

Polysaccharides from marine sources consist of different macromolecules with different biological properties [12, 13]. Marine polysaccharides can be classified mainly into marine animal, marine plant, and marine microbial polysaccharides based on their source [14]. Elsakhawy et al. reported that the microbial populations, including bacteria, fungi, and algae, represent marine microbial polysaccharides [15]. These microbial populations play an essential role in energy cycling between the surface and the bottom in the sea by aggregation of small molecules, causing them to sink. They are bio-active agents which are useful in antitumor, antiviral, antioxidant, anticoagulant, food, and feed applications [15, 16]. The primary source of marine polysaccharides is marine algae. Seaweeds are considered multicellular marine algae and a significant source of polysaccharides [17]. The polysaccharides extracted from marine algae (seaweeds) are agar alginate, carrageenans, and laminarin. These are known as cell wall structural and storage polysaccharides [18, 19]. The most notable structural polysaccharide extracted from marine animal sources is chitin. Skeletons of crustaceans and cartilaginous fish tissues are principal sources of chitin. Cell walls of fungi and nematodes are other sources of chitin. Table 1.2 shows the most representative structural polysaccharides from marine sources.

## *1.1.5 Mucosubstances*

Kiernan reported that mucosubstance includes all glycans and glycoconjugates found in tissue sections [20]. In general, mucosubstances are the substances formed from the mucilage, mucus, or slime. There are two types, mucopolysaccharides and mucoproteins. Mucopolysaccharides are acidic or aminated polysaccharides formed from

1.1 Introduction

**Table 1.2** Typical example of structural polysaccharides from marine sources [15–19]

| Source | Polysaccharides | Functional groups | Occurrence | Pharmaceutical uses |
|---|---|---|---|---|
| Seaweed | Alginate $(C_6H_8O_6)_n$ | Polyanionic $(COO^-)$ | Brown algae | Stabilizing agent |
| | Agar $(C_{24}H_{38}O_{19})$ | Neutral | Red algae | Microbiology and food industry |
| | Carrageenans $(C_{24}H_{36}O_{25}S_2)$ | Polyanionic $(SO_4^-)$ | Red algae | Applications in the food and pharmaceutical industries are due to their stabilizing, gelling, thickening, antiviral, antitumor, anti-inflammatory properties |
| Animal | Chitin $(C_8H_{13}O_5N)_n$ | Acetylamino group | Crustacean shell, fungi | Chitosan manufacturing |
| | Chondroitin sulfate $(C_{13}H_{21}NO_{15}S)$ | Polyanionic $(SO_4^-$, sulfate) | Mammalian and non-mammalian marine animal | Anticoagulant applications in pharmaceutical industries |
| | Hyaluronic acid $(C_{14}H_{21}N_{11})_n$ | Hydrated polyanionic | The cartilage of marine animals | Various applications in pharmaceutical industries |

galactose, mannose, uric acid, and sugar derivatives found in plants and animals. Mucopolysaccharides bind proteins in the cell walls and connective tissues, providing lubrication in ligaments and tendons. Hyaluronic acid is a mucopolysaccharide present in the extracellular fluid of animal tissues, synovial fluid, and cerebrospinal fluids. Hyaluronic acid occurs in cementing material between animal and inside cell coats. On the other hand, mucoprotein or glycoprotein is conjugated monosaccharides form mucus. Mucoprotein is found in the stomach, nasal secretion, and intestine. They are antibacterial and protective in function.

## *1.1.6 Polysaccharides as a Human Nutrition Source*

The three main polysaccharides (i.e., starch, cellulose fibers, and glycogen) obtained from terrestrial plant, animal, and marine sources provide energy and dietary fibers for human consumptions. Starch from plant foods, including cereals, legumes, and tubers (especially the potato), provides the primary glucose source in the human diet [21]. The dietary fibers are indigestible carbohydrates of plant origin. They encompass a range of divergent compounds of non-starch carbohydrate and lignin that are easily fermented by the colonic microbiota in the small and large intestines of humans [22]. Turner and Lipton [23] reported that dietary fiber contributes no calories to our diet. However, the metabolites released by the bacteria in the colon provide energy to humans and other mammals. Marine polysaccharides are good natural products for medicinal and dietary applications [16]. For instance, marine polysaccharide seaweeds have health-promoting molecules and materials such as dietary fiber; ω-3 fatty acids; essential amino acids; and A, B, C, and E vitamins [24]. Table 1.3 shows the list of polysaccharides and food examples.

## 1.1.7 Chitosan Market

Chitin/chitosan derivative's potential uses in food and pharmaceutical industries require quality end products. For example, ultra-pure chitosan production for food and pharmaceutical applications requires strict quality control from raw material extraction to the final product than the technical grade chitosan. From an economic point of view, this control is costly, and the vast majority of the samples belong to the low-end market. The prices of chitin/chitosan and their derivatives vary from a few dollars to hundreds of dollars per kilogram, depending on the polymer's quality. For example, the low-end chitosan product for wastewater treatment and agricultural uses is around 9 USD/Kg [27]. The price of the pharmaceutical-grade chitosan depends on the quality of the product. The global chitin and chitosan markets are expected to grow significantly in the upcoming years. This is due to applications in various industries, such as water treatment, pharmaceuticals, medical, food, beverage, and cosmetic industries [28]. According to the Grand View Research report, the global chitosan market size was estimated at USD 6.8 billion in 2019 [29]. Figure 1.8 shows the estimated uses of chitosan in water treatment, food, and pharmaceutical industries.

The chitosan market is expected to grow globally to USD 17.84 billion by 2025 [29]. The increase in cost of chitosan is primarily due to the lack of raw material availability [29].

Figure 1.9 shows that the demand for chitosan is expected to increase while suppliers remain limited. Consequently, the industrial applications for chitin and chitosan derivatives with high purity products may inhibit industry growth.

## 1.1.8 Summary

Polysaccharides can be termed as storage polysaccharides, structural polysaccharides, and mucosubstances. The contribution of polysaccharides in maintaining the life cycle of plant and animal is well-known. For example, structural polysaccharides play an essential role in forming the structural framework of the cell walls in plants and the skeleton of animals.

Polysaccharides, from terrestrial plants and animals and marine sources, are good natural products for medicinal and dietary applications. The most notable structural polysaccharide from marine source is chitin and chitosan. Chitin and chitosan and

**Table 1.3** List of polysaccharides and description of sources

| Polysaccharides | Description | Composition | Kcal/g | Food source |
|---|---|---|---|---|
| *Digestible* | | | | |
| Starch | Starch has a large number of glucose units joined by glycosidic bonds. Most green plants produce this polysaccharide as energy storage | Glucose | 4.2 | Cereal grains and their products (bread, pasta, pastries, cookies), potatoes, tapioca, yam, legumes |
| Dextrin | Dextrins are a group of low-molecular-weight carbohydrates produced by the hydrolysis of starch | Glucose | 3.8 | Artificially produced food additives |
| Glycogen | Glycogen forms glucose on hydrolysis | Glucose | 4 | Shellfish, animal liver |
| *Dietary fibers* | | | | |
| Cellulose | Cellulose is exceedingly insoluble and hard to digest by a human enzyme | Glucose | 0 | The cell walls of the plants contain cellulose as a vital component. Cellulose forms about 25% of the fiber in grains and fruit. It forms about 33% of fibers in vegetables and nuts. Most of the fiber in cereal bran is cellulose |
| Hemicellulose | Hemicellulose is a polysaccharide that contains sugars other than glucose. Associated with cellulose in cell walls and present in both water-soluble and insoluble forms | Arabinose + Xylose | 0 | Forms about 33% of the fiber in vegetables, fruits, legumes, and nuts. The primary dietary sources are cereal grains |

(continued)

**Table 1.3** (continued)

| Polysaccharides | Description | Composition | Kcal/g | Food source |
|---|---|---|---|---|
| Polydextrose | The average degree of polymerization for polydextrose is 12. It is synthesized from glucose and sorbitol. It gets partially fermented in the colon (~50% in humans). Polydextrose has bulking and prebiotic properties | Glucose | 1.2 | Polydextrose is a food additive. Polydextrose, for example, is used in energy products as a bulking agent to replace sugars and to provide texture to foods. Its contribution to energy is low at just 1.2 kcal/g |
| Beta-glucan | Glucose polymers, unlike cellulose, have a branched structure enabling them to form viscous solutions | Glucose | ~2 | It is a major component of cell wall material in oats and barley. It is present only in small amounts in wheat |
| Pectin | Pectin is a polysaccharide. Its main component is galacturonic acid, which is a sugar acid derived from galactose. It is soluble in hot water, and it forms gels on cooling | Various monosaccharides | 3.3 | Pectin is found in cell walls and intracellular tissue of fruits and vegetables. Pectin represents 15–20% of the fiber in vegetables, legumes, and nuts. Sugar beet and potatoes are also sources of pectin |
| Inulin | Inulin comprises of three to ten sugar units. It occurs naturally in plants. It is consumed as foods, mainly vegetables, cereals, and nuts | Fructose + Glucose | 1–2 | The primary dietary sources of inulin are onions, chicory, and Jerusalem artichokes |
| Gum and mucilage | Gums are hydrocolloids derived from plant exudates. Mucilage is present in the cells of the outer layers of seeds of the plantain family, for example, psyllium. Gum and mucilage are used as gelling agents, thickeners, stabilizers, and emulsifying agents | Various monosaccharides | | Gums: plant exudates (gum arabica and tragacanth), seeds (guar and locust beans), and seaweed extracts (agar, carrageenans, alginates). Mucilage: psyllium seed husk |

(continued)

## 1.1 Introduction

**Table 1.3** (continued)

| Polysaccharides | Description | Composition | Kcal/g | Food source |
|---|---|---|---|---|
| *Marine polysaccharide* | | | | |
| Alginate | Alginate is present in brown algae's cell walls, as algenic acid's calcium, magnesium, and sodium salts. It consists of a linear copolymer of β-(1–4) linked d-mannuronic acid and β-(1–4) linked l-guluronic acid unit | Galactose | | Food additives derived from marine algae |
| Carrageenan | Carrageenans are linear sulfated polysaccharides extracted from red edible seaweeds | Galactose | | Food additives derived from marine algae |
| Chitin and chitosan | Chitin is a homopolymer of 2-acetamido-2-deoxy-β-D-glucopyranose. Some of the glucopyranose residues are in the deacetylated form as 2-amino-2-deoxy-β-D-glucopyranose. When chitin is deacetylated by more than 50%, it is referred to as chitosan. Chitosan is soluble in dilute acids | Glucosamine | | Dietary supplements derived from shells of crustaceans |

Copyright © 2008 Buttriss, J. L. and Stokes, C. S. © 2002 British Nutrition Foundation, Nutrition Bulletin, 33, 186–200, 2008. Reproduced from Ref. [25] with permission. https://www.nutrientsreview.com/carbs/polysaccharides.html. Reproduced from Ref. [26] Online

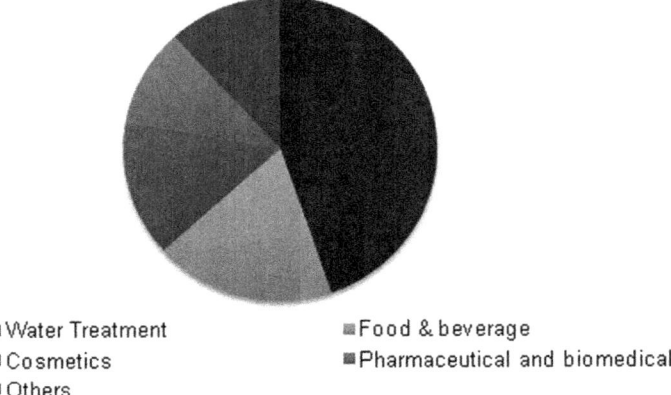

**Fig. 1.8** Uses of chitin/chitosan derivatives in various industries. Copyright © 2021 Grand View Research, Inc. https://www.grandviewresearch.com/industry-analysis/global-chitosan-market. Reproduced from Ref. [29] with permission

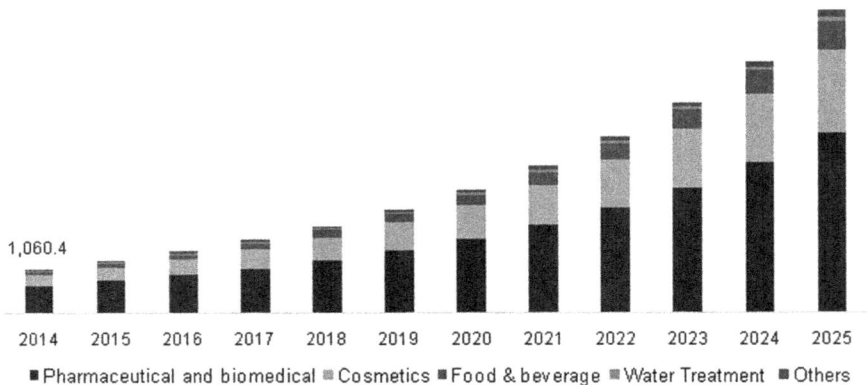

**Fig. 1.9** US market revenue for chitosan. Copyright © 2021 Grand View Research, Inc. https://www.grandviewresearch.com/industry-analysis/global-chitosan-market. Reproduced from Ref. [29] with permission

their derivatives are used in various applications. The quality of chitosan products regulates its market price and their type of applications.

## References

1. Percival E (1979) The polysaccharides of green, red, and brown seaweeds: their basic structure, bio-synthesis and functions. Br Phycol J 14:103–117
2. Mishaki A (1994) Structural aspects of some functional polysaccharides. In: Nishimary K, Doi

E (eds) Food hydrocolloids: structure, properties and functions. Plenum Press, New York, pp 1–19
3. Willfahrt A, Steiner E, Hötzel J et al (2019) Printable acid-modified corn starch as non-toxic, disposable hydrogel-polymer electrolyte in supercapacitors. Appl Phys A 125:474. https://doi.org/10.1007/s00339-019-2767-6
4. (a) Lin N, Huang J, Chang PR et al (2011) Preparation, modification, and application of starch nanocrystals in nanomaterials: a review. J Nanomater 13p. https://doi.org/10.1155/2011/573687. Article ID 573687. (b) Seung D (2020) Amylose in starch: towards an understanding of biosynthesis, structure and function. New Phytol 228:1490–1504. https://doi.org/10.1111/nph.16858
5. Zeeman SC, Tiessen A, Pilling E et al (2002) Starch synthesis in Arabidopsis. Granule synthesis, composition, and structure. Plant Physiol 129:516–529
6. Manners DJ (1991) Recent developments in our understanding of glycogen structure. Carbohyd Polym 16(1):37–82
7. Singh AV (2011) Biopolymers in drug delivery: a review. Pharmacol Online 1:666–674
8. Wertz LJ, Bedue O, Mercier PJ (2010) Cellulose science and technology. EPFL Press, Boca Raton, USA
9. Zablackis E, Huang J, Mullerz B, Darvill AC, Albersheim P (1995) Characterization of the cell-wall polysaccharides of Arabidopsis thaliana leaves. Plant Physiol 107:1129–1138
10. Chinga-carrasco G, Miettinen A, Hendriks L et al Structural characterization of kraft pulp fibers and their nanofibrilated materials for biodegradable composite applications, Nanocomposites and polymers with analytical methods, Dr. Cuppoletti J (ed) ISBN: 978-953-307-352-1, InTech, Available from: http://www.intechopen.com/books/nanocomposites-andpolymers-with-analytical-methods/structural-charcaterisation-of-kraft-pulp-fibers-and
11. Nejati Hafdani F, Sadeghinia N (2011) A review on application of chitosan as a natural antimicrobial. Int J Med, Health, Biomed, Bioeng Pharm Eng 5(2)
12. Piontek J, Lunau M, Handel N, Borchard C, Wurst M, Engel A (2010) Acidification increases microbial polysaccharide degradation in the ocean. Biogeosciences 7:1615–1624
13. Bajpai VK, Rather IA, Lim J, Park Y-H (2014) Diversity of bioactive polysaccharide originated from marine source. Indian J Geo-Mar Sci 43:10
14. Wang W, Wang S-X, Guan H-S (2012) The antiviral activities and mechanisms of marine polysaccharides: an overview. Mar Drugs 2012 10:2795–2816. https://doi.org/10.3390/md10122795
15. Elsakhawy TA, Sherief FA, Abd-El-Kodoos RY (2017) Marine microbial polysaccharides: environmental role and applications (an overview). Env Biodiv Soil Secur 1:61–70
16. Fedorov SN, Ermakova SP, Zvyagintseva TN, Stoni VA (2013) Anticancer and cancer preventive properties of marine polysaccharides: some results and prospects. Mar Drugs 11:4876–4901. https://doi.org/10.3390/md11124876
17. Cardoso MJ, Costa RR, Mano JF (2016) Marine origin polysaccharides in drug delivery systems. Mar Drugs 14:34
18. Vera J, Castro J, Gonzalez A, Moyenne A (2011) Seaweed polysaccharides and derived oligosaccharides stimulate defense responses and protection against pathogens. Mar Drugs 9:2514–2525
19. Silva TH, Alves A, Ferreira BM, Oliveira JM et al (2012) Materials of marine origin: a review on polymers and ceramics of biomedical interest. Int Mater Rev 57:276–306
20. Kierman JA (2010) Carbohydrate histochemistry. In: Education guide-special stains and H&E, pp 75–92
21. Ellis PR, Rayment P, Wang Q (1996) A physico-chemical perspective of plant polysaccharides in relation to glucose adsorption, insulin secretion and entero-insular axis. Proc Nutr Soc 55:881–898
22. Mišurcová L et al (2012) Health benefits of algal polysaccharides in human nutrition. In: Henry J (ed) Advances in food and nutrition research, vol 66. Academic Press, Burlington, pp 75–145
23. Turner ND, Lupton JR (2011) Dietary fiber. Adv Nutr 2(2):151–152

24. Rajapakse N, Kim S-K (2011) Chapter-2 Nutritional and digestive health benefits of seaweed. Adv Food Nutr Res 64:17–28
25. Buttriss JL, Stokes CS (2008) Dietary fiber and health: an overview. Nutr Bull 33:186–200
26. www.nutrientsreview.com/carbs/polysaccharides.html
27. Aranaz I, Mengiber M, Acosta N, Heras A (2015) Chitosan: a natural polymer with potential industrial applications. A Sci J COMSATS-Sci Version 21(1&2):41–50
28. http://www.marketresearchstore.com/report/chitosan-market-z82339
29. https://www.grandviewresearch.com/industry-analysis/global-chitosan-market

# Chapter 2
# Preparation of Chitin and Chitosan

**Abstract** This chapter provides information on chitin formation chemistry, extraction of chitin, and chitosan preparation processes. In addition, the relationship between structural and biological properties of chitosan, different chitin and chitosan preparation processes, and their limitations are discussed.

## 2.1 Chitin and Chitosan

Chitin, a polysaccharide, is the second most abundant natural material next only to cellulose. Chitin derives its name from the Greek word for envelope. Chitosan is prepared by deacetylation of chitin, generally in a concentrated sodium hydroxide solution. The chitin and chitosan are the basic organic nitrogen-containing structural polysaccharides that occur naturally in lower animals and fungi. In animals, it occurs as the principal constituent in the exoskeleton of crustacea, mollusks, and insects. Crustacean shell, the most available source of chitin, consists mainly of protein (30–40%), mineral salts (30–50%), chitin (13–42%), and also carotenoids as pigment [1]. The cell walls of fungi can also be used as source materials for chitin and chitosan [2]. For example, the chitin content in the fungal cell wall may vary from 2 to 42% (dry basis) in yeast and Euascomycetes, respectively [3]. However, chitin in fungi is associated with other polysaccharides that makes its extraction difficult [4]. Other sources of chitin include algae, protozoa, Mollusca, Annelida, Arthropoda, bacteria, marine diatoms, insects, and vertebrates. Their uses and applications range from agricultural uses to medical applications. Table 2.1 lists the applications of chitin and chitosan.

Unlike other naturally occurring polysaccharides such as cellulose, which are acidic, chitin and chitosan are highly basic. Also, the amino groups in chitosan make it more versatile than cellulose. Figure 2.1 shows the molecular structure of chitin and chitosan.

The occurrence and applications of chitin, chitosan, and their derivatives are the subjects of a book by Pariser and Lombardi [5]. The rapid pace at which scientists and technologists worked with this new raw material for the past 40–50 years shows the multifaceted uses of chitin and chitosan. Starting with one or two patents in

**Table 2.1** Various applications of chitin and chitosan

| Area of applications | Example |
|---|---|
| Agriculture | A defensive mechanism in plants<br>Stimulation of plant growth<br>Seed coating<br>Frost protection<br>Time release of fertilizers<br>Nutrients into soil |
| Food industry | Antimicrobial agent<br>– Bactericidal<br>– Fungicidal<br>– Measure of mold contamination in agricultural commodities<br>Edible film industry<br>– Controlled moisture transfer between food and the surrounding environment<br>– Controlled release of antimicrobial substances<br>– Controlled release of antioxidants<br>– Controlled release of nutrients, flavors, and drugs<br>– Reduction of oxygen partial pressure<br>– Controlled rate of respiration<br>– Temperature control<br>– Controlled enzymatic browning in fruits<br>– Reverse osmosis membranes<br>Additive<br>– Clarification and deacidification of fruits and beverages<br>– Natural flavor extender<br>– Texture controlling agent<br>– Emulsifying agent<br>– Food mimetic<br>– Thickening and stabilizing agent<br>– Color stabilization<br>Nutritional quality<br>– Dietary fiber<br>– Hypocholesterolemic effect<br>– Livestock and fish feed additive<br>– Reduction of lipid absorption<br>– Production of single-cell protein<br>– Antigastritis agent<br>– Infant feed ingredient<br>– Recovery of solid materials from food processing wastes<br>– Affinity flocculation<br>– Fractionation of agar<br>– Enzyme immobilization<br>– Encapsulation of nutraceuticals |
| Water and wastewater treatment | Coagulating/flocculating agent<br>Heavy metal ions removal |
| Cosmetics and toiletries | Skincare |

(continued)

## 2.1 Chitin and Chitosan

**Table 2.1** (continued)

| Area of applications | Example |
|---|---|
| Biomedical | Anticoagulation, tissue engineering, wound dressing, orthopedic, drug delivery |
| Biotechnology | Enzyme immobilization, gene delivery |
| Textile | Fibers for textile |

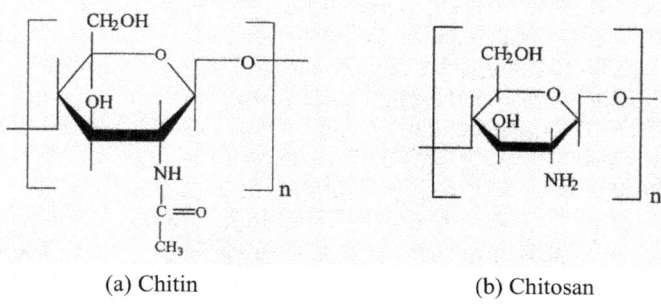

(a) Chitin  (b) Chitosan

**Fig. 2.1** Chemical structure of **a** chitin and **b** chitosan

1978–80, the number of patents on chitosan has grown to nearly 700 in 1996–2001. During the periods 1991–96 and 1996–2001, the number has increased from 180 to 680. This is a remarkable growth in the number of patents since 1996. It indicates the increased interest in chitosan and its applications in various fields. These recent advances have made chitin a precious raw material and have increased the value of chitin. Chitin is insoluble but biodegradable. Both soluble and insoluble versions of chitosan have been prepared.

Isolation of chitin started with the work of Braconnot around 1811 when he isolated a material from mushrooms and called it fungine. Later in 1823, Odier [Odier, A. Mem. Soc. Hist. Nat. Paris, 1, 29–42 (1823)] named it chitin [meaning tunic in Greek as it has a resemblance to fabric]. Chitin has a crystalline structure [6] with an unusual property, including toughness, biodegradability, and relative inertness. Chitin, a (1-4)-linked 2-acetamido-2-deoxy-β-D-glucan, is insoluble in water and most common solvents. The deacetylation of its N-acetylglucosamine units increases the hydrophilicity of chitin. This deacetylation produces chitosan, a (1-4)-linked 2-amino-2-deoxy-β-D-glucan, and it is the most important derivative of chitin. Chitin is poly[β-(1-4)-2-acetamido-2-deoxy-D-glucopyranose] and chitosan is poly[β-(1-4)-2-acetamido-2-amino2deoxy-D-glucopyranose]. Three different forms—α-chitin, a tightly arranged antiparallel form, β-chitin tightly arranged parallel form, and γ-chitin a mixed arrangement—are known to exist [7, 8]. Chitin is a high molecular mass polymer linked by [β-(1-4)-] bonds. Structural studies reveal that chitosan is closer to α-chitin. Ogawa [9] has also studied chitosan's conformational properties by X-ray diffraction and $^{13}$C NMR techniques. Chitosan and chitin represent a continuum of copolymers. Chitosan is insoluble in mineral acids but soluble in organic acids, and it contains a reactive hydroxyl (OH) and an amine (NH$_2$) group. Because of amino groups, chitosan is soluble in aqueous acidic media through protonation. It forms viscous solutions to produce gels in various forms, e.g., beads, membranes, coatings, fibers, and sponges. The amino and the hydroxyl groups are easily altered chemically to improve the mechanical properties of chitosan.

The most abundant sources of chitin are shellfish waste and other microbial sources. Mahmoud et al. [10] have reviewed the production of chitin from these sources. Dutta and coworkers [11] have outlined a general procedure for treating crustacean shells to produce chitin and chitosan. Merzendorf and Zimoch [12] discuss the formation of chitin from trehalose [the primary hemolymph sugar in most insects]. They also provide details of the importance of the regulation of chitin synthases and degradation. The insect cuticles are biocomposites with chitin fibers embedded in a protein matrix. Zhang et al. [12a] (as reported by Merzendorf and Zimoch [12]) provide the necessary steps involved in the formation of chitin (Fig. 2.2). Chitins are non-toxic, odorless, and biocompatible in animal tissues [13]. They represent a renewable source of natural biodegradable polymers. The difference between chitin and chitosan depends on the degree of deacetylation. Chitosan refers

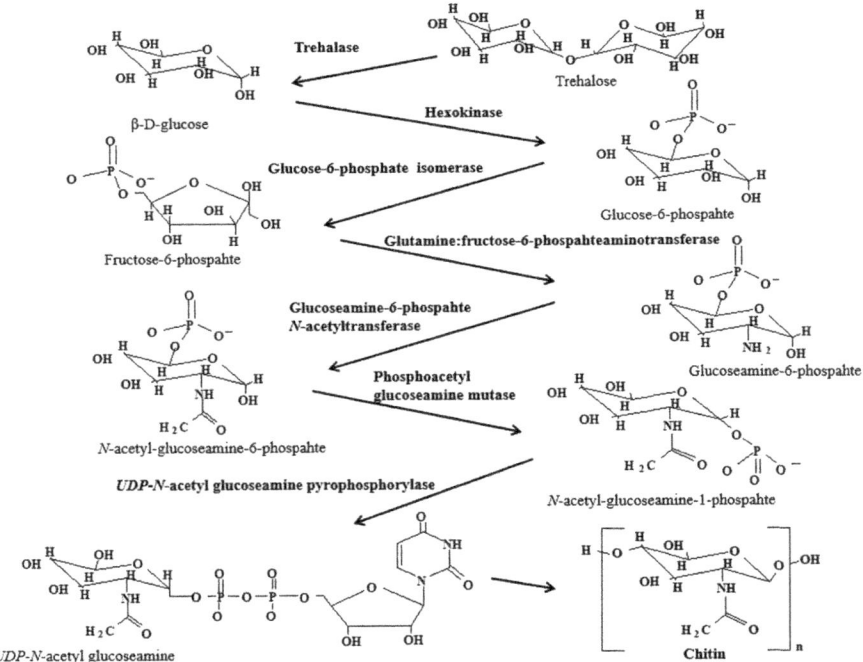

**Fig. 2.2** Formation of chitin. Copyright: © 2021 Zhang, X., Yuan, J., Li, F., Xiang, J. Licensee MDPI, Basel, Switzerland. This article is an open access article distributed under the terms and conditions of the Creative Commons Attribution (CC BY) license. Reproduced from Ref. [12a]

to deacetylation above 50%. Muzzarelli pioneered work on chitin and chitosan [14, 15]. The authors discuss methods of deacetylation, structure, properties, binding characteristics, applications, and a host of other things.

Most of the chitin produced today is from crustaceans such as shellfish, oysters, crab, squid, and shrimp. The other source used is a microbial source. Chitin extraction by chemical and biological methods is discussed in the review articles [16, 17]. The chemical method uses acid and bases, whereas the biological method uses microorganisms to extract chitin. Chen et al. [18] discuss the microstructure and strength of the chitin fiber. The book on Chitin and Chitosan in Fungi by Peter [19] discusses the chemical structure, detection, biosynthesis, production, degradation of chitin, and chitosan. Chitosan is used in water treatment, cosmetic and drug manufacturing, food additives, semi-permeable membranes, and in the development of biomaterials. Tharanathan and Kittur [20] have written an authoritative review of the potential of chitin. Khor [21] has reviewed the applications of chitin and chitosan as a biomaterial. Khor [21] narrates the chemistry and structural properties of chitin, biomedical applications of chitin, use of chitin as a biomaterial, use in tissue engineering, and procedures in getting chitin and its derivatives approved by the regulatory authority. Jayakumar et al. [22] have reviewed different methods such as free radical, radiation, enzymic and cationic graft copolymerization onto chitosan, the influence of grafting parameters such as grafting percentage and grafting efficiency, and the properties of grafted chitosan. This review also screens the current applications of graft copolymerized. Chitosans for drug delivery, tissue engineering, antibacterial, biomedical, metal adsorption, and dye removal are also discussed. Rafat and Sahl [23] have reviewed the antimicrobial activity of chitosan. Xia et al. [24] reviewed the biological activities of chitosan and chitooligosaccharides for hypocholesterolemic, antimicrobial, immunostimulating, antitumor, and anticancer effects. Shahidi and Abuzaytoun [25] have reviewed the structure, solubility, preparation, applications, and health effects of chitin and chitosan.

## 2.2 Chitosan Preparation Methods

### 2.2.1 Introduction

An interest in chitin has increased due to its biocompatibility, biodegradability, and lack of toxicity and allergenicity. Both chitin and chitosan are available commercially in industrial grade. Numerous derivatives of chitin and chitosan have been developed, but most of them are still on the laboratory scale. Large-scale commercial use is yet to be determined. Chitosan has also been cross-linked with a variety of chemicals to modify its physical and chemical properties. The starting raw material for any chitin or chitosan-based products is chitin, which is extracted or isolated from natural sources. Approximately 70% of marine capture fisheries are utilized for processing, and a considerable quantity of the processed catch remains as waste [26]. So far, the primary commercial sources of chitin are extracted from marine crustaceans and fish industries wastes. Therefore, the production of chitin is an environmentally safe use of waste from shellfish and fish processing industries [27–29]. The sources of chitin include marine invertebrates, insects, and fungi as the conjugated form with proteins [30–34]. It is also present in the cell wall of plants and the cuticles of animals. Also, chitins in animal tissues are frequently calcified, such as in the case of shellfish. Table 2.2 lists the content of chitin depending on the sources of marine catch (crustacean).

The primary source of industrial-grade α-chitin is from shrimps and crabs [35]. The β-chitin is prepared mainly from squid pens [8, 36]. Due to the higher reactivity [37–40] and higher affinity toward solvents [41] of β-chitin, it is increasingly used in industrial applications. Cauchie [42] (as reported by Synowiecki and Al-Khateeb [31]) noted that the chitin content of crustacean order might vary depending on where they live: freshwater or seawater and is given in Table 2.3.

The extraction efficiency of chitin from these sources depends both on protein and mineral contents. The chemical composition of these sources (Table 2.4) [31, 43, 44, 45, 46] determines the amount of acid or alkali added and the time of exposure. In the USA, crab and shrimp are mainly used for chitin extraction, and about 39,000 tons are available annually [47].

Commercial chitin is extracted from crustacean waste using chemical processes. The chemical process removes inorganic matter and hydrolyzes the protein content of the waste. When chitin is deacetylated more than 50%, the resultant polymer is termed chitosan. Chitosan is soluble in dilute acids. Chitosan preparation depends on the type and source of chitin besides other factors. Wan et al. [48] reviewed different methods of preparation, characterization, and applications of chitosan. Rinaudo [49] discusses the morphology, identification, characterization, processing, and sources of chitin. The author also discusses the preparation and chemical modifications of chitosan and its applications. Tan et al. [50] used waste mycelia to prepare chitosan with 84% deacetylation and viscosity-average molecular mass of $1.4 \times 10^5$. De Santos et al. [51] characterized three commercially available chitosans using different techniques. They used proton NMR, IR spectroscopy, conductometric titration, thermogravimetry, and differential scanning calorimetry to characterize the samples of chitosan. A solid-phase method, different from the usual liquid phase, has been described by Gao et al. [52]. In this method, chitosan was prepared using chitin and solid alkali in a single-screw extruder at a given temperature. Potentiometric titration gel permeation chromatography, IR spectroscopy, and carbon-13 magnetic resonance spectroscopy were used to characterize the chitosan product. Domard and Chaussard [53] prepared high molecular mass chitosan. The authors studied the kinetics of deacetylation and the energy of activation as a function of NaOH concentration. Large-scale production of chitin and chitosan is primarily from the discarded crab and shrimp shells, prawn waste, and crustacean shells. According to Kumar, the production of 1 kg of 70% deacetylated chitosan from chitin (obtained from shrimp shells) requires 6.3 kg of HCl and 1.8 kg of NaOH, in addition to nitrogen and water (1400 L) [54].

**Table 2.2** Chitin content of various sources

| Crustaceans [30] | Chitin content (%) |
|---|---|
| *Nephro* (Lobster) | 69.8[a] |
| *Euphausia superb* (Krill) | 24[a] |
| *Homarus* (Lobster) | 60–70[a] |
| *Crangon crangon* (Shrimp) | 17.8[a] |
| *Lepas* (Goose barnacle) | 58.3[a] |
| *Chionoecetes opilio* (Crab) | 26.6[a] |
| **Insects [30]** | |
| *Blatella* (Cockroach) | 18.4[a] |
| *Coleoptera* (Ladybird) | 27–35[a] |
| *Diptera* | 54.8[a] |
| *Pieris* (Butterfly) | 64.0[a] |
| *Bombyx* (Silk worm) | 44.2[a] |
| *Galleria* (Wax worm) | 33.7[a] |
| **Microorganism species [17, 30]** | |
| *Aspergillus niger* | 42.0[b] |
| *Penicillium notatum* | 18.5[b] |
| *Penicillium chrysogenum* | 19.5–42[b] |
| *Saccharomyces gutulata* | 2.3[b] |
| *Mucor rouxii* | 9.4[b] |
| *Aspergillus phoenicis* | 23.7[b] |
| *Histoplasma capsulatum* | 25.8–26.4[b] |
| *Trichoderma virdis* | 12.0–22.0[b] |
| *Blastomyces dermatidis* | 13.0[b] |
| *Paracoccidioides brasiliensis* | 11.0[b] |
| *Neurospora crassa* | 8.0–11.9[b] |
| *Tremella mesenterica* | 3.7[b] |
| *Lactarius vellereus* | 19.0 |

Copyright © 2015 Elsevier Ltd., Trends in Food Science & Technology, Volume 48, February 2016. Reproduced from Ref. [30] with permission. Copyright © 2015 Faculty of Food Technology and Biotechnology, University of Zagreb, W. Arbia et al., Food Technol. Biotechnol. 51(1)12–25 (2013); ISSN 1330-9862. https://hrcak.srce.hr/file/146860, Creative Common CC BY license. Reproduced from Ref. [17]

[a]Chitin % compared to the dry mass of exoskeletons
[b]Chitin % relative to the dry mass of mycelium

**Table 2.3** Chitin content of crustacean order from fresh and seawater [31, 42, 43]

| Crustacean order | Chitin content as % body mass of crustacean from | | |
|---|---|---|---|
| | Fresh water | Sea water | References |
| Copepoda | 12.4 | 5.8 | [31] |
| | (10.8–13.9) | (3.1–8.6) | [42] |
| Amphipoda | – | 7.3 | [31] |
| | (8.3 ± 0.2) | (5.8–9.4) | [42] |
| Cladocera | 4.9 | 12.2 | [31] |
| Cladocera | 16–17 | – | [43] |
| Anostraca | 2.2 | 1.5 | [31] |
| Decapoda | – | 8.8 | [31] |
| | – | (5.5–11.2) | [42] |
| Branchiopoda | (1.5–6.4) | (0.9–12.2) | [42] |

**Table 2.4** Composition of crustacean shell wastes on a dry basis

| Source | Species | Protein (%) | Chitin (%) | Ash (%) | Lipid (%) | References |
|---|---|---|---|---|---|---|
| Shrimp | Penaeus notialis | 42.5 | 26.08 | 19.5 | 5.32 | [27] |
| | Pandalus borealis | 41.9 | 17.0 | 34.2 | 5.2 | [31] |
| | Metapenaeus affinis | 44 | 18.0 | 22.8 | 7.3 | [32] |
| | Penaeus monodon | 49.6 | 13.5 | 21.9 | 6.3 | [32] |
| | Xiphopenaeus kroyeri | 39.4 | 19.9 | 32.0 | 3.8 | [32] |
| | Litopenaeus vannamei (abdominal) | 25.9 | 43.8 | 27.9 | 2.4 | [32] |
| | Litopenaeus vannamei (head) | 40.7 | 23.2 | 22.2 | 13.9 | [32] |
| | Crangon crangon | 40.6 | 17.8 | 27.5 | 9.9 | [46] |
| Crab | Portunus pelagicus | 25.98 | 18.83 | 36.34 | 4.54 | [32] |
| | Scylla serrata | – | 16.8 | 37.7 | – | [32] |
| | Callinectes amnicola | 45.7 | 19.36 | 23.8 | 4.95 | [27] |
| | Callinectes sapidus Rathbun | 45.0 | 11.0 | 28.0 | 1.0 | [28] |
| | Chinoecetes opilio | 29.2 | 26.6 | 40.6 | 1.3 | [45] |
| Lobsters | Thenus orientalis | 26.0 | 20.4 | 38.2 | 3.0 | [32] |
| | Panulirus argus | 14.0 | 16.0 | 54.0 | 1.1 | [32] |
| Squids | Loligo chenisis | 57.2 | 34–37 | 1.37 | 1.9 | [32] |
| Crawfish | Procamborus clarkit | 29.8 | 13.2 | 46.6 | 5.6 | [31] |
| Krill | Euphausia superba | 41.0 | 24.0 | 23.0 | 11.6 | [44] |

A Russian patent [55] describes a method of preparing chitosan. The stock containing chitin is decolorized using hypochlorite and then calcined before deproteinization with NaOH. The extraction of chitin and chitosan from shrimp (*Metapenaeus monoceros*) has been described by Naznin [56]. Naznin reported the use of 30% commercial-grade HCL acid in the demineralization step and 1.5 N NaOH in the deproteinization step yielded a good quality of chitosan. Yamanami et al. [57] described a method to produce low molecular mass chitin powder to be used for making certain coating materials, plastics, and synthetic fiber. Li et al. [58] described a microwave method to prepare chitosan with a high degree of deacetylation (i.e., close to 95%). They reported chitosan prepared using the microwave method had a molecular mass varying from $3.8 \times 10^5$ to $2 \times 10^6$. They also reported that the molecular mass of the chitosan was inversely proportional to the degree of deacetylation. Although α-chitin is the most common form, the tightly arranged parallel β-chitin is useful for specific applications. Microwave technology has also been used by Cao et al. for the deacetylation of chitin [59]. They have used statistical methods to study the variables on the degree of deacetylation and molecular mass of the chitosan produced. The optimum conditions cited are reaction time under microwave treatment of 20 min, the concentration of NaOH 50%, soaking time of ethanol 2.5 h, and concentration of ethanol 80%. Chitin was soaked in ethanol before subjecting it to the microwave technique. Saito and Kosakada [60] have prepared β-chitin using squid tendons. They have prepared β-chitin with varying degrees of crystallinity and purity. Hammer and Carr [61] described several methods of production of chitin and chitosan. The authors used both chemical and genetically modified methods to produce chitin synthase, chitin deacetylase, and glutamine-fructose-6-phosphate aminotransferase. Optimization of reaction times is an essential factor in preparing chitosan, as is evidenced by the work of Yen and Mau [62]. The authors show that the average molecular mass of fungal chitosan prepared from shiitake stipes decreased with increased reaction times. They also studied the patterns of the chitosan synthesized using differential scanning calorimetry (DSC) and scanning electron microscopy (SEM). A new method called forced penetration has been used by Zhang et al. [63] to prepare chitosan of relatively high molecular mass. This method appears to be a rapid method of preparing chitosan from chitin materials. The preparation is carried out in a vacuum, which reduced the reaction time. The authors have characterized the product using various techniques such as powder XRD, FTIR, NMR, and SEM.

## 2.2.2 *Extraction of Chitin*

In the skeletal tissue of crustacean shells, protein and chitin combine to form a matrix through a covalent bond, which is then extensively calcified to yield hard shells [17, 64]. Chitin can be extracted from the shells using either the chemical or the biological (enzymatic) process. In the biological process, the reaction time to deproteinization is longer than the chemical process [65]. This makes the chemical process more attractive for the commercial production of chitin. The major steps involved in these processes are described below.

### 2.2.2.1 Chemical Process

The basic steps for chitin isolation from crustacean shells are as follows:
- Demineralization (acid removal of calcium carbonate).
- Deproteinization (removal of proteins).
- Depigmentation (removal of pigment).

Demineralization of source material (crustacean shells) starts with the removal of minerals such as calcium carbonate ($CaCO_3$) using a hydrochloric acid bath of 2.5% HCl solution. In this process, calcium carbonate reacts with hydrochloric acid and decomposes to form water-soluble calcium salt with the release of carbon dioxide ($CO_2$), as shown in the following equation [66]:

$$2HCl + CaCO_3 \rightarrow CaCl_2 + H_2O + CO_2 \uparrow . \tag{2.1}$$

Other minerals in the shellfish wastes also decompose to soluble salts in the presence of acid. The ash content of the demineralized shell indicates the effectiveness (i.e., strength) of the demineralization process [67]. The products from the demineralization of the shell contained 31–36% ash.

Deproteinization usually involves alkaline treatment [68]. In this process, the disruption of the chemical bonds between protein and chitin occurs, and protein (albumen) ultimately decomposes to water-soluble amino acids [16, 69]. In addition to deproteinization, partial deacetylation and depolymerization of chitin may occur due to alkali treatment [64]. In the chemical method, the order of the two steps of chitin extraction, namely demineralization and deproteinization, can also be reversed [16, 64, 67]. However, the extracted chitin structure remains intact and stable when demineralization steps occur first [70]. Table 2.5 provides the composition of the crawfish shell and the wholemeal. As can be seen from the table, calcium is the major constituent of the mineral portion of the shell. The other minerals are present in trace quantities. Acids such as 90% formic acid, 22 and 37% hydrochloric acid [71, 72], acetic acid [71], and sulfuric acid [71] have also been suggested for the demineralization step. The physical and chemical properties of the finished product depend on the temperature, duration, the concentration of the chemicals, as well as the concentration and the size of the crushed shells.

Some shells may also contain lipids and pigments (Table 2.5). The pigment in the crustacean shells forms complexes with chitin. Pigment residue may be removed from chitin by extraction at room temperature. A hydrocarbon solvent such as acetone, chloroform, ethyl acetate, ethanol, or mild oxidizing treatment carotenoids (e.g., astaxanthin) can be used for this extraction. A decolorizing step is sometimes incorporated to obtain white chitin. This step can also influence the characteristics of the

## 2.2 Chitosan Preparation Methods

**Table 2.5** Typical chemical analysis of wholemeal and shells of crawfish

| Composition | Wholemeal | Shell |
|---|---|---|
| Crude protein, % | 35.8 | 16.9 |
| Fat, % | 9.9 | 0.6 |
| Fiber, % | 16.5 | 23.6 |
| Chitin, % | 15.9 | 23.5 |
| Ash, % | 38.1 | 63.6 |
| Minerals | 12.3 | 24.8 |
| – Ca, % | 0.8 | 1.0 |
| – P, % | 1.0 | 0.1 |
| – K, % | 0.2 | 0.3 |
| – Mg, % | 545 | 200 |
| – Mn, ppm | 78 | 180 |
| – Fe, ppm | | |
| Astaxanthin, ppm | 78 | 108 |

Copyright © 1989, American Chemical Society, J Agric Food Chem 37(3):575–579. Reproduced from Ref. [73] with permission

chitin molecule. Decolorization agent includes sodium hypochlorite solution [74], absolute acetone [75], 3% hydrogen peroxide [76], and ethyl acetate [77]. Figure 2.3 shows the steps involved in the extraction of chitin from crawfish shells.

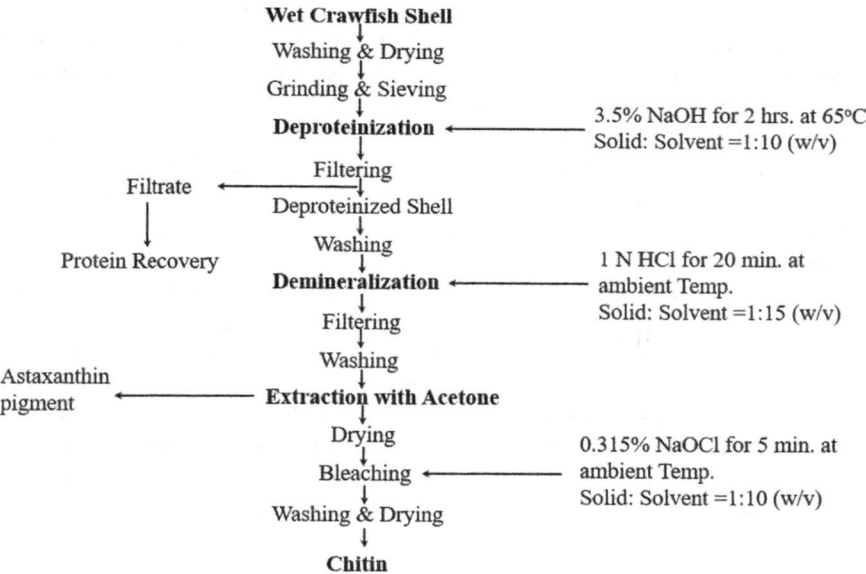

**Fig. 2.3** Steps in the extraction of chitin from crawfish. Copyright © 1989, American Chemical Society, J Agric Food Chem 37(3):575–579; Reproduced from Ref. [73] with permission

## 2.2.2.2 Enzymatic Process

Chitin isolated by chemical processes contains lysinoalanine and other amino acid derivatives. Another issue with the chemical process is the depolymerization of chitin. Ethylenediamine-tetra-acetic acid (EDTA) has been suggested instead of HCl for the removal of minerals [78, 79]. However, various environmental issues regarding the disposal of EDTA may prevent its use in the future. Various microorganisms are used for enzymatic deproteinization of crustacean shells for the production of chitin. However, in this method, crustacean shells are first demineralized and then deproteinated using a protease. Enzyme preparations with chitinase, chitosanase, and lysozyme are primarily used to hydrolyze chitin and chitosan [80]. For example, chitinases are capable of converting chitins and chitosans to low molecular mass products. The enzyme chitinases can break up the crystalline chitin structure by randomly hydrolyzing $\beta$-1,4 bonds between the sugar units [81, 82]. Horn et al. [82] reported that the degradation of chitin or cellulose is mono-, di-, and trisaccharides such as N-acetyl glucosamine and N, N'-di- acetylchitobiose, which are typical reaction end products. Beier and Bertilsson [83] reported bacteria as a significant mediator of chitin-degrading enzymes in nature. They summarize different mechanisms and the main steps involved in chitin degradation at a molecular level. They also discuss the coupling of community composition to measured chitin hydrolysis activities and substrate uptake.

Earlier attempts to use proteolytic enzymes to deproteinize crustacean waste include that of Takeda and Abe [84], Takeda and Katsuura [85], and Broussignac [86]. Enzymes used in these studies included tuna proteinase at pH 8.6 and 37.8 °C, papain at pH 5.5–6.0 at 37.5 °C, and a bacterial proteinase at pH 7.0 and 60 °C. However, it took over 60 h to deproteinate about 90% of the protein from the shell. Broussignac [86] recommended the use of gapain, pepsin, or trypsin for the deproteinization of crustacean shells to prepare chitin. However, Shimahara and Takiguchi [87] used bacterial protease from *Pseudomonas maltophili* to deproteinate crustacean shells and reported that after 24 h, the protein content remained in the shells was only about 1%.

Bustos and Healy studied [88] the deproteinization of prawn shell waste using different proteolytic microorganisms. The microorganisms tested were: *Pseudomonas maltophilia, Bacillus subtilis, Streptococcus faecium, Pediococcus pentosaseus*, and *Aspergillus oryzae*. All microorganisms in a medium that contained prawn shells and 0.2% phosphate buffer had good cell growth. Over 80% deproteinization of demineralized prawn shells was achieved with inoculant BAFP 202 for all the microorganisms except *P. maltophilia*. Chitin obtained by the microbial method showed a higher molecular weight than did chitin prepared by a chemical process.

Wang and Chio [89] conducted deproteinization tests by fermenting shrimp and crab shell powder using *Pseudomonas aeruginosa K-187* [89]. In the liquid-phase fermentation process, 55% of protein removal from shrimp and crab shell powder (SCSP) occurs after seven days, whereas 48 and 61% protein removal from shrimp shells and shrimp heads occurs after five days of liquid-phase fermentation. In the case of solid-phase fermentation, 68 and 46% protein was removed from crab shell powder and acid-treated powder, respectively, after ten days of fermentation. It was further observed that 82% protein was removed from shrimp shells, and 81% of the protein was removed from shrimp heads after five days of solid-phase fermentation.

Synowiecki and Al-Khateeb [46] provided an outline (Fig. 2.4) of a process for the extraction of chitin from demineralized shrimp shells using a commercially available protease, Alcalase from *Bacillus licheniformis*. The isolated chitin contained about 4.5% protein impurities.

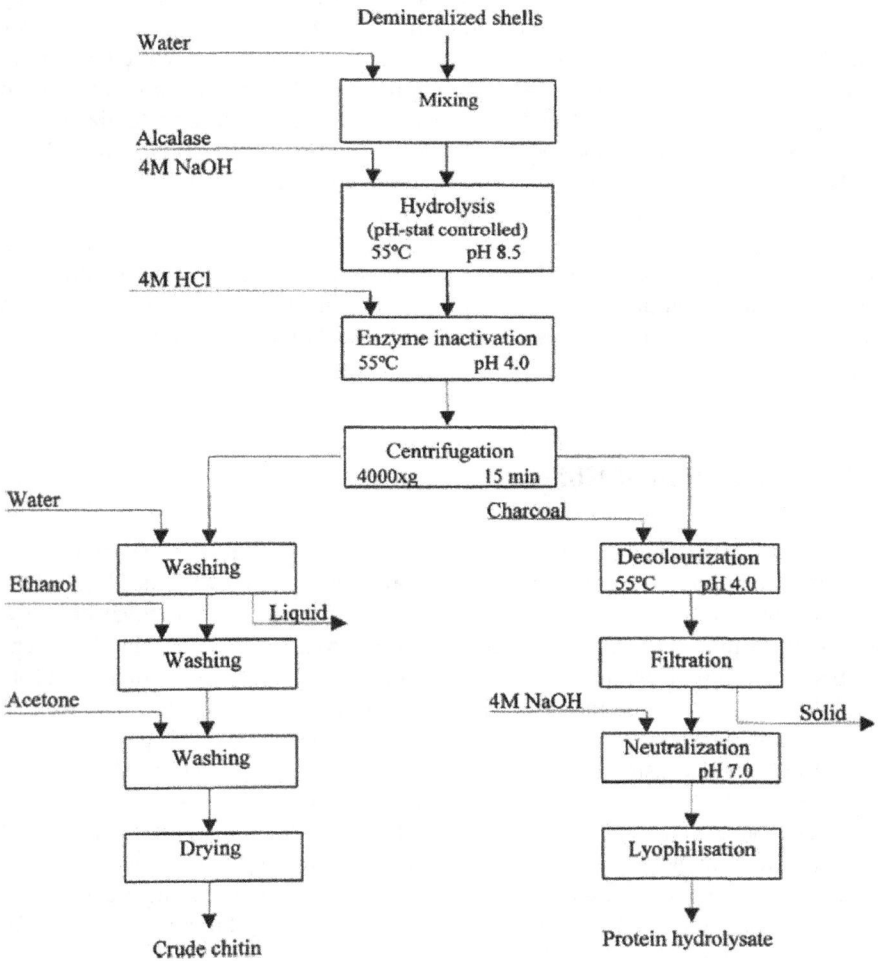

**Fig. 2.4** Flowsheet for isolation of chitin from demineralized shrimp shells by an enzymatic process. Copyright © 1999 Elsevier Science Ltd. Food Chem., 68:147–152, 2000. Reproduced from Ref. [46] with permission

Other microorganisms have been investigated for deproteinization of crustacean shell wastes including *Lactobacillus plantarum* [90], *Bacillus subtilis* [91, 92], *Lactobacillus paracasei* [93], *Lecanicillium fungicola* [94], and *Penicillium chrysogenum* [95]. Most of the processes to extract chitin start with demineralized shells. However, several researchers [90, 96] have investigated the microbial demineralization processes. Hall and Silva [96] used lactic acid bacterial fermentation of shrimp waste for demineralization with added carbohydrate sources such as cassava or molasses. Rao et al. [90] used organic acids and salt [97] as nutrients for the microorganisms. However, Jung et al. [98] suggested that a one-step combined cofermentation process using a lactic acid-producing bacteria and protease-producing bacteria may be used to extract chitin directly from shell wastes. They used *Lactobacillus paracasei subsp. tolerans* KCTC-3074, a lactic acid-producing bacterium, and *Serratia marcescens* FS-3, a protease-producing bacterium, to extract chitin from red crab shell wastes. A 97% demineralization and 53% deproteinization on day seven had occurred. Oh et al. [99] used *Pseudomonas aeruginosa* F722 to study the microbial demineralization and deproteinization of crab shell waste to extract chitin. For carbon sources, various concentrations of glucose were used. At the optimal temperature of 30 °C, demineralization was 92% and deproteinization was 63%. Figure 2.5 shows the results of their studies. Both the demineralization and deproteinization depend on temperature since all the microorganisms have an optimal growth temperature. Also, the solid-to-liquid ratio had a dramatic effect on the process.

## 2.3 Preparation of Chitosan

Chitosan is usually obtained by the deacetylation of chitin of greater than 50% under alkaline conditions. Deacetylation degree represents the proportion of N-acetyl-D-glucosamine units to the total number of units. It refers to the removal of acetyl groups from the chitin. During deacetylation, the hydrolysis of acetamide groups occurs when chitin is treated with 40% sodium hydroxide at 100 °C for 1–3 h [100, 101]. From a chemical point of view, either acid or alkali treatment can be used to remove acetamide groups from chitin [16]. However, the N-acetyl groups (acetamide) cannot be removed using acidic reagent without hydrolysis of chitin [67]. This is because the glycosidic (covalent) bonds are repeated monomers of chitin linked together. These bonds are susceptible to acid. In acid treatment, the hydrolysis of the glycosidic linkage occurs. This hydrolysis involves both the protonation of the glycosidic oxygen and the addition of water to yield the reducing sugar end groups [65]. Therefore, the alkaline method is preferred over the acid treatment method for the deacetylation of chitin [102].

Kurita [103] reported that the chitin samples have different degrees of deacetylation (i.e., 0.05–0.15). The degree of deacetylation depends on the origin and the mode of isolation of chitin samples. Table 2.6 shows the different processes of the deacetylation of chitin. The deacetylation of the chitin sample can be performed either at

## 2.3 Preparation of Chitosan

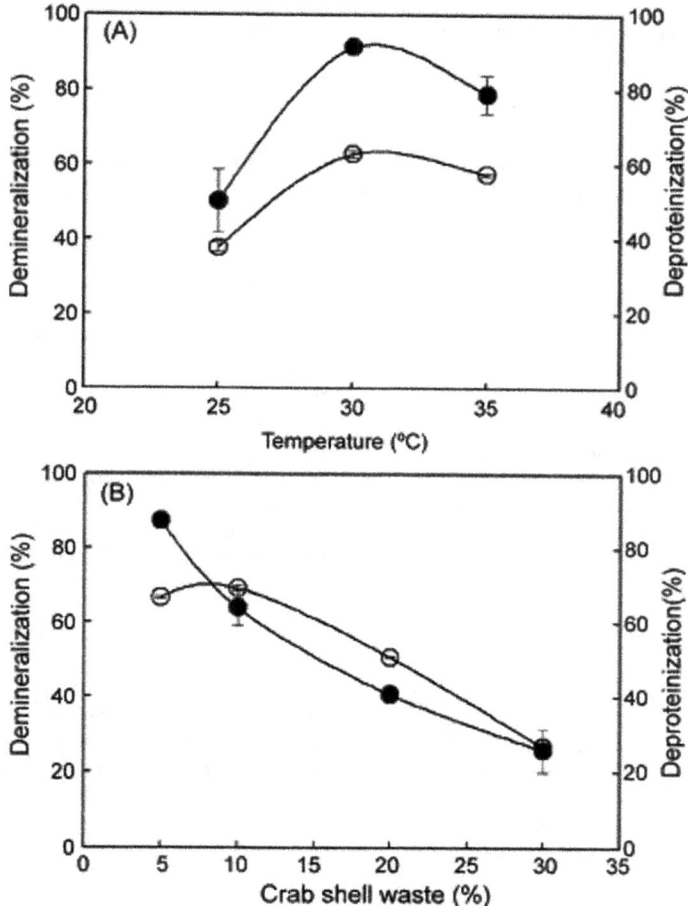

**Fig. 2.5** Effects of temperature (a) and the amount of crab shell waste (b) on demineralization (●) and deproteinization (○) after fermentation with crab shell waste for seven days with *Pseudomonas aeruginosa* F722. Copyright © 2007 Elsevier Ltd. Process Biochemistry, 42(7):1069–1074, 2007. Reproduced from Ref. [99] with permission

room temperature or at elevated temperature. At room temperature (25 °C), chitin is dispersed in NaOH solution for 3 h or more, followed by dissolution in crushed ice around 0 °C. This process is also called a homogeneous method of deacetylation [16]. The process performed at high temperature is called a heterogeneous process in which chitin is treated with concentrated NaOH solution at 100 °C or higher to remove acetyl groups. Here, 2-amino-3-deoxy-β-D-glycopyronase (deacetylated) unit is created from 2-acetamido-2deoxy-β-D-glucopyranose (acetylated) unit depending on the concentration of NaOH, temperature, and reaction time [104]. For instance, chitosan

**Table 2.6** Preparation of chitosan from chitin using different processes

| Raw material | Chitin | | |
|---|---|---|---|
| Deacetylation process | Heterogeneous | Homogeneous | Enzymatic |
| Reagent | 40–50% NaOH | 40–50% NaOH | Enzyme |
| Process temperature | 100–150 °C | 25 °C | – |
| Reaction time | 1 to a few hours | For 3 h followed by dissolution in crushed ice at approximately 0 °C | The molecular weight of chitosan decreased after the glycosidic bonds were broken by this method |
| Reaction product | Chitosan | Chitosan | Chitosan |
| Advantage | Chitosan produced has a higher degree of deacetylation up to 85–90% | This process produces deacetylation with acetyl groups uniformly distributed along the chain of chitosan structure | This is an alternative process to prepare novel chitosan oligomers. This process minimizes adverse chemical modification and promotes the biological activities of chitosan |
| Disadvantage | Solubility and degree of aggregation of chitosan vary in aqueous solution due to irregular distribution of N-acetyl-D-glucosamine residue<br>– Low molecular weight<br>– Environmental pollution | This process produces soluble chitosan with only 48–55% degree of deacetylation<br>– Low molecular weight<br>– Environmental pollution | The enzyme is not effective on insoluble chitosan. The degree of deacetylation is <10%. Cost is high of the non-reusable enzymes for batch production systems |

Copyright © 2015 Younes, I., Rinaudo, M., Licensee MDPI, Basel, Switzerland. Creative Commons Attribution CC BY license, Mar. Drugs 2015, 13(3):1133–1174; https://doi.org/10.3390/md1303 1133, Reproduced from Ref. [16]

with ~85–99% degree of deacetylation is achieved by the heterogeneous process, and 48–55% degree of deacetylation is achieved by the homogeneous process [16].

## 2.4 Chitin and Chitosan Properties

Chitosan is the deacetylated form of chitin with a degree of deacetylation greater than 50%. In general, chitin is a high molecular mass polymer linked by [β-(1-4-] bonds.

## 2.4 Chitin and Chitosan Properties

Structural studies reveal that chitosan is closer to α-chitin. Chitosan and chitin represent a continuum of copolymers. The amino and the hydroxyl groups of chitosan can be easily altered chemically, which improves the mechanical properties of chitosan [105, 106]. Compared to cellulose and other naturally occurring polymers, chitin and chitosan are suitable resource materials for many applications due to their non-toxicity, biocompatibility, and biodegradability properties. For instance, chitosan is used in water treatment, textiles, food, and pharmaceutical industries [105–109]. Biopolymer chitosan has drawn particular attention in the scientific community due to its biological and structural properties. The degree of deacetylation (DD) and molecular weight (MW) are considered structural (physicochemical) parameters of chitosan. The DD and MW represent the proportion of deacetylated units in chitosan, depending mainly on the source of chitin and chitosan preparation process. Biodegradability is one of the most important biological properties of chitosan. Other biological properties, such as biocompatibility and bio-adhesiveness of chitosan, allow its use in various medical applications [107]. Also, antifungal, antibacterial, antioxidant, and antimicrobial are considered critical biological properties of chitosan for their applications in biomedical, food, and pharmaceutical industries. Table 2.7 shows the intrinsic properties of chitosan.

Several studies focused on the physicochemical properties of chitosan and their effects on the biological properties to understand the structure-property-activity relationship of chitosan products [108–111]. Table 2.8 shows typical examples of chitosan properties and their relationship to the structural (physicochemical) properties.

Chitin is highly hydrophobic and is insoluble in water and most organic solvents due to its acetylated amine groups. It is soluble in hexafluoroisopropanol, hexafluoroacetone, chloroalcohols in conjunction with aqueous solutions of mineral acids [11]. Chitosan is insoluble in neutral or alkaline pH conditions of the water. It is reported that the chitin with less than 40% degree of deacetylation becomes completely insoluble in water due to the following reasons [111]:

1. Formation of numerous intermolecular H-bonds that occur between alcohol, amide, and ether functionalities distributed on the repeating units all along the polymer chains.
2. The hydrophobic interactions due to the presence of the methyl groups of the acetamide functions and –CH and –$CH_2$ groups of the glucosidic rings.

Furthermore, the compact α-, β-, and γ- conformations formed through hydrogen bonding make chitosan insoluble in water [112]. The units cell of α-chitin contains disaccharide sections of two chains with full intramolecular C(3)–OH...O(5) and intermolecular C(2)NH...O=C(7) hydrogen bonding and O(6′)H...O(7)/O(6)H...O(6′) intra-/intermolecular hydrogen bonds [113]. The hydrogen bonding in the chitosan structure is responsible for the insolubility of chitin in water and the formation of chitosan fibrils. The interchain hydrogen bonding between hydroxyl groups within chitosan makes it paracrystalline and non-porous.

The solubility of chitosan is an essential parameter as a higher percentage of chitosan solubility in an aqueous solution is desirable for medical applications.

Table 2.7 Intrinsic properties of chitosan

| Physical–chemical properties | Polyelectrolyte (at acidic pH) | Biological properties |
|---|---|---|
| • Linear amino-polysaccharide with high nitrogen content | • Cationic biopolymer with high charge density (one positive charge per glucosamine residue) | Biocompatible |
| • Rigid D-glucosamine structure | • Chelating and complexing properties | • Safe |
| • Numerous reactive groups | • Flocculant agent | • Biodegradable to normal body constituent's natural polymer |
| • High crystallinity | • Entrapment and adsorption properties | • Binds to mammalian and microbial cells aggressively |
| • Hydrophilicity | • Filtration and separation | • Renovated effect on connective gum tissue |
| • Capacity to form hydrogen bonds | • Film-forming ability | Bioactivity |
| • Weak base (powerful nucleophile, pKa ~6.3) | • Adhesive | • Hemostatic (causes stop bleeding) |
| • Soluble in dilute acidic aqueous solutions | | • Accelerates the formation of osteoblast responsible for bone formation |
| • Insoluble in water and organic solvents | | • Fungistatic |
| • Ionic conductivity | | • Spermicidal |
| | | • Anticholesteremic |
| | | • Anticancer or antitumor |
| | | • Antiacid and antiulcer |
| | | • Central nervous system depressant |
| | | • Immunoadjuvant |

Copyright © All rights are reserved by EM El Kady, Creative Commons open access CC BY license, World J Agri & Soil Sci. 3(1):2019. WJASS.MS.ID.000553. https://doi.org/10.33552/WJASS.2019. 03.000553. Reproduced from Ref. [10]. Copyright © 2014 King Saud University. Production and hosting by Elsevier B.V. Arabian Journal of Chemistry, 2017, 10:S3826–S3839. Reproduced from Ref. [107] with permission. Copyright© 2020 Jiménez-Gómez, C. P., and Cecilia, J. A., Licensee MDPI, Basel, Switzerland. Creative Commons Attribution (CC BY) license, Molecules 2020, 25:3981. https://doi.org/10.3390/molecules25173981. Reproduced from Ref. [107a]

Chitosan is soluble in nearly all dilute aqueous acidic solutions. For example, the degradation of chitosan structure will take place under acidic conditions such as acetic acid ($CH_3COOH$) and also in hydrochloric (HCl) and nitric ($HNO_3$) acids at a concentration of 1% or less, whereas the parent chitin is insoluble in most organic solvents. Table 2.9 shows the solubility of chitin and chitosan in different solvents.

**Table 2.8** Relationship between structural and biological properties of chitosan

| Property | Structural property | Relationship |
|---|---|---|
| Analgesic | DD | The analgesic property of chitosan is directly proportional to its DD |
| Antimicrobial | DD, MW | The antimicrobial property of chitosan is directly proportional to DD and MW |
| Antioxidant | DD, MW | The antioxidant property of chitosan is directly proportional to DD and inversely proportional to MW |
| Biodegradability | DD, MW | The biodegradability property of chitosan is inversely proportional to DD and MW |
| Biocompatibility | DD | The biocompatibility property of chitosan is directly proportional to DD |
| Crystallinity | DD | The crystallinity of chitosan is inversely proportional to MW [crystallinity increases with the decrease of MW] |
| solubility | DD | The solubility of chitosan is directly proportional to DD of chitosan |
| Viscosity | DD | The viscosity of chitosan is directly proportional to DD of chitosan |

Copyright © 2009 Bentham Science Publishers Ltd., Current Chemical Biology, 3(2009):203–230. Reproduced from Ref. [1] with permission. Copyright © 2011 Elsevier Ltd., Progress in Polymer Science, 36:981–1014, 2011. Reproduced from Ref. [110] with permission

**Table 2.9** Solubility of chitin and chitosan

| Chitin | Chitosan |
|---|---|
| 5% LiCl in dimethylacetamide, diethylacetamide | Dilute aqueous organic or mineral acids below pH 6.5 |
| LiCl in N-methyl-2pyrrolidone | Dimethylsulfoxide |
| $CaCl_2 \cdot 2H_2O$-saturated methanol | p-Toluene sulfonic acid |
| Hexafluoroisopropyl alcohol | 10-Camphorsulfonic acid |
| Heaxafluoroacetone sesquihydrate | |
| The mixture of 1,2-dichloroethane and trichloroacetic acid (35:65) | |
| A fresh saturated solution of lithium thiocyanate | |

Copyright © 2008 Elsevier Ltd., Reactive and Functional Polymers 68:1013–1051, Reproduced from Ref. [114] with permission

The soluble–insoluble phase transition for chitosan occurs at its pKa value (acid dissociation constant) around pH between 6 and 6.5. Therefore, the free amino groups of chitosan get less protonated in the pH range of 6.0–6.5. Wang et al. [115] determined the dissociation constants of chitosan in aqueous solutions by the potentiometric titration. Katchalsky's equation described below is used to describe the potentiometric behavior [115]:

$$pK_a = p^H + \log\left[\frac{1-\alpha}{\alpha}\right] = pK_{\text{int}} - \frac{\varepsilon \Delta\varphi(\alpha)}{KT}. \tag{2.2}$$

The definition of $\varepsilon$, $k$, and $T$ is the unitary charge, Boltzmann's constant, and the absolute temperature, respectively. For the rodlike model, $\Delta\varphi(\alpha)$ represents the difference between the surface potentials of the polyion and the reference state at a distance ($\alpha$) from the axis.

Navarro et al. [116] used the Henderson–Hasselbalch equation to determine acid–base properties and chitosan deprotonation. The pKa value of chitosan varied with ionic strength and the salt concentration in the solution [117]. Moreover, in solution, the degree of deacetylation influences the balance of hydrophobic interaction and hydrogen bonding on chitosan, thus significantly affecting pKa values [118]. Chitosan starts precipitating when the pH of the solution increases more than the pKa value of chitosan. One of the possible explanations of this phenomenon is the change of molecular conformation in chitosan leading to the formation of precipitates. The changes of pKa value occur due to the formation of intermolecular hydrogen bond involving $-NH_2$ group [119]. This precipitate of hydrogen-bonded chitosan may have a similar structure to those found in chitin. The degree of ionization of the precipitated chitosan is independent of temperature, but depends on the concentration of deacetylated monomer and salt [120]. The pKa value is highly dependent on the degree of deacetylation (DD) [121]. The solubility of chitosan is dependent on both the degree of deacetylation (DD) and the method of deacetylation used to extract chitosan from chitin [121]. This is because the stiffness of the chitosan polymer chain increases with the decrease of DD. The stiffness results from the amplified steric hindrance from the acetylated unit. Hsu et al. [122] reported that the degree of deacetylation affects the hydrophilicity of the chitosan films [122]. Chitosan sample with a higher degree of deacetylation tends to dissolve in an acidic solution through the protonation of amine ($-NH_2$) groups. Amine groups come from the glucosamine unit in the chitosan structure. In a low-density aqueous ionic environment, as pH is reduced (acidic pH), chitosan's molecular conformation change occurs. The glycosidic linkages of chitosan are susceptible to acid, which leads to the degradation of chitosan in acidic solution [123]. The protonation of amino groups of glucosamine units contributes to the disruption of hydrogen bonding and the solvation of cationic sites. Subsequently, the solubilization of chitosan occurs when the balance between solvent/polymer and polymer/polymer interactions becomes favorable [114]. To a certain extent, the solubility of chitosan depends on the degree of deacetylation and molecular weight. The molecular weight affects the rate of degradation and mechanical properties. No et al. [124] reported that the degree of deacetylation of chitosan must be at least 85% complete to achieve the desired solubility of chitosan.

Chitosan is also susceptible to the oxidative attack, which leads to chain repulsion, diffusion of ions inside the gel, and dissociation of secondary interaction. The rate of the oxidative attack increases with temperature, the concentration of peroxide, and degree of deacetylation [125]. Yin et al. [126] reported that hydroxyl radical plays an essential role in the oxidative degradation of chitosan. They also suggested that the oxidative degradation of chitosan solution by peroxide can be controlled with the pH

## 2.4 Chitin and Chitosan Properties

of the solution. Protons will scavenge hydroxyl radicals in solution; therefore, the higher acidity of the solution will cause to slow down the degradation of chitosan. Chitosan, like other biopolymers, is very sensitive to degradation induced by the hydrolytic, thermal, photoionic, ultrasonic, and gamma radiation [127]. This degradation is due to chitosan undergoing the chain scission reaction. The degradation of chitosan in the liquid phase by the gamma radiation is due to the formation of reactive hydroxyl radicals from the radiolysis of water molecules. The degradation process usually begins with random splitting of β (1 → 4)-glycosidic bonds (depolymerization) followed by N-acetyl linkage (deacetylation) [128]. Maznah et al. [129] reported that hydroxyl radicals could attack the (1 → 4) glycosidic linkages resulting in the breaking the chitosan chain. The $H^+$ and $OH^-$ radicals formed by radiolysis during irradiation of water accelerate the molecular chain scission of chitosan. Possible mechanisms of chitosan degradation processes are as follows (Fig. 2.6).

Dash et al. [110] reported that the rate of degradation and mechanical properties of chitosan is related to its degree of deacetylation and length of the chain

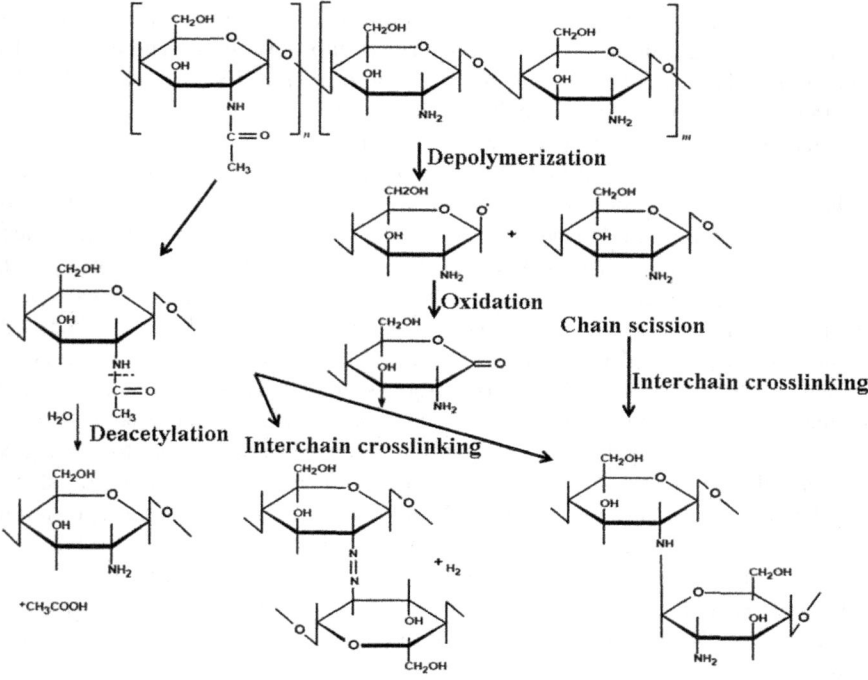

**Fig. 2.6** Possible degradation mechanisms of chitosan. Copyright © 2015 Szymanska, E. and Winnicka, K., Licensee MDPI, Basel, Switzerland. Creative Commons Attribution CC BY license, *Mar. Drugs* 2015, 13(4):1819–1846. https://doi.org/10.3390/md13041819. Reproduced from Ref. [130]

(Mw of polymer chain). Clayer et al. [131] have studied the structure–property relation of some polysaccharide gels. They found the same general chemical structure for the series of linear copolymers of linked β (1 → 4) glucosamine, and N-acetylglucosamine. They also found that the concentration of polymer must be above a critical value to balance between hydrophobic and hydrophilic interactions, and the gelation must occur simultaneously everywhere in the medium.

The viscosity of the chitosan solution plays a significant role in quality control in the industrial uses of chitosan. The intrinsic viscosity of the polymer is related to the molecular weight of the polymer. The weight average molecular weight (MW) of native chitin exceeds millions. The commercial chitosan products are available in three different molecular weight ranges: low molecular weight, medium molecular weight, and high molecular weight. The desired molecular weight is obtained by the chemical or enzymatic processes for the various applications in food and pharmaceutical activities [109]. It is reported that chitosan with a molecular weight less than 220,000 formed a random coil conformation, whereas chitosan with a molecular weight higher than 220,000 formed a more compacted molecule [132]. Chitosan with a high molecular weight possesses more intramolecular hydrogen bonds and more even charge distribution. The amino groups of chitosan in an aqueous medium get protonated, and the polymer is transformed into polycation. The electrostatic repulsion of charged groups affects the chain conformation [133]. Kasaai et al. [134] reported that a higher value of the degree of acetylation results in a rigid conformation, leading to a higher degree of expansion of chitosan. The number of positive charges of chitosan will be higher at the low pH of the solvent. This leads to a higher degree of expansion of chitosan due to electrostatic repulsion. The viscosity of the chitosan solution depends on several factors like DD value, average molecular weight, concentration, ionic strength, pH, and temperature [112]. In an aqueous solution, the polymer conformation and the polymer to solvent interactions depend on the number of positive charges ($NH_3^+$) on chitosan and the pH and ionic strength of the solvent [134]. Morris et al. [135] reported that chitosan has conformation close to the semi-flexible coil or rigid rod-type structure. They also reported a large number of possible conformations which could fall into either of these two categories [135]. The viscosity of the chitosan solution is linearly proportional to the degree of deacetylation of chitosan. At a higher degree of deacetylation, chitosan shows an extended conformation with a more flexible chain due to the charge repulsion in the molecule [136]. The viscosity of the chitosan solution is also affected by factors such as concentration and temperature. As the chitosan concentration in the solution increases and the temperature decreases, the viscosity of the chitosan solution increases. The DD also plays a significant role in affecting the molecular weight of chitosan. A lower DD leads to a higher molecular weight [137].

Chitosan is semi-crystalline. The degree of crystallinity is a function of the degree of deacetylation [138]. Crystallinity is maximum for both chitin (i.e., 0% deacetylated) and fully deacetylated (i.e., 100%) chitosan. Table 2.7 shows that the crystallinity of chitosan is related to the physicochemical properties of chitosan. Figure 2.7 shows the typical XRD patterns for pure chitosan and chitosan–iron

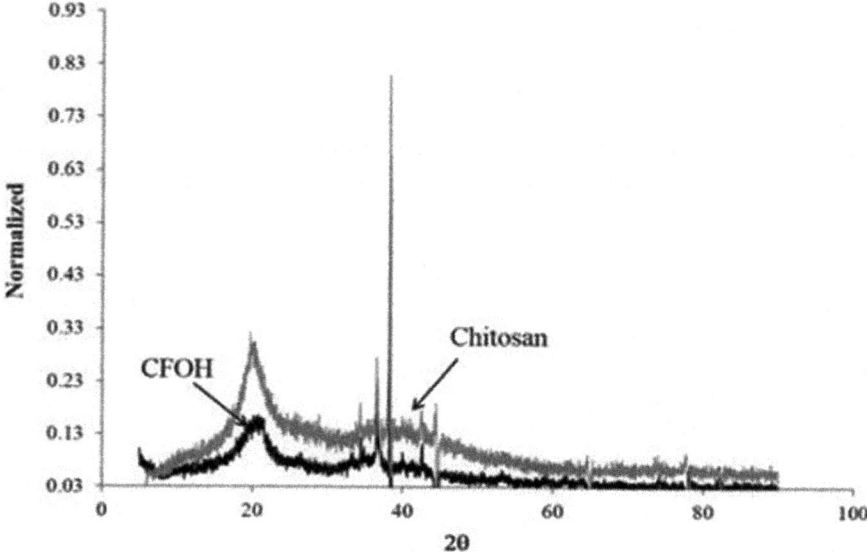

**Fig. 2.7** XRD pattern for chitosan and chitosan–iron-based (CFOH) bead. Copyright © 2014, Taylor & Francis, Separation Science and Technology, 49:2863–2877, 2014. Reproduced from Ref. [106] with permission

(CFOH) beads. The chitosan sample showed a diffraction peak near 20°, indicative of chitosan's relatively regular crystal lattices (110, 040) [139].

Ogawa et al. investigated the crystal diffraction pattern of chitosan using X-ray diffraction measurement [140]. Chitosan membrane samples were prepared using chitosan with different molecular weights in their study. They reported that chitosan membrane samples show hydrated and anhydrous crystal diffraction patterns at approximately 10° ($2\theta$) and 15° ($2\theta$). The chitosan sample with a low molecular weight showed a strong reflection ring for anhydrous crystal at 15° ($2\theta$), and its crystallinity increased with the decreasing chitosan molecular weight. At the same time, the diffraction pattern changed from anhydrous to hydrated crystal with an increase in the molecular weight of the chitosan sample. The degree of crystallinity also depends on the degree of deacetylation and the molecular weight (MW) of chitosan. The higher the degree of deacetylation of chitosan means low acetyl groups, increasing the degree of crystallinity but decreasing the molecular weight. Therefore, the chitosan chain becomes more flexible than the chitosan with a lower degree of deacetylation [141, 142]. Foster et al. [143] studied the relationship between crystallinity and the degree of deacetylation (DD) of the chitosan. They observed that the crystallinity of chitosan film increased linearly with the increase of DD of chitosan sample. Changing the DD of the chitosan also changed other essential properties such as MW, ash content, protein content, and crystallinity [144]. The main structural parameters of the chitosan influencing the physical and biological properties are its molecular weight and degree of deacetylation. The degree of deacetylation identifies the biopolymer as chitin or chitosan. The molecular weight determines the viscosity and rate of degradation [145].

The biodegradability and biocompatibility are critical biological properties of chitosan (Table 2.8) for medical and biomedical uses. These properties are related to the cationic nature of chitosan. The cationic nature of chitosan results from the presence of free protonated amino groups and depends on the degree of deacetylation (DD). However, molecular weight (MW) also impacts biological properties [145]. For instance, the degradation of chitosan is affected by the length of the chain. The length of the chain depends on the MW of chitosan [1]. Chitin and chitosan degrade before melting. A significant interest in chitin and chitosan arises because the behavior and chemical characteristics of these two substances are similar to those of lysozymes. Lysozyme is an enzyme present in the human body. Chitosan is biodegradable as it is quickly metabolized by specific enzymes, especially lysozymes [108]. The chemical properties of chitosan related to hemostasis include molecular weight, the extent of ionization, degree of acetylation, and degree of crystallinity. The ability to bind with tissues is a function of these parameters [146]. The properties such as biocompatibility, biodegradability, and mucoadhesion have expanded the potential uses of chitin and chitosan in biomedical applications [114].

Mireless-DeWitt [112] reported that the quality and properties of chitosan products such as purity, viscosity, MW, DD vary significantly due to several factors associated with the manufacturing process. Based on the structural properties and purity, the applications of chitosan can be classified, (1) technical grade (mostly used for agriculture and water treatment), (2) pure grade (mostly used for food and cosmetic applications), and (3) ultra-pure grade (used in biopharmaceutical industry).

Commercial- or technical-grade chitosan will often have a random distribution of the remaining N-acetyl glucosamine units after deacetylation. Also, the glucosamine residues in the chitosan structure are polydispersed. Therefore, the molecular weight of most technical-grade chitosan is identified as average molecular weight instead of distinct molecular weight [147]. In general, commercial- or technical-grade chitosan is classified based on its molecular weight range. The molecular weight, molecular weight distribution, degree of deacetylation, and purity level of commercial-grade chitosan may vary due to its raw material source [130].

The application of chitosan and its derivatives as new physiological materials has been investigated due to their antitumor, immunoenhancing, antimicrobial, and hypocholesterolemic properties [148]. Chitosan has been studied for various biomedical applications such as artificial skin, absorbable surgical sutures, wound-healing accelerators, tissue engineering, drug delivery, and medical devices. Szymanska and Winnicka [130] reported that there are hardly any chitosan-based pharmaceutical products available commercially (except for hemostatic dressing, wound-healing and nutraceutical products). They identified the following factors that affect the stability of chitosan in the product (Table 2.10).

Uses of chitosan in the medical field require chitosan with ultra-pure chemical, toxicological, physical, morphological, and mechanical properties. Chitosan with ultra-pure quality is termed as medical-grade chitosan and is sought to prepare biomedical products and medical devices. Struszczyk and Struszczyk [149] reported

## 2.4 Chitin and Chitosan Properties

**Table 2.10** Factors affecting chitosan stability

| Factors | | |
|---|---|---|
| Internal | Environmental | Processing |
| • Purity level<br>• Molecular weight<br>• Polydispersity index<br>• Degree of deacetylation<br>• Moisture content | • Humidity<br>• Temperature | • Involve acidic dissolution<br>• Sterilization<br>• Thermal processing<br>• Physical method (centrifugation and compression force) |

Copyright © 2015 Szymanska, E. and Winnicka, K., Licensee MDPI, Basel, Switzerland. Creative Commons Attribution CC BY license, *Mar. Drugs* 2015, 13(4):1819–1846. https://doi.org/10.3390/md13041819. Reproduced from Ref. [130]

the following crucial aspects of chitin and its derivatives that affect the safety and performance of medical devices:

1. Purity (medical chitosan processes the chitin under clean-room conditions in a series of verifiable purification steps).
2. Reproductivity of chitin sources.
3. Absence of European Standards regulating requirements for chitin and its derivatives in the design of medical devices.
4. Compliance with the European Standards [EN ISO 22442-1/2/3 Standards].
5. Sterilization of medical devices containing chitin or its derivatives and critical process may affect their biocompatibility.

Table 2.11 shows the chitosan and chitosan hydrochloride properties recommended by the European Pharmacopeia 6.0 and the United States Pharmacopeia 34-NF 29.

Several researches have reviewed the preparation and characterization methods for chitosan for pharmaceutical and biomedical applications [148, 150, 151]. Marques et al. [150] stressed the need of accurate information about MW, DD, viscosity, and purity as characterization parameters for medical-grade chitosan. In addition, it is reported that chitosan sample must pass bioburden (<100 cfu/g), pyrogenicity, and bacterial endotoxin ($\leq$20 EU/g, endotoxin unit) test in order to meet primary requirements for medical applications [152–154].

Skaugrud et al. [155] reported that one of the main requirements to address regulatory issues related to pharmaceutical applications of biopolymer (chitosan) based materials is the proper characterization of the physicochemical (structural) properties of biomaterial. The regulatory issues related to pharmaceutical or biomedical uses of chitosan are given in Table 2.12. The purity (ash content), moisture, proteins, endotoxin, and heavy metals content are essential parameters for human food and medical applications. These parameters play a crucial role in determining the mechanism and the speed of polymer degradation [156]. For instance, the environmental moisture content is responsible for a plasticizing or swelling effect of solid or semi-solid polymer assemblies [130] that can alter chitosan's the physicochemical and biological properties. For pharmaceutical use, the endotoxin content of the

Table 2.11 Chitosan properties recommended by the European Pharmacopeia 6.0 and the United States Pharmacopeia 34-NF 29

| Parameters | Acceptance criteria | |
| --- | --- | --- |
| | Chitosan hydrochloride (Eur. Ph. 6.0) | USP 34-NF 29 |
| The appearance of solid product | White or almost white fine powder | Not determined |
| Degree of deacetylation | 70–95% | 70–95% |
| Distribution of molecular weight | Not determined | 0.85–1.15 |
| pH of 1% (g/mL) solution | 4.0–6.0 | Not determined |
| Loss on drying | Not determined | $\leq 5\%$ |
| Insoluble/impurities | $\leq 0.5\%$ | $\leq 1.0\%$ |
| Heavy metals | $\leq 40$ ppm | $\leq 10$ ppm |
| Iron | Not determined | $\leq 10$ ppm |
| Sulfated ash | $\leq 1\%$ | Not determined |
| Protein | Not determined | $\leq 0.2\%$ |
| Microbiological contaminants | Not determined | Absence of *Pseudomonas aeruginosa* and *Staphylococcus aureus* |
| Aerobic microbials | Not determined | $10^3$ CFU |
| Molds and yeasts | Not determined | $10^2$ CFU |

Copyright © 2015 Szymanska, E. and Winnicka, K.; Licensee MDPI, Basel, Switzerland. Creative Commons Attribution CC BY license, *Mar. Drugs* 2015, 13(4):1819–1846; https://doi.org/10.3390/md13041819. Reproduced from Ref. [130]

Table 2.12 Regulatory issues [130, 150, 155]

| Characterization | Product specification (purity, viscosity, MW, DD)<br>Product performance<br>• Quality<br>• Efficacy<br>Product stability |
| --- | --- |
| Manufacture | Good manufacturing practice (GMP)<br>• Reproducibility of the scientific data<br>• Validity of the scientific data<br>Documentation (DMF: Drug Master File) |
| Toxicology and safety | Risk assessment studies<br>• To evaluate possible adverse immune responses that affect both safety and efficacy of the product<br>Application-specific studies |

polymer is required to be reduced to an acceptable level. Also, the biopolymer must be manufactured in compliance with good manufacturing practice (GMP) guidelines. Primary toxicological data and polymer properties are documented in Drug Master Files (DMF).

## 2.5 Summary

Chitin can be extracted from crustacean waste using both chemical and biological (enzymatic) processes. Chitin production using a chemical process requires a shorter reaction time than the biological process. Whereas the biological process minimizes adverse chemical modification and promotes the biological activities of chitosan, the higher cost of non-reusable enzymes is one of the drawbacks of biological process. The natural polymer chitosan is the deacetylation form of chitin. The deacetylation of chitin can be carried out using either acid or alkali treatment methods. The alkaline method of deacetylation is preferred over acid treatment method as the N-acetyl groups (acetamide) cannot be removed using acidic reagent without hydrolysis of chitin. Chitosan is a polysaccharide of considerable interest because of its abundance and unusual combination of structural and biological properties. Notable properties of chitosan including biodegradability, biocompatibility, and inertness are making it potential candidate for many applications in medical, biomedical, cosmetics, agricultural, and food industries.

## References

1. Amine R, Tarek C, Hassane E et al (2021) Chemical proprieties of biopolymers (chitin/chitosan) and their synergic effects with endophytic Bacillus species: unlimited applications in agriculture. Molecules 26:1117. https://doi.org/10.3390/molecules26041117
2. Fernando LD, Dickwella Widanage MC, Penfield J et al (2021) Structural polymorphism of chitin and chitosan in fungal cell walls from solid-state NMR and principal component analysis. Front Mol Biosci 8:727053. https://doi.org/10.3389/fmolb.2021.727053
3. Abo Elsoud MM, El Kady EM (2019) Current trends in fungal biosynthesis of chitin and chitosan. Bull Natl Res Cent 43:59. https://doi.org/10.1186/s42269-019-0105-y
4. Silva TH, Alves A, Ferreira BM, Oliveira JM, Reys LL, Ferreira RJF, Sousa RA, Silva SS, Mano JF, Reis RL (2012) Materials of marine origin: a review on polymers and ceramics of biomedical interest. Int Mater Rev 57:276–306
5. Praiser ER, Lombardi DP (1989) Chitin source book: a guide to research literature. Wiley, Chichester
6. Roberts GAF (1992) Chitin chemistry. The Macmillan Press Ltd., London
7. Prasad N, Ramakrishnan C (1972) Structure and conformation of chitins, 1. Mathematics of the rigid body refinement procedure as applied to the structure of α-chitin. Ind J Pure Appl Phys 10:501–505
8. Gardner KH, Blackwell J (1975) Refinement of the structure of β-chitin. Biopolymers 14:1581–1595
9. Ogwa K (2005) Conformational diversity of chitosan. J Metals, Mater Miner 15(1):1–5

10. Mahmoud MG, El Kady EM, Asker MS (2019) Chitin, chitosan and glucan, properties and applications. World J Agri Soil Sci 3(1). WJASS.MS.ID.000553. https://doi.org/10.33552/WJASS.2019.03.000553
11. Dutta PK, Ravikumar MNV, Dutta J (2002) Chitin and chitosan for versatile applications. J Macomolecular Sci Part C C42:307–354
12. Merzendorfer H, Zimoch L (2003) Chitin metabolism in insects: structure, function and regulation of chitin synthases and chitinases. J Exptl Biol 206:4393–4412. (a) Zhang X, Yuan J, Li F, Xiang J (2021) Chitin synthesis and degradation in crustaceans: a genomic view and application. Mar Drugs 19:153. https://doi.org/10.3390/md19030153
13. Borzacchiello A, Ambrosio L, Netti PA et al (2001) Chitosan-based hydrogels: synthesis and characterization. J Mater Sci Mater Med 12:861–864
14. Muzzareli RAA (1977) Chitin. Pergamon of Canada, Toronto
15. Muzzarelli RAA (1973) Natural chelating polymers. Pergamon of Canada, Toronto
16. Younes I, Rinaudo M (2015) Chitin and chitosan preparation from marine sources. Structure, properties and applications. Mar Drugs 13:1133–1174
17. Arbia W, Arbia L et al (2013) Chitin recovery using biological methods. Food Technol Biotechnol 51(1):12–25
18. Chen B, Peng X, Cal C, Dong Q (2006) Research on the hybrid fiber-reinforced microstructure of chafer cuticle. Solid State Phenom 111:135–138
19. Peter MG, Chitin and chitosan in Fungi. Biopolymer online, Wiley-VCH Verlag GmbH & Co. https://doi.org/10.1002/3527600035.bpol6005
20. Tharanathan RN, Kittur FS (2003) Chitin-the undisputed biomolecule of great potential. Critical Rev Food Sci Nutr 43:61–87
21. Khor E (2001) Chitin: fulfilling a biomaterial promise. Elsevier, New York
22. Jayakumar R, Prabhakaran M, Reis RL, Mano JF (2005) 3B's research group-biomaterials, biodegradables & biomimetics. Carbohyd Polym 62(2):142–158
23. Rafat D, Sahl HG (2009) Chitosan and its antimicrobial potential, a critical review. Microb Biotechnol 2(2):186–201
24. Xia W, Liu P, Zhang J, Chen J (2011) Biological activities of chitosan and chitooligosaccharides. Food Hydrocolloids, 2592:170–179
25. Shahidi F, Abuzaytoun R (2005) Chitin, chitosan, and co-products: chemistry, production, applications, and health effects. Adv Food Nutr Res 49:93–135
26. Bruck WM, Slater JW, Carney BF (2010) Chitin and chitosan from marine organisms. In: Kim S-K (ed) Chitin, chitosan, oligosaccharides and their derivatives: biological activities and applications. CRC Press, pp 11–23
27. Olafadehan OA, Amoo KO, Ajayi TO et al (2021) Extraction and characterization of chitin and chitosan from *Callinectes amnicola* and *Penaeus notialis* shell wastes. J Chem Eng Mater Sci 12(1):1–30, January–June 2021, https://doi.org/10.5897/JCEMS2020.0353, Articles Number: B4E8A7966131, ISSN 2141-6605
28. Webster C, Onokpise O, Abazinge M et al (2014) Turning waste into usable products: a case study of extracting chitosan from blue crab. Am J Environ Sci 10(4):357–362; https://doi.org/10.3844/ajessp.2014.357.362. (a) Hossain MS, Iqbal A (2014) Production and characterization of chitosan from shrimp waste. J Bangladesh Agril Univ 12(1):153–160
29. Cheba BA (2011) Chitin and chitosan: marine biopolymers with unique properties and versatile applications. Glob J Biotechnol Biochem 6(3):149–153
30. Hamed I, Ozogul F, Regenstein JM (2016) Industrial applications of crustacean by-products (chitin, chitosan, and chitooligosaccharides): a review. Trends Food Sci Technol 48:40–50
31. Synowiecki J, Al-Khateeb NA (2003) Production, properties and some new applications of chitin and its derivatives. Crit Rev Food Sci Nutr 43(2):145–171
32. Tan HW et al (2022) Potential economic value of chitin and its derivatives as major biomaterials of seafood waste, with particular reference to Southeast Asia. J Renew Mater 10(4):909–938; JRM 10(4) 2022. https://doi.org/10.32604/jrm.2022.018183. ISSN: 2164-6341 (online)

33. Tolaimate A, Desbrieres J, Raíz M, Alagui A (2003) Contribution to the preparation of chitins and chitosans with controlled physico-chemical properties. Polymer 44:7939–7952
34. Ashford NA, Hattis D, Murray AE (1977) Industrial prospects for chitin and protein from shellfish wastes. MIT Sea Grant Report MITSG 77-3, MIT, Cambridge
35. Minke P, Blakwell J (1978) The structure of α-chitin. J Mol Biol 120(2):167–181
36. Kurita K, Tomita K, Tada T, Ishii S, Nishimura SI, Shimoda K (1993) Squid chitin as a potential alternative chitin source: deacetylation behavior and characteristic properties. J Polym Sci, Part A: Polym Chem 31(2):485–491
37. Kurita K, Ishii S, Tomita K, Nishimura SI, Shimoda K (1994) Reactivity characteristics of squid β-chitin as compared with those of shrimp chitin: high potentials of squid chitin as a starting material for facile chemical modifications. J Polym Sci Part A: Polym Chem 32(6):1027–1032
38. Kurita K, Tomita K, Ishii S, Nishimura SI, Shimoda K (1993) β-chitin as a convenient starting material for acetolysis for efficient preparation of $N$-acetylchitooligosaccharides. J Polym Sci, Part: A Polym Chem 31(9):2393–2395
39. Sannan T, Kurita K, Iwakura Y (1976) Studies on chitin, 2 Effect of deacetylation on solubility. Makromol Chem 177(12):3589–3600
40. Lee VFP (1974) Solution and shear properties of chitin and chitosan. PhD thesis, Université de Washington, Seattle
41. Austin PE, Castle JE, Albisetti CJ (1989) Chitin and chitosan. In: Skjak-Braek, Anthonsen GT, Sandford P (eds) Elsevier, Essex, p 749
42. Cauchie HM (1997) An attempt to estimate crustacean chitin production in the hydrosphere. In: Diamond A, Roberts GAF, Varum KM (eds) Advances in chitin science, Lyon Jacques Andre Publisher, pp 32–38; (a) Cauchie H-M (2002) Chitin production by arthropods in the hydrosphere. Hydrobiologia 470:63–96
43. Kaya M, Baran T, Saman I et al (2014) Physicochemical characterization of chitin and chitosan obtained from resting eggs of C. Quadrangula (Branchiopoda: Cladocera: Daphniidae). J Crustac Biol 34(2):283–288. https://doi.org/10.1163/1937240X-00002221
44. Naczk M, Synowiecki J, Sikorski ZE (1981) The gross chemical composition of Antarctic krill shell waste. Food Chem 7:175–179
45. Shahidi F, Synowiecki J (1991) Isolation and characterization of nutrients and value added products from snow crab (Cinoecetes opilio) and shrimp (Pandalus borealis) processing discards. J Agric Food Chem 39:1527–1532
46. Synowiecki J, Al-Khateeb NA (2000) The recovery of protein hydrolysate during enzymatic isolation of chitin from shrimp Crangon crangon processing discards. Food Chem 68:147–152
47. Knorr D (1991) Recovery and utilization of chitin and chitosan in food processing waste management. Food Technol 45:114–122
48. Wan Y, Creber KAM, Peppley B, Bui VT (2002) Recent Res Dev Appl Polym Sci 1(Pt. 1):17–36
49. Rinaudo M (2006) Chitin and chitosan: properties and applications. Prog Polym Sci 31(7):603–632
50. Tan T, Wang B, Chen P (2001) Production of chitosan by biological method. Huagong Jinzhan 20(9):4–5, 21
51. De Santos JE, Soares JP, Dockal ER, Campana F, Sergio P, Cavalheiro E (2003) Polimeros: Ciencia e Tecnologia 13(4):242–249
52. Gao L-P, Du Y-M, Zhang D-B, Shi X-W, Zhan H-Y, Song W-H (2003) J Nat Sci 8(4):1156–1160
53. Domard A, Chaussard G (2002) New approach in the study of the production of chitosan from squid pens: kinetics, thermodynamic and structural aspects. Adv Chitin Sci 5:1–5. ISBN: 974-229-412-7
54. Kumar MNV (2000) A review of chitin and chitosan applications. React Funct Polym 46:1–27
55. Voronov AF, Nifant'ev NE, Polle SN, Abramov VA, Benkogenov AV, Isadskaya GB, Mar'ina VF, Zyrzin AP (2001) (Proizvodstvennoe Ob'edinenie "Aleksinskii Khimicheskii Kombinat", Russia). Russia Application: RU 2000-112383 20000519

56. Naznin R (2005) Extraction of chitin and chitosan from shrimp (Metapenaeus monoceros) shell by chemical method. Pak J Biol Sci 8(7):1051–1054
57. Yamanami T, Tsuchida S, Ieta K, Seki M, Kobayashi S, Ise H, Kobayashi T (2006) Low-molecular-weight chitin powder. Jpn Kokai Tokkyo Koho, JP 2006348093 A 20061228
58. Li Q-X, Song B-Z, Yang Z-Q, Fan H-L (2006) Preparation of chitosan with high viscosity-average molecular weight and high degree of deacetylation by rapid interim microwave heating method. Guocheng Gongcheng Xuebao 6(5):789–793. Squid tendon-derived β-chitin-based substances, their powders, and manufacture thereof by deproteinization
59. Cao J, Dai Y, Wang H, Wang Y (2005) Research on chitosan preparation by microwave deacetylation of chitin. Shipin Kexue (Beijing, China) 26(11):120–125
60. Saito Y, Kosakada K (2006) Squid tendon-derived β-chitin-based substances, their powders, and manufacture thereof by deproteinization. Jpn Kokai Tokkyo Koho, 7 p. JP 2006321961 A 20061130
61. Hammer PE, Carr B (2006) Methods for production of chitin/chitosan with transgenic organisms and for in vitro production of chitin/chitosan. PCT Int Appl 140p. WO 2006124779 A2 20061123
62. Yen M-T, Mau J-L (2007) Physico-chemical characterization of fungal chitosan from shiitake stipes. Food Sci Technol 40(3):472–479
63. Zhang Y, Xue C, Li Z, Zhang Y, Fu X (2006) Preparation of half-deacetylated chitosan by forced penetration and its properties. Carbohyd Polym 65(3):229–234
64. Elieh-Ali-Komi D, Hamblin MR (2016) Chitin and chitosan: production and application of versatile biomedical nanomaterials. Int J Adv Res (Indore) 4(3):411–427
65. Einbu A (2007) Characterization of chitin and a study of its acid-catalysed hydrolysis. PhD thesis submitted to Faculty of Natural Sciences and Technology, Department of Biotechnology, Norwegian University of Science and Technology, Trondheim, April 2007
66. Mohan K, Muralisankar T, Jayakumar R et al (2021) A study on structural comparisons of a-chitin extracted from marine crustacean shell waste. Carbohyd Polym Technol Appl 2:100037
67. Fernandex-Kim S-O (2014) Physicochemical and functional properties of crawfish chitosan as affected by different processing protocols. MS thesis, Louisiana State University and Agricultural and Mechanical College, December 2014
68. Einbu A, Naess SN, Elgsaeter A, Varum KM (2004) Solution properties of chitin in alkali. Biomacromol 5:2048–2054
69. Puvvada YS, Vankayalapati S, Sukhvasi S (2012) Extraction of chitin from chitosan from exoskeleton of shrimp for application in the pharmaceutical industry. Int Curr Pharm J 1(9):258–263
70. Aranaz I, Mengibar M, Harris R, Panos I, Miralles B, Acosta N, Galed G, Heras A (2009) Functional characterization of chitin and chitosan. Curr Chem Biol 3:203–230
71. No HK, Meyers SP (1997) Preparation of chitin and chitosan. In: Muzzarelli RAA, Peter MG (eds) Chitin handbook. European Chitin Society, pp 475–489
72. Synowiecki J, Sikorski ZE, Naczk M (1981) The activity of immobilized enzymes on different krill chitin preparations. Biotechnol Bioeng 23:2211–2215
73. No HK, Meyers SP, Lee KS (1989) Isolation and characterization of chitin from crawfish shell waste. J Agric Food Chem 37(3):575–579
74. Blumberg R, Southall CL, Van Rensburg NJ, Volckman OB (1951) South African fish products. XXXII-The rock lobster: a study of chitin production from processing wastes. J Sci Food Agric 2:571
75. Kamasastri PV, Prabhu PV (1961) Preparation of chitin & glucosamine from prawn shell waste. J Sci Znd Res 20D:466
76. Brine CJ, Austin PR (1981) Chitin isolates: species variation in residual amino acids. Comp Biochem Physiol 70B:173
77. Brzeski MM (1982) Concepts of chitin/chitosan isolation from Antartic Krill *(Euphausia superba dana)* shells on a technical scale. In: Hirano S, Tokura S (eds) Proceedings of the second international conference on chitin and chitosan. The Japan Society of Chitin and Chitosan: Sapporo, Japan, p 15

78. Austin PR, Brine CJ, Castle JE, Zikakis JP (1981) Chitin: new facets of research. Science 212:749–753
79. Roberts G (1997) Chitosan production routes and their role in determining the structure and properties of the product. In: Domard A, Roberts GAF, Varum KM (eds) Advances in chitin science. Jaques Andre Publisher, Lyon, pp 22–31
80. Jung W-J, Park R-D (2014) Bioproduction of chitooligosaccharides: present and perspectives. Mar Drugs 12:5328–5356. https://doi.org/10.3390/md12105328
81. Aguila EMD, Gomes LP, Andrade CT et al (2012) Bopcatalytic production of chitosan polymers from shrimp shells, using a recombinant enzyme produced by *Pichia pastoris*. Am J Mol Biol 2:341–350. https://doi.org/10.4236/ajmb.2012.24035 Published Online October 2012 http://www.SciRP.org/journal/ajmb/
82. Horn SJ, Sikorski P, Cederkvist JB et al (2006) Cost and benefits of procesivity in enzymatic degradation of recalcitrant polysaccharides. PNAS 103(48):18089–18094
83. Beier S, Bertilsson S (2013) Bacterial chitin degradation-mechanisms and ecophysiological strategies. Front Microbiol 4, Article 149, June 2013. www.frontiersin.org. https://doi.org/10.3389/fmicb.2013.00149
84. Takeda M, Abe E (1962) Isolation of crustacean chitin. Decalcification by sodium ethylenediaminetetraacetate and enzymic hydrolysis of incidental proteins. Norinsho Suisan Koshuso Kenkyu Hokoku 11:399–406
85. Takeda M, Katsuura H (1964) Purification of king crab chitin. Suisan Daigaku Kenkyu Hokoku 13:109–116
86. Broussignac P (1968) Chitosan, a natural polymer not well known by the industry. Chim Ind Genie Chim 99:1241–1247
87. Shimahara K, Takiguchi Y (1988) Methods in enzymology. In: Wood WA, Kellogg ST (eds), vol 161. Academic Press, New York, pp 417–423
88. Bustos RO, Healy M (1994) Microbial deproteinisation of waste prawn shell. In: Institution of Chemical Engineers Symposium Series Institution of Chemical Engineers, Rugby, England, pp 13–15
89. Wang S-L, Chio S-H (2007) Deproteinization of shrimp and crab shell with the protease of Pseudomonas aeruginosa K-187. Enzyme Microb Technol 22(7):1069–1074
90. Rao MS, Munoz J, Stevens WF (2000) Critical factors in chitin production by fermentation of shrimp biowaste. Appl Microbiol Biotechnol 54:808–813
91. Yang JK, Shih IL, Tzeng YM, Wang SL (2000) Production and purification of protease from a *Bacillus subtilis* that can deproteinize crustacean wastes. Enzyme Microbiol Technol 26:406–413
92. He H, Chen X, Sun C, Zhang Y, Gao P (2006) Preparation and functional evaluation of oligopeptide-enriched hydrolysate from shrimp (*Acetes chinensis*) treated with crude protease from *Bacillus* sp. SM98011. Biores Technol 97:385–390
93. Shirai K, Guerrero I, Huerta S, Saucedo G, Castillo A, Gonzalez RO et al (2001) Effect of initial glucose concentration and inoculation level of lactic acid bacteria in shrimp waste ensilation. Enzyme Microbiol Technol 28:446–452
94. Laura RC, Marìa del Carmen MC, Huerta S, Revah S, Shirai K (2006) Enzymatic hydrolysis of chitin in the production of oligosaccharides using *Lecanicillium fungicola* chitinases. Process Biochem 41:1106–1110
95. Patidar P, Agrawal D, Banerjee T, Patil S (2005) Optimisation of process parameters for chitinase production by soil isolates of *Penicillium chrysogenum* under solid substrate fermentation. Process Biochem 40:2962–2967
96. Hall GM, Silva S (1992) Lactic acid fermentation of shrimp (*Penaus monodon*) waste for chitin recovery. In: Brine CJ, Sandford PA, Zikakis JP (eds) Advance in chitin and chitosan. Elsevier Applied Science, London, pp 633–668
97. Rao MS, Guyot JP, Pintado J, Stevens WF (2002) Improved conditions for lactobacillus fermentation of shrimp waste into chitin. In: Scchiva K, Chandrkrachang S, Methacanon P, Peter MG (eds) Advance in chitin science, vol V. Bangkok, Thailand, pp 40–44

98. Jung WJ, Jo G-Y, Kuk JH et al (2007) Extraction of chitin from red crab shell waste by cofermentation with *Lactobacillus paracasei* KCTC-3074 and *Serratia marcescens* FS-3. Carbohydr Polym 68:746–750
99. Oh K-T, Kim Y-J, Nguyen VN, Jung W-J, Park R-D (2007) Demineralization of crab shell waste by *Pseudomonas aeruginosa* F722. Process Biochem 42(7):1069–1074
100. Badawy MEI, Rabea EI (2011) A biopolymer chitosan and its derivatives as promising antimicrobial agents against plant pathogens and their applications in crop protection. Int J Carbohydr Chem, 29p, Article ID 460381
101. Domard A, Rinaudo M (1983) Preparation and characterization of fully deacetylated chitosan. Int J Biol Macromol 5:49–52
102. Mincea M, Negrulescu A, Ostafe V (2012) Preparation, modification, and applications of chitin nanowhiskers: a review. Rev Adv Mater Sci 30:225–242
103. Kurita K (2001) Controlled functionalization of the polysaccharide chitin. Prog Polym Sci 26:1921–1971
104. Azevedo EP (2011) Aldehyde functionalized chitosan and cellulose: chitosan composite: application as drug carriers and vascular by-pass grafts. PhD thesis, University of Iowa
105. Jabeen S, Saeed S, Kausar A, Muhammad B, Gul S, Farooq M (2016) Influence of chitosan and epoxy cross-linking on physical properties of binary blends. Int J Polym Anal Charact 21:2
106. Hasan S, Ghosh A, Race K, Schreiber R Jr, Prelas MA (2014) Dispersion of FeOOH on chitosan matrix for simultaneous removal of As(III) and As(V) from drinking water. Sep Sci Technol 49:2863–2877
107. Crini G, Morin-Crini N, Fatin-Rouge N et al (2017) Metal removal from aqueous media by polymer-assisted ultrafiltration with chitosan. Arab J Chem 10:S3826–S3839; (a) Jiménez-Gómez CP, Cecilia JA (2020) Chitosan: a natural biopolymer with a wide and varied range of applications. Molecules 25:3981. https://doi.org/10.3390/molecules25173981
108. Kristiansen A, Varum KM, Grasdalen H (1998) the interaction between highly de-N-acetylated chitosan and lysozyme from chicken egg white studied by 1H-NMR spectroscopy. Eur J Biochem 15, 251(1–2):335–342
109. Kumirska J, Weinhold MX, Thoming J, Stepnowski P (2011) Biomedical activity of chitin/chitosan-based materials-influence of physicochemical properties apart from molecular weight and degree of N-acetylation. Polymers 3:1875–1901
110. Dash M, Chiellini F, Ottnbrite RM, Chiellini E (2011) Chitosan-a versatile semi-synthetic polymer in biomedical applications. Prog Polym Sci 36:981–1014
111. Abbas AOM Chitosan for biomedical applications. PhD thesis, University of Iowa, http://ir.uiowa.edu/etd/771
112. Mireless-DeWitt CA (1994) Complex mechanism of chitosan and naturally occurring polyanions. MS thesis, Oregon State University, February 28
113. Struszczyk MH (2002) Chitin and chitosan. Polimery 47(5):316–326
114. Mourya VK, Inamdar NN (2008) Chitosan-modifications and applications: opportunities galore. React Funct Polym 68:1013–1051
115. Wang QZ, Chen XG, Liu N et al (2006) Protonation constants of chitosan with different molecular weight and degree of deacetylation. Carbohyd Polym 65:194–201
116. Navarro R, Guzman J, Saucedo I, Revilla J, Guibal E (2003) Recovery of metal ions by chitosan: sorption mechanisms and influence of metal speciation. Macromol Biosci 3(10):552–561
117. Rinaudo M, Pavlov G, Desbrieres J (1999) Influence of acetic acid concentration on the solubilization of chitosan. Polymer 40:7029–7032
118. Wang QZ, Chen XG, Liu N, Wang SX, Liu CS, Meng XH, Liu CG (2006) Protonation constants of chitosan with different molecular weight and degree of deacetylation. Carbohyd Polym 65:194–201
119. Park JW, Choi K-H, Park KK (1983) Acid-base equilibria and related properties of chitosan. Bull Korean Chem Soc 4(2):68–71

120. Fillon D, Lavertu M, Buschmann MD (2007) Ionization and solubility of chitosan solutions related to thermosensitive chitosan/glycerol-phosphate systems. Biomacromol 8:3224–3234
121. Cho YW, Jang J, Park CR, Ko SW (2000) Preparation and solubility in acid and water of partially deacetylated chitins. Biomacromol 1:609–614
122. Hsu S-H, Whu SW, Tsai C-L, Wu Y-H, Chen H-W, Hsieh K-H (2004) Chitosan and scaffold materials: effect of molecular weight and degree of deacetylation. J Polym Res 11:141–147
123. Einbu A (2007) Characterisation of chitin and a study of its acid-catalysed hydrolysis. PhD (Doctor of Philosophy) thesis, Norwegian University of Science and Technology. ISBN ISBN 978-82-471-1643-2 (electronic ver.)
124. No HK, Meyers SP (1995) Preparation and characterization of chitin and chitosan-a review. J Aquat Food Prod Technol 4(2):27–52
125. Macquarrie DJ, Hardy JJE (2005) Application of functionalized chitosan in catalysis. Ind Eng Chem Res 44:8499–8520
126. Yin X, Zhang X, Lin Q et al (2004) Metal-coordinating controlled oxidative degradation of chitosan and antioxidant activity of chitosan-metal complex, ARKIVOC (9):66–78. https://quod.lib.umich.edu/a/ark/5550190.0005.910
127. Mucha M, Bialas S (2013) Thermal and photochemical stability of chitosan doped by nano-silver. J Chitin Chitosan Sci 1(3):235–239
128. Chatelet C, Damour O, Domard A (2001) Influence of the degree of acetylation on some biological properties of chitosan films. Biomaterials 22:261–268
129. Maznah M, Muhammad IN, Norzita Y, Norhashidah T, Zahid A (2014) Degradation of chitosan by gamma ray with presence of hydrogen peroxide. Adv Nucl Res Energy Dev. AIP Conf Process 1584:136–140
130. Szymanska E, Winnicka K (2015) Stability of chitosan—a challenge for pharmaceutical and biomedical applications. Mar Drugs 13:1819–1846
131. Clayer A, Vachoud L, Viton C, Domard A (2003) Atypical polysaccharide physical gels: structure/property relationships. Macromol Symp 200:1–8
132. Mucha M, Pawlak A (2002) Complex study on chitosan degradability. Polimery 47:7–8
133. Halabalova V, Simek L, Mokrejs P (2011) Intrinnsic viscosity and conformational parameters of chitosan chains. Rasayan J Chem 4(2):223–241
134. Kasaai MR, Arul J, Charlet G (2000) Intrinsic viscosity-molecular weight relationship for chitosan. J Polym Sci: Part B: Polym Phys 38:2591–2598
135. Morris GA, Castile J, Smith A, Adams GG, Harding SE (2009) Macromolecular conformation of chitosan in dilute solution: a new global hydrodynamic approach. Carbohyd Polym 76:616–621
136. Pedroni VI, Schulz PC, Gschaider ME, Andreucetti N (2003) Chitosan structure in aqueous solution. Colloid Polym Sci 282:100–102
137. El-hefian EA, Yahaya AH, Misran M (2009) Characterisation of chitosan solubilized in aqueous formic and acetic acids, Maejo. Int J Sci Technol 3(03):415–425
138. Senel S, McClure SJ (2004) Potential applications of chitosan in veterinary medicine. Adv Drug Deliv Rev 56:1467–1480
139. Wang YH, Wu YA, Wen D (2006) Biodegradable polylactide/chitosan blend membranes. Biomacromol 7(4):1362–1372
140. Ogawa K, Yui T, Miya M (1992) Dependence on the preparation procedure of the polymorphism and crystallinity of chitosan membranes. Biosci, Biotech, Biochem 56(6):858–862
141. Li Q, Dunn ET, Grandmaison EW (1977) Application and properties of chitosan. In: Goosen MFA (ed) Application of chitin and chitosan. Technomic Publishing Co. Lancaster
142. Ahmad LO, Permana D, Wahab SH, Sabarwati LO, Ramadhan AN, Improved chitosan production from tiger shrimp shell waste (*Penaeus mondon*) by multistage deacetylation method and effect of bleaching. Adv Environ Geol Sci Eng. www.wseas.us/e-library/conferences/2015/Salerno/EG/EG-49.pdf
143. Foster LJR, Ho S, Hook J, Basuki M, Marcal H (2015) Chitosan as a biomaterial: influence of degree of deacetylation on its physicochemical, material and biological properties. PLoS ONE 10(8):1–22. https://doi.org/10.1371/journal.pone.0135153

144. Yuan Y, Chesnutt BM, Haggard WO, Bumgardner JD (2011) Deacetylation of chitin: material characterization and in vitro evaluation via albumin adsorption and pre-osteoblastic cell cultures. Materials 4:1399–1416. https://doi.org/10.3390/ma4081399
145. Aranaz I, Alcántara AR, Civera MC et al (2021) An overview of its properties and applications. Polymers 13:3256. https://doi.org/10.3390/polym13193256
146. Whang HS, Kirsch W, Zhu YH, Yang CZ, Hudson SM (2005) Hemostatic agents derived from chitin and chitosan. J Macromol Sci 45:309–323
147. Viorel MR (2004) Composite materials made of chitosan and nanosized apatite; Preparation and physicochemical characterization. PhD thesis, University of Potsdam
148. Zhang J, Xia W, Liu P et al (2010) Chitosan modification and pharmaceutical/BIOMEDICAL applications. Mar Drugs 8:1962–1987. https://doi.org/10.3390/md8071962
149. Struszczyk MH, Struszczyk KJ (2007) Medical application of chitin and its derivatives. Polish Chitin Society, Monograph XII, pp 139–147
150. Marques C, Som C, Schmutz M, Borges O, Borchard G (2020) How the lack of chitosan characterization precludes implementation of the safe-by-design concept. Front Bioeng Biotechnol 8:165. https://doi.org/10.3389/fbioe.2020.00165
151. Ahmed TA, Bader M, Aljaeid BM (2016) Preparation, characterization, and potential application of chitosan, chitosan derivatives, and chitosan metal nanoparticles in pharmaceutical drug delivery. Drug Des, Dev Therapy 2016:10:483–507
152. Hein S, Ng CH, Chandrkrachang S et al (2001) A systematic approach to quality assessment system of chitosan. Chitin and chitosan: chitin and chitosan in life science, Yamaguchi, pp 327–335
153. Struszczyk MH, Struszczyk KJ (2007) Medical applications of chitin and its derivatives, progress on chemistry and application of chitin and its derivatives. In: Jaworska M (ed), vol XII, ISSN 1896-5644, pp 139–148
154. Struszczyk MH (2006) Global requirements for medical applications of chitin and its derivatives. Polish Chitin Society, Monograph XI, pp 95–102
155. Skaugrud O, Hagen A, Borgersen B, Dornish M (1999) Biomedical and pharmaceutical applications of alginate and chitosan. Biotechnol Genet Eng Rev 16:23–40
156. Aranaz I, Mengiber M, Acosta N, Heras A (2015) Chitosan: a natural polymer with potential industrial applications. A Sci J COMSATS-Sci Version 21(1&2):41–50

# Chapter 3
# Chitosan Characterization

**Abstract** Chitosan is one of the most abundant natural polymers synthesized by the deacetylation of chitin. Chitin is a major component of crustacean shells and is available in abundance in nature. The physical and chemical properties of chitin and chitosan are discussed in this chapter.

## 3.1 Introduction

Chitin is a high molecular mass polymer and is insoluble in water like cellulose and has low reactivity. It is a white, stiff, and inelastic nitrogenous polysaccharide. Chitosan is a deacetylated derivative of chitin. Both chitin and chitosan are commercially important because of the high percentage of nitrogen (5–8% depending on the degree of deacetylation) compared to cellulose (1.25%). This high nitrogen content makes chitin and chitosan good chelating agents. The many applications of chitosan are due to its excellent properties such as biocompatibility, biodegradability, non-toxicity, and adsorption properties. The properties of chitin and chitosan are dependent on their ability to form polyoxysalts, films, and chelated metal ions.

Techniques to characterize the surface morphology and intermolecular interactions of chitin and chitosan are as follows:

1. Morphology of the surface using scanning electron microscopy (SEM).
2. Energy-dispersive spectroscopy (EDS) X-ray microanalysis.
3. Transmission electron microscopy (TEM).
4. Thermogravimetric analysis (TGA).
5. Surface area and surface charge analysis.
6. Moisture sorption study.
7. Tensile strength measurement.
8. Diffuse reflective UV-spectroscopy (DRUV).
9. X-ray photoelectron spectroscopy (XPS).
10. Electron spin resonance (ESR) spectroscopy.

The physical characterization of chitin and chitosan is included in Sect. 3.1 of this chapter.

1. Molecular weight.
2. Degree of deacetylation.
3. Swelling behavior and solubility.
4. Percentage of moisture.
5. Percentage of ash.

### 3.1.1 Molecular Weight

The molecular weight (MW) of a polymer determines the strength of its fiber and the viscosity of the solution. The degree of deacetylation (DD) significantly affects the MW of chitosan [1]. The MW of chitosan decreases with the increase in its DD value. The intrinsic viscosity of chitin and chitosan is related to their molecular weight [2, 3]. The viscosity of polymer solution is a direct measure of the hydrodynamic volume of the polymer molecules. The hydrodynamic volume depends on the molecular size or the chain length and the molecular weight [4]. The molecular weight of chitin and chitosan can be determined by methods such as chromatography, light scattering, and viscometry [5–7]. The light scattering method provides absolute values for MW, but the technique is complicated, and sometimes, the data are not easy to interpret [8]. Viscometry and gel permeation chromatography (GPC) are the two most commonly used methods for determining the molecular weight of a polymer. The viscosity-average molecular weight is determined by measuring the relative viscosity with an Ostwald viscometer of buffer solutions formed with acetic acid/sodium acetate at various concentrations. The MW can be estimated from the intrinsic viscosity using the Mark–Houwink–Sakurada equation [9]:

$$[\eta] = kM_v^a, \qquad (3.1)$$

where $M_v$ is the viscosity-averaged molecular weight of the polymer, $k$ and $a$ are constants depending on the polymer and the solvent systems used and temperature. The measurement of intrinsic viscosity depends on the hydrodynamic volume of the macromolecule. The hydrodynamic volume is a function of the MW conformation properties and the polymer–solvent interaction. The polymer molecules often do not assume the shape of an extended straight chain in solution and are present in a coil form [4]. These coils or aggregates offer resistance for mobility and deter the flow of molecules imparting viscosity. Rinaudo reported that various solvents and chitosan mixtures could be used to calculate the molecular weight (MW) of chitosan [10]. The viscosity and the molecular weight for a given polymer–solvent system are virtually related to each other. Table 3.1 shows Mark–Houwink–Sakurada parameters for chitosan solutions reported by Knaul et al. [11].

## 3.1 Introduction

**Table 3.1** Published Mark–Houwink–Sakurada constants for chitosan

| Solvent | T (°C) | %DDA | K (mL/g) | a | $M_v$ range ($\times 10^4$) | Method |
|---|---|---|---|---|---|---|
| 0.2 M CH$_3$COOH/0.1 M NaCl/0.4 M Urea | – | – | $8.93 \times 10^{-2}$ | 0.71 | 1.38 | SD |
| 0.1 M CH$_3$COOH/0.2 M NaCl | 25 | Derived chitosan | $3.04 \times 10^{-5}$ | 1.26 | – | Absorbance |
| 0.17 M CH$_3$COOH/0.47 M NaCl | 25 | 90–100 | 1.115 | 0.147 | 1.5–0.16 | SD |
| 2% CH$_3$COOH/0.2 M CH$_3$COOH + dichloroacetic acid | 25 | 82–88 | $1.38 \times 10^{-2}$ | 0.85 | 1.5–0.16 | SD |
| 0.33 M CH$_3$COOH/0.3 M NaCl | 21 | 80–82 | $3.41 \times 10^{-3}$ | 1.02 | 5.2–13.3 | SD |
| 0.1 M CH$_3$COOH/0.2 M NaCl | 25 | ≈80 | $1.81 \times 10^{-3}$ | 0.93 | 63–0.48 | |
| 0.2 M CH$_3$COOH/0.1 M CH$_3$COONa | 30 | 69<br>84<br>91<br>100 | $0.104 \times 10^{-3}$<br>$1.424 \times 10^{-3}$<br>$6.589 \times 10^{-3}$<br>$16.80 \times 10^{-3}$ | 1.12<br>0.96<br>0.88<br>0.81 | 25.1–1.94<br>1.94 | LS |
| 0.3 M CH$_3$COOH/0.2 M CH$_3$COONa | 25 | 98<br>88.5<br>79 | 0.082<br>0.076<br>0.074 | 0.76<br>0.76<br>0.76 | 20.3–29.0 | GPC/MALL LS/LALLS |

(continued)

Table 3.1 (continued)

| Solvent | T (°C) | %DDA | K (mL/g) | a | $M_v$ range ($\times 10^4$) | Method |
|---|---|---|---|---|---|---|
| 0.02 M $CH_3COOH$/0.1M$CH_3COONa$ and chitosan chloride | 20 | 100<br>85<br>40 | Log $k_{0.1}$ = −0.427−3.821$F_A$ | $a_{0.1}$ = 0.6169 + 0.759$F_A$ | 1.5–31.0 | OM |
| 0.2 M $CH_3COOH$/0.2 M $CH_3COONa$ | 25 | 42<br>89 | | 1.14<br>0.521 | | SD |
| 1% $CH_3COOH$ | 30 | | 0.0474 | 0.723 | 20.46–65.74 | LS |
| 0.5 M $CH_3COOH$/0.5 M $CH_3COONa$ | 25 | 69.8 | 0.119 | 0.59 | 274–11.5 | GPC/LS |
| 0.3 M $CH_3COOH$/0.2 M $CH_3COONa$ | 25 | | Plots but no values provided | | | SEC, LS |

Copyright © National Research Council of Canada, Nov. 1998. Can J Chem 76:1699–1706, 1998. Reproduced from Ref. [11] with permission
DDA—degree of deacetylation; GPC—gel permeation chromatography; HPLC—high-performance liquid chromatography; LS—light scattering; LALLS—low-angle laser light scattering; OM—osmometry; SD—sedimentation and diffusion; SEC—size exclusion chromatography

## 3.1 Introduction

### 3.1.2 Degree of Deacetylation (DD) and Dissociation Constant of Chitosan

#### 3.1.2.1 Degree of Deacetylation (DD) of Chitosan

The degree of deacetylation is one of the structural parameters contributing to the physical and biological properties of chitosan. The degree of deacetylation identifies the biopolymer as chitin or chitosan. The DD of chitosan depends mainly on the source of the chitin and chitosan preparation process. In the case of chitosan, DD must be above 50%.

Conductometric titration is a relatively reliable and straightforward approach to estimate DD of chitosan [12, 13]. By this titration method, the $-NH_2$ fraction in the polymer can be obtained by dissolution of chitosan in the presence of a small excess of HCl. This is followed by the neutralizing the protonated $-NH_2$ groups by NaOH. Finally, a titration curve is prepared to show the pH of the chitosan solution and the volume of NaOH added. Figure 3.1 shows an example of a titration curve for pH vs. volume of NaOH (mL).

The degree of deacetylation DD is calculated using the equation given below [14]:

$$DD = \frac{203.2}{42.0 + \frac{1000 M}{C_{NaOH}(\Delta V)}}. \quad (3.2)$$

In Eq. 3.2, $\Delta V$ is the difference between the maxima of the first derivative, which corresponds to the equivalence point of the excess HCl ($V_1$) in the solution, and the protonated amino groups ($V_2$), determined. $M$ denotes the mass of chitosan, whereas

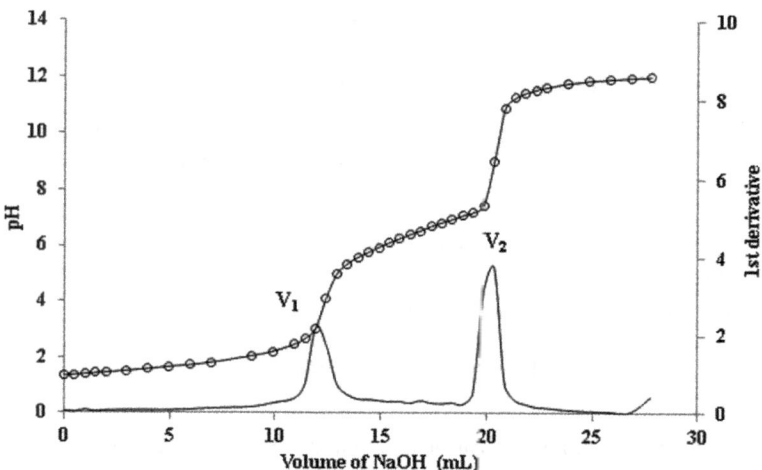

**Fig. 3.1** Typical potentiometric titration curves of crustacean shells-based chitosan samples. Unpublished work by authors

203.2 is the molecular weight of the acetylated unit, and 42.0 is the difference between the molecular weights of the acetylated and deacetylated units in Eq. 3.2.

Figure 3.1 shows a typical NaOH versus pH plot for chitosan samples. Two inflection points of the curve were calculated using the first derivative value vs. volume of NaOH. Figure 3.1 shows the first inflection point ($V_1$) at a pH about 3.03 and the second ($V_2$) at pH 9.01. The difference between the two points is $\Delta V$, representing the amount of base required to neutralize the weak acid ($-NH_3^+$) in the solution. The degree of deacetylation (DD) of chitosan is calculated from the first and second inflection points of the titration curve. The number of amine groups (mM) in chitosan samples can be calculated using the following equation:

$$NH_2 = \left[\left(\left(\frac{1}{161.18}\right) * DD\right) + \left(\left(\frac{1}{203.19}\right) * DA\right)\right] * M. \quad (3.3)$$

In the above equation, 161.18 (g/mole) represents the molar mass of degree of deacetylation (DD), 203.19 (g/mole) represents the molar mass for DA (degree of acetylation) of chitosan, and M (gram) represents the amount of chitosan in solution.

### 3.1.2.2 Dissociation Constant of Chitosan

The dissociation constant of a polyelectrolyte mainly depends on the degree of dissociation. Several studies focused on determining the dissociation constants of chitosan in aqueous solutions by potentiometric titration [15, 16]. The following equilibrium reaction describes the degree of ionization of amine groups of chitosan in solution [17]:

$$R - NH_2 + H_3O^+ \leftrightarrow R - NH_3 + H_2O. \quad (3.4)$$

Therefore, the dissociation constant of protonated chitosan may be defined as:

$$K_a = \frac{[R - NH_2][H_3O^+]}{[R - NH_3^+]} \text{ and } pK_a = -\log K_a. \quad (3.5)$$

The potentiometric behavior is well represented by the Katchalsky–Spitnik equation [18]. In the Katchalsky–Spitnik equation, the electrical free energy change occurs during the neutralization of a polyelectrolyte [19]:

$$pH = pK_a + n \log(\alpha/(1 - \alpha)), \quad (3.6)$$

where $\alpha$ is the degree of neutralization calculated from the titration curve at each pH value, the term n is an empirical parameter related to the free energy change during titration. At $\alpha = 0.5$, the pKa value is equal to the pH of the chitosan solution.

## 3.1 Introduction

Two inflection points of the titration curve (Fig. 3.1) provide information about the buffer capacity of the chitosan. The pKa value is calculated from the first derivative versus pH plot from the potentiometric titration (Fig. 2a). The second derivative method is also used to estimate the pKa value of the chitosan samples. In the second derivative method, the pKa value is estimated from the point where the value of $\Delta^2 pH/\Delta V^2$ is zero (Fig. 2a). The pKa value of the chitosan sample can determine by inspection of the second derivative plot [13].

Figure 2b shows the degree of ionization versus pH of the solution, and Fig. 2c shows the degree of neutralization, $\alpha$ versus pH. The degree of neutralization value $\alpha$ is obtained at each pH value from the titration curve. The pKa value is obtained from the point where the value of $\alpha$ is 0.5. Figure 2d shows the plot using the Katchalsky–Spitnik equation (Eq. 3.6). The pKa and $n$ value can be calculated from the intercept and slope of the straight line of pH versus $\log(\alpha/1-\alpha)$ (Fig. 2d). The pKa value of a typical chitosan sample is in the range of 6–6.5 [20].

Chitosan starts precipitating at pH > pKa value during titration. This precipitation reduces the concentration of chitosan in the solution. The precipitates may produce an error in calculating the degree of deacetylation using Eq. 3.2.

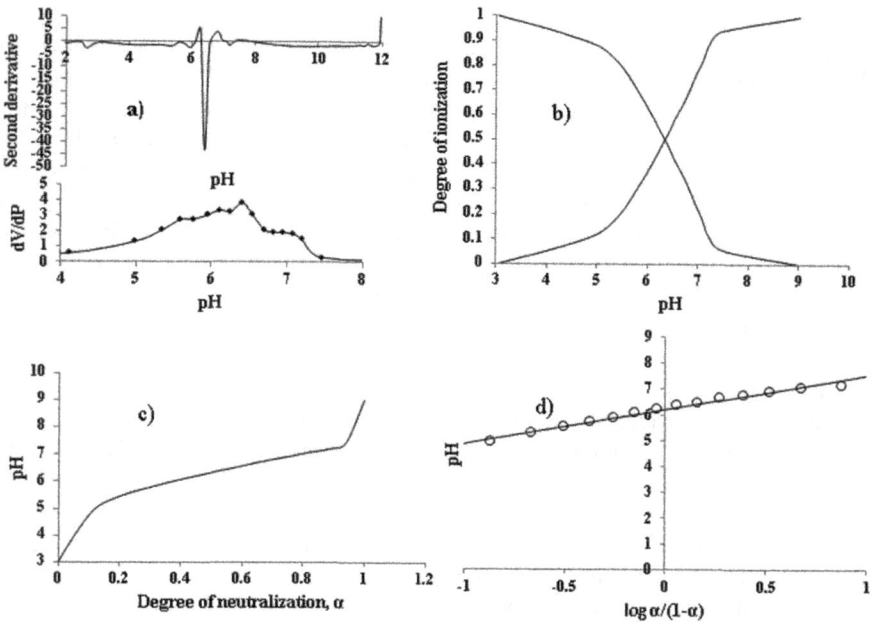

**Fig. 3.2** Potentiometric titration of chitosan sample for typical pKa value calculation using **a** first and second derivative, **b** degree of ionization, **c** degree of neutralization, and **d** Katchalsky–Spitnik equation. Unpublished work by authors

## 3.1.3 Swelling Behavior and Solubility

3.1.3.a Swelling of chitosan

The swelling behavior of chitosan is a function of time, temperature, and pH values. The degree of swelling of the chitosan sample is calculated from the following equation:

$$\text{Degree of swelling} = \left(\frac{W_s - W_d}{W_d}\right). \tag{3.7}$$

$W_s$ and $W_d$ are sample weights at the equilibrium swelling and dry state, respectively. The weight of the dry state is obtained by placing the lyophilized (freeze-dried) samples under a vacuum overnight. The swelling characteristics of chitosan hydrogel depend on the ionizable groups present within the gel structure [21]. The $-NH_2$ groups of chitosan are protonated in the acidic pH range. The swelling behavior of chitosan in deionized water can be attributed to the high repulsion of the $-NH_3^+$ groups. At a pH higher than 6, the carboxylic acid groups are ionized. The electrostatic repulsive forces between the charge (COO-) cause an increase in swelling [22, 23].

3.1.3.b Solubility of chitosan

The solubility of chitosan is an essential parameter as a higher percentage of chitosan solubility in an aqueous solution is desirable for medical applications. Chitosan is soluble in an organic acid solution. It is insoluble in water above pH 5.5 because of its compact α-, β-, and γ- conformations formed through hydrogen bonding, as mentioned in Chap. 2. Chitosan is soluble in acetic acid ($CH_3COOH$), hydrochloric (HCl), and nitric ($HNO_3$) acids at a concentration of 1% or less. A chitosan sample with a higher degree of deacetylation tends to dissolve in an acidic solution due to the protonation of amine ($-NH_2$) groups. The pKa value of chitosan is in the range of 6–6.5. However, in a low-density aqueous ionic environment, the molecular conformation change occurs in chitosan as pH is reduced (acidic pH). The glycosidic linkages of the chitosan are susceptible to acid, leading to the degradation of chitosan in acidic solutions. To a certain extent, the solubility of chitosan depends on the degree of deacetylation and molecular weight. No et al. reported that the degree of deacetylation of chitosan must be at least 85% to be completely soluble in acidic solutions [24].

A typical chitosan solubility measurement approach is described here. Approximately 10 mg chitosan and 1% (v/v) HCl 1% solution are mixed using a magnetic stirrer until a homogeneous solution appears. A microprocessor pH meter was used to measure the pH. The insoluble portion was removed by filtration using Whatman filter paper 22 μm and weighed. The solubility of chitosan in the solution is calculated using the following equation [25]:

$$S = \left[\left(\frac{10 - W_1}{10}\right)\right] \times 100\%, \tag{3.8}$$

## 3.1 Introduction

**Table. 3.2** Transmittance of modified chitosan materials

| Carboxylic acid | The stoichiometric ratio of crosslinking agent (%) | Results | Transmittance (%) |
|---|---|---|---|
| Succinic acid | 25 | Opalescent | 76 |
| | 50 | Precipitate | |
| | 100 | Precipitate | |
| Malic acid | 25 | Opalescent | 85 |
| | 50 | Opalescent | 80 |
| | 100 | Opalescent | 71 |
| Tartaric acid | 25 | Clear | 99 |
| | 50 | Clear | 92 |
| | 100 | Opalescent | 88 |
| Citric acid | 25 | Opalescent | 38 |
| | 50 | Opalescent | 38 |
| | 100 | Precipitate | |

Copyright © 2005, American Chemical Society, Biomacromolecules 6(5):2521–2527. Reproduced from Ref. [26] with permission

where $W_1$: the weight of the undissolved chitosan (mg) and $S$: solubility (%).

Bodner et al. [26] evaluated the solubility of chitosan nanoparticles in deionized water. They used carboxylic acid as a crosslinking agent. At a lower crosslinker concentration, the solubility of colloid dispersion is more significant. This is due to the protonation of free amino groups of the chitosan chain. The solution was either clear or opalescent [26]. Table 3.2 summarizes the transmittance of chitosan nanoparticles obtained by crosslinking various percentages of the carboxylic acids.

Bodner et al. [26] also observed that the solubility of the chitosan nanoparticles is related to two factors. These two factors are the hydrophilic character of the crosslinking agents and the ratio of free amino groups to total amino groups of the chitosan chain. Cho et al. [27] investigated the viscoelastic properties of the chitosan system in solution and gel states of glycerophosphates (GP) and polymer concentrations. Figure 3.3 represents gelation temperature as a function of glycerophosphate and polymer concentrations. The region below the surface is the solution state, while the above is the gel. Cho et al. observed that the phase transition is on edge between concentration-induced and heat-induced gelation. The chitosan solutions showed a liquid-like behavior at low β-glycerophosphate (GP) and chitosan (C) content but deviated from the solution state at higher concentrations [27].

### *3.1.4 Percentage of Moisture (Loss on Drying)*

The gravimetric method can determine the loss of moisture on drying of the chitosan samples. Different amounts of chitosan, between 0.1 and 0.3 g, are weighted with an accuracy of ±0.0002 in dry vials of known mass. In this process, the water mass loss is determined by drying the sample by placing it in the oven at ~100 °C for 48 h

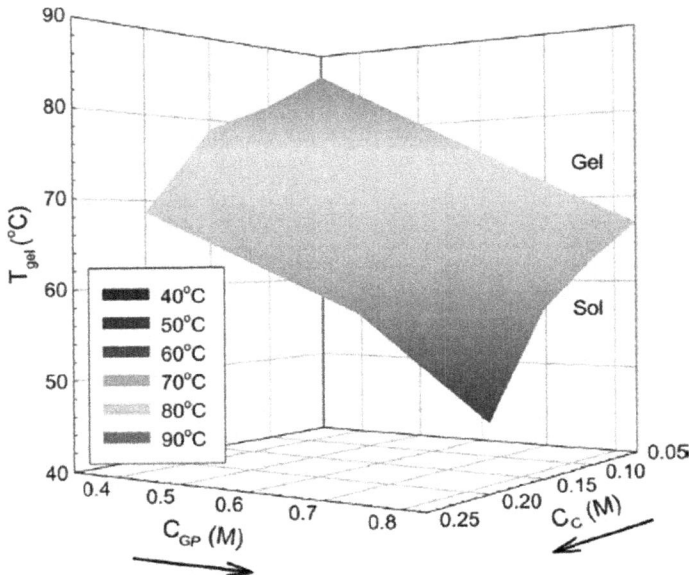

**Fig. 3.3** Phase diagram for $T_{gel}$ under various β-glycerophosphate and chitosan concentrations. Copyright © 2005 Elsevier Ltd. Food Hydrocolloids 20:936–945. Reproduced from Ref. [27] with permission

(until constant weight). The samples are weighed before and after drying. The water mass will be the difference between the weights of the wet and oven-dried samples as per the following equation:

$$\%\text{loss on drying} = \frac{(\text{Wet weight} - \text{Dry weight})}{\text{Dry weight}} \times 100. \tag{3.9}$$

## 3.1.5 Percentage of Ash

Ash content indicates the effectiveness of the demineralization step for removing calcium carbonate from the chitosan. A high-grade quality of chitosan should have less than 1% ash content. Some residual ash of chitosan may affect their solubility, consequently contributing to lower viscosity, or can affect other vital characteristics of the final product [24].

The ash content of the chitosan samples can be determined using the gravimetric method. Approximately 2.0 g of chitosan is placed into a previously baked, cooled, and tarred crucible. The sample is heated for 4 h in a muffle furnace preheated to 650 °C. The crucible is allowed to cool in the furnace to less than 200 °C. The crucible is then placed into a desiccator with a vented top. The percentage of ash value is calculated using the following equation:

$$\text{Ash\%} = \frac{\text{(Weight of residue, g)}}{\text{(Sample weight, g)}} \times 100. \quad (3.10)$$

## 3.2 Morphology of the Chitosan Using SEM, EDS X-Ray Analysis, and TEM

### 3.2.1 Scanning Electron Microscopy (SEM) Analysis

Scanning electron microscopy (SEM) can be used to investigate the surface morphology of the chitosan sample [28]. Using high-current heating lanthanum hexaboride or tungsten filament produces an electron beam. The beam can be accelerated using a potential of up to 30,000 V. When the beam focuses on a sample, electrons strike the sample and generate different signals such as backscattered electrons (BSE), secondary electrons (SE), and X-ray photons. The BSE signal generated during the elastic scattering of electrons in the beam retains almost all of their initial energy during the operation. SE electrons are very low in energy. They are ejected from the surface of the target material after the interaction with electrons of the primary beam. They can easily escape the sample. On the other hand, X-rays are generated inside the target when electrons jump from higher states to lower ones to fill the vacancies created by ejected electrons from K, L, and M shells. An electron from a higher energy state (shell) will move to a lower energy state emitting its excess energy in the form of X-ray photons, which are discrete and characteristics of the specific element that emitted it. The sample for SEM analysis must be free from contaminants that might interfere with electron movement. A sample is placed on aluminum stabs using double-sided carbon tape. Depending on the experiment, the sample is coated with approximately 20 nm of gold or carbon using a sputter coater to ensure the electric and heat conductance of the specimen. Generally, samples are sputter coated with gold for electrical contacts. If the gold peaks interfere with the X-ray analysis, carbon is used for coating the sample. Carbon-coating is not usually used for taking SEM micrographs due to the charging of the sample. In the case of chitosan, carbon-coating did not interfere with the SEM analysis. The SEM images shown in Fig. 3.4 are generated using backscattered electrons at 20 kV accelerating voltage. The SEM analysis is performed using the AMRAY 1600 T SEM machine utilizing a LaB6 crystal with a maximum accelerating voltage of 20 kV. Electron micrographs from the field emission scanning are recorded on an S4700 electron micrograph (Hitachi, Japan) operating at 120 kV.

Figure 3.4 shows a typical example of an SEM micrograph for chitosan and chitosan-based material. The SEM of chitosan showed no particular shape, instead

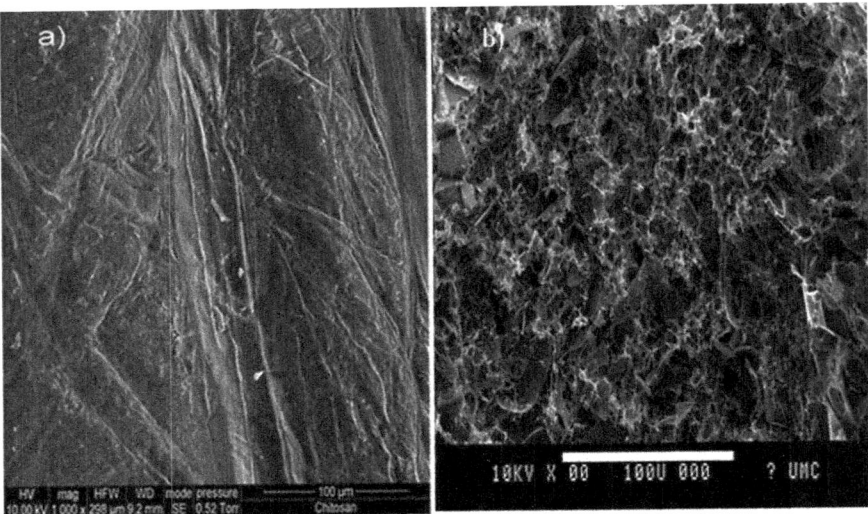

**Fig. 3.4** SEM micrograph of the surface morphology of **a** chitosan flake and **b** cross section of chitosan-coated perlite beads. Copyright © 2007 Elsevier B.V., J. Hazardous Material, 152(2):826–837, 2008. https://doi.org/10.1016/j.jhazmat.2007.07.078. Reproduced from Ref. [29] with permission

it appeared like flakes (Fig. 3.4a). Figure 3.4b shows the SEM of the cross section of chitosan-based material. The inside of the material is porous. Figure 3.4b shows that chitosan-based material's surface morphology and texture—are entirely different from that of chitosan flake.

## 3.2.2 Energy-Dispersive Spectroscopic (EDS) X-Ray Microanalysis

During EDS analysis, X-rays are directed to a liquid nitrogen-cooled semiconductor detector where an electrical signal is generated proportional to the energy of X-rays striking it. A multichannel analyzer then processes these pulses. X-ray intensity (counts) versus energy (keV) is plotted. The EDS microanalysis is a rapid process, and it provides simultaneous multielement analysis of the full X-ray spectrum. It does not show the higher-order X-ray lines. The sample preparation method for EDS is similar to that used for SEM analysis, except that carbon instead of gold is used for coating.

The EDS X-ray microanalysis can be performed on the same chitosan sample used for the SEM micrograph. The EDS microanalysis is used for the elemental analysis of the chitosan sample (Fig. 3.5). The carbon, oxygen, and nitrogen peaks occur at

## 3.2 Morphology of the Chitosan Using SEM, EDS X-Ray Analysis, and TEM

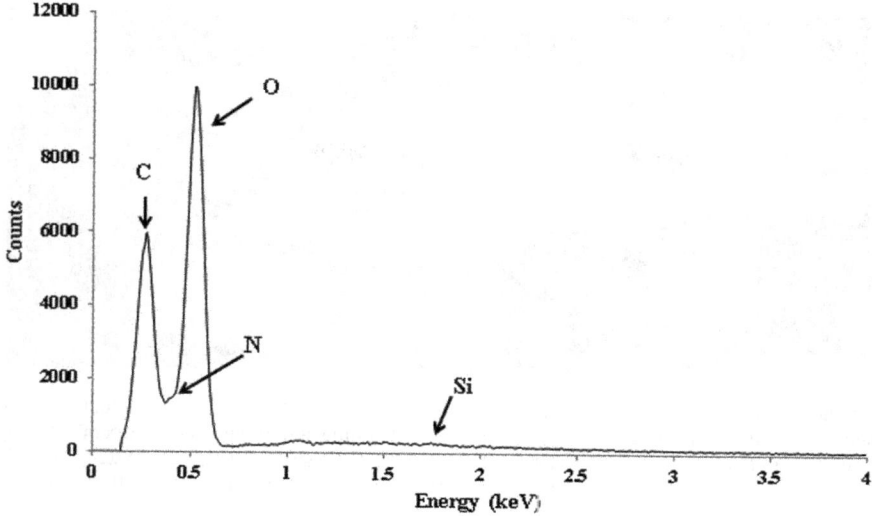

**Fig. 3.5** EDS X-ray microanalysis of chitosan sample. Unpublished work by authors

0.3 keV, 0.36, and 0.5 keV, respectively [28]. Carbon, oxygen, and nitrogen are the main components of chitosan (Fig. 3.5).

EDS X-ray microanalysis can be used for elemental analysis of the chitosan sample. Table 3.3 shows that the chitosan sample contains 23.53% C, 12.68% N, and 63.71% O element estimated from EDS X-ray microanalysis. The EDS X-ray microanalysis shows a peak for silicon. During chitosan sample analysis, silicon may come from the chitosan sample or the coating process.

**Table 3.3** EDS analysis of C, O, and N percentages in chitosan sample

| Element line | K ratio | Weight % | Weight % error |
|---|---|---|---|
| C K | 0.31 | 23.53 | ±0.19 |
| N K | 0.1 | 12.68 | ±0.86 |
| O K | 0.58 | 63.71 | ±0.46 |
| Si K | 0.00 | 0.08 | ±0.02 |
| Si L | – | – | – |
| Total | | 100 | |

Unpublished work by authors

**Fig. 3.6** Typical example of transmission electron micrograph (TEM) of chitosan-coated perlite beads exposed to copper ion (Philips EM 430 STEM equipped with an energy-dispersive X-ray spectroscopy (EDS). Copyright © 2007 Elsevier B.V., J. Hazardous Material, 152(2):826–837, 2008. https://doi.org/10.1016/j.jhazmat.2007.07.078. Reproduced from Ref. [29] with permission

## 3.2.3 Transmission Electron Micrograph (TEM)

The electron beam of TEM travels through the target material. The differences in contrast in the image are due to the density variation of the sample. Chitosan sample is prepared by embedding in Epon-Spurr resin and microwaved under vacuum at 250 W for 3 min, then placed in an oven for 24 h at 60 °C to polymerize [28]. Samples are then cut on a Leica Ultracut UCT ultramicrotome at 65–85 nm thick. Sections are then placed on carbon-coated copper grids for viewing in the TEM. The Philips TEM operates at 300 keV. Figure 3.6 shows the transmission electron microscopic (TEM) micrograph of chitosan-coated perlite bead exposed to copper ions in solution.

## 3.2.4 Thermogravimetric Analysis (TGA) of Chitosan

Thermogravimetric analysis (TGA) can be used to determine the water content of chitosan and to determine the temperature at which the chitosan would burn out. This information is helpful in determining the thermal stability of chitosan (Fig. 3.7). The mass loss of the sample is recorded as a function of temperature. Figure 3.7a shows thermogravimetric analysis (TGA) of chitosan performed using a TGA (TA Instruments) analyzer with flowing nitrogen at 200 mL/min. Approximately 20 mg of chitosan is heated in the temperature range from 30 to 600 °C in an open alumina crucible at a set heating rate. The TGA measures the amount and rate of weight loss of the sample as it is heated. Thermogravimetric analysis of chitosan provides complementary information on changes in its composition as heating progresses under controlled conditions. The heating rate in this analysis (Fig. 3.7a) is set at

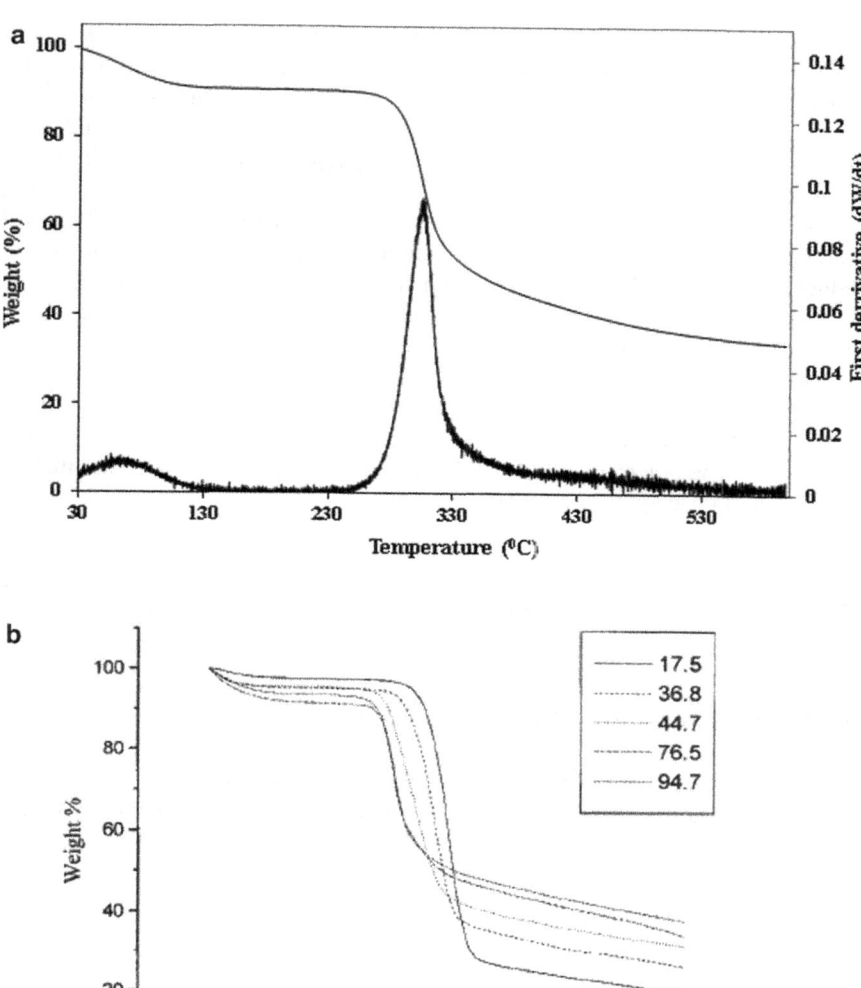

**Fig. 3.7 a** Typical thermogravimetric analysis (TGA) of chitosan samples under nitrogen atmosphere. Unpublished work by authors, **b** TGA curves of deacetylated products (DD = 17.5, 36.8, 44.7, 76.5, and 94.7%). A Perkin Elmer TGA 7 thermogravimetric analyzer measures the thermal weight loss of deacetylated products of chitin. Copyright © 2004 Wiley Periodicals, Inc. Journal of Applied Polymer Science 93:2416–2422. Reproduced from Ref. [32] with permission

10 °C min$^{-1}$ under a nitrogen stream. Based on the TGA profile, the dehydration of chitosan occurs in the range of 30–105 °C with a weight loss of about 8%. Water is present in macromolecules in three different forms: weakly surface-bound water, tightly bound to hydrophilic groups, and crystalline-frozen water embedded in the chitosan matrix [30, 31]. The weight loss in this temperature range occurs due to moisture vaporization. The anhydrous chitosan further decomposed in the second step, with a weight loss of ~33.5% at 305 °C. Finally, it is burnt entirely at 600 °C with additional weight loss of 24.5%. The remaining 34% is assumed to be combustion products of the chitosan at 600 °C. The first derivative of weight loss also shows one broad peak representing the slow dehydration process. Another sharp peak at 305 °C represents the decomposition of the chitosan sample.

Chen et al. reported thermogravimetric analysis data of β-chitin with different degrees of deacetylation (Fig. 3.7b) [32]. Two regimes of weight loss can be observed in these TGA curves. One weight loss regime at 50–150 °C is due to moisture vaporization. The other regime of weight loss at 250–350 °C is due to the thermal degradation of deacetylated products of β-chitin.

## 3.2.5 Determination of Surface Area and Surface Charge of Chitosan

### 3.2.5.1 Determination of Chitosan Surface area

The total surface area of the bead was determined from the adsorption of N$_2$ at 77 K using the Brunauer–Emmett–Teller (BET) method. The following BET equation is described by Hines et al. [33]:

$$\frac{P}{N(P_0 - P)} = \frac{1}{N_m C} + \frac{(C-1)P}{N_m C P_0}, \qquad (3.11)$$

where

$P$ = adsorbate gas pressure,
$P_0$ = saturation pressure of adsorbate gas at the system temperature,
$N$ = equilibrium amount of gas adsorbed,
$N_m$ = amount of gas adsorbed in monolayer,
$C$ = dimensionless constant.

The amount of adsorbed gas corresponding to the monolayer coverage ($N_m$) is calculated from Eq. 3.11. Then, the total surface area per unit mass of the sample can be calculated from the knowledge of the effective cross-sectional area of the adsorbate molecule using Eq. 3.12.

## 3.2 Morphology of the Chitosan Using SEM, EDS X-Ray Analysis, and TEM

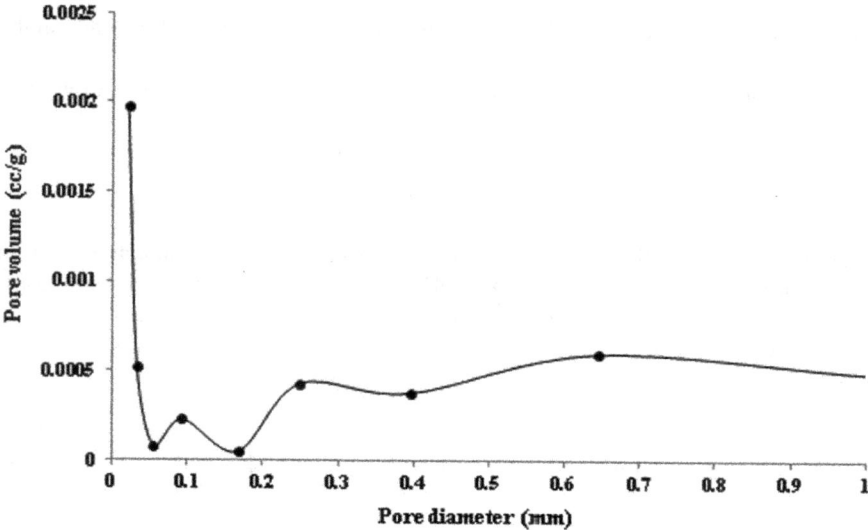

**Fig. 3.8** Pore size and pore volume distribution of chitosan flake. Unpublished by authors

$$S_g = \frac{N_m A_0 \alpha}{W_s}, \quad (3.12)$$

where

$A_0$ = Avogadro's number,
$\alpha$ = effective cross-sectional area of adsorbate molecule (16.2 Å$^2$ for nitrogen),
$W_s$ = mass of the sample,
$S_g$ = total surface area per unit mass of the sample.

The equilibrium nitrogen adsorption data at the liquid nitrogen temperature in the pressure range of 0.05–0.30 atm (0.74–4.41 psi) are used to calculate the surface area. Figure 3.8 shows a typical example of pore volume and pore diameter of chitosan. Figure 3.8 was produced through the nitrogen adsorption/desorption at 77 K using a Micromeritics ASAP 2010 surface analyzer. For the information shown in Fig. 3.8, the BET surface area of chitosan is 1.36 m$^2$/g. For the chitosan flakes, the average pore diameter and pore volume are 0.004 μm and 1 × 10$^{-3}$ cm$^3$/g, respectively.

### 3.2.5.2 Surface Charge

The surface charge of the chitosan can be determined by a standard potentiometric titration method in the presence of an asymmetric electrolyte, e.g., sodium nitrate. The magnitude and the surface charge can be measured at the point of zero charge (PZC). The pH of the PZC for a given surface depends on the relatively basic and acidic properties of the solid. The surface charges of beads can be determined in the following manner.

The surface charge (C m$^{-2}$) can be calculated using the following equation [34]:

$$\sigma_0 = \frac{(C_a - C_b + [\text{OH}^-] - [\text{H}^+])\text{F}}{Sa}, \quad (3.13)$$

where $C_a$ and $C_b$ are the concentration of acid or base required to reach a point on the titration curve in mol/L, [H$^+$] and [OH$^-$] are H$^+$ and OH$^-$ ion concentrations, respectively. [H$^+$] and [OH$^-$] are calculated from the pH of the solution with appropriate corrections using the activity coefficient. The activity coefficient is calculated using the Davis equation [35]:

$$\log \gamma_i = -0.5109 \left( \frac{\sqrt{I}}{1 + \sqrt{I}} - 0.3I \right), \quad (3.14)$$

$$I = 0.5 \sum_i C_i Z_i^2, \quad (3.15)$$

where $\gamma_i$ is the activity coefficient of component $i$ of concentration $C_i$ and ionic charge $Z_i$ and $I$ is the ionic strength (mol/L). F is Faraday constant (96,400 C/mol), and $S$ is the specific surface area of beads measured by the BET method using N$_2$ as adsorbate at 77 K.

Figure 3.9 shows a typical surface charge analysis of chitosan-based adsorbent material. The PZC value of the chitosan-coated perlite bead is 8.5, as shown in

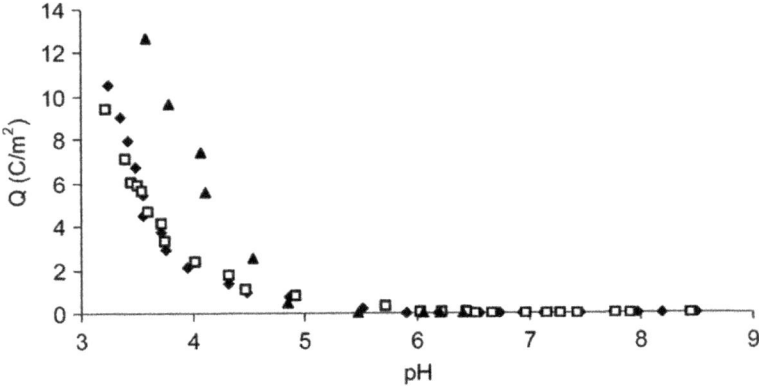

**Fig. 3.9** Typical example of surface charge of chitosan-coated perlite beads in the presence of CdCl$_2$: (◆) 0.1 M NaNO$_3$, (□) 0.05 M NaNO$_3$, and (▲) cadmium solution [20]. Copyright © 2006, American Chemical Society, Ind Eng Chem Res 45:5066. Reproduced from Ref. [20] with permission

Fig. 3.9, which is similar to the value reported by Jha et al. for chitosan flake [36]. However, Udaybhaskar et al. reported a PZC value in the range of 6.2–6.8 for pure chitosan [37]. The surface charge of the chitosan-coated perlite bead is almost zero in the pH range of 6–8.5. The pKa value of perlite is determined to be ~7. The protonation of the beads sharply increased at the pH range of 3–4.5, making the surface positive. At pH <3.5, the difference between the initial and equilibrium pH is not significant due to the complete protonation of chitosan. At higher pH (4.5–8.5), the surface charge of the bead slowly decreased due to the slow protonation of chitosan on the bead. The surface charge of the bead could be due to the modification of chitosan when coated on perlite, making the bead amphoteric in nature.

### 3.2.6 Moisture Sorption onto Chitosan

The biodegradable property of chitosan has increased its interest in food and pharmaceutical industries. During food dehydration, the moisture content plays a vital role in the electrochemical and biological activity and the biodegradation of food products [38, 39]. Polysaccharides usually have a strong affinity for water. In a solid state, these macromolecules may have disordered structures that can be hydrated. Water molecules form complex structures with the hydrophilic groups of polymer chains through interhydrogen bonding [40, 41]. The macromolecular structure and permeability characteristics of most biodegradable films change with alterations in the surrounding atmosphere's relative humidity (RH). The temperature influences the mobility of the molecules of water in the films. Temperature also affects the dynamic equilibrium between the vapor and the adsorbent phase. A plasticizer is an essential hydrophilic material used in food packaging [42]. In most hydrophilic films, water functions as a plasticizer, and the relative humidity of the environment determines water adsorption and desorption on film [43]. Plasticization by water molecules was reported to affect the glass transition temperature ($T_g$) of chitosan film [31]. The $T_g$ of chitosan decreases with an increase in moisture content in the film. An increase in moisture content modifies the performance and biodegradability of the film [30, 31, 44]. The water sorption capacity of chitosan-based materials is analyzed by water sorption isotherm and water activity ($a_w$) at a constant temperature and pressure [40, 44–46]. Water activity is defined as the partial vapor pressure of water in a chitosan-based material at a specific temperature divided by the partial vapor pressure of water in the atmosphere at the same temperature. The water activity $a_w$ becomes the basic safety and quality controlling factor in preserving the moisture-sensitive material. The following equation is used to obtain the water activity:

$$a_w = \frac{\text{Equilibrium relative humidity}(\%)}{100}. \tag{3.16}$$

The moisture sorption isotherm of chitosan-based material is described by different mathematical models given in Table 3.4. Aguirre-Loredo et al. [30] reported

Table. 3.4 Isotherm models for moisture sorption data fitting [46–50]

| Model | Equation | Parameters |
|---|---|---|
| BET | $X = \dfrac{x_m C a_w}{(1-a_w)(1-a_w+C a_w)}$ | $x_m$ = monolayer water content<br>$C$ = energy difference between the upper layers and the monolayer |
| GAB | $X = \dfrac{x_m C k a_w}{(1-k a_w)(1-k a_w+C k a_w)}$ | $x_m$ = monolayer water content<br>$C$ = energy difference between the upper layers and the monolayer<br>$K$ = degree of freedom of water molecules |
| Henderson | $X = \left[ -\ln \dfrac{(1-a_w)}{A} \right]^{\frac{1}{B}}$ | $A, B$ = empirical parameters |
| Halsey | $X = \left[ \dfrac{A}{\ln(1/a_w)} \right]^{\frac{1}{B}}$ | $A, B$ = empirical parameters |
| Oswin | $X = A \left( \dfrac{a_w}{1-a_w} \right)^{B}$ | $A, B$ = empirical parameters |
| Iglesias-Chirife | $X = A \left[ \dfrac{a_w}{(1-a_w)} \right] + B$ | $A, B$ = empirical parameters |
| Flory–Huggins | $X = A \exp(B a_w)$ | $A, B$ = empirical parameters |

that not all models apply to the full scale of water activity that ranges from 0 to 1. They described that the moisture adsorption isotherms belong to three regions. The region of $a_w < 0.2$ describes water adsorption in the monolayer region. The region corresponding to $a_w$ 0.2–0.7 describes additional layers of water adsorption. The region corresponding to $a_w > 0.7$ describes the condensation of water in the pores of the material, followed by the dissolution of the material. Suryatem et al. [46] reported that the isotherm is more rapid in the initial stages ($a_w$ 0.00–0.20). They also reported that a less moisture is absorbed, while $a_w$ increases toward 0.56 [46].

Among all equations shown in Table 3.4, the Guggenheim, Anderson, and de Boer (GAB) equation has a theoretical basis, while other models are empirical or semi-empirical [47]. The GAB model has been widely used to describe equilibrium moisture sorption isotherms [39, 48]. The significant advantages of the GAB model are as follows [49, 50]:

1. It has a theoretical background refined from the BET theory.
2. It describes the sorption behavior of nearly all foods from zero to 0.9 $a_w$.
3. It has a relatively simple mathematical form with three parameters.
4. All the parameters have physical meanings in terms of the sorption process.
5. It can describe some temperature effects of isotherms by the Arrhenius-type equation.

## 3.2.7 Mechanical Properties of Chitin and Chitosan

The stress–strain behavior of polymeric materials depends on various parameters, such as molecular characteristics, microstructures, and temperature. Tensile properties describe the force required to elongate to the breakpoint of the material. The stress and strain at the breakpoint are called ultimate strength and elongation at break. The tensile strength and elongation at the breakpoint of the polymeric material are calculated using the following equations [4]:

$$\text{Tensile strength} \left(\frac{N}{mm^2}\right) = \frac{\text{Breaking force (N)}}{\text{Cross} - \text{sectional area of sample (mm}^2)}, \quad (3.17)$$

$$\text{Elongation at break } (\%) = \frac{\text{Increase in length at breaking point (mm)}}{\text{Original length (mm)} \times 100\%}. \quad (3.18)$$

Mushi et al. [51] describe specific cases of chitin and chitosan film at various porosities. They described the stress–strain relations and failure mechanisms of the film.

## 3.2.8 Diffuse Reflective UV-Vis (DRUV) Spectroscopy

Diffuse reflective UV-vis spectroscopy can provide important characteristic information about the electronic transition of chitosan before and after a crosslinking reaction. The polymer chromophore absorbs light at characteristic wavelengths in the UV-vis region. Diffuse reflective UV-vis (DRUV) spectroscopy is applied to study covalent and non-covalent interactions of chitosan. Figure 3.10 shows a typical example of diffuse reflective UV-vis (DRUV) spectra of pure chitosan and chitosan crosslinked with an iron oxide sample. The UV-is spectrum of pure chitosan showed maximum absorption at 300 nm (Fig. 3.10). The absorption spectrum of the chitosan–iron sample changes with the development of a broad absorption band at 300 nm, and the shoulder peak at 300 nm for chitosan disappeared. Figure 3.10 also shows absorption spectra for chitosan crosslinked with iron oxide known as CFOH. The absorption intensity for the CFOH sample is higher than chitosan. The hydrated iron oxide shows a broad peak at 266 nm and another in the visible range at ~480 nm. The chromophore responsible for the absorption band of chitosan at 300 nm is replaced by new chromophores generated by the crosslinking of chitosan with iron oxide. In the CFOH bead, the new chromophore shows a wide band with a maximum in the UV region, close to 300 nm. The new chromophore also indicates a small shoulder peak at higher wavelengths, reaching the visible region. The lower intensity of the CFOH sample compared to the hydrated iron oxide is attributed to the charge transfer of $Fe^{3+}$ in hydroxo-complex species [52]. The spectrum obtained after the subtraction

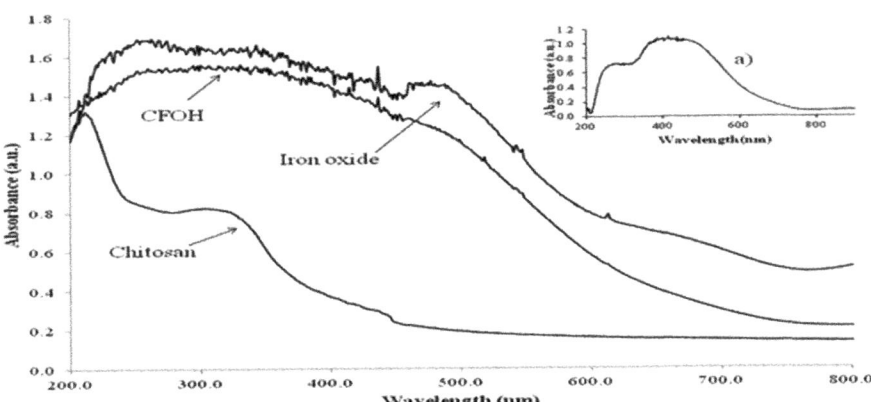

**Fig. 3.10** Typical example of diffuse reflective UV-vis (DRUV) spectra of chitosan, chitosan–iron composite (CFOH), and hydrated iron oxide. **a** DRUV spectrum after subtraction of pure chitosan spectrum from that of chitosan–iron composite sample. Copyright © 2014, Taylor & Francis. https://doi.org/10.1080/01496395.2014.949774. Reproduced from Ref. [53] with permission

of the pure chitosan spectrum from the CFOH spectrum shows a new absorption band at 460 nm. This absorption band is due to the formation of a covalent bond between chitosan and iron.

### 3.2.9 X-Ray Photoelectron Spectroscopy (XPS) Analysis

Photoelectron spectroscopy, also known as electron spectroscopy for chemical analysis (ESCA), is extremely useful for chemical analysis of a variety of compounds. Essentially, the X-ray photoelectron spectroscopy is used for the direct study of the valence shell [54]. The primary chemical information provided by XPS is the core binding energies used to identify elements of chemical compounds. A typical example of the XPS experimental data is obtained by using a KRATOS model AXIS 165XPS spectrometer with non-monochromatic Mg X-rays (hv = 1253.6 eV), which are usually used as the excitation source at a power of 240 W. The spectrometer is equipped with an eight-channel hemispherical detector. The energy of 5–160 eV was used during the analysis of the samples. Each sample can be exposed to X-rays for the same period of time and intensity.

In XPS analysis, a survey scan ensures that the elemental composition at the surface is representative of the entire sample. The higher resolution utility scans also provide the atomic concentration of C, N, and O in the sample. Figure 3.11 shows the peak positions of carbon, oxygen, and nitrogen obtained by the XPS for chitosan flake. In Fig. 3.11, the C-1s peak appears at 284.3 eV with an FWHM of 3.27 eV (full width at maximum height). Figure 3.11a shows two peaks, one for C–N (284.3 eV) and the other for C–C (283 eV) linkage. From XPS spectra (Fig. 3.11),

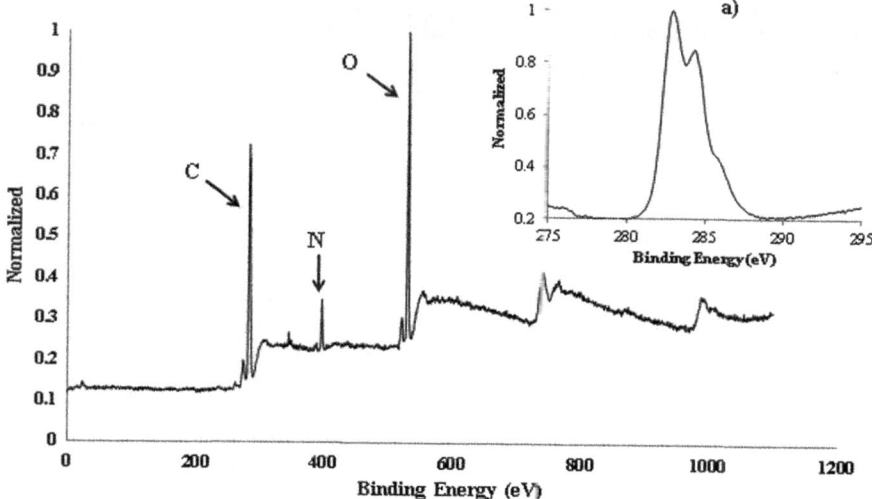

**Fig. 3.11** XPS survey scan spectra of chitosan **a** C-1s peak of chitosan. Copyright © 2006, American Chemical Society, Ind Eng Chem Res 45:5066. Reproduced from Ref. [20] with permission

it appears that the oxygen-containing group (O–1s) in chitosan flake shows a peak at 530.5 eV. Kurmaev et al. [55] reported that the primary contribution to the O–1s spectra comes from the functional groups in chitosan structure (–OH, –O– in the ring and possibly –O– between the rings). The –NH$_2$ groups in chitosan are considered as the active binding sites. The N–1s in chitosan flake shows a peak at 397.5 eV (FWHM 1.87 eV), attributed to the amino groups in chitosan. Table 3.5 shows the surface elemental concentration of C, N, and O as determined from peak/area ratios after correction using the experimentally determined sensitive factor (±5%).

## 3.2.10 Electron Spin Resonance (ESR) Spectroscopy

Electron spin resonance (ESR) or electron paramagnetic resonance (EPR) measurement provides evidence of chemical species with one or more unpaired electrons (electronic spin state and oxidation state of metal) crosslinked with chitosan samples. The ESR spectra are measured at the microwave power of 6.35 mW and the amplitude of modulation of 2.0 G in the range of 2000–4000 G (5 scans with the sweep time of 42 s) or 100–4600 G (2 scans at the sweep time of 84 s). Figure 3.12 shows a typical example of the ESR spectra of the chitosan-copper sample.

Webster et al. [58] investigated different metals crosslinked with chitosan, and their oxidation state by ESR is given in Fig. 3.13.

**Table 3.5** Atomic concentration and assignment of spectral bands of C, N, and O on chitosan flake as obtained from XPS data

| Sample | Element | Binding energy (eV) | Atomic fraction (%) | Assignment | References |
|---|---|---|---|---|---|
| Chitosan flake | C 1s | 284.6 | 10 | C–C or contaminated carbon | [56] |
|  | C 1s | 286.2 | 44.5 | C–N or C–O or C–O–C |  |
|  | C 1s | 287.9 | 8 | C=O or C–O–C |  |
|  | Total C 1s | 531.8 | 62.5 | – |  |
|  | O 1s | 532.8 | 3.5 | C–O or C–O–C or chemisorbed water or stoichiometric hydroxide |  |
|  | O 1s | 399.1 | 26.5 | >C–O or O–H or bound $H_2O$ |  |
|  | Total O 1s |  | 30 | – |  |
|  | N 1s |  | 7 | N |  |
| Chitosan flake | C 1s | 283 | 67.28 | C–N and C–C | [29] |
|  | O 1s | 530.5 | 27.06 | C–O or C–O–C or O–H |  |
|  | N 1s | 397.5 | 5.1 | N |  |

Copyright © 2002 Published by Elsevier Ltd. Water Res. 2002, 36:3699. Reproduced from Ref. [56] with permission; Copyright © 2007 Elsevier B.V. https://doi.org/10.1016/j.jhazmat.2007.07.078. Reproduced from Ref. [29] with permission

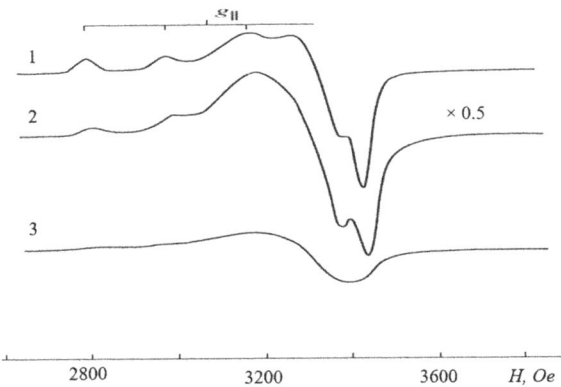

**Fig. 3.12** ESR spectra of different starting Cu/chitosan samples taken at 20 °C: (1) coprecipitated 0.5% Cu/chitosan, (2) coprecipitated 1.5% Cu/chitosan, (3) 0.5% Cu/chitosan prepared by adsorption methods. Copyright © 1969, MAIK "Nauka/Interperiodica", Kinetics and Catalysis 44(6):793–800. Reproduced from Ref. [57] with permission

| M | CS | | |
|---|---|---|---|
| | Oxidation State | Outer Shell electrons | ESR Signal |
| Cr | ? | ? | − |
| Mn | 2+ | $d^5$ | + |
| Fe | 3+ | $d^5$ | + |
| Co | 2+ | $d^6$ | − |
| Ni | 2+ | $d^8$ | − |
| Cu | 2+ | $d^9$ | + |
| Zn | 2+ | $d^{10}$ | − |

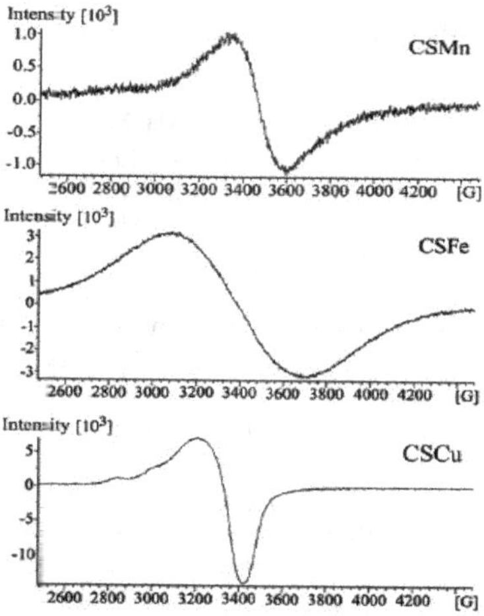

Fig. 3.13 Mn, Fe, and Cu complexed with both chitosan showed strong ESR signals. Oxidation states and outer shell electrons are summarized: M = metal; d = d-shell electrons; − indicates no ESR signal; + indicates ESR signal is observed. Copyright © 2007 Elsevier Ltd., Carbohydrate Research, 342, 1189–1201, (2007). Reproduced from Ref. [58] with permission

## 3.3 Summary

The deacetylation of chitin to chitosan yields a linear chain of glucosamine, making a weaker hydrogen bond network in the biopolymer and altering the chemical and physical properties of chitin. The main parameters that influence the physical and chemical properties of chitosan are its molecular weight and degree of deacetylation. The molecular weight determines the viscosity and the strength of its fiber (rate of degradation). The degree of deacetylation identifies the biopolymer as chitin or chitosan and plays a significant role in affecting the molecular weight of chitosan. Scanning electron microscopy (SEM) and transmission electron microscopic (TEM) analysis provide surface morphological information of chitosan. The moisture content in chitosan samples can play a key role in electrochemical and biological activity and their stability regarding hydrolysis and biodegradation. Thermogravimetric analysis (TGA) of chitosan samples provides essential information about the thermal stability of chitosan. The energy-dispersive spectroscopy (EDS) X-ray microanalysis estimates the presence of elements in the chitosan sample. The primary chemical information provided by the X-ray photoelectron spectroscopy (XPS) is based on core binding energies used to identify elements of chemical compounds. The diffuse reflective UV-vis spectroscopy (DRUV) provides important characteristic information about the electronic transition of chitosan before and after the crosslinking reaction. Electron spin resonance (ESR) provides evidence of chemical species with one or more unpaired electrons (electronic spin state and oxidation state of metal) that crosslinked with chitosan samples.

# References

1. El-hefian EA, Yahaya AH, Misran M (2009) Characterization of chitosan solubilized in aqueous formic and acetic acids, Maejo. Int J Sci Technol 3(03):415–425
2. Szymanska E, Winnicka K (2015) Stability of chitosan—a challenge for pharmaceutical and biomedical applications. Mar Drugs 13:1819–1846
3. Khan TA, Peh KK, Ch'ng HS (2002) Reporting degree of deacetylation values of chitosan: the influence of analytical methods. J Pham Pharmaceut Sci 5(3):205212. www.ualberta.ca/~csps
4. Chattopadhyay DP, Inamdar MS (2010) Aqueous behavior of chitosan. Int J Polym Sci 2010:7p. Article ID 939536. https://doi.org/10.1155/2010/939536
5. Morris GA, Castile J, Smith A, Adams GG, Harding SE (2009) Macromolecular conformation of chitosan in dilute solution: a new global hydrodynamic approach. Carbohyd Polym 76:616–621
6. Laka M, Chernyavskaya S (2006) Preparation of chitosan powder and investigation of its properties. Proc Estonian Acad Sci Chem 55(2):78–84
7. Dahmane EM, Taourirte M, Eladlani N, Rhazi M (2014) Extraction and characterization of chitin and chitosan from Parapenaeus longirostris from Moroccan local sources. Int J Polym Anal Charact 19(4):342–351
8. Badawy MEI, Rabea EI (2011) A biopolymer Chitosan and its derivatives as promising antimicrobial agents against plant pathogens and their applications in crop protection. Int J Carbohydr Chem 2011:29p. Article ID 460381
9. Kasaai MR (2007) Calculation of Mark–Houwink–Sakurada (MHS) equation viscometric constants for chitosan in any solvent-temperature system using experimental reported viscometric constants data. Carbohyd Polym 68:477–488
10. Rinaudo M (2006) Chitin and chitosan: properties and applications. Prog Polym Sci 31:603–632
11. Knaul JZ, Kasaai MR, Bui VT, Creber KAM (1998) Characterization of deacetylated chitosan and chitosan molecular weight review. Can J Chem 76:1699–1706
12. Qiang Z, Adams C (2004) Potentiometric determination of acid dissociation constants (pKa) for human and veterinary antibiotics. Water Res 38:2874–2890
13. Wang QZ, Chen XG, Liu N, Wang SX, Liu CS, Meng XH, Liu CG (2006) Protonation constants of chitosan with different molecular weight and degree of deacetylation. Carbohyd Polym 65:194–201
14. Sarkar K, Srivastava R, Chatterjee U, Kundu PP (2011) Evaluation of chitosan and their self-assembled nanoparticles with pDNA for the application in gene therapy. J Appl Polym Sci 121:2239–2249
15. Navarro R, Guzman J, Saucedo I, Revilla J, Guibal E (2003) Recovery of metal ions by chitosan: sorption mechanisms and influence of metal speciation. Macromol Biosci 3(10):552–561
16. Rinaudo M, Pavlov G, Desbrieres J (1999) Solubilization of chitosan in strong acid medium. Int J Polym Anal Charact 5:267–276
17. Pillai CKS, Willi P, Sharma CP (2009) Chitin and chitosan polymers: chemistry, solubility and fiber formation. Prog Polym Sci 34:641–678
18. Katchalsky A, Shavit N, Eisenberg H (1954) Dissociation of weak polymeric acids and bases. J Polym Sci: Part A; Polym Chem 13(68):69–84
19. Park JW, Choi K-H, Park KK (1983) Acid-base equilibria and related properties of chitosan. Bull Korean Chem Soc 4(2):68–71
20. Hasan S, Krishnaiah A, Ghosh TK, Viswanath DS, Boddu VM, Smith ED (2006) Adsorption of divalent cadmium (Cd (II)) from aqueous solutions onto chitosan-coated perlite beads. Ind Eng Chem Res 45:5066–5077
21. Yazdani-Pedram M, Tapia C, Retuert J, Arias JL (2003) Synthesis and unusual swelling behavior of combined cationic/non-ionic hydrogels based on chitosan. Macromol Biosci 3:577–581
22. Ray M, Pal K, Anis A, Banthia AK (2010) Development and characterization of chitosan based polymeric hydrogel membranes. Des Monomers Polym 13(3):193–206

23. Felinto MCFC et al (2007) The swelling behavior of chitosan hydrogels membranes obtained by UV- and γ-radiation. Nucl Inst Methods Phys Res B 265:418–424
24. No HK, Meyers SP (1995) Preparation and characterization of chitin and chitosan—a review. J Aquat Food Prod Technol 4(2):27–52
25. Xie Y, Liu X, Chen Q (2007) Synthesis and characterization of water-soluble chitosan derivate and its antibacterial activity. Carbohyd Polym 69:142–147
26. Bonder M, Hartmann JF, Borbely J (2005) Preparation and characterization of chitosan-based nanoparticles. Biomacromol 6:2521–2527
27. Cho J, Marie-Claude Heuzey M-C, Bégin A, Carreau PJ (2006) Chitosan and glycerophosphate concentration dependence of solution behavior and gel point using small amplitude oscillatory rheometry. Food Hydrocolloids 20:936–945
28. Hasan S (2005) Development of materials for the removal of metal ions from radioactive and non-radioactive waste streams. PhD thesis, University of Missouri, Columbia
29. Hasan S, Ghosh TK, Viswanath DS, Boddu VM (2008) Dispersion of chitosan on perlite for enhancement of copper (II) adsorption capacity. J Hazard Mater 152(2):826–837
30. Aguirre-Loredo RY, Adriana Inés Rodriguez-Hernandez AI, Gonzalo Velazquez G (2017) Modelling the effect of temperature on the water sorption isotherms of chitosan films. Food Sci Technol, Campinas, 37(1):112–118
31. Dhawade PP, Jagtap RJ (2012) Characterization of the glass transition temperature of chitosan and its oligomers by temperature modulated differential scanning calorimetry. Adv Appl Sci Res 3(3):1372–1382
32. Chen C-H, Wang F-Y, Ou Z-P (2004) Deacetylation of Chitin. I. Influence of the deacetylation conditions. J Appl Polym Sci 93:2416–2422
33. Hines AL, Ghosh TK, Loyalka SK, Richard CW Jr (1993) Indoor air quality and control. PTR Prentice-Hall Inc., Englewood Cliffs, New Jersey, pp 242–243
34. Davranche M, Lacour S, Bordas F, Bollinger J-C (2003) An easy determination of the surface chemical properties of simple and natural solids. J Chem Educ 80(1):76–78
35. Hasan S, Ghosh TK, Viswanath DS, Loyalka SK, Sengupta B (2007) Preparation and evaluation of fuller's earth beads for removal of cesium from waste streams. Sep Sci Technol 42:717–738
36. Jha IN, Iyengar LR, Prabhakararao AVS (1988) Removal of cadmium using chitosan. J Environ Eng 114:962
37. Udayabhaskar P, Iyengar L, Prabhakar R (1990) Hexavalent chromium interaction with chitosan. J Appl Polym Sci 39:739
38. Mucha M, Ludwiczak S, Kawińska M (2005) Kinetics of water sorption by chitosan and its blends with poly(vinyl alcohol). Carbohydr Polym 62:42–49
39. Kabalan T, Clément YBY, Françoisea KA, Mathiasb OK (2008) Determination and modelling of moisture sorption isotherms of chitosan and chitin. Acta Chim Slov 55:677–682
40. Julkapli NM, Akil HM, Ahmad Z (2012) Thermal properties of 4,4-oxydiphathalic anhydride chitosan filled chitosan bio-composites. J Therm Anal Calorim 107:365–376
41. Chang PR, Jian R, Yu J, Ma X (2010) Fabrication and characterization of chitosan nanoparticle/plasticized starch composite. Food Chem 120(3):730–740
42. Ibrahim N, Tajaddodi Talab K, Spotar S, Kharidah M, Rosnita AT (2013) Desorption isotherm model for a Malaysian rough rice variety (MR219). Pertanika J Trop Agric Sci 36(2):189–198
43. Saberi B, Vuong QV, Chockchaisawasdee S, Golding JB, Scarlett CJ, Stathopoulos CE (2016) Water sorption isotherm of pea starch edible films and prediction models. Foods 5(1):2–18. https://doi.org/10.3390/foods5010001; www.mdpi.com/journal/foods
44. Quyen DTM, Adisak J, Rachtanapun P (2012) Relationship between solubility, moisture sorption isotherms and morphology of chitosan/methylcellulose films with different carbendazim content. J Agric Sci 4(6)
45. Mucha M, Wańkowicz K, Ludwiczak S, Balcerzak J (2006) Water sorption by biodegradable chitosan/polylactide composites. Polish Chitin Society, Monograph XI, pp 41–51
46. Suriyatem R, Rachtanapun C, Raviyam P et al (2015) Investigation and modeling of moisture sorption behavior of rice starch/carboxymethyl chitosan blend films. In: Global conference on polymer and composite materials (PCM 2015). IOP Conf Ser: Mater Sci Eng 87:012080

47. Rosa GS, Moraes MA, Luiz AA, Pinto LAA (2010) Moisture sorption properties of chitosan. LWT—Food Sci Technol 43:415–420
48. Blahovec J, Yanniotis S (2010) 'GAB' generalized equation as a basis for sorption spectral analysis. Czech J Food Sci 28:345–354
49. Andrade RD, Lemus R, Perez CE (2011) Models of sorption isotherms for food: uses and limitations, Vitae, Revista De La Facultad De Quomica Farmac…UTICA ISSN 0121-4004/ISSNe 2145-2660. Volumen 18 n˙mero 3, a Òo 2011.
50. Van den Berg C (1984) Description of water activity of foods for engineering purposes by means of the GAB model of sorption. In: McKenna BM (ed) Engineering and foods, vol 1. Elsevier, New York, pp 311–321
51. Mushi NE, Utsel S, Berglund LA (2014) Nanostructured biocomposite films of high toughness based on native chitin nanofibers and chitosan. Front Chem 2, Article 99
52. Belver C, Banares-Muroz MA, Vicenta MA (2004) Fe-seponite pillar and impregnated catalysts I. Preparation and characterization. Appl Catal B. Environ 50:101–112
53. Hasan S, Ghosh A, Race K, Schreiber R Jr, Prelas MA (2014) Dispersion of FeOOH on chitosan matrix for simultaneous removal of As (III) and As(V) from drinking water. Sep Sci Technol 49:2863–2877
54. Seigbahn K (1974) Electron spectroscopy: an outlook. Electron Spectrosc 5:58–97
55. Kurmaev EZ, Shin S, Watanabe M et al (2002) Probing oxygen and nitrogen bonding sites in chitosan by X-ray emission. J Electron Spectrosc Relat Phenom 125:133
56. Dambis L, Vincent T, Guibal E (2002) Treatment of arsenic containing solutions using chitosan derivatives: uptake mechanism and sorption performances. Water Res 36:3699
57. Kramareva NV, Finashina ED, Kucherov AV, Kustov LM (2003) Copper complexes stabilized by chitosan: peculiarities of the structure, redox, and catalytic properties. Kinet Catal 44(6):793–800
58. Webstar A, Halling MD, Grant DM (2007) Metal complexation of chitosan and its glutaraldehyde cross-linked derivative. Carbohyd Res 342:1189–1201

# Chapter 4
# The Structural Difference Between Chitin and Chitosan

**Abstract** Chitin is poly[β-(1-4)-2-acetamido-2-deoxy-D-glucopyranose], and chitosan is poly[β-(1-4)-2-acetamido-2-amino2deoxy-D-glucopyranose]. Chitin is available in three different crystalline polymorphic forms: α-chitin, β-chitin, and γ-chitin. The polymeric chains of α-chitin are arranged in a tightly antiparallel fashion. In β-chitin, the chains of monomer are tightly arranged in parallel to each other, whereas γ-chitin has a mixed arrangement of both α- and β-chitin. Chitosan is closer to α-chitin. Based on the spectroscopic analysis, comprehensive intermolecular structural information related to chitin and chitosan has been discussed.

## 4.1 Molecular Structure of Chitin

Chitin is a polymer of N-acetyl-D-glucosamine. Its structural formula is multiple acetyl glucosamine residues (chitobiose), linked together by 1-4-β-glucosidic bonds. Chapter two described the general chemical structures of chitin and chitosan. Structural studies reveal that the marine chitin has three reactive chemical groups on its structure, and they are primary (C-6), secondary (C-3) hydroxyl groups on each repeat unit, and acetamide (C-2) groups [1]. Both hydroxyl and acetamide groups in chitin are connected through strong inter- and intramolecular hydrogen bonds in the same or neighboring chain. This makes chitin paracrystalline and non-porous [2]. Chitin is available in three different crystalline polymorphic forms: α-chitin, β-chitin, and γ-chitin. All three forms exist naturally. Figure 4.1 shows the details of the chemical bonding of the chitin with an intermolecular hydrogen bond. The hydrogen bonding between molecules (neighboring sugar ring of chitin) determines the α-, β-, and γ-chitin structures [3–5]. The α-chitin is the most abundant form of structural marine polysaccharide extracted from crabs, lobsters, krill, and shrimp shells (6). The β-chitin is mostly derived from *Loligo* squid pens and tubeworms (7). The γ-chitin exists in the cocoon fibers of the *Ptinus* beetle and the stomach of *Loligo* (8).

Blackwell [9] studied the structure and various physicochemical properties of α- and β-chitin and noted that the parallel chitin chains are in bonded piles or sheets linked together by N–H...O=C hydrogen bonds through the amide groups (Fig. 4.2).

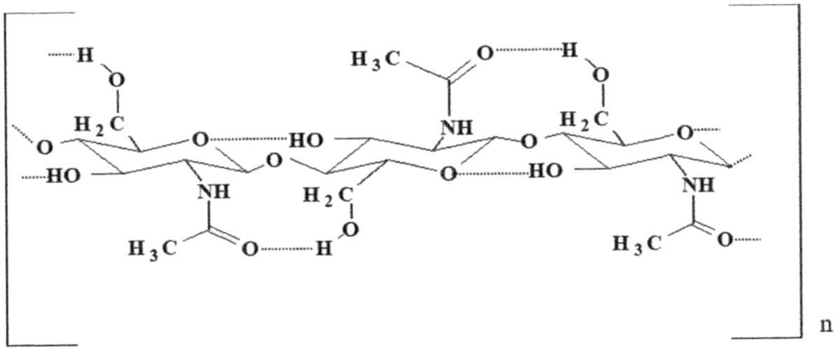

**Fig. 4.1** Chemical structure of chitin. Dotted lines indicate hydrogen bonding between the neighboring sugar rings of chitin. Copyright © 2020 Jiménez-Gómez, C.P., and Cecilia, J. A. Licensee MDPI, Basel, Switzerland. Creative Commons Attribution (CC BY) license, Molecules 2020, 25, 3981; https://doi.org/10.3390/molecules25173981. Reproduced from Ref. [5]

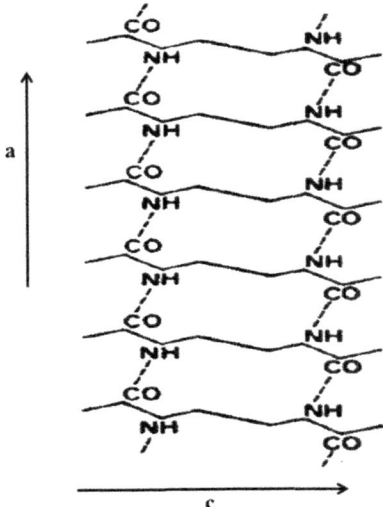

**Fig. 4.2** Pile or sheet of chitin chains viewed along the fiber axis: (a) hydrogen bond direction, (c) side-chain direction. Copyright © 1969, John Wiley and Sons, Biopolymers 7:281–298, 1969. Reproduced from Ref. [9] with permission

The α, β, and γ forms of chitin differ only on the arrangement of piles of chains. Piles are arranged alternately antiparallel in α-chitin, whereas they are all parallel in β-chitin [10–12]. For γ-chitin, they follow a parallel and antiparallel pattern. Figure 4.3 shows the arrangement of the α-chitin, β-chitin, and γ-chitin structures.

Carlstrom [13] studied the structure of α-chitin in more detail and proposed a buckled chain structure arranged in an orthorhombic cell. Figure 4.4 shows his proposed structure. He claimed that the proposed structure could adequately account for the chemical and physical properties of α-chitin. Prasad and Ramakrishnan [14] presented a similar structure later for α-chitin.

4.1 Molecular Structure of Chitin

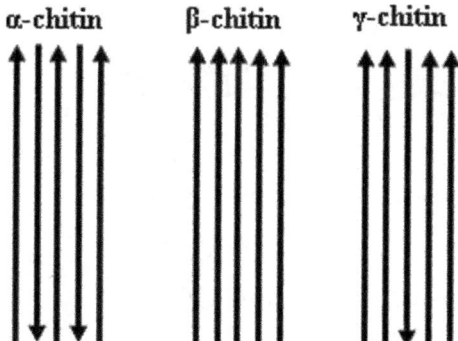

**Fig. 4.3** Arrangement of the α-chitin, β-chitin, and γ-chitin structure. Copyright © 2008 Elsevier Ltd. Reactive & Functional Polymers, 68, 1013–1051, 2008. Reproduced from Ref. [12] with permission

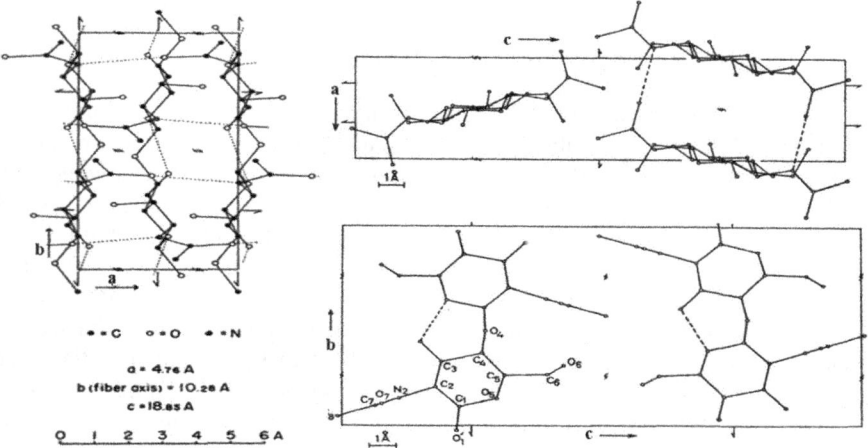

**Fig. 4.4** Projection of the proposed model for α-chitin. Copyright © Copyright, 1957, by The Rockefeller Institute for Medical Research, J Biophys Biochem Cytol. 1957 Sep 25; 3(5): 669–683. Reproduced from Ref. [13] with permission

Kameda et al. [15] studied the hydrogen bonding structure and stability of α-chitin. They found that α-chitin contains different types of hydrogen bonding with the C=O groups. Exclusive inter- and intramolecular hydrogen bonding occurs between the amide (NH) group and the C (6)–OH groups (Fig. 4.5).

Muthukrishnan et al. [16] reported that the carbonyl group (C7) and the side-chain amino group contribute to the extraordinary stability of α-chitin compared to cellulose. Kameda et al. further added that approximately 60% of carbonyl groups exclusively contribute to the intermolecular hydrogen bonding. About 40% of carbonyl groups contribute to the combination of intermolecular and intramolecular hydrogen

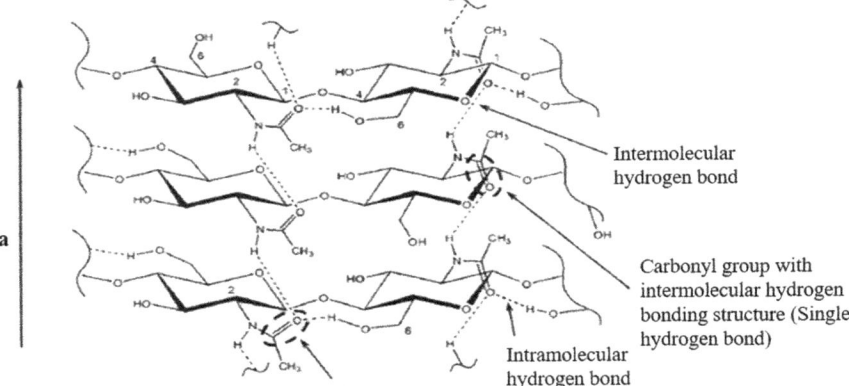

**Fig. 4.5** Hydrogen bonding structure for α-chitin. Copyright © 2005 WILEY-VCH Verlag GmbH & Co., Macromolecular Bioscience. 5:103–106, 2005. Reproduced from Ref. [15] with permission

bonding [15]. The crystallinity of α-chitin is more than 80% [17]. The polymeric chains of α-chitin are tight antiparallel, leading to a mostly orthorhombic crystalline form [18]. The extensive hydrogen bonds among the hydroxyl and acetamide groups make a highly ordered crystal structure in α-chitin. The low solubility properties of chitin mostly come from the densely packed crystalline structure of α-chitin.

Several researchers have studied the structure of β-chitin [19–22]. Lotmar and Picken [19] first suggested the β-chitin form as a well-oriented fiber with unit cell dimensions of $a = 9.32$ Å, $b = 10.17$ Å, and $c = 22.15$ Å in notopodial chaetae and squid *Loligo*. A critical review of infrared data of β-chitin by Gardner and Blackwell [20] indicates that the –CH$_2$OH group is hydrogen bonded. It was suggested that the structure contains the $O^{6'}$–H ... $O^7$ intrasheet hydrogen bond, leading to the following structure shown in Fig. 4.6.

The tightly arranged chains of the monomer of β-chitin give it a monoclinic crystal symmetrical structure, whereas γ-chitin is a mixed arrangement of both α- and β-chitin [8, 21, 22]. Muzzarelli et al. reported that the crystal structure of β-chitin lacks hydrogen bonds along the b-axis (Fig. 4.6). Therefore, it is more susceptible than α-chitin to intracrystalline swelling and acid hydrolysis, even at low acid concentrations [24]. Table 4.1 shows the properties of α- and β-chitin.

The β- and γ-chitin provide physiological functions other than support in chitin structure [23]. The β-chitin has weak intermolecular forces [19, 26] and exhibits higher modification reactivity and higher affinity for solvents than α-chitin [25–29]. This phenomenon makes β-chitin susceptible to swelling in water and dissolution in formic acid. It is noteworthy that chitosan prepared from β-chitin exhibits high reactivity compared to α-chitin [30] and significant bactericidal activity [31].

4.1 Molecular Structure of Chitin

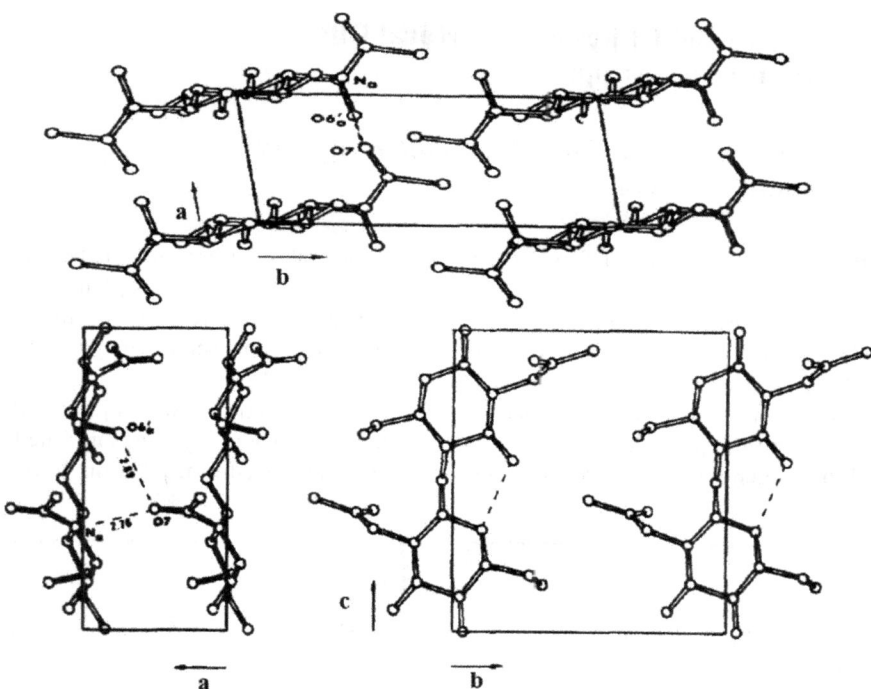

**Fig. 4.6** Proposed models for β-chitin. The structure contains N–H. O=C and $O^{6'}$–H ...$O^7$ inter-sheet hydrogen bonds. Copyright © 1975 John Wiley & Sons, Inc., Biopolymers 14:1581–1595, 1975. Reproduced from Ref. [20] with Permission

**Table 4.1** Typical properties of α-chitin and β-chitin [3–8. 21–24]

| α-chitin | β-chitin |
|---|---|
| Swells very little in water | Swells in water |
| α-chitin cannot be converted to β-chitin | β-chitin can be converted to α-chitin (treatment with strong acid) |
| It gives moderately sharp IR bands | It gives sharp IR bands (pognophore and diatom spines, but not sharp in *Loligo* β-chitin) |
| Moderately crystalline | *Loligo* β-chitin has a crystalline structure highly crystalline structure |
| No change with humidity (XRD pattern) | Spacing changes with humidity (XRD pattern) |

These results suggest the high potential of chitosan derived from β-chitin as a novel functional biopolymer.

## 4.2 Chitin and Chitosan: Structural Differences from Spectroscopic Analysis

### 4.2.1 X-Ray Diffraction (XRD) Pattern of Chitin and Chitosan

The structure of different forms (α, β, and γ) of chitin differs only in how the polymeric chains form crystallographic structures with various crystal planes [32]. Figure 4.7 shows the XRD spectra of α- and β-chitin and their corresponding chitosan [33]. The X-ray diffraction patterns of the α-chitin and the corresponding hydrolyzed chitosan showed strong reflections at 2θ around 9°–10° and 2θ of 20°–21° and minor reflections at 2θ values at 26.4° and higher [33]. The β-chitin showed a peak at 2θ = 9.9°, and the peak decreased significantly in its corresponding chitosan. The β-chitin is much more amenable to N-deacetylation than the α-form. γ-Chitin, having

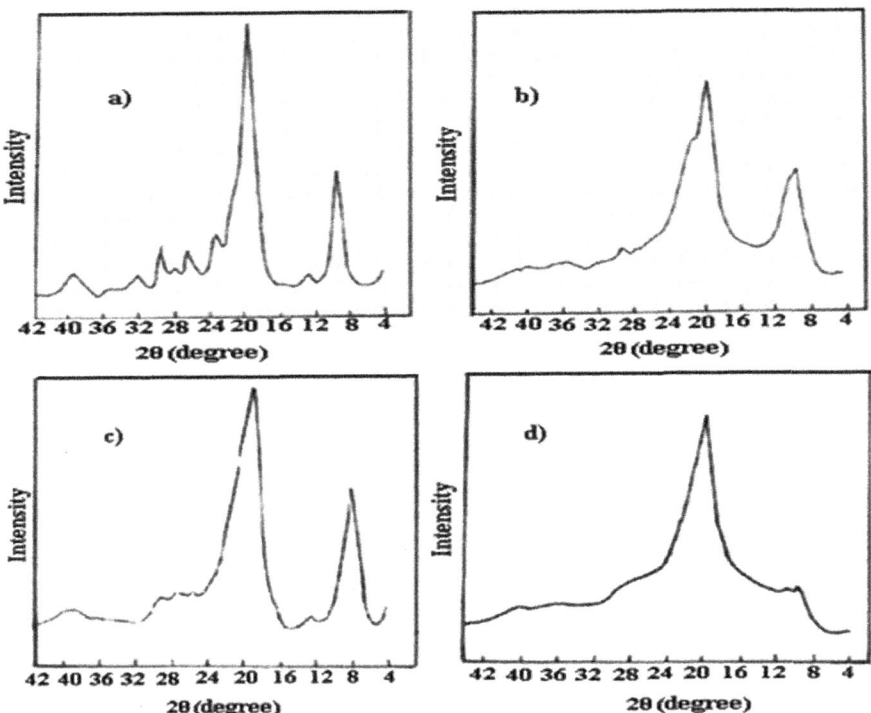

**Fig. 4.7** X-ray diffraction pattern of **a** α-chitin, **b** corresponding chitosan, **c** β-chitin, and **d** corresponding chitosan sample. Copyright © 2007 Elsevier Ltd., Bioresource Technology 99:1359–1367, 2008. Reproduced from Ref. [33] with permission

## 4.2 Chitin and Chitosan: Structural Differences ...

an antiparallel and parallel structure, was similar in its X-ray diffraction patterns to α-chitin [34].

Several studies [35a, 35b], as noted by Svezdova and Uzov [35c], reported the crystal structure of α- and β-chitin (Fig. 4.8) [35]. They reported that anhydrous β-chitin has a similar molecular sheet structure to α-chitin in the ac plane. However, these sheets having the same polarity are stacked because of hydrophobic forces to form a parallel structure. Thus, no strong hydrogen bond exists between the sheets along the b-axis. When the α-chitin structure is aligned in an antiparallel fashion, it has a stronger intermolecular hydrogen bonding. The β-chitin can incorporate small

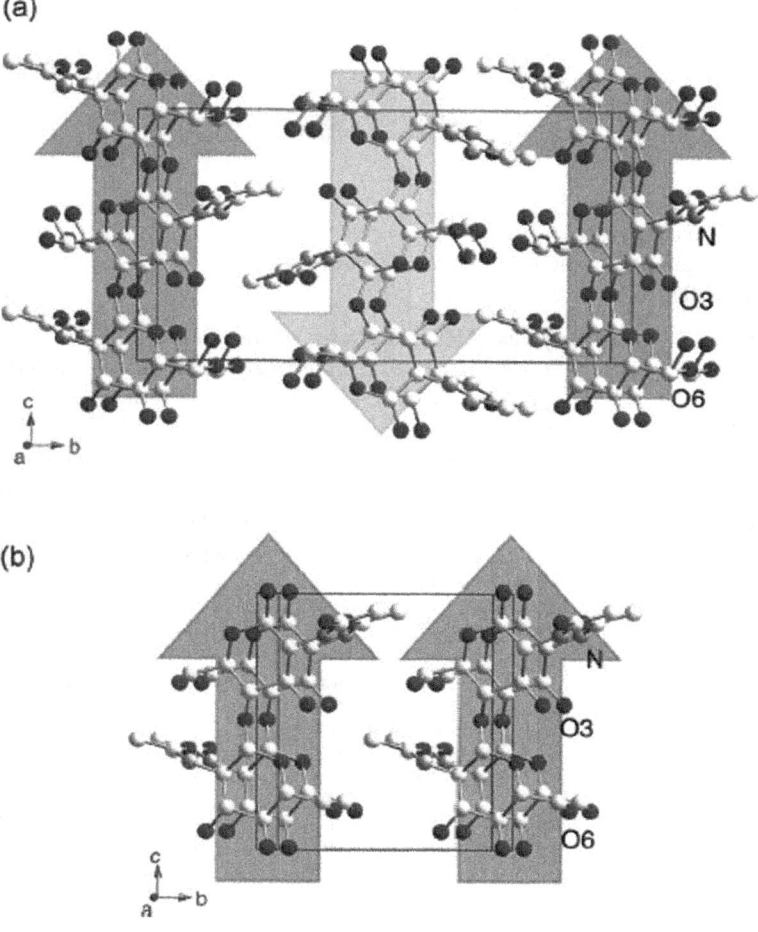

**Fig. 4.8** Crystal structure of **a** α-chitin and **b** β-chitin. Copyright © 2011, American Chemical Society, Macromolecules, 44, 950–957, 2011; Copyright © 2009, American Chemical Society, Biomacromolecules, 10, 1100–1105, 2009; Management and Education, http://www.conference-burgas.com/maevolumes/vol8/BOOK%204/b4_11.pdf. Reproduced from Ref. [35] with permission

molecules such as water and amine due to the lack of hydrogen bonds between its sheets.

The crystallinity index (CrI) indicates the crystal state of chitin and chitosan with various crystal planes. The crystallinity indices are determined by the following equation [36]:

$$\text{Crystallinity (CrI, \%)} = \frac{I_{110} - I_{am}}{I_{110}} \times 100 \quad (4.1)$$

where $I_{110}$ = the maximum intensity of diffraction peaks at (110) plane at $2\theta = {\sim}19°$, $I_{am}$ = the intensity of amorphous diffraction regions at $2\theta = {\sim}12.6°$.

Among various crystal lattice types, the variable d is the distance between atomic layers in a crystal. The distance $d$ is calculated using Bragg's law

$$d\,(\text{Å}) = \frac{\lambda}{2 \sin \theta} \quad (4.2)$$

where

$d$    plane spacing,
$\lambda$    the wavelength of the incident X-ray beam,
$\theta$    the incidence or reflection angle.

Figure 4.9 shows the typical XRD pattern for chitin and chitosan. The sample is scanned from 0 to 80° ($2\theta$) at a 1°/min speed. The chitin spectrum exhibits broad peaks at $d = 0.34$, 0.46, 0.50, and 1.09 nm with a shoulder peak at $d = 0.71$ nm, whereas chitosan shows the spectral peaks at $d = 0.45$, 0.98, and 2.93 nm. Chitosan sample shows a diffraction peak near 20°, indicative of relatively regular crystal lattice (110, 040) of chitosan. The degree of deacetylation (DD) of chitosan impacts the XRD peak intensity for different crystal planes [38, 38a]. Table 4.2 shows the XRD parameters of chitin and chitosan with different degrees of deacetylation.

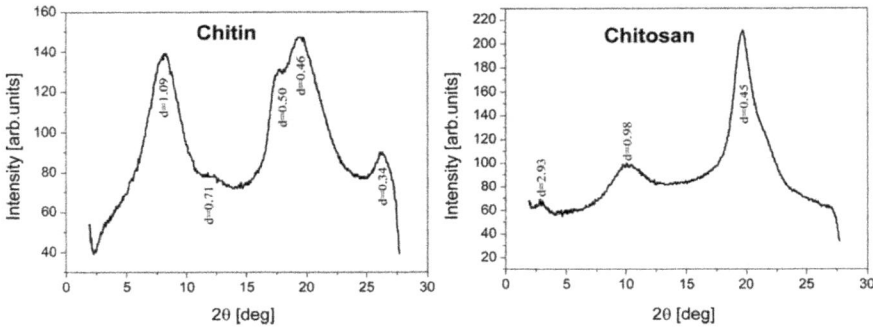

**Fig. 4.9** X-ray diffraction spectra of chitin and chitosan. Copyright © 2004 Elsevier Ltd., Carbohydrate. Polym. 56:137–146, 2004. Reproduced from Ref. [37] with permission

**Table 4.2** Typical XRD parameters of chitin and chitosan with different DA and DD

| Sample | $2\theta$ (°) | $d$-spacing (Å) | Relative intensity (%) | DA (%) | DD (%) |
|---|---|---|---|---|---|
| Chitin | 9.39, 19.22, 20.73, 23.41, 26.39 | 9.41, 4.61, 4.28, 3.79, 3.37 | 100, 94.2, 38.1, 21.9, 28.7 | 83.1 | 16.9 |
| Chitin (crab shell), from commercial source | 9.4, 9.5, 30.73, 23.4, 26.45 | 9.4, 4.55, 4.29, 3.79, 3.37 | 39.5, 100, 40.1, 23.0, 37.5 | 91.3 | – |
| Chitin (crab shell) from the Black Sea | 9.54, 19.6, 22.9, 23.4 | 9.54, 4.53, 3.88, 3.8 | 36.8, 100, 22.1, 20.51 | 90.76 | – |
| Chitosan (shrimp) from commercial source | 9.06, 20.0, 23.78, 26.85 | 9.76, 4.43, 3.74, 3.33 | 100, 92.8, 37.4, 35.1 |  | 59.4 |
| Chitosan (shrimp) from commercial source | 10.65, 20.20 | 8.29, 4.39 | 58.9, 100 |  | 87.0 |
| Chitosan (shrimp shell) from the Black Sea | 9.3, 20.2 | 9.5, 9.39 | 94.12, 100 | – | 84.0 |
| Chitosan (crab shell), from commercial source | 9.3, 20.2, 21.4 | 9.5, 4.39, 4.15 | 20.63, 100, 77.8 | – | 86.53 |
| Chitosan (crab shell) from the Black Sea | 9.2, 20.5 | 9.6, 4.33 | 52.1, 100 | – | 79.94 |
| Chitosan (shrimp) from commercial source | 11.91, 20.35 | 7.42, 4.36 | 59.9, 100 | – | 92.8 |

Copyright © 2005 Elsevier Ltd., Carbohydrate Research, 340, 1914–1917, 2005; Management and Education, http://www.conference-burgas.com/maevolumes/vol8/BOOK%204/b4_11.pdf. Reproduce from Refs. [35, 38a] with permission

## 4.2.2 Differential Scanning Calorimetry (DSC)

Differential scanning calorimetry (DSC) is a powerful analytical technique. The physical and chemical changes in the polymer are measured as a function of temperature using the DSC technique. The DSC technique provides fingerprint information about chitin and chitosan's thermal properties and their derivatives under air or inert environment. These thermal properties include specific heat, glass transition temperature $T_g$, and melting point $T_m$. For example, the glass transition temperature $T_g$ of the polymer provides information about the temperature region where the polymer turns from hard glassy material to soft rubbery material. The information about the glass transition temperature $T_g$ is used to estimate polymer–polymer miscibility, whereas the melting temperature $T_m$ is used to study polymer–polymer blend compatibility [38].

The DSC measurement is generally conducted with a cyclic heating and cooling cycle. About 2–10 mg sample is placed into a covered aluminum sample holder. An empty sample holder is used as a reference. The measurement runs should be performed in duplicate. The sample is first heated at 10 °C/min from room temperature to 110 °C under dry nitrogen purge (10–100 mL/min). This step should be followed by cooling to room temperature to evaporate the moisture. The sample is then reheated from room temperature to the desired temperature (e.g., 600 °C, same as TGA) at a heating 10 °C/min. Figure 4.10 shows the DSC analysis curves of chitin. Jang et al. [34] reported that chitin has a broad but weak endothermic peak around 50–140°, which is due to the evaporation of the binding water in chitin [34]. They observed that the exothermic peaks of α-chitin, β-chitin, and γ-chitin appeared at 330, 220, and 300 °C, respectively, as shown in Fig. 4.10a. This order is due to the α-chitin having a very rigid crystalline structure because of intersheet and intrasheet hydrogen bonding. The β-chitin has a relatively weak intermolecular force due to intrasheet interaction. Crystalline structure is the reason for a higher exothermic peak in α-chitin than β-chitin. Since γ-chitin has an antiparallel and parallel structure, an exothermic peak appears between α-chitin and β-chitin.

Figure 4.10b shows both endothermic and exothermic peaks for various chitin and chitosan samples. Polysaccharide like chitin and chitosan has a strong affinity for water. As shown in Fig. 4.10b, the endothermic peaks correspond to the loss of water. In chitin, the initial water molecules are associated with hydrophilic hydroxyl groups. The water molecules form a complex structure with the chitosan polymer chains through interhydrogen bonds [39]. The macromolecule, such as chitin and chitosan, may have disordered structures that can be easily hydrated [40]. In chitin and its N-deacetylated analogs, the endothermic peak area increased with an increase in N-deacetylation. Therefore, the endotherm related to bound water evaporation reflects chemical and molecular changes during N-deacetylation [41].

The exothermic peak, which appears in the temperature range between about 250 and 350 °C, corresponds to the decomposition of the polymer. The degradation of the chitosan chains causes the exothermic peak at around 290 °C [42]. The glass transition ($T_g$) temperature of pure chitosan is at 200 °C. The crosslinked chitosan–polyacrylonitrile copolymer blend has higher $T_g$, therefore, better thermal stability when compared with chitosan alone [38].

4.2 Chitin and Chitosan: Structural Differences … 89

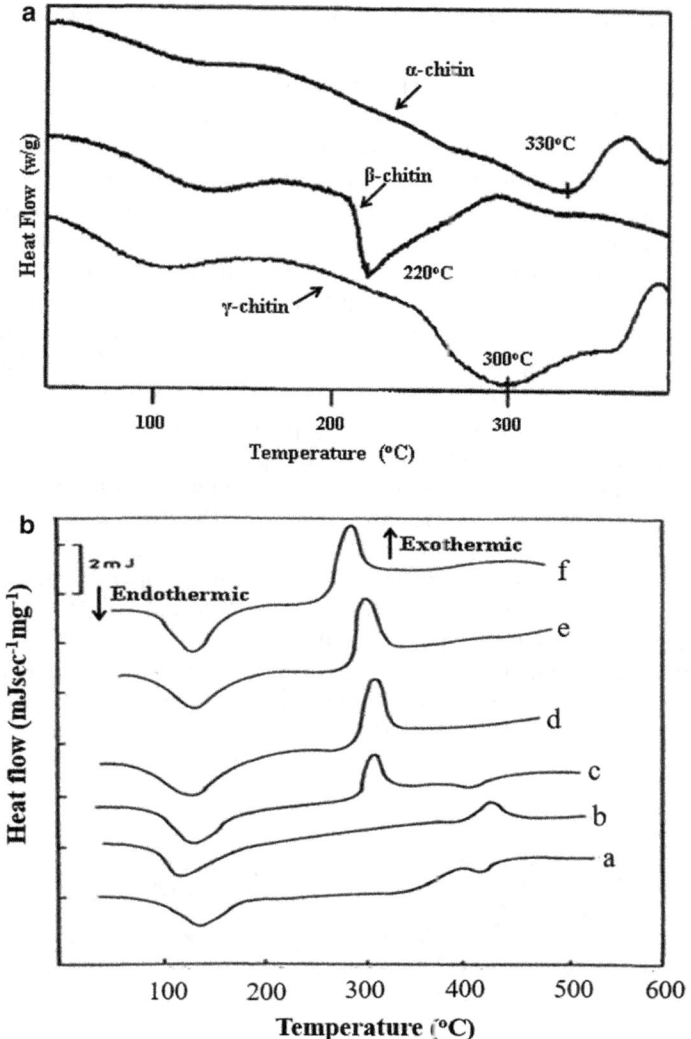

**Fig. 4.10 a** DSC thermogram of α-chitin, β-chitin, and γ-chitin. Copyright © 2004, John Wiley and Sons, Journal of Polymer Science: Part A: Polymer Chemistry 42:3423–3432, Reproduced from Ref. [34] with permission. **b** DSC thermograms of (a) chitin, (b) chitosan (%DD = 25), (c) chitosan (%DD = 48), (d) chitosan (%DD = 79), (e) chitosan (%DD = 84), and (f) chitosan (%DD = 89). Copyright © 2002 Elsevier Science Ltd., Carbohydrate Polymer 49:185–193, 2002. Reproduced from Ref. [41] with permission

## 4.2.3 Fourier Transform Infrared (FTIR) Spectroscopy

Fourier transform infrared spectra (FTIR) provides molecular fingerprint information of polymer. FTIR spectroscopy is a potent tool to analyze the molecular structure of chitin and chitosan. Because of the different hydrogen bonds [44], Cardenas et al. [44] reported that the α- and β-chitin could be distinguished by FTIR spectroscopy. In the FTIR spectra, α-chitin, β-chitin, and γ-chitin showed a doublet, a singlet, and a semi-doublet at the amide I band (Fig. 4.11). The more apparent spectral difference is the frequency of the vibration modes of amide I in the region 1660–1620 $cm^{-1}$. In α-chitin, two absorption peaks appear at 1660 and 1620 $cm^{-1}$, respectively. In the β-chitin, only one band appears at 1650 $cm^{-1}$. Jang et al. reported that the absorption band of hydrogen bonding between the carbonyl groups (–C=O) of amide I and amide II (–NH–) appears at 1660 $cm^{-1}$. The absorption band of hydrogen bonding between the –$CH_2OH$ side chain and the carbonyl group (–C=O) appears at 1620 $cm^{-1}$ [34]. Therefore, the amide I band splits in the FTIR spectrum of α-chitin. These results show that α-chitin is characterized by intersheet and intrasheet hydrogen bonding. The α-chitin has the structure of an antiparallel chain. The absorption band for amide I is split at 1660 and 1620 $cm^{-1}$. Cardenas et al. reported that half of the carbonyl groups are bonded through hydrogen bonds to the amino group inside the same chain (C=O...H–N) responsible for the vibration mode at 1660 $cm^{-1}$ [43]. The other half produces the same bond plus another with the group –$CH_2OH$ from the side chain. This additional bond with the –$CH_2OH$ group causes a decrease in the amide I band at 1627 $cm^{-1}$. These interchain bonds are responsible for the high chemical stability of the α-chitin structure. Also, β-chitin has intrasheet hydrogen bonding by a parallel

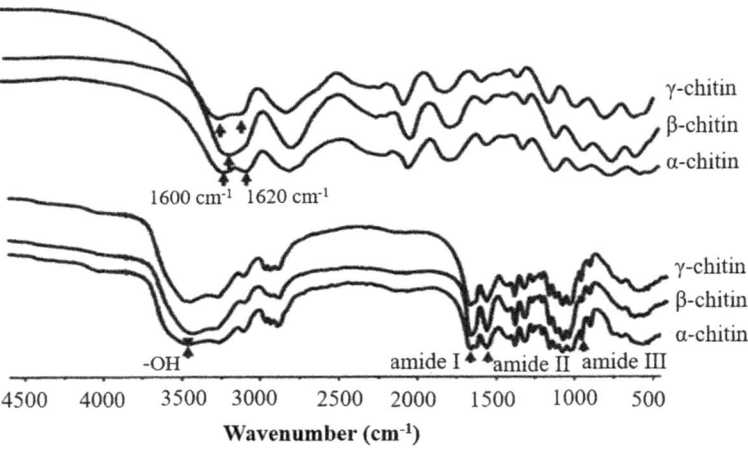

**Fig. 4.11** FTIR spectra of amide band of α-chitin, β-chitin, and γ-chitin. Copyright © 2004, John Wiley and Sons, Journal of Polymer Science: Part A: Polymer Chemistry 42:3423–3432. Reproduced from Ref. [34] with permission

chain, and the absorption band for amide I is a single peak at 1650 cm$^{-1}$. The crystalline structure of γ-chitin, being antiparallel, is similar to that of α-chitin. Therefore, the amide I absorption band of γ-chitin is closer to that of α-chitin. Corresponding to the region of OH and NH (3600–3000 cm$^{-1}$) groups, the α-chitin exhibits a more detailed structure than the β-chitin. This is due to the different packing arrangements of the macromolecules [43]. The shoulder peak in the α-chitin spectrum at 3579 cm$^{-1}$ is due to the intramolecular hydrogen bond involving the OH (6)…O=C. Such a bond is missing in the β-chitin.

Table 4.3 lists the standard wavelength of the main bands obtained for chitin and chitosan and their derivatives [44]. Chitosan exhibits peaks corresponding to the stretching O–H and N–H groups and C–H stretching vibration in –CH and –CH$_2$,

**Table 4.3** IR wavelength of functional groups of chitosan and its derivatives

| Sample | Peak position (cm$^{-1}$) | Peak type | Assignment |
|---|---|---|---|
| Characteristic saccharide structure | 905<br>1153–1158 | Medium<br>Strong | Aliphatic aldehyde<br>Primary or secondary alcohol |
| Characteristic primary amine | 1580–1650<br>3450 | Strong<br>Broad | NH$_2$ deformation vibration<br>NH$_2$ symmetric stretching vibration |
| Characteristic chitin (secondary amide) | 1322–1325<br>1515–1570 (1554)<br>1649–1655 | Weak<br>Medium<br>Strong | Amide III: OH and CH deformation<br>Amide II: N–H deformation and C–N stretching vibration<br>Amide I: C=O stretching vibration |
| Other characteristic peaks for chitin and chitosan | 1083<br>1263–1267<br>1377–1383<br>1420<br>1700–1725<br>2800–2900<br>3200–3600<br>1409<br>1500–1625<br>1548, 1560 | Strong<br>Weak<br>Strong<br>Medium<br>Strong<br>Weak<br>Strong<br>Strong<br>Weak | C–O stretching vibration<br>CH wag (ring) vibration<br>CH$_3$ deformation (bend) vibration<br>OH and CH deformation (ring)<br>Aliphatic carboxylic acid dimer<br>C–H stretch vibration<br>OH stretch vibration<br>COOH C–O stretch and O–H deformation<br>NH$_3^+$ deformation vibration Symmetric and symmetric deformation |
| Chitosan hydrogel | 1510–1570<br>1573<br>1640–1690<br>1643<br>1500–1625<br>1548, 1560 | Weak<br>Weak<br>Medium | C–N<br>C=N(imines) Schiff's base<br>NH$_3^+$ deformation vibration Symmetric and asymmetric deformation |

Copyright © 2004 Society of Chemical Industry. Polym. Int, 53, 911–918, 2004. Reproduced from Ref. [44] with permission

respectively (Fig. 4.11). The absorption bands at around 2921 and 2877 cm$^{-1}$ are due to C–H symmetric and asymmetric stretching. The complicated nature of the absorption spectrum in the 1650–1500 cm$^{-1}$ region suggests the IR spectra include amine (–NH$_2$) and carbonyl (–CONHR) band [45, 46]. The peaks at 1563 cm$^{-1}$ and 1661 cm$^{-1}$ represent amide II and amide I, respectively, for chitosan. The peaks confirm the presence of residual N-acetyl groups at around 1645 cm$^{-1}$ (C=O stretching of amide I) and 1325 cm$^{-1}$ (C–N stretching of amide III) [47]. Kumariska et al. [48] reported that the amide I for β-chitin shows a peak at 1630 cm$^{-1}$. For α-chitin, peaks appear at 1660 cm$^{-1}$ and 1630 cm$^{-1}$ due to the influence of hydrogen bonding or the presence of an enol form of the amide moiety [48]. In the region of 1350 to 1450 cm$^{-1}$, the bands indicated alkane C–H symmetric bending vibration in –CHOH [49]. In the region of 1000–1200 cm$^{-1}$, the bands at 1066 and 1028 cm$^{-1}$ correspond to C–O stretching.

Figure 4.12 shows the typical IR spectrum of different functional groups present in chitin and chitosan. Following the FTIR technique, the ratio of the amide II band's absorbance to that of the band at 3450 cm$^{-1}$ can be used to calculate the degree of deacetylation of chitosan [52, 53]. The absorbance at 1655 and 3450 cm$^{-1}$ is used to calculate the DD according to the following equation [54]:

$$\text{DD} = \left(1 - \frac{A_{1655}}{A_{3450}}\right) \times \frac{1}{1.33} \times 100 \qquad (4.3)$$

where

DD         the deacetylation degree,
$A_{1655}$ cm$^{-1}$    absolute heights of absorption bands of amide groups,

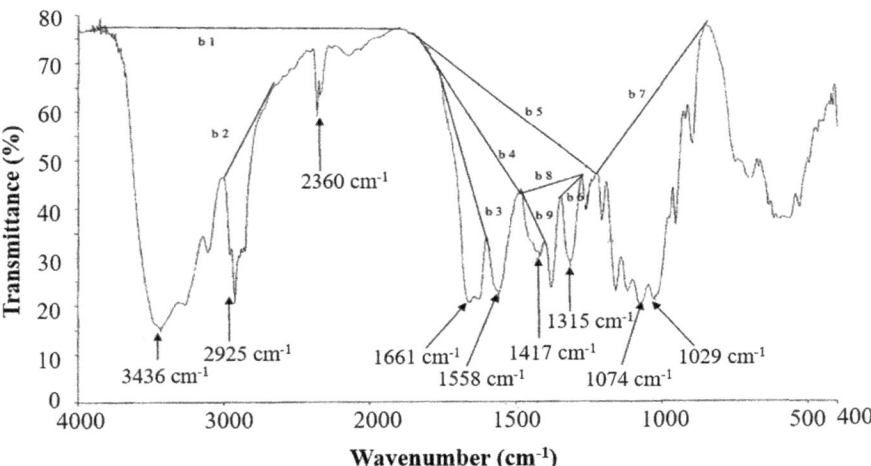

**Fig. 4.12** Typical IR spectra of chitin. 2001 WILEY-VCH Verlag GmbH, Macromolecular Symposia, 2001, 168, 1–20. Reproduced from Ref. [50] with permission

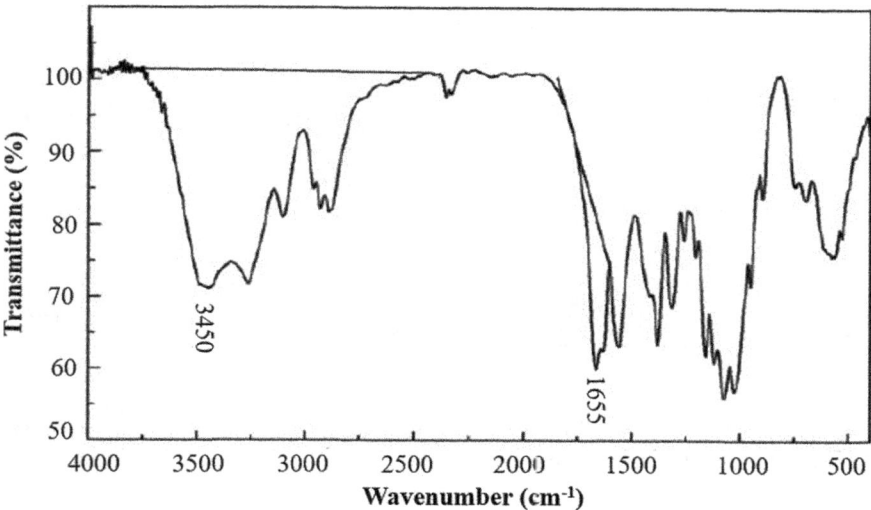

**Fig. 4.13** Typical IR spectrum of chitosan showing the two baselines for calculating the amide I band absorbance for the ratio A1655/A3450. Copyright © 1992 Published by Elsevier B.V., Int. J Biol Macromol 14(3):166–169, 1992. Reproduced from Ref. [51] with permission

$A_{3450}$ cm$^{-1}$  absolute heights of absorption bands of hydroxyl groups.

The factor 1.33 represents the value of the ratio of A1655/A3450 for fully N-acetylated chitosan (Fig. 4.13).

### 4.2.4 Nuclear Magnetic Resonance (NMR) Spectroscopy

Nuclear magnetic resonance spectroscopy data provide information on the chemical structure. It also provides information on the complex conformational equilibria of polysaccharides in solution. Cunha et al. [55] reported that the NMR data reflect primarily short range through bond and space interactions via the Nuclear Overhauser Effect (NOE) or local perturbations to electronic shielding (chemical shifts). The average amino group content of chitin and chitosan directly correlates to the degree of deacetylation. It is measured using $^1$H liquid-state NMR, solid-state $^{13}$C NMR, and $^{15}$N NMR technique [50, 56]. The $^{13}$C solid-state NMR appears to be the most reliable for evaluating the acetyl content of chitin and chitosan. It does not require the solubilization of the polymer. However, it needs a high level of purification of the studied material [50]. In the NMR spectra, α-chitin and β-chitin are distinguished, but γ-chitin is nearly the same as that of α-chitin (Fig. 4.14).

Jang et al. reported that solid-state CP–MAS $^{13}$C NMR spectra of C3 and C5 carbon atoms in α-chitin and β-chitin are different because of the differences in the hydrogen bonding forces [34]. The C3 and C5 spectra of β-chitin form one

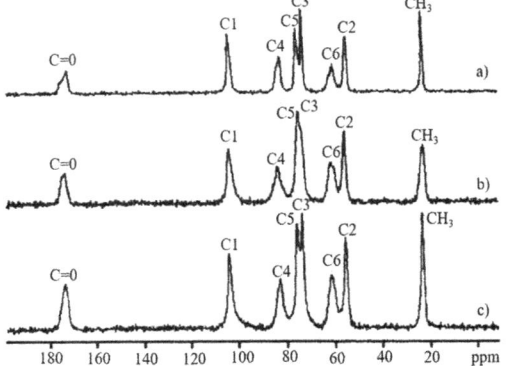

**Fig. 4.14** Solid-state CP/MAS NMR spectra of **a** α-chitin, **b** β-chitin, and **c** γ-chitin. Copyright © 2004, John Wiley and Sons, Journal of Polymer Science: Part A: Polymer Chemistry 42:3423–3432. Reproduced from Ref. [34] with permission

single resonance, whereas two separate resonances occur in α-chitin [57]. Different H bond interactions could be identified between the proton donor (amine from the glucosamine unit) and acceptor groups [58]. The NMR spectra of C3 and C5 carbon atoms for α-chitin around 73 and 75 ppm (Fig. 4.14) depicted the antiparallel structure of α-chitin with intersheet and intrasheet hydrogen bonding. The spectra for β-chitin show a singlet around 74 ppm for C3 and C5 carbon atoms due to parallel molecular structure. The same chemical shift $\delta = 74$ ppm for C3 and C5 indicates the high structural homogeneity for β-chitin [59]. Although the spectra of α and β-chitin are nearly identical, it is not easy to differentiate them by solid-state $^{13}$C NMR (Fig. 4.15) [60]. The peaks corresponding to C3 and C5 carbon atoms of γ-chitin occur at 73 and 75 ppm, respectively, because γ-chitin forms intersheet hydrogen bonding. The degree of deacetylation value of chitosan depends on the ratio of the signal intensities of methyl on acetyl groups ($I_{CH_3}$) and the ring carbon ($I_C$) obtained from the solid-state $^{13}$C-NMR spectrum. The following equation can be used to estimate the degree of deacetylation values of chitosan [61]:

$$\text{DD}(\%) = \left(1 - \frac{I_{CH_3}}{(I_{C1} + I_{C2} + I_{C3} + I_{C4} + I_{C5} + I_{C6})/6}\right) \times 100\%. \quad (4.4)$$

Solid-state $^{15}$N NMR is less sensitive than $^{13}$C NMR. The $^{15}$N NMR is an excellent tool to identify the crystallinity of the sample [60]. Figure 4.15 shows the $^{15}$N NMR spectra of α-chitin and chitosan samples. The $^{15}$N NMR provides only two signals related to the amino group and the *N*-acetylated group, as shown in Fig. 4.15. Alvarenga reported that the amide, amine, and ammonium nitrogen atoms are displayed at $\delta = \sim 101, 0,$ and 13 ppm, respectively [62]. The degree of acetylation is calculated as

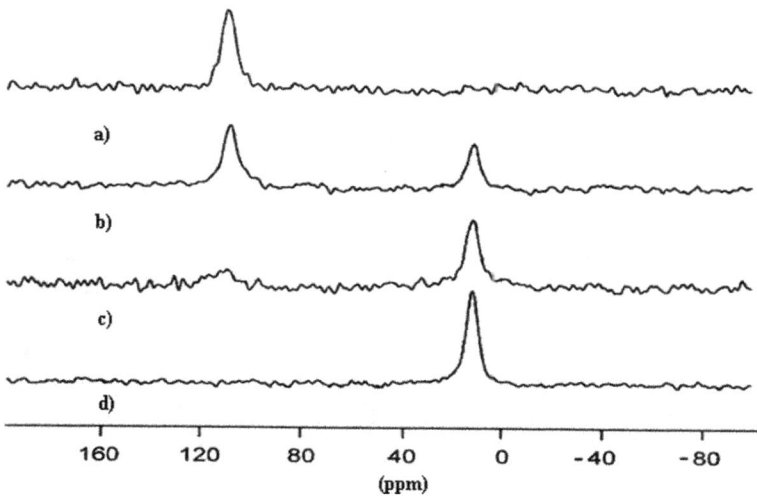

**Fig. 4.15** $^{15}$N NMR spectra of **a** α-chitin, **b** homogeneous partially re-acetylated chitosan, and **c** and **d** heterogeneous commercial chitosan. Copyright © 2006 Elsevier Ltd., Prog Polym Sci 31:603–632, 2006. Reproduced from Ref. [60] with permission]

$$DD\% = \left(1 - \frac{A_{101}}{(A_{101} - A_0)}\right) \times 100 \quad (4.5)$$

where $A_{101}$ and $A_0$ are the integral areas of peaks at $\delta = 101$ and 0 ppm, respectively.

The $^1$H NMR is a powerful tool for chemical identification, but it typically quantifies protons and is not always accurate in bond identification [63]. This technique is limited to the samples that are soluble in the solvent. The degree of acetylation is calculated from the integral ratio between the protons of the acetyl group (GlcNAc) and N-glucosamine (GlcN) groups [64]. The technique is beneficial for the soluble part of chitosan. Typical chitosan sample preparation for the $^1$H NMR technique is given by Czechowska-Biskup et al. [65]. A typical chitosan solution in a deuterated (deuterated chloride) aqueous acid DCl/D$_2$O is prepared by dissolving 0.05 g chitosan in 6 cm$^3$ D$_2$O and 1 cm$^3$ deuterium chloride. This solution is prepared in an NMR tube. An additional amount of D$_2$O and DCl (deuterated chloride) can be added to the solution to make a final volume of 10 cm$^3$. The pH of the chitosan solution should be approximately 4.0. The sample requires freeze-drying using D$_2$O (99.9%) to exchange labile protons by deuterium atoms. Figure 4.16 shows a typical $^1$H NMR spectrum of a chitosan sample.

The degree of acetylation of chitosan is calculated using the methyl signal area and the summation of H2, H3, H4, H5, H6, and H6′ signals in the $^1$H NMR spectrum, according to the following formula [62]:

$$\%DA = \left(\frac{2 \times A_{CH_3}}{A_{H_2-H_6}}\right) \times 100. \quad (4.6)$$

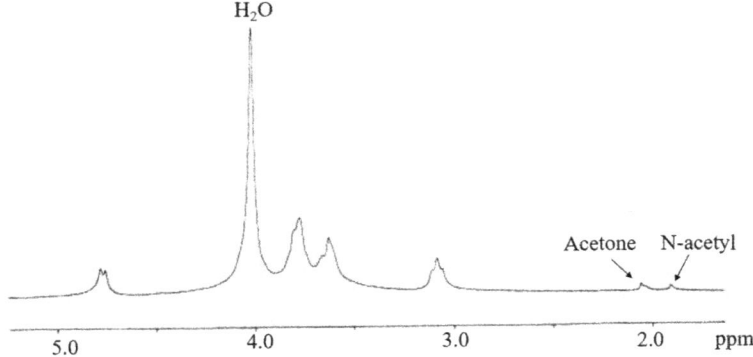

**Fig. 4.16** 1H NMR spectrum of chitosan sample (%acetyl 15%). Copyright © 1987 Published by Elsevier B.V., Int J Biol Macromol 9(4):233–237, 1987. Reproduced from Ref. [66] with permission

**Table 4.4** Chemical shifts (ppm) of chitosan

| Chitosan | | | |
|---|---|---|---|
| $^1$H NMR | | $^{13}$C NMR | |
| Proton | δ (ppm) | Carbon | δ (ppm) |
| H-1 | 4.77 | C-1 | 97.23 |
| H-2 | 3.07 | C-2 | 55.80 |
| H-3 | 3.80 | C-3 | 69.83 |
| H-4 | 3.77 | C-4 | 77.09 |
| H-5 | 3.63 | C-5 | 74.55 |
| H-6 | 3.66<br>3.77 | C-6 | 60.28 |
| MeO-3 | – | MeO-3 | – |
| MeO-6 | – | MeO-6 | – |
| (Me)$_3$ N-2 | – | (Me)$_3$N-2 | – |
| ACNH-2 | 1.91 | | |

Copyright © 1987 Published by Elsevier B.V., Int J Biol Macromol 9(4):233–237, 1987. Reproduced from Ref. [66] with permission

A few HCl drops are added to D$_2$O during the solubilization of polymer [60, 66] to avoid the overlap of acetic acid and acetyl group signals. Table 4.4 shows a typical chemical shift for $^1$H and $^{13}$C NMR. All the $^{13}$C signals have been assigned, and their chemical shifts are given in Table 4.4. The low N-acetylation degree (~5%) is not identifiable from the $^{13}$C NMR spectrum. It appears on the $^1$H NMR spectrum in which the protons of the acetyl group resonate at δ = 1.91 ppm (Fig. 4.16) [66].

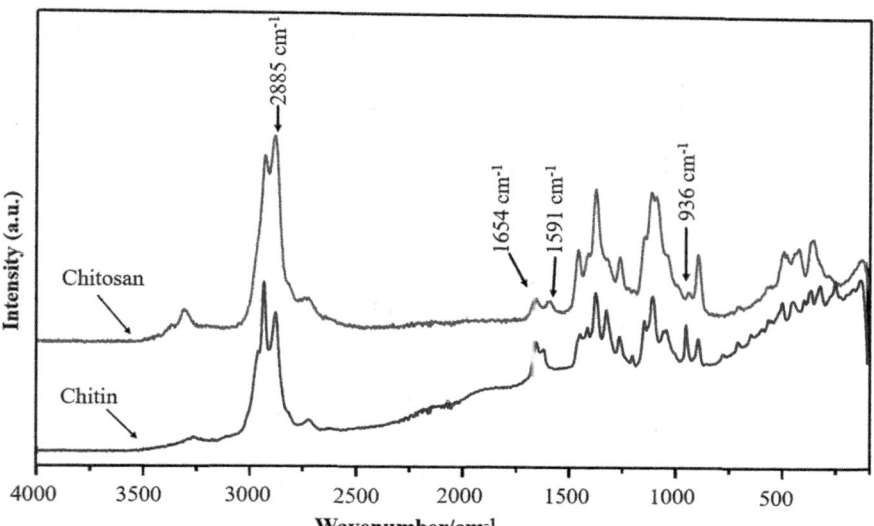

**Fig. 4.17** Typical Raman spectra of chitin. Copyright © 2014 Elsevier B.V. All rights reserved, Spectrochimica Acta Part A: Molecular and Biomolecular Spectroscopy 134 (2015) 114–120. Reproduced from Ref. [71] with permission

### 4.2.5 Raman Spectroscopy

Raman spectra are analyzed for fingerprint and group frequency peaks. The scans are collected from 2000 to 200 cm$^{-1}$ with a step size of 8 cm$^{-1}$. Depending on the solid-phase sorbate concentration and background fluorescence, the scan time ranged from less than a minute to a maximum of 15 min. The scans should be collected at the low-energy laser output. This prevents the local burning of the chitosan sample. The C–C, C–H, and C–N bonds of the chitosan samples can be determined by Raman spectroscopic analysis. In Raman spectroscopy, the peak for amine and alcohol is easily distinguishable as the N–H stretch is stronger than the O–H stretch [67]. Figure 4.17 shows the typical Raman spectra of chitin. The functional groups of chitin and chitosan and their characteristic frequency identified by Raman spectroscopy are given in Table 4.5 [68–71].

## 4.3 Summary

The hydrogen bonding between molecules (neighboring sugar ring of chitin) determines the α-, β-, and γ-chitin structures. The structure of different forms (α, β, and γ) of chitin differs only in how the polymeric chains form crystallographic structures with various crystal planes. Due to the extensive hydrogen bonds among hydroxyl

**Table 4.5** Typical Raman excitation frequencies for functional groups present in chitosan and chitin [68–71]

| Sample | Wavenumber (cm$^{-1}$) | Assignments | Reference |
|---|---|---|---|
| Chitosan | 470 | $\nu$ (C–C(=O)–C) | [69] |
| | 1000 | $\nu$ (C–H) | |
| | 1800 | $\delta$ (C=CCOOR) $\delta$ (C=O) | |
| | 2630 | $\delta$ (CH) rings | |
| | 3250 | $\nu$ (NH$_2$) | |
| Chitosan | 2850–2960 | Alkane: C–C, C–H (methylene) | [68, 70] |
| | 3350–3400 3270–3330 | Amine: NH$_2$ (primary) | |
| | 3600–3200 | Alcohol: C–OH | |
| | 1100 | Ether: C–O–C | |
| Chitosan | 2885 | $\nu$ (CH$_2$) | [71] |
| | 1654 | $\nu$ (CO) | |
| | 1591 | $\delta$ (NH$_2$) | |
| | 936 | $\delta$ (CN) | |
| Chitin | 1260 (Amide III) | Monosubstituted amide: NH–C=O–CH$_3$ | [68, 70] |
| | 3350–3310 | Amine: NH (secondary) | |
| | 1460 (bending) 1375 (deformation) | Methyl, CH$_3$ | |

$\nu$ = stretching, $\delta$ = in-plane bending vibration.

and acetamide groups, the crystal structure of α-chitin is highly ordered (crystalline orthorhombic form) than β-chitin. Weak intermolecular forces characterize β-chitin, and it has higher affinity for solvents than α-chitin.

Chitosan is semi-crystalline. The degree of crystallinity is a function of the degree of deacetylation. Crystallinity is maximum for both chitin (i.e., 0% deacetylated) and fully deacetylated (i.e., 100%) chitosan. The spectroscopic analysis provides fingerprint information about the structural difference between chitin and chitosan. For instance, α- and β-chitin could be distinguished by the Fourier transform infrared (FTIR) spectroscopy because of the different hydrogen bonds. The DSC technique provides fingerprint information about chitin and chitosan's thermal properties and their derivatives under air or inert environments, whereas Raman spectra are analyzed for fingerprint and group frequency peaks. In the NMR spectra, the chemical structure of α-chitin and β-chitin is distinguished, but γ-chitin is nearly the same as that of α-chitin.

# References

1. Chopin N, Guillory X, Weiss P, Bideau JL, Colliec-Jouault S (2014) Design polysaccharides of marine origin: chemical modifications to reach advanced, versatile compounds. Curr Organ Chem 18(7):867–895
2. Roberts GAF (1992) Chitin chemistry. Macmillan, London
3. Kaya M, Mujtaba M, Ehrlich H et al (2017) On chemistry of γ-chitin. Carbohyd Polym 176:177–186
4. Roy CJ, Salaun F, Giraud S et al (2017) Solubility of chitin: solvents, solution behaviors and their related mechanisms. In: Xu Z (ed) Solubility of polysaccharide, pp 109–127. https://doi.org/10.5772/intechopen.71385
5. Einbu A (2007) Characterization of chitin and a study of its acid-catalysed hydrolysis. Ph.D. thesis submitted to Faculty of Natural Sciences and Technology, Department of Biotechnology, Norwegian University of Science and Technology, Trondheim
6. Arbia W, Arbia L et al (2013) Chitin recovery using biological methods. Food Technol Biotechnol 51(1):12–25
7. Inairo A, Di Giosa M, Fermani S et al (2014) Customizing properties of β-chitin in squid pen (Gladius) by chemical treatments. Mar Drugs 12:5979–5992
8. Jung J (2013) New development of β-chitosan from jumbo squid pens (*Dosidicus gigas*) and its structural, physicochemical, and biological properties. Ph.D. dissertation, Oregon State University, 29 May 2013
9. Blackwell J (1969) Structure of β-chitin or parallel chain systems of poly-β-(1→4)-N-acetyl-D-glucosamine. Biopolymers 7:281–298
10. Bakshi PS, Selvakumar D, Kadirvelu K, Kumar NS (2020) Chitosan as an environmentally friendly biomaterial—a review on recent modifications and applications. Int J Biol Macromol 150:1072–1083
11. Aranaz I, Mengibar M, Harris R, Panos I, Miralles B, Acosta N, Galed G, Heras A (2009) Functional characterization of chitin and chitosan. Curr Chem Biol 3:203–230
12. Mourya VK, Inamdar NN (2008) Chitosan modifications and its applications: opportunities galore. React Funct Polym 68:1013–1051
13. Carlstrom D (1957) The crystal structure of alpha-chitin (poly-N-acetyl-D-glucosamine). J Biophys Biochem Cytol 3(5):669–683
14. Prasad J, Ramakrishnan C (1972) Studies on structure and conformation of chitins.1. Mathematics of rigid body refinement procedure as applied to structure of alpha-chitin. Ind J Pure Appl Phys 10:501–505
15. Kameda T, Miyazawa M, Ono H, Yoshida M (2005) Hydrogen bonding structure and stability of α-chitin studied by 13C solid-state NMR. Macromol Biosci 5:103–106
16. Muthukrishnan S, Merzendorfer H, Arakane Y, Yang Q (2016) Chitin metabolic pathways in insects and their regulation. In: Cohen E, Moussian B (eds) Extracellular composite matrices in arthropods. Springer International Publishing Switzerland. https://doi.org/10.1007/978-3-319-40740-1_2
17. Fan Y, Saito T, Isogai A (2008) Preparation of chitin nanofibers from squid pen beta-chitin by simple mechanical treatment under acid conditions. Biomacromolecules 7:1919–1923
18. Minke R, Blackwell J (1978) The structure of alpha chitin. J Mol Biol 120:167–181
19. Lotmer W, Picken LER (1950) A new crystallographic modification of chitin and its distribution. Experientia 6:58–59
20. Gardner KH, Blackwell J (1975) Refinement of structure of β-chitin. Biopolymers 14:1581–1595
21. Fabritius H, Sachs C, Rabbe D, Nikolov S, Friak M, Neugebauer J (2011) Chitin in the exoskeletons of Arthropoda: from ancient design to novel material science. In: Gupta MS (ed) Topics in geobiology, Springer Science+ Business Media, B.V.
22. Fabritius H, Sachs C, Triguero PR, Rabbe D (2009) Influence of structural principles on the mechanics of a biological fiber-based composite material with hierarchical organization: the exoskeleton of the lobster *Homarus americanus*. Adv Mater 21:391–400

23. Muzzarelli RAA (1977) Chitin. Pergamon Press, Oxford, Great Britain, pp 1–309
24. Muzzarelli RAA, El-Mehtedi M, Mattiolo-Belmonte M (2014) Emerging biomedical applications of nano-chitins and nano-chitosans obtained via advanced eco-friendly technologies from marine resources. Mar Drugs 12:5468–5502. https://doi.org/10.3390/md12115468
25. Mincea M, Negrulescu A, Ostafe V (2012) Preparation, modification, and applications of chitin nanowhiskers: a review. Rev Adv Mater Sci 30:225–242
26. Rudal KM (1963) The chitin/protein complexes on insect cuticles. Adv Insect Physiol 1:257–313
27. Kurita K, Tomita K, Tada T, Ishii S, Nishimura S, Shimoda K (1993) J Polym Sci, Part A: Polym Chem 31:485
28. Kurita K, Tomita K, Ishii S, Nishimura S, Shimoda K (1993) J Polym Sci, Part A: Polym Chem 31:2393
29. Kurita K, Ishii S, Tomita K, Nishimura S, Shimoda K (1994) J Polym Sci, Part A: Polym Chem 32:1027
30. Kurita K, Tomita K, Tada T, Nishimura S, Ishii S (1993) Polym Bull 30:429
31. Shimojoh M, Masaki K, Kurita K, Fukushima K (1996) Nippon Nogeikagaku Kaishi 70:787
32. Jooyeoun J (2013) Development of β-chitosan from jumbo squid pens (*Dosidicus gigas*) and its structural, physicochemical, and biological properties. Ph.D. (Doctor of Philosophy) thesis, Oregon State University
33. Abdou ES, Nagy KSA, Elsabee MZ (2008) Extraction and characterization of chitin and chitosan from local sources. Biores Technol 99:1359–1367
34. Jang M-K, Kong B-G, Jeong Y-I et al (2004) Physicochemical characterization of α-chitin, β-chitin, and γ-chitin separated from natural resources. J Polym Sci, Part A: Polym Chem 42:3423–3432
35. (a) Nishiyama N, Noshiki Y, Wada M (2011) X-ray structure of anhydrous β-chitin at 1 Å resolution. Macromolecules 44:950–957; (b) Sikorski P, Hori R, Wada M (2009) Revisit of α-chitin crystal structure using high resolution X-ray diffraction data. Biomacromolecules 10:1100–1105; (c) Zvezdove D, Uzov C (2012) Determination of the degree of deacetylation of chitin and chitosan by X-ray powder diffraction. Manage Educ VIII (4):85–89. http://www.conference-burgas.com/maevolumes/vol8/BOOK%204/b4_11.pdf
36. Kaya M, Lelešius E, Nagrockaitė R et al (2015) Differentiation of chitin content and surface morphologies of chitins extracted from male and female grasshopper species. PLoS ONE. https://doi.org/10.1371/journal.pone.0115531
37. Muzzarelli C, Francescangeli O, Tosi G, Muzzarelli RAA (2004) Susceptibility of dibutyryl chitin and regenerated chitin fibers to deacetylation and depolymerization by lipases. Carbohyd Polym 56:137–146
38. Acharyulu GRT, Sudha PN (2013) Physico-chemical characterization of cross-linked chitosan-polyacrylonitrile polymer blends. Der Pharmacia Lettre 5(2):354–363; (a) Zhang Y, Xue C, Xue Y et al (2005) Determination of the degree of deacetylation of chitin and chitosan by X-ray powder diffraction. Carbohyd Res 340:1914–1917
39. Julkapli NM, Akil HM, Ahmad Z (2012) Thermal properties of 4,4-oxydiphathalic anhydride chitosan filled chitosan bio-composites. J Therm Anal Calorim 107:365–376
40. Prashanth KVH, Kittur FS, Tharanathan RN (2002) Solid-state structure of chitosan prepared under different N-deacetylating conditions. Carbohyd Polym 50:27–33
41. Kittur FS, Prashanth KVH, Udaysankar K, Tharanathan RN (2002) Characterization of chitin, chitosan and their carboxymethyl derivatives by differential scanning calorimetry. Carbohyd Polym 49:185–193
42. El-Hafian EA, Elgannoudi ES, Mainal A, Yahaya AH (2010) Characterization of chitosan in acetic acid: rheological and thermal studies. Turk J Chem 34:47–56
43. Cardenas G, Cabrera G, Taboada E, Miranda P (2004) Chitin characterization by SEM, FTIR, XRD, and 13C Cross polarization/Mass Angle Spinning NMR. J Appl Polym Sci 93:1876–1885
44. Wang T, Turhan M, Gunasekaran S (2004) Selected properties of pH-sensitive, biodegradable chitosan-poly (vinyl alcohol) hydrogel. Polym Int 53:911–918

45. Pakula M, Bimiak S et al (1998) Chemical and electrochemical studies of interactions between Iron (III) ions and an activated carbon surface. Langmuir 14(11):3082–3089
46. Grant J, Cho J et al (2006) Self-assembly and physicochemical and rheological properties of a polysaccharide-surfactant system formed from the cationic biopolymer chitosan and non-ionic sorbitan esters. Langmuir 22(9):4327–4335
47. Queiroz MF, Melo KRT, Sabry DA, Sasaki L, Rocha AO (2015) Does the use of chitosan contribute to oxalate kidney stone formation? Mar Drugs 13:141–158
48. Kumirska J, Czerwicka M, Kaczyński Z et al (2010) Application of spectroscopic methods for structural analysis of chitin and chitosan. Mar Drugs 8:1567–1636. https://doi.org/10.3390/md8051567
49. Guinesi LS, Cavalheiro ETG (2006) The use of DSC curves to determine the acetylation degree of chitin/chitosan samples. Thermochim Acta 444:128–133
50. Brugnerotto J, Desbrières J, Heux L, Mazeau K, Rinaudo M (2001) Overview on structural characterization of chitosan molecules in relation with their behavior in solution. Macromol Symp 168:1–20
51. Baxter A, Dillon M, Taylor KDA, Roberts GAF (1992) Improved method for I.R. determination of the degree of N-acetylation of chitosan. Int J Biol Macromol 14:166–169
52. Yaghobi N, Mirzadeh H (2004) Enhancement of chitin's degree of deacetylation by multistage álcali treatments. Iran Polym J 13(2):131–136
53. Hussain MR, Imam M, Maji TK (2013) Determination of degree of deacetylation of chitosan and their effect on the release behavior of essential oil from chitosan and chitosan-gelatin complex microcapsules. Int J Adv Eng Appl 6(4):4–12
54. Kurita K (2001) Controlled functionalization of the polysaccharide chitin. Prog Polym Sci 26:1921–1971
55. Cunha RA, Soares TA, Rusu VH et al (2012) The molecular structure and conformational dynamics of chitosan polymers: an integrated perspective from experiments and computational simulations. In: The Complex World of Polysaccharides, Chapter 9. Intech, pp 229–256. https://doi.org/10.5772/51803
56. Lyalina T, Zubareva A, Lopatin S, Zubov V, Sizova S, Svirshchevskaya E (2017) Correlation analysis of chitosan physicochemical parameters determined by different methods. Organ Med Chem Int J 1(3):555562
57. De Velde V, Kiekens P (2004) Structure analysis and degree of substitution of chitin, chitosan and dibutyrylchitin by FT-IR spectroscopy and solid-state 13C NMR. Carbohyd Polym 58:409–416
58. Pereira AGB, Muniz EC, Hsieh Y-L (2015) $^1$H NMR and $^1$H–$^{13}$C HSQC surface characterization of chitosan–chitin sheath-core nanowhiskers. Carbohyd Polym 123:46–52
59. Chen C-H, Wang F-Y, Ou Z-P (2004) Deacetylation of β-chitin. I. Influence of the deacetylation conditions. J Appl Poly Sci 93:2416–2422
60. Rinaudo M (2006) Chitin and chitosan: properties and applications. Prog Polym Sci 31:603–632
61. Hsiao H-Y, Tsai C-C, Chen S, Hsieh B-C, Chen RLC (2004) Spectrophotometric determination of deacetylation degree of chitinous materials dissolved in phosphoric acid. Macromol Biosci 4:919–921
62. De Alvarenga ES (2011) Characterization and properties of chitosan. In: Elnashar M (ed) biotechnology of biopolymers. InTech (2011). ISBN 978-953-307-179-4. http://www.intechopen.com/books/biotechnology-of-biopolymers/characterization-and-properties-of-chitosan
63. Buschmann MD, Merzouki A, Lavertu M, Thibault M, Jean M, Darras V (2013) Chitosans for delivery of nucleic acids. Adv Drug Deliv Rev. https://doi.org/10.1016/j.addr.2013.07.005
64. Badawy MEI, Rabea EI (2011) A biopolymer chitosan and its derivatives as promising antimicrobial agents against plant pathogens and their applications in crop protection. Int J Carbohyd Chem 2011, Article ID 460381, 29 p
65. Czechowska-Biskup R, Jarosinska D, Rokita B, Ulanski P, Rosiak JM (2012) Determination of degree of deacetylation of chitosan-comparison of methods. Prog Chem Appl Chitin Its Derivatives XVII:5–20

66. Domard A, Gey C, Rinaudo M, Terrassin C (1987) 13C and 1H N.M.R. Spectroscopy of chitosan and N-trimethyl chloride derivatives. Int J Biol Macromol 9:233–237
67. Babatunde EO, Ighalo JO, Akolo SA et al (2020) Investigation of biomaterial characteristics of chitosan produced from crab shells. Mat Int 2:0303–0310. https://doi.org/10.33263/Materials23.303310
68. Martin R, Hild S, Walther P, Ploss K, Boland W, Tomaschko K-H (2007) Granular chitin in the epidermis of nudibranch molluscs. Biol Bull 213:307–315
69. Eddya M, Tbib B, El-Hami K (2020) A comparison of chitosan properties after extraction from shrimp shells by diluted and concentrated acids. Heliyon 6:e03486. https://doi.org/10.1016/j.heliyon.2020.e03486
70. Dreyer EC (2006) Characterization of electrodeposited chitosan films by atomic force microscopy and Raman spectroscopy. MS (Master of Science) Thesis, University of Maryland
71. Zajac A, Hanuza J, Wandas M, Dymińska L (2015) Determination of N-acetylation degree in chitosan using Raman spectroscopy. Spectrochim Acta Part A: Mol Biomol Spectrosc 134:114–120

# Chapter 5
# Preparation and Application of Chitosan Derivatives

**Abstract** Chitin and chitosan are natural, nitrogen-containing polysaccharides. They can be further functionalized through reaction into various gels, membranes, and fibers. Thus far, the materials' mechanical and chemical properties have been largely unexploited for industrial application. There needs to be further research in utilizing these materials for broader applications, including a study on the sustainability aspects of their use.

## 5.1 Physicochemical Crosslinking of Chitosan

Crosslinked polymers are used in engineering due to their higher stability at elevated temperatures and mechanical deformation resistance than their parent polymer [1]. Chitosan is modified through crosslinking to improve its physical or chemical properties for several medical, biomedical, and wastewater treatment applications. Various chemical agents are used to crosslink chitosan in an acidic, basic, or neutral environment. The amino and hydroxyl groups present in the chitosan structure can be crosslinked to form bridges between polymeric chains [2]. Crini reported that the crosslinking process drastically reduces segment mobility in the polymer, and several chains interconnect by formation of new interchain linkages [3]. The chitosan crosslinking process can be a physical or a chemical crosslinking process.

The physical crosslinking method is a simple process. It does not require the use of an organic event or high temperature, and no chemical interaction is involved [4]. Ionic and thermal crosslinking is a typical example of physical crosslinking methods. A network of ionic bridges or polyelectrolyte complex networks can form between negatively charged crosslinking agents and the positively charged chitosan chains [5, 6]. Ionic crosslinking involves the formation of a strong but reversible electrostatic link [5, 6]. For example, poly (acrylic acid), a negatively charged polyelectrolyte macromolecules, can form a polyelectrolyte complex with chitosan. This chitosan-polyelectrolyte complex has the necessary physicochemical properties for biomedical applications [7, 8]. In the ionic crosslinking process, the crosslinking of chitosan with an anionic crosslinking agent such as sodium sulfate or tripolyphosphate (TPP) forms a hydrogel. The hydrogel or ionic gelation is the product of this physical crosslinking process. Another example of a physical crosslinking process is the thermal crosslinking of chitosan with citric acid providing insolubility property [4]. Varhosaz and Alinagri reported that the crosslinking process of sulfate and citrate with chitosan is much faster than that of TPP due to their smaller molecular size [9]. They also reported that the TPP/chitosan crosslinked beads possessed better mechanical strength. The force required to break the TPP/chitosan crosslinked beads was approximately ten times higher than sulfate/chitosan or citrate/chitosan beads. The main disadvantage of physical crosslinking is the difficulty in controlling the crosslinking process; therefore, it is tough to obtain a desired crosslinking [10].

Chemical crosslinking is free radical, condensation (cationic/anionic), UV radiation, or small-molecule crosslinking [11]. Chemical crosslinking of chitosan creates permanent networks between chitosan's chains. Crosslinking agents are either homo-bifunctional or hetero-bifunctional reagents with identical or non-identical reactive groups [12]. Glutaraldehyde ($C_5H_8O_2$) and dimethyl adipimidate ($C_8H_{16}N_2O_2$) are widely used as homo-bifunctional crosslinking agents in biomedical and pharmaceutical industries. 4-azidobenzoic ($C_7H_5N_3O_2$) acid is a typical example of a hetero-bifunctional crosslinking agent used in tissue engineering [13]. Kitagawa et al. [14] reported that the hetero-bifunctional agent has an advantage over the homo-bifunctional agents for the improved enzyme labeling of hormones and drugs. A polymer network is formed based on the crosslinker's covalent or ionic bonds [15]. In the case of covalent bonding, counterions are diffused into the polymer. The crosslinking agent reacts with the polymer by forming covalent bonds. The crosslinker is either intermolecular (long range) or intramolecular (short range) [16]. The crosslinking process involves apolycondensation reaction with the polymer. Multifunctional monomers provide for chain growth and network formation. The concentration of crosslinking agent and crosslinking time affects the covalent crosslinking reaction. However, the covalent crosslinker concentration, e.g., glutaraldehyde, dominates the reaction [6].

Homogeneous crosslinked materials are easy to prepare. They are insoluble in acidic or alkaline media and organic solvents. Glutaraldehyde ($C_5H_8O_2$) is the most widely studied homo-bifunctional crosslinking agent among all crosslinking agents. It is a five-carbon molecule terminated at both ends by aldehyde groups. Glutaraldehyde is mainly available as acidic aqueous solutions (pH 3.0–4.0) ranging in concentration from 2 to 70% (w/v) [17]. It is soluble in water, alcohol, and organic solvents, and it can be used in acidic, neutral, or basic environments as a crosslinking agent.

**Fig. 5.1** Structural formula of glutaraldehyde

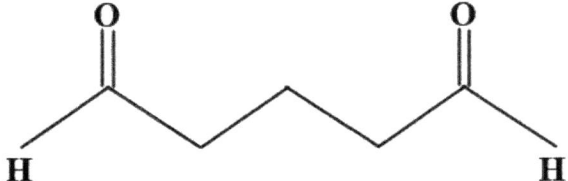

## 5.1 Physicochemical Crosslinking of Chitosan

Figure 5.1 shows the chemical structure of the bifunctional crosslinking agent glutaraldehyde. Glutaraldehyde can be present in the solution in different forms such as free aldehyde, mono- and dehydrated monomeric glutaraldehyde, monomeric, and polymeric cyclic hemiacetals. The glutaraldehyde form depends on the pH, temperature, and other solution conditions [17, 18]. An increase in temperature leads to a substantial increase in the content of aldehyde [19].

The crosslinking reactions are influenced by the size and type of crosslinking agent and chitosan functional groups in chitosan. It is reported that the chemical modification of chitosan improves properties without changing any fundamental skeleton. Crosslinked chitosan becomes more resistant to shear, high temperature, and low pH than pure chitosan [20]. Crosslinking also reduces the amount of crystalline domain in the polysaccharide. The reaction of glutaraldehyde with the amine groups produces covalent crosslinking through a Schiff base reaction. In the acidic reaction condition, the protonation of amine groups of chitosan does not affect the crosslink reaction. However, the concentration of glutaraldehyde affects the properties of the crosslinked product [21].

Several aldehydes can react with different functional groups, such as carboxyl, hydroxyl, and amide. The pKa value of chitosan is 6.5 [22]. At low pH value, the amine groups of chitosan become protonated, determining its solubility in water. More free amino groups are available for crosslinking reaction when solution pH is near or lower the chitosan's pKa value. Glutaraldehyde reacts rapidly with amine groups of chitosan during crosslinking through Schiff's base reaction. This reaction occurs due to the nucleophilic attack by the nitrogen of the amino groups of chitosan on the carbon of the glutaraldehyde. This reaction displaces the oxygen of the aldehyde resulting in the C=N bond [23]. Equation 5.1 describes the possible reaction mechanism for the Schiff base reaction:

$$\text{[structure with CHO groups]} + 2R-NH_2 \longrightarrow \text{[structure with N-R / CH groups]} \tag{5.1}$$

It is reported that the concentration of glutaraldehyde affects the physical properties of the crosslinked chitosan product [24, 25]. Kil'deeva et al. [19] reported that the crosslinked rigidity increases when the glutaraldehyde/$NH_2$ ratio is above 0.4 mol/mole. Rithe et al. [18] reported that a higher concentration of glutaraldehyde leads to a higher degree of crosslinking, making the polymer chains closer to each other, further decreasing their ability to absorb water. Hence, chitosan crosslinked with glutaraldehyde does not swell. It is also reported that the crosslinking reaction of glutaraldehyde with chitosan in acetic acid solution proceeds with a high rate of reaction. Stronger acid has a longer gelling time than weaker acid (e.g., acetic acid), except multibasic acids, such as citric acid [24]. Citric acid has a higher ionic strength than acetic acid. The greater ionic strength reduces the activity of protonated chitosan and thereby reducing the gelation rate.

Kil'deeva et al. [19] further reported that the product of the aldol reaction forms a pH of 5.6 during crosslinking between chitosan and glutaraldehyde. The product formation from an aldol reaction takes place at pH > 7.2 [19]. Berger et al. [2] reported that dialdehyde, such as glyoxal and glutaraldehyde, is the most common crosslinker, but it is considered toxic. Glyoxal is a known mutagenic substance. Glutaraldehyde is neurotoxic, but its fate in the human body is not fully understood [2].

Silva et al. [21] studied glutaraldehyde crosslinked chitosan membranes using simulated body conditions. They concluded that the chitosan membrane could be adequate for biomedical applications based on cytotoxicity screening and cell culture tests. Epichlorohydrin, ethylene glycol diglycidyl ether, and formaldehyde are also investigated as crosslinkers for chitosan [25–27]. Ngah et al. [25] studied chitosan crosslinked with glutaraldehyde, epichlorohydrin, and ethylene glycol diglycidyl ether to obtain adsorbent material insoluble in acidic and basic solutions. It is reported that the crosslinking of polymer with epichlorohydrin maintains the cationic amino function and improves the mechanical properties of the material [26]. Figure 5.2 shows the typical chemical crosslinked products of chitosan with different crosslinking agents.

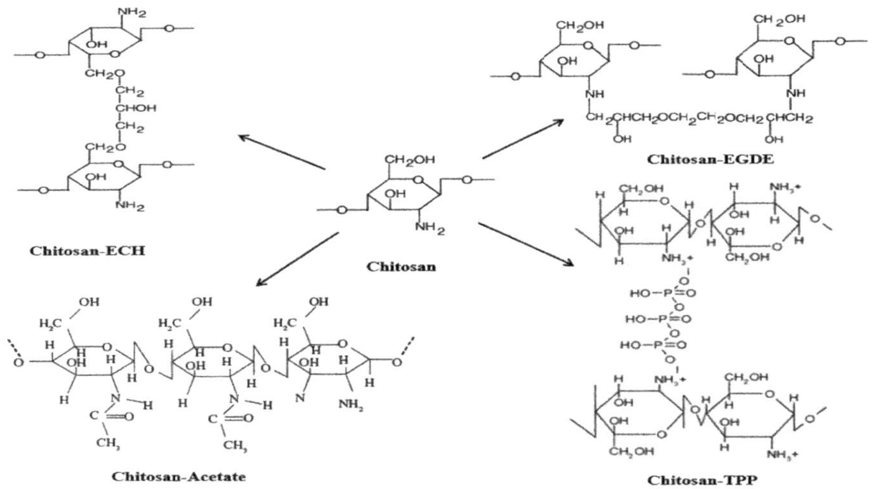

**Fig. 5.2** Chitosan crosslinked with acetate, epichlorohydrin (ECH), ethylene glycol diglycidyl ether (EGDE), and tripolyphosphate (TPP). Copyright © 2002 Elsevier Science B.V., Reactive & Functional Polymers 50:181–190, 2002; © 1992 John Wiley & Sons, Inc. Journal of Polymer Science 30(1):2187–2193, 1992; https://www.scielo.br/j/mr/a/bBbcqRHcxjcWtdj8ND wdLMm/?format=pdf&lang=en, Materials Research 10(4):347–352, 2007. Reproduced from the Refs. [25–27] with permission

## 5.2 Radiation-Induced Crosslinking

In radiation-induced crosslinking, the chemical changes in chitosan occur due to irradiation. The process does not require any external heat or catalyst. The extent of radiation-induced reaction only depends on the polymer network structure [28]. Sabharwal et al. [29] reported that natural polymers' radiation processing causes chain scission reactions to occur when natural polymers are exposed to high-energy radiation [29]. It is reported that the irradiation of chitosan yields lower viscosity and results in chain scission of chitosan [30, 31]. When the radiation from the source interacts with a polymer material, the polymer material absorbs its energy [32]. Active species such as free radicals are produced, which initiates various chemical reactions. The $H^+$ and $OH^-$ radicals formed by radiolysis during irradiation of water accelerate the molecular chain scission of chitosan. The reaction between the $H^+$ and $OH^-$ free radical and chitosan molecules leads to the rapid degradation of chitosan in an aqueous solution [33]. Pengfei et al. [34] reported radiation-induced grafting of styrene onto chitin and chitosan. They reported that the grafting reaction was promoted in the presence of methanol, and oxygen delayed the grafting reaction but did not inhibit it altogether. Yoksan et al. [35] reported that γ-irradiation reduces the molecular weight of chitosan and its chemical structure retains its functionalities as demonstrated by its activity. They observed that the degradation of chitosan by γ-rays is most useful for the amorphous structure.

Chitosan can also be crosslinked using UV radiation without any chemical additives at ambient temperature. Mane et al. [11] reported that polymerization using UV radiation is the safest and clean way of polymerization since it cannot deteriorate the polymer properties. As a result, the polymer retains its biocompatibility [11]. Table 5.1 shows typical applications of different chitosan crosslinked products obtained using different crosslinking processes.

## 5.3 Water-Soluble Chitosan Derivatives

Chitosan is soluble in 0.5–1% acetic, hydrochloric, nitric, perchloric, and phosphoric acid solution. Chitosan precipitates in solution by several acids, such as sulfuric, phosphotungstic, iodomercuric, iodobismuthic, molybdic, picric, and tannic acids [37]. Qin [38] categorized organic solvents for chitosan depending on their viscosity, as given in Table 5.2.

This solubility property of chitosan limits its specific application in biomedical and pharmaceutical fields. For instance, chitosan applications are restricted in biological fields due to its insolubility in water [39–41]. Chitosan has a hydrated, dehydrated, and nanocrystalline structure. The extensive intramolecular and intermolecular hydrogen bonding caused by the acetamido or primary amino group residues plays a major role in its insolubility in water [42, 43]. The growing demand for chitosan in the biomedical and pharmaceutical sectors has increased the interest to

**Table 5.1** Typical example of crosslinked chitosan and their uses

| Cross-linker | Application |
|---|---|
| ECH | Protein binding |
| ECH, formaldehyde | Protein binding |
| ECH | Flavonoids recovery |
| Oxalic acid | Transdermal drug delivery system |
| Glyoxal | Hydrogel for site-specific antibiotics delivery in the stomach |
| N,N'-methylene bisacrylamide | Biocompatible hydrogel for controlled delivery of amoxicillin |
| Citrate | Film for controlled release of riboflavin |
| Sulfate | Film for improvement of sustained release of a drug |
| Tripolyphosphate | Gel beads for controlled release of piroxicam |
| GLA | Recovery of Cu (II), Ni (II), Zn (II), Pd (II), Au (III), Pt (IV), and Mo (VI) from the aqueous solution |
| EDGE | Recovery of Cu (II) and Pb (II) from the aqueous solution |
| ECH | Recovery of Cu (II), Ni (II), Cr (VI), Pb (II), Cd (II), Co (II), and Zn (II) from the aqueous solution |
| Formaldehyde | Recovery of Ag (II) and Cd (II) from the aqueous solution |

Copyright © 2003 Published by Elsevier B.V. European Journal of Pharmaceutics and Biopharmaceutics, 57, 19–34, (2004); Copyright © 2013 International Research Journal of Pharmacy, 4(2), 2013, 45–50. Reproduced from the Refs. [2, 36] with permission

*GLA* Glutaraldehyde; *ECH* Epichlorohydrin; *EGDE* Ethylene glycol diglycidyl ether

**Table 5.2** Organic solvents for chitosan

| Solvents | Properties | Solubility of chitosan |
|---|---|---|
| Solution (2 M) of acetic, citric, formic, lactic, malic, malonic, pyruvic, or tartaric acid | Slightly non-Newtonian solution (viscosity of solution changes with applied shear) | No clearly defined solubility limit |
| Solution (2 M) of dichloroacetic acid or oxalic acid | Very non-Newtonian solution | Chitosan solution forms gels when allowed to stand for a long period |
| Benzoic acid (0.041 M), Salicylic acid (0.036 M), and sulfonic acid (0.052 M) | – | A slight degree of solubility |
| Dimethyl formamide (DMF), Dimethyl sulfoxide (DMSO), and pyridine | – | Non-solvent (non-soluble) |

Copyright © 2021 Xue, C., Wilson, L.D. Licensee MDPI, Basel, Switzerland. Creative Commons Attribution (CC BY) license. J. Compos. Sci. 2021, 5, 160. https://doi.org/10.3390/jcs5060160. Reproduced from Ref. [38a]; Ph.D. Thesis, University of Leeds, UK, April 1990. https://etheses.whiterose.ac.uk/11299/1/531880.pdf. Reproduced from Ref. [38b] online

improve chitosan's solubility in aqueous solutions. Kubota and Eguchi [44] cited literature to prepare water-soluble chitosan. This preparation involved changing morphology, molecular weight, and the crystalline structure of chitosan. Chemical modification of chitosan structure is also an important step to improve its solubility in water under neutral conditions [45]. The water solubility of chitosan can be improved by reducing its molecular weight by physical (e.g., shear force and ultrasonic variants), chemical, and enzymatic methods [46]. Qin et al. [47] reported an enzymatic process to prepare water-soluble low molecular weight chitosan. The enzymatic process minimizes the alteration of the chemical nature of the water-soluble chitosan product. The low molecular weight chitosan obtained by the enzymatic process has higher solubility compared to that obtained by the physical and chemical processes [46, 47]. Nevertheless, the cost of the enzymes (e.g., chitosanase, lysozyme) limits the enzymatic process for the commercial production of water-soluble chitosan. As mentioned in chapter four, chitosan possesses an amino group ($-NH_2$) at the C2 position and the primary and secondary hydroxyl groups ($-OH$) at C6 and C3 positions of its structure. The amino groups at the C2 position are nucleophilic. The nitrogen in the amino groups of chitosan structure has a lone pair electron making it more nucleophilic under neutral conditions than the hydroxyl groups of chitosan. The polysaccharides are modified by the esterification and etherification reactions using the hydroxyl groups on their structure [48]. Buschmann et al. [49] reported that the hydroxyl on C3 is less reactive. The amine groups at the C2 position and hydroxyl functional groups at the C6 position make chitosan one of the most modifiable polymers. The chemical modification of chitosan allows amine and hydroxyl or both site-specific chemical reactions by attaching hydrophilic functional groups to the chitosan chain without changing its fundamental skeleton and keeping the original physicochemical and biochemical properties [50, 51]. Several studies investigated the chemical modification of chitosan using its reactive amino and hydroxyl groups [42, 48, 52–54] to increase chitosan's solubility in aqueous solution and organic solvents. Figure 5.3 shows the chemical structure of different chitosan derivatives. The chitosan derivatives are prepared by modifying amino functional groups of chitosan through alkylation, acylation, and quaternization [40, 41, 48]. Typical examples of site-specific chemical modifications of chitosan and their applications are included in Table 5.3.

## 5.4 Overall Synthesis Process for Water-Soluble Chitosan Derivatives

### 5.4.1 Alkylation of Chitosan

Solubility in water at neutral or basic pH of chitosan is improved primarily by specific attachment of carbohydrates to the 2-amino functional groups achieved by Maillard reaction or further reductive alkylation of Schiff bases [55]. The Maillard

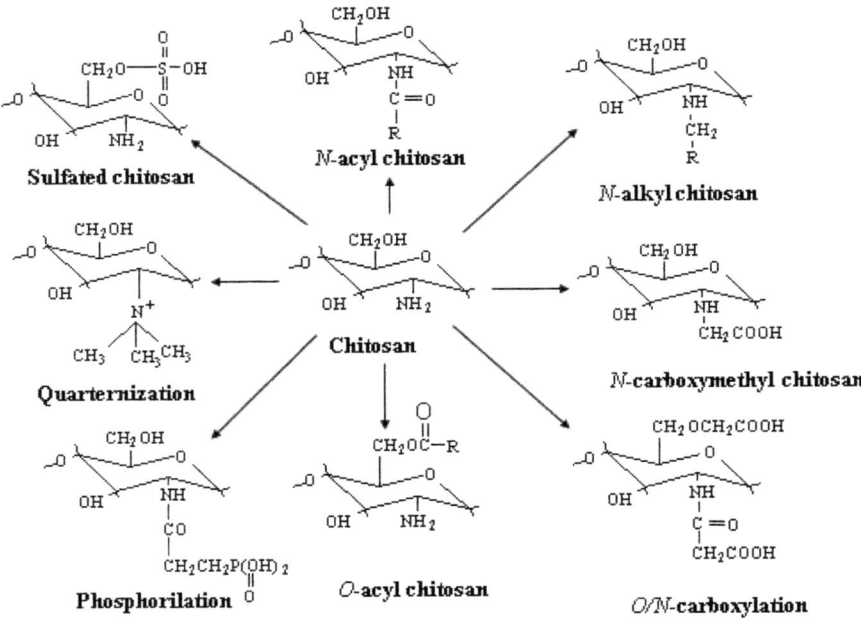

**Fig. 5.3** Chemical structure of typical chitosan derivatives. Copyright © 2019 Brasselet, C., Pierre, G., Dubessay, P., et al. Licensee MDPI, Basel, Switzerland. Creative Commons Attribution (CC BY) license. Applied Sciences, 2019, 9 (7), 33 p. https://doi.org/10.3390/app9071321. Reproduced from Ref. [40]; Copyright © 2018 Zhao, D. et al. Licensee MDPI, Basel, Switzerland. Creative Commons Attribution (CC BY) license. Polymers 2018, 10, 462; https://doi.org/10.3390/polym1 0040462. Reproduced from Ref. [41]; Copyright © 2020 Wang, W., Meng, Q., Li, Q., et al. Licensee MDPI, Basel, Switzerland. Creative Commons Attribution (CC BY) license, J. Mol. Sci. 2020, 21, 487; https://doi.org/10.3390/ijms21020487. Reproduced from Ref. [48]

reaction is a well-known chemical reaction between an amino acid and reducing sugar, usually requiring heat. N-alkylation modification of chitosan introduces alkyl or carboxymethyl groups into the chitosan structure [56]. The grafting tends to occur on the primary amine groups of chitosan following Schiff base reaction to form secondary imin (aldimines and ketimines) [39, 50, 55, 57]. The reaction product (secondary imin) is further hydrogenated to obtain N-alkylchitosan using reducing agents [50, 55]. The reducing agents used are sodium/potassium borohydride ($NaBH_4$/$KBH_4$) or sodium cyanoborohydride ($NaBH_3CN$). The schematic flow diagram of N-alkyl derivatives of chitosan is given in Fig. 5.4 [50].

Equation 5.2 shows the overall reaction mechanisms for N-alkyl derivatives of chitosan. Most of these reactions proceed smoothly in a binary solvent mixture of aqueous acetic acid and methanol. The solubility of chitosan derivatives is affected by inter- and intramolecular hydrogen bonds. Apart from this, the solubility of chitosan and N-substitute chitosan derivatives in the acidic region would be caused by the

## 5.4 Overall Synthesis Process for Water-Soluble Chitosan Derivatives

protonation of the amino group changing from $-NH_2$ to $-NH_3^+$. Hafdani and Sadeghinia [43] reported that the N-alkylation reaction of chitosan, at the C2 position, with the disaccharides enhanced the solubility of chitosan. Lactose, maltose, and cellobiose are the disaccharides used in this study. The solubility in the basic pH region would change the carboxyl groups to carboxylate ions ($-COOH$ to $-COO-$) [58].

**Table 5.3** Chitosan derivatives and their potential applications

| Derivatives | Reaction pathways | Product/solubility | Applications |
|---|---|---|---|
| N-alkylation | Chitosan undergoes Schiff base reaction with aldehydes and ketones to convert into N-alkyl derivatives | N-alkyl or sugar modified chitosan | Biomedical, pharmaceuticals, cosmetics, and antibacterial coatings |
| Carboxylation | Introduction of carboxyl groups onto the amino groups of chitosan | Water-soluble chitosan, amphoteric polyelectrolytes | Carboxymethyl chitosan is used for biomedical applications. Moisture retainer |
| Hydroxyalkyl | The reaction of chitosan with epoxide and glycidol | Hydroxyalkyl chitosan improves the water solubility of chitosan | Emulsifier, cosmetics |
| N-acylation | Reaction between chitosan and acid anhydride or acyl halide | Acyl chitosan | Biomedical, pharmaceutical, biomedicine, biolabeling, and biosensor |
| O-acylation | O-acyl chitosan is prepared from acyl chlorides in the presence of methanesulfonic acid | Acyl chitosan | Biodegradable-coating materials, emulsifier |
| S-chitosan | Acylation, addition, and substitution reactions are used to obtain sulfur-containing chitosan derivatives | Sulfated chitosan | Anticoagulant and hemagglutination inhibition activities. Other applications include antisclerotic, antiviral, anti-HIV, antibacterial, antioxidant, and enzyme inhibition activities |
| P-chitosan | The reaction of chitosan with phosphorous pentoxide in the presence of acid | Phosphorylated chitosan | Water soluble, and it can be used for biomedical applications. P-chitosan has bactericidal, osteoinductive, and metal chelating properties. It can promote tissue regeneration |

(continued)

**Table 5.3** (continued)

| Derivatives | Reaction pathways | Product/solubility | Applications |
|---|---|---|---|
| Quaternization | Quaternization of chitosan is usually carried out in basic media using alkyl halides | Quaternized chitosan derivatives; soluble in the solvent at physiological pH | Cationic nature chitosan, moisture retention, bio-adhesives, antimicrobial, and permeation enhancing effect |

Copyright © 2019 Brasselet, C., Pierre, G., Dubessay, P., et al. Licensee MDPI, Basel, Switzerland. Creative Commons Attribution (CC BY) license. Applied Sciences, 2019, 9 (7), 33 p. https://doi.org/10.3390/app9071321. Reproduced from Ref. [40]; Copyright © 2018 Zhao, D. et al. Licensee MDPI, Basel, Switzerland. Creative Commons Attribution (CC BY) license. Polymers 2018, 10, 462; https://doi.org/10.3390/polym10040462. Reproduced from Ref. [41]; Copyright © 2020 Wang, W., Meng, Q., Li, Q., et al. Licensee MDPI, Basel, Switzerland. Creative Commons Attribution (CC BY) license, J. Mol. Sci. 2020, 21, 487; https://doi.org/10.3390/ijms21020487. Reproduced from Ref. [48]

**Fig. 5.4** Schematic of the N-alkyl chitosan preparation process. Copyright © 2011 Cellulose Chem. Technol., 45(9–10), 2011, 619–625. Reproduced from Ref. [50] with permission of Romanian Academy Publishing House, the owner of the publishing rights

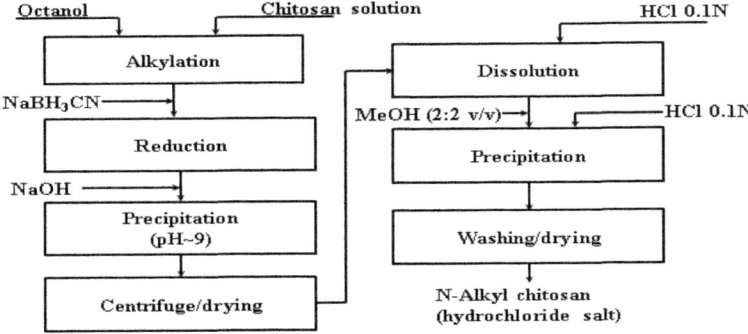

(5.2)

Similarly, chitosan derivatives are prepared by introducing chemical substitutes into the chitosan structure [50]. Hydroxyl alkyl, carboxyalkyl, and carboxymethyl alkyl are typical examples of these chitosan derivatives. The presence of bulky substituents like alkyl or carboxymethyl groups in chitosan structure weakens the chitosan structure by altering its hydrogen bonds [52]. Therefore, N-alkylchitosan

## 5.4 Overall Synthesis Process for Water-Soluble Chitosan Derivatives

derivatives swell in the water despite the hydrophobicity of the alkyl chain [59], which ultimately causes chitosan soluble in both organic and aqueous solvents.

### 5.4.1.1 N-Acylation of Chitosan

N-acylation is one of the many types of N-substituted chitosan derivatives that is highly soluble in a wide pH range. The acylation of chitosan is more versatile than the N-alkylation process as it introduces hydrophobic moieties at amino, alcohol, or both residues [52]. The process involves an addition/elimination type reaction mechanism between N-amino functional groups of chitosan and an acid anhydride, or acyl halide [52, 60]. This process restores the amide functionality of the N-amino groups. The N-acylation is mainly done by acid chlorides or acid anhydride, which introduces the amido group ($CONH_2$ group in amides that replace the hydroxyl group in a carboxyl group) to the chitosan nitrogen. The acylation of chitosan can be achieved in a homogeneous reaction between chitosan and acid anhydrides (linear or cyclic) or acyl halide in acetic acid/methanol solutions [43, 60]. Methanol is used to help solvate the anhydride in the medium. Moyura and Majumder [61] reported that the acylation reactions are carried out frequently in aqueous media with acetic acid/methanol, pyridine, pyridine/chloroform, trichloroacetic acid/dichloroethane, and ethanol/methanol. The amino group, hydroxyl group, or both chitosan groups react in the chitosan acyl synthesis process. Therefore, the acylation reaction can be controlled at the expected sites, i.e., on either amino or hydroxyls, or both. The introduction of hydrophobic branches also endows the polymers with a better soluble range than chitosan itself. Equation 5.3 shows the overall acylation reaction mechanism:

(5.3)

### 5.4.1.2 O-Acylation of Chitosan

Following the O-acylation method, the nucleophilic N-amino groups of chitosan are protonated to prevent them from further reaction. The hydroxyl groups of chitosan are preferentially substituted. However, the experimental conditions can influence O-substitution. Introducing a hydrophobic moiety with an ester linkage into chitosan has two benefits [61]:

1. Hydrophobic groups contribute solubility in organic solvents.
2. The ester linkage is hydrolyzed by the enzyme (glycosidases degrade the glycoside linkage of chitosan derivatives).

The most common process of getting an O-acylated chitosan derivative is the protonation of the amine group of chitosan with methanesulfonic acid [62]. Methanesulfonic acid worked as a suitable solvent for chitosan and also an efficient catalyst for the esterification reaction. As a result, hydroxyl groups become active in nucleophilic-substitution reactions. Therefore, during the preparation of O-acyl chitosan, hydrophobic moiety (acyl chloride) preferentially reacts with hydroxyl groups in the presence of methanesulfonic acid ($MeSO_3H$). The introduction of acyl chloride (hydrophobic moiety) with hydroxyl groups (ester linkage) contributes to the organo-solubility of chitosan [63]. After the O-acyl chitosan is prepared, the protection of amino groups of chitosan can be removed for further N-acylation reaction to obtain N, O-acylated chitosan [64]. Equation 5.4 shows the overall reaction mechanism for N, O-acylated chitosan:

$$(5.4)$$

## 5.4.2 Sulfated Chitosan

Sulfated chitosan is water soluble and is an anionic derivative of chitosan. Its applications include antiviral, anticoagulant, antimicrobial, and osteogenic activities [65]. To prepare sulfated chitosan, sulfating reagents such as sulfuric acid and sulfur trioxide are used. Other sulfating agents used include sulfur-based reaction media such as sulfur trioxide/pyridine, sulfur trioxide/trimethylamine, sulfur trioxide/sulfur dioxide, and chlorosulfonic acid–sulfuric acid [61]. Vongchan et al. [66] suggested that using reagent sulfuric acid, chlorosulfonic acid, and sulfur trioxide alone could cause degradation of the product. Combining Lewis base with these sulfating reagents can prevent degradation during reaction [66]. Pires et al. [67] reported the preparation of sulfating complex through drop by drop addition of $HClSO_3$ (chlorosulfonic acid) to DMF (N, N-dimethylformamide) under continuous stirring. In a typical sulfated chitosan synthesis process, chitosan and DMF mixture is stirred for 12 h at room temperature to obtain solvated chitosan. The solvated chitosan is then mixed with a

## 5.4 Overall Synthesis Process for Water-Soluble Chitosan Derivatives

previously prepared sulfating complex, and the reaction proceeds at room temperature for 5 h with stirring. The final mixture can be neutralized by 20% (m/v) NaOH and precipitated with methanol in an ice bath. Equation 5.5 shows the overall reaction mechanism:

$$\text{Chitosan} \xrightarrow{\text{Methanol}} N\text{-sulfobenzoyl chitosan} \quad (5.5)$$

### 5.4.3 Phosphorylation of Chitosan

Phosphorylated chitosan is prepared with the reaction of chitosan and phosphoric acid in the presence of a catalyst. Chitosan modified with phosphoric groups exhibits cytocompatibility, bio-absorbability, and osteoinductiveness [68]. Several studies reported the preparation of phosphorylated chitosan [67, 69, 70]. The various applications of these phosphorylated chitosans in the biomedical and pharmaceuticals field have also been reported. Joykumar reported that the introduction of phosphonic or phosphonate groups onto chitosan structure increases chitosan's specific physical properties [51]. These properties include chelating ability and solubility. Phosphorylation of chitosan can be performed through three different reaction media, namely in the presence of $H_3PO_4$/urea, $H_3PO_4$/$Et_3PO_4$/$P_2O_5$, or $P_2O_5$/$CH_3SO_3H$ [71]. Phosphorylation of chitosan with phosphorus pentoxide provides a high degree of substitution (DS) in the presence of methane sulfonic acid as a catalyst. Phosphorylated chitosan with a higher degree of substitutions forms a polyion complex. This makes it insoluble in water. However, the phosphorylated chitosan with a low DS value is soluble in water [51, 68]. In a typical synthesis process, phosphorous pentoxide is added to the mixture of chitosan and methane sulfonic acid. The mixture is then stirred at 0–5 °C for 2–3 h. Ether is used to precipitate the product. The product is then centrifuged to separate from the aqueous phase. The precipitate is then washed with ether and acetone and dried. Equation 5.6 shows the overall reaction for the synthesis of phosphorylated chitosan:

$$\text{[chitosan structure with CH}_2\text{OH, OH, NH}_2\text{]} \xrightarrow[\text{CH}_3\text{SO}_3\text{H, 0-5 °C}]{P_2O_5} \text{[phosphorylated/sulfated chitosan]} \quad (5.6)$$

## 5.4.4 Quarternized Chitosan

In the chitosan quaternization, nucleophilic alkylation occurs at the primary amino group at the C2 position of chitosan. Quaternary ammonium substituents also react at the primary amino group at the C2 position [43, 53]. Quarternization of chitosan is the most used synthesis route, rendering chitosan with a permanent positive charge [72]. The amino groups of chitosan are quarternized to synthesize the derivatives of chitosan. It is reported that polymers containing quaternary ammonium groups are cationic polyelectrolytes with permanent charges, differing from polymeric amines, which become charged in acidic media only [73]. The presence of the quaternary ammonium charges in the chitosan backbone makes it a cationic characteristic, which is independent of the solvent pH [74]. For instance, the introduction of $CH_3$ moiety in the amino groups of chitosan structure generates a chitosan derivative with permanent cationic characteristics and is soluble in the solvent at physiological pH [43, 74]. Quaternary salts of chitosan can be prepared with different degrees of substitution, mainly via methyl iodide or dimethyl sulfate synthetic route [75]. The typical overall reaction mechanism for quaternary chitosan preparation is as follows (Eq. 5.7):

$$\text{chitosan} \xrightarrow[\text{HAOH}]{(CH_3)_2SO_4} \text{N,N,N-trimethyl chitosan}$$

$$\text{chitosan} \xrightarrow[\text{NaOH, NMP}]{CH_3I, NaI} \text{[quaternized chitosan with I}^-\text{]} \xrightarrow{\text{Ion-exchange}} \text{[quaternized chitosan with Cl}^-\text{]} \quad (5.7)$$

## 5.5 Preparation of Chitosan-Based Hydrogel

### 5.5.1 Introduction and Background

Hydrogels are crosslinked polymeric materials that swell in water to an equilibrium value and are insoluble in it. It is a complex network of crosslinked hydrophilic polymers capable of retaining large amounts of water or biological fluids. The unique property of higher water absorption by the hydrogels is due to the three-dimensional polymeric links between the monomers. The links between the monomers occur through various types of chemical bonds, such as hydrogen bonding, van der Waals interactions, and chemical bridges. Development of these synthetic "superabsorbing" polymers capable of absorbing and holding large amounts of water can be traced to the early 1950s and recently has seen tremendous applications in pharmaceutical, personal care, and cosmetic applications. These hydrogels can absorb water up to 1000 times their weight and are used in diapers and sanitary napkins. The hydrophilic sites present in the polymeric network of the hydrogel are responsible for water absorption from aqueous media, thus forming a hydrogel structure [76–78]. The amino, carboxyl, and hydroxyl groups are the notable hydrophilic sites in the polymeric network of the hydrogel. It is reported that the water absorption capacity of a hydrogel is dependent on the chemical structure and crosslinking density of polymers and environmental conditions [79]. The polymeric hydrogel can contain free, semi-bound, and bound water molecules, as shown in Fig. 5.5.

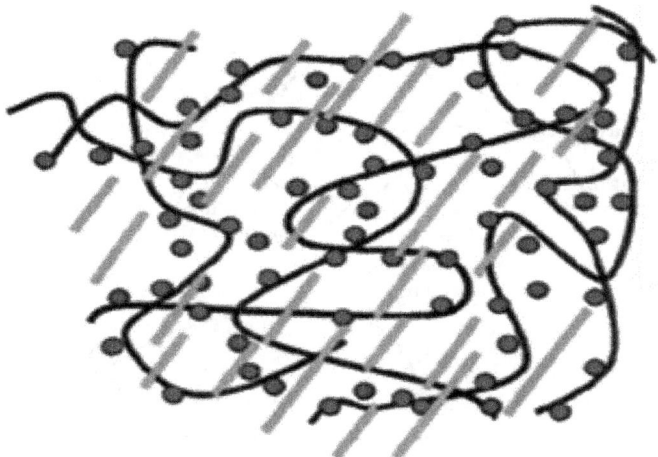

——Free water  ●Bound water  ●Semi-bound water

**Fig. 5.5** Schematic of water molecules in polysaccharide-based hydrogel. Copyright © 2012 Pasqui, D.; De Cagna, M.; Barbucci, R; Creative Common CC BY license, Polymers 4(3):1517–1534, 2012, https://doi.org/10.3390/polym4031517. Reproduced from Ref. [80]

The amount of free water molecules is dependent on the hydrogel structure, and there is no bond between free water and polymer functional groups. Free water ultimately influences the swelling ratio of hydrogel [81]. Semi-bound or bound water can form weak interactions with functional groups in the polymeric chain. Hydrogen bonding between polymeric chains and water molecules forms bound water [82].

In general, the preparation of polymer hydrogels uses two main processes: physical and chemical crosslinking processes [79]. In a review paper, Calo and Khutoryaniskiy [83] discussed the preparation of hydrophilic polymer-based hydrogels and their biomedical applications. The hydrogels preparation using physical and chemical crosslinking processes was also discussed. In physical crosslinking of hydrogel preparation, molecular entanglements, ionic strength, electrostatic, hydrophobic, and hydrogen bonding forces between the polymer chains play a critical role in forming the network.

Hydrogel obtained from physical crosslinking can be dissolved in a solvent by changing the pH, temperature, or ionic strength of the solution. This type of hydrogel is relatively weak. It has reversible links formed from the temporary associations between chains [84]. The formation of a more robust and more extensive intermolecular association in the hydrogel is possible through the chemical covalent crosslinking as they are permanent and irreversible [85]. A dual-network hydrogel is prepared by combining both physical and chemical crosslinking processes [86]. The three crosslinks depicted in Fig. 5.6 represent possible crosslinked hydrogel structures. Figure 5.6a shows an ideal network with tetrahedral covalent links. Figure 5.6b, c shows the molecular entanglements and multifunctional junctions in the hydrogel structure.

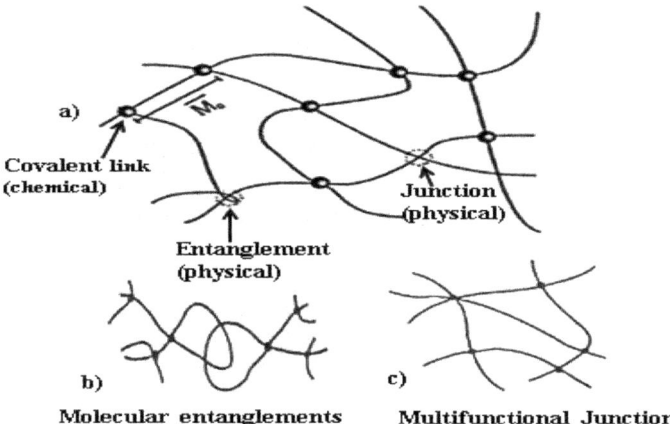

**Fig. 5.6** Schematic of a hydrogel network structure: **a** Tetrahedral covalent links, **b** molecular entanglements, and **c** multifunctional junction. Copyright © 2015 Elsevier B.V., Materials Science and Engineering C 57:414–433, 2015. Reproduced from Ref. [82] with permission

## 5.5 Preparation of Chitosan-Based Hydrogel

The chemical crosslinking process generally requires at least one difunctional, small molecular weight crosslinking agent. That agent usually links two longer molecular weight chains through its di- or multifunctional groups [87]. Chai et al. [88] reported that the molecular weight of the polymer chain affects the degree of crosslinking between the neighboring crosslink points. Chemical bonding and type of monomer are the two leading factors contributing to a hydrogel structure. In most cases, hydrogels can be amorphous, or semicrystalline, and hydrogen bonded [89]. The amorphous hydrogels are composed of randomly arranged chains. The semicrystalline gels contain dense regions of ordered macromolecules. Three-dimensional networks are usually formed in hydrogen-bonded hydrogels [89]. The following characteristics categorize the polymers used to prepare hydrogel: monomer composition, mechanical and structural integrity, overall ionic charge, and basic structure [90]. For example, a hydrogel synthesized using only one monomer is called a homopolymer. Similarly, copolymer hydrogels are composed of two different kinds of monomers. Ullah et al. [82] classified hydrogels depending on their properties, sources of raw material, method of preparation, and the nature of crosslinking. Properties considered were swelling ability, ionic charge, and the rate of biodegradation (Fig. 5.7).

There is considerable interest in the research and development of polymer-based "smart gels." These "smart gels" respond to minute changes in their environment, such as temperature, pH, and ionic strength. For example, some acrylic-based gels can absorb water at 25 °C and desorb at 35 °C. These gels are also known as "smart

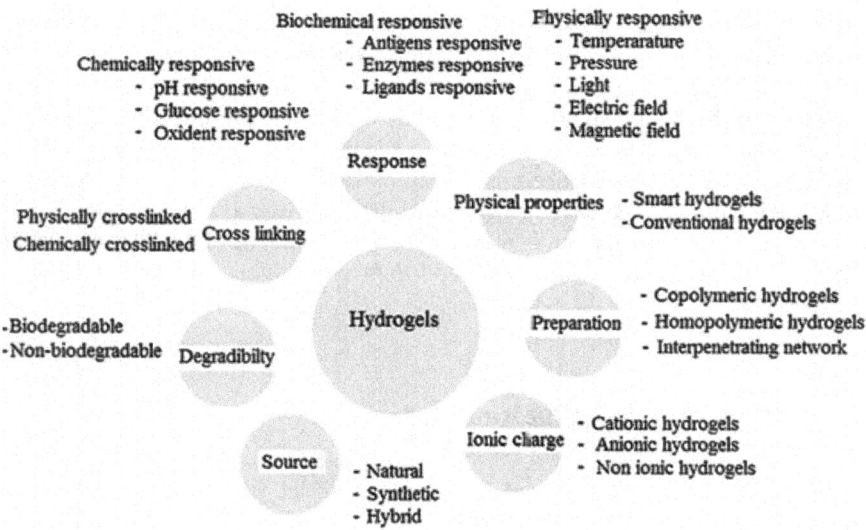

**Fig. 5.7** Classification of polymer hydrogels based on their properties. Copyright © 2015 Elsevier B.V., Materials Science and Engineering C 57:414–433, 2015. Reproduced from Ref. [82] with permission

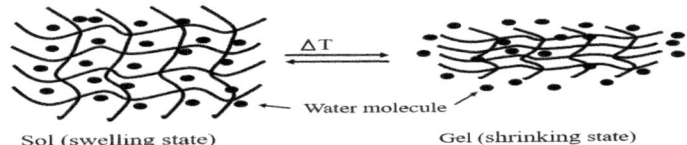

**Fig. 5.8** Schematic of a thermosensitive hydrogel in response to a temperature change. Copyright © 2020 Wang, J., Gao, S., Tian, J., et al. Licensee MDPI, Basel, Switzerland. Creative Commons Attribution (CC BY) license, Water 2020, 12, 692; https://doi.org/10.3390/w12030692. Reproduced from Ref. [92]

gels" because of their memory to return to their original state when the initial environmental conditions (pH, temperature, ionic strength) are regained [91]. When a dry or partially swollen gel is placed in a solvent, causing the crosslinked network expands, and hydrogel swelling occurs (Fig. 5.8). This reaction is thermodynamic. Elastic and hydrophobic retractive forces counterbalance the swelling force, causing the network to expand within the polymer gel. The point at which these two forces reach equilibrium is known as the swollen state. The gels can be reused several times without appreciable loss of water absorption capacity. The temperatures at which a gel expels the absorbed water and collapses are called the critical temperatures.

Several natural and synthetic polymers have been studied in hydrogel research [93]. The most common hydrogels are prepared by direct crosslinking of polymers, which are soluble in water. The water-soluble synthetic polymers are poly (acrylic acid), poly (vinyl alcohol), poly(vinylpyrrolidone), poly (ethylene glycol), and polyacrylamide. The water-soluble natural polysaccharides such as chitosan, agarose, alginate, and hyaluronan are also used to form hydrogels [83, 93]. Several research activities have been published on chitosan-based gel and hydrogel. These published research studies reported various chitosan-based hydrogels and their applications in numerous biomedical applications, as well as food and pharmaceutical industries [79, 94–96]. Potential biomedical (e.g., extended drug delivery system, enzyme encapsulation), food protein separation, and analytical applications of the thermosensitive hydrogels are extensively researched, and their end products are currently being developed [94].

## *5.5.2 Preparation of Chitosan Hydrogel*

Chitosan hydrogel can be formed by the physical and chemical crosslinking of chitosan macromolecules [2]. The preparation of a chitosan hydrogel by physical crosslinking starts with the solubilization of the macromolecules in an acidic or alkaline aqueous solvent (Fig. 5.9). For the acidic solvent, the gelation process starts with the protonation of amine groups and the entanglement of chitosan macromolecules. Intermolecular hydrogen bond interactions are the essential gelling mechanism in alkaline systems [97]. In the chemical crosslinking process, the formation of hydrogel

## 5.5 Preparation of Chitosan-Based Hydrogel

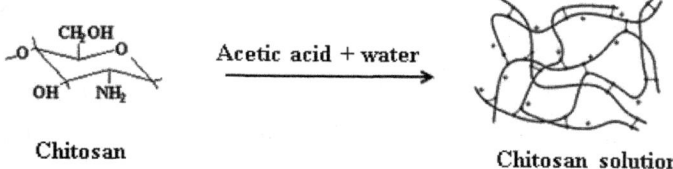

**Fig. 5.9** Schematic of a chitosan solution in acetic acid. Copyright © 2003 Published by Elsevier B.V. European Journal of Pharmaceutics and Biopharmaceutics 57(1):19–34, 2004. Reproduced from Ref. [2] with permission

occurs through the covalent crosslinking of chitosan macromolecules. The formation of bonds in chitosan macromolecules in the hydrogel network shows irreversibility (covalent link) for the chemical crosslinking process. It is reversible for the physical crosslinking process. Hydrogels are in different forms, namely copolymers (or self-crosslinked), blends (hybrid), and interpenetrating networks (IPNs) [2, 93, 98]. Copolymeric hydrogels are prepared from two or more different monomer species. The self-crosslinked hydrogels are prepared from two structural units, and they may or may not belong to the same chitosan polymeric chain. For the hybrid hydrogel, crosslinking occurs between the structural unit of chitosan and other polymeric chains [2]. Ahmad reported that IPNs consist of two independent crosslinked synthetic and natural polymer components, which are contained in a network form [98]. At the same time, the semi-IPNs hydrogel is a combination of crosslinked and non-crosslinked polymers.

In the case of chitosan, a non-reacting polymer was added to the chitosan solution before crosslinking. This addition leads to a crosslinked chitosan hydrogel with a semi-interpenetrating network (semi-IPNs) [2, 98]. Further crosslinking of semi-IPNs hydrogel with additional polymer creates two entangled crosslinked networks, thus forming a full-IPNs. The properties and microstructure of IPNs can differ from their corresponding semi-IPN [2, 93]. The schematics of different states of the hydrogel are shown in Figs. 5.9 and 5.10.

Chitin and chitosan have been chemically modified to yield hydrogel structure. Chitosan-modified gels are prepared using nitration, phosphorylation, sulfation, xanthation, acylation, hydroxylation, Schiff's base formation and alkylation, and graft polymerization processes. The positively charged amino groups of chitosan have the potential to attract biomolecules [92, 99]. These biomolecules include peptides, antibodies, and proteins. In recent years, thermosensitive injectable systems have gained special attention due to their ability to deliver therapeutic agents, molecules, or cells [100]. Chitosan-based hydrogels possess thermodynamically active functional groups on polymer chains that make them sensitive to stimuli, and this temperature-sensitive property has numerous biomedical applications [94, 101]. For example, chitosan glycerol phosphate-based hydrogel with thermosensitive properties can be prepared at body temperature following a sol–gel transition. This thermosensitive hydrogel has great potential in tissue engineering [85, 96]. Figure 5.11 shows the thermally gelling chitosan system. The hydrogel is prepared by neutralizing highly

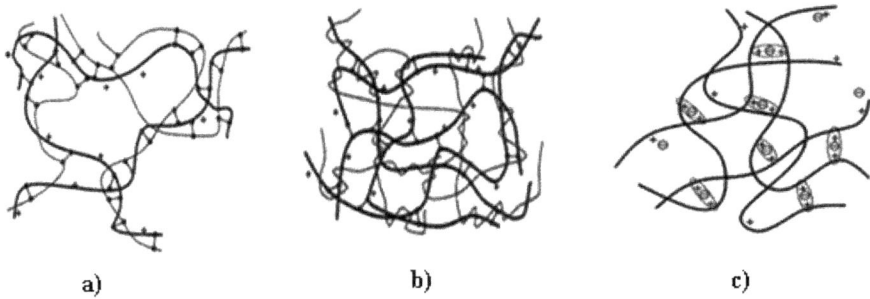

**Fig. 5.10** Structure of chitosan hydrogel: **a** hybrid polymer network, **b** semi-interpenetrating network, and **c** ionic crosslinking of chitosan (●—● covalent crosslinker + the positive charge of chitosan, ▬▬ chitosan, ▬▬ additional polymer, ⊖ charged ionic crosslinker, and ⬯ ionic interaction). Copyright © 2003 Published by Elsevier B.V. European Journal of Pharmaceutics and Biopharmaceutics 57(1):19–34, 2004. Reproduced from Ref. [2] with permission

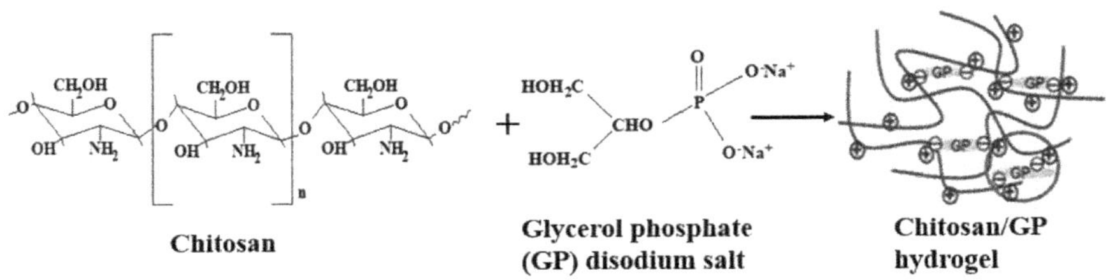

**Fig. 5.11** Overall reaction mechanism of chitosan/GP hydrogels. Copyright © 2012 Kim, G.O., Kim, N., Kim, D.Y. et al. Creative Commons CC BY license, Molecules 17(12):13, 704–13, 711, 2012. Reproduced from Ref. [105]

deacetylated chitosan solutions with β-glycerol phosphate at physiological pH and temperature [102]. Han et al. [103] reported that chitosan solution displayed a sol–gel phase transition in a pH and temperature-dependent manner and formed an endothermic hydrogel after subcutaneous injection into a mouse in the presence of β-GP. Several studies reported that the thermogelling chitosan/β-glycerol phosphate (GP) solution is especially attractive as an injectable scaffold for cartilage tissue engineering [79, 92, 94, 96]. Nonetheless, these systems also exhibit weaknesses for specific applications, such as low mechanical strength, chemical stability, prolonged gelation time, and uncontrollable physiological degradability. The toxicity of crosslinking agents presents the major drawback in using polymers as injectable matrices or as in situ-forming polymer scaffolds since their seepage into body fluids can be catastrophic even at low concentrations.

The crosslinking agent, genipin, has been successfully used as a substitute for toxic crosslinking agents. Genipin is commonly used in herbal medicine and as a food dye. It is a naturally occurring material capable of binding with biological tissues and biopolymers. Genipin can be used as an alternative to dialdehyde. Moura et al. [104] have focused on modifying the chitosan structure by crosslinking with genipin.

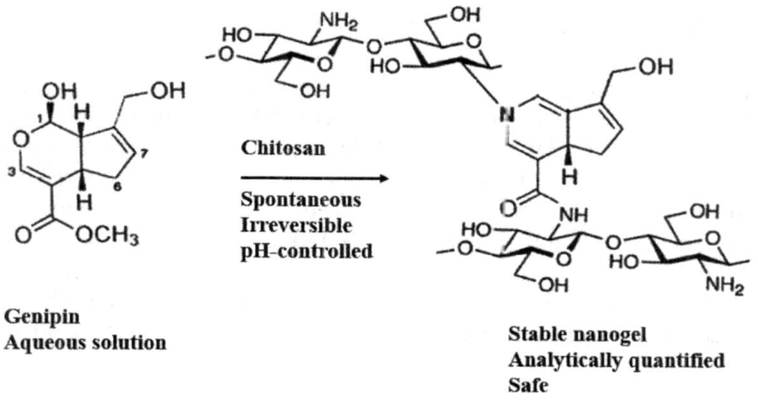

## 5.5 Preparation of Chitosan-Based Hydrogel

**Fig. 5.12** Genipin crosslinked chitosan hydrogel. Copyright © 2015 Muzzarelli, R.A.A., El Mehtedi, M., Bottegoni, C. et al. Creative Common CC BY license, Mar Drugs 13:7314–7338, 2015. https://doi.org/10.3390/md13127068. Reproduced from the Ref. [106]

This modification is effective for cells and bioactive molecule encapsulation [104]. Figure 5.12 shows the chemical structure of chitosan crosslinked with genipin.

Muzzarelli et al. reported that chitosan crosslinked with genipin is safe, spontaneous, and has irreversible reactions [106]. Dimida et al. [107] reported that crosslinking with genipin stabilizes the chitosan-genipin hydrogel structure because a chemical bridge is formed between the side backbone of chitosan. Lai [108] demonstrated that the genipin (GP) crosslinked chitosan is compatible with human retinal pigment epithelial cells. They reported that genipin crosslinked chitosan samples improved the preservation of corneal endothelial cell density and possessed better anti-inflammatory activities, indicating the benefit of the genipin crosslinker.

### 5.5.3 Chitosan-Based "Smart" Hydrogels

Chitosan and inorganic polymer-based mixed hydrogels have been reported as smart gels for biomedical and pharmaceutical use [109–111]. The operating principle of smart gels is dependent on materials whose volume changes abruptly. Smart gels display a volume phase transition (VPT) when an appropriate stimulus is applied [112]. Poly(N-isopropyl acrylamide), known as PNIPAm, can be crosslinked to create gels that undergo a continuous reversible volume phase transition at around 33 °C. A typical synthesis process for chitosan smart gel is described in the next section.

### 5.5.3.1 Chitosan Copolymer Smart Gel Synthesis

In a typical smart gel synthesis process, chitosan is dissolved in a 1% acetic acid solution. $N$-isopropyl acrylamide (97%) (NIPA) and $N,N'$-methylene bisacrylamide (>99.5%) (BIS) are added to the chitosan solution under continuous stirring with $N_2$ flow in an ice bath for 30 min. Ammonium persulfate (98+%) (APS) and $N, N,N',N'$-tetramethylene diamine (TEMED) are then added to the above chitosan mixture. The mixture is then stirred briefly and poured into a Teflon frame between two glass plates to polymerize at room temperature. Finally, the gel is immersed in deionized water at least four times at minimum intervals of 1 h for cleaning. Equation 5.8 shows the overall reaction mechanism of chitosan-PNIPAm hydrogel preparation [113]:

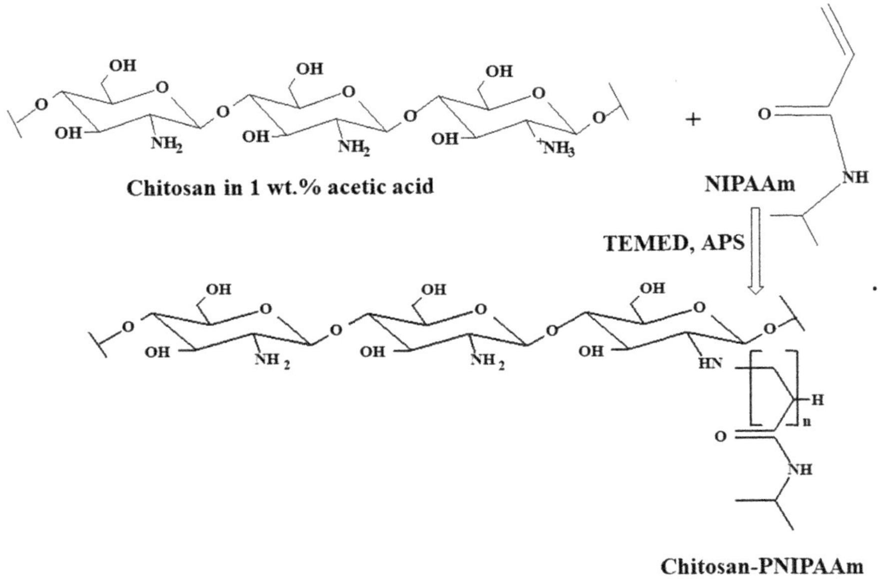

Chitosan in 1 wt.% acetic acid    NIPAAm

TEMED, APS

Chitosan-PNIPAAm
(5.8)

Poly($N$-isopropyl acrylamide) (PNIPAm) is a non-toxic and one of the most interesting stimuli-responsive polymers. The physical stimuli response behavior of a hydrogel occurs due to the effect of temperature, electric or magnetic field, light, pressure, and sound. The chemical stimuli include pH, solvent composition, ionic strength, and molecular species [98, 114]. Bao et al. [115] reported that PNIPAm has hydrophilic amide groups and hydrophobic isopropyl groups in its side chains (Fig. 5.13). PNIPAm is soluble in low polarity organic solvents due to the side-chain groups. In the case of solubility in water, it can further undergo a coil to globule transition [115, 116]. The hydrophilic/hydrophobic state of the gel can be changed by varying the temperature below or above the lower critical solution temperature [111, 114]. The lower critical solution temperature is also known as LCST, where the system will be miscible below the LCST. In PNIPAm hydrogel, the hydrogen bond

## 5.6 Chitosan Nanoparticles

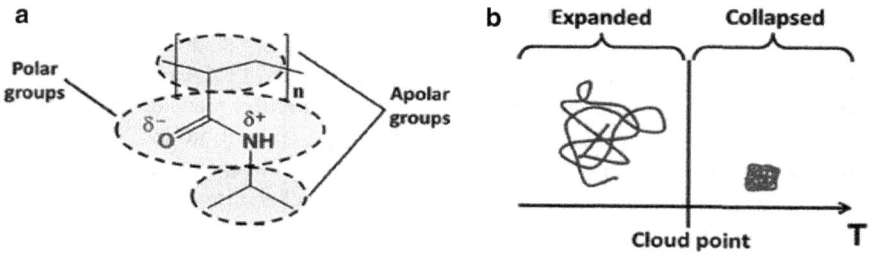

**Fig. 5.13 a** Chemical formula of PNIPAm and **b** representation of the volume phase transition between the coil (left) and globular (right) hydrogel conformation. Copyright © 2017 Lanzalaco, S., and Armelin, E. Creative Common CC BY license, Gels 3(4):36, 2017. Reproduced from the Ref. [114]

occurs between the water molecules and hydrophilic groups, and the bond becomes stronger below LCST [109]. Above the LCST, the phase transition of the hydrogel network may occur. This makes the gel hydrophobic and insoluble, thus causing gel formation [88, 109, 110, 117]. The phase transition above LCST is reported to involve the breakage of intermolecular hydrogen bonds with the water molecules. It creates intramolecular hydrogen bonds among the dehydrated amide groups [111]. PNIPAm-based thermosensitive hydrogels are known to undergo a volume phase transition at or above the lower critical solution temperature (LCST) [117].

In a proper environment, PNIPA gels are also capable of releasing liquid through a reversible discontinuous volume phase transition [118, 119]. The thermoreversible property of gel is used in solute extraction and separation, controlled drug release, manufacture of artificial organs, and enzyme immobilization [120].

In recent years, polymer blending has become a method for providing polymeric materials with desirable properties for practical applications. Heavy metals ions are not biodegradable. They are usually removed physically or chemically from the contaminated water. Chemical precipitation, membrane filtration, ion exchange, and adsorption processes are generally used to remove heavy metal ions from the contaminated water [121]. Hydrogel absorption proposes an economically as well as environmentally sound way of ultimately solving an age-old problem.

## 5.6 Chitosan Nanoparticles

The nanostructure of particles can promote orientations or conformations of the reactants and reaction intermediates, thereby facilitating the activation of particle bonds in the molecules by the active sites [122]. Chitosan nanoparticles are subnanosized colloidal structures with size ranges from 1 to 1000 nm [123, 124]. They are used as drug manufacturing components in nanomedicine. Chitosan nanoparticles have a long shelf life [125, 126] and a larger activated surface-to-volume ratio than other physical forms of chitosan materials [127]. It is reported that the chitosan nanoparticle has a high affinity for macromolecules [128], increasing its potential as a carrier for the therapeutic drug delivery system [129]. Wang et al. [130] discussed the applications of chitosan nanoparticles in nanomedicine. It can carry many drugs, including gene drugs, protein drugs, anticancer chemical drugs, and antibiotics. The possible routes of drug administration include oral, nasal, intravenous, and ocular. [130].

The mucoadhesive properties of chitosan nanoparticles improve drug absorption and bioavailability. These mucoadhesive properties and the high surface-to-volume ratio of nanoparticles allow extended drug contact with the mucosal layer [131]. In a review article, Elgadir et al. [132] discuss mucoadhesive, permeation enhancing, gene expression, gelling, and anionic drug delivery properties of chitosan. This review also captures the understanding of drug delivery systems, particularly in cases where chitosan drug-loaded nanoparticles are applied. Several researchers reviewed the published research work on chitosan nanoparticles, including their preparation methods, characteristics, modification, in vivo metabolic processes, and applications [130, 133–135]. Ahmed and Aljaeid [131] reported that the selection of the preparation method for nanoparticles depends upon the particle size, the stability of the active constituents, the residual toxicity present in the final product, and the kinetics of the drug release profile. Bellich et al. [136] discuss the different methods for preparing and characterizing chitosan micro and nanoparticles. The pharmaceutical applications of these particles in drug delivery are also discussed. Two techniques (top-down and bottom-up) are used to develop micro and nanoparticles as drug carriers. The top-down techniques are usually employed for successive breaking down of bulk particles to smaller particles using mechanical, chemical means, or other forms of energy. The top-down techniques include emulsion diffusion, salting out, nanoprecipitation, and emulsion evaporation [134]. Chemical synthesis is an excellent example of a bottom-up technique that usually starts with a monomer. Therefore, it allows nanostructures to be generated from individual atoms or molecules capable of self-assembling complex structures [137]. The bottom-up nanoparticle preparation process includes ionic gelation, reverse micellar, polyelectrolyte complexation, and coacervation [128]. Ahmed et al. [131] reported the crosslinking and drying techniques for preparing chitosan micro/nanoparticles loaded with thermosensitive or less stable substances. These substances include proteins, peptides, hormones, vaccines, plasmid DNA, and antigens.

Chitosan contains abundant amino and hydroxyl groups, which enable nanoparticle formulation via bottom-up techniques such as physical and chemical crosslinking [138]. Figure 5.14 shows a typical chemical crosslinking reaction mechanism for chitosan nanoparticles.

In chemical crosslinking, glutaraldehyde is widely used as a crosslinking molecule in covalent formulations [139]. Glutaraldehyde acts as a bridge between two glucosamine units of chitosan to form a Schiff base that provides a final product with an irreversible rigid network structure and high resistance to dissolution [2]. The chemical crosslinking agents often are toxic. Therefore, for food and pharmaceutical use, the preparation of chitosan nanoparticles using a physical crosslinking process is usually preferred over the chemical crosslinking process [124, 126, 136,

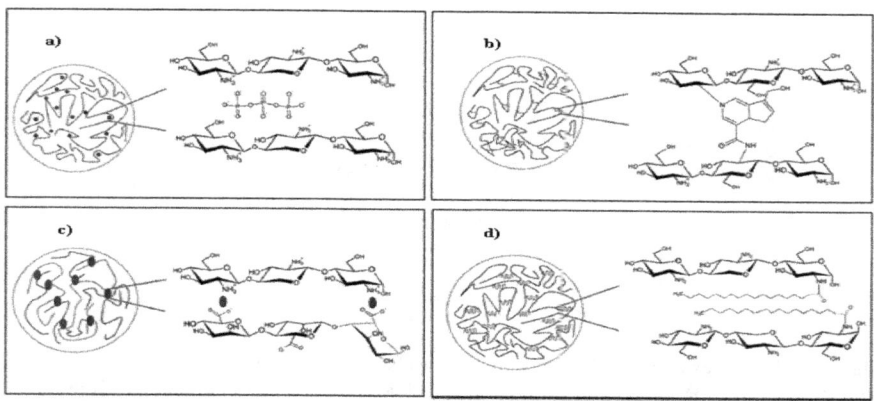

**Fig. 5.14** Chitosan nanoparticle preparation mechanism: **a** ionic crosslinking, **b** covalent crosslinking, **c** polyelectrolyte complexation (PEC), and **d** self-assembly. Copyright © 2016 Bellich, B., D'Agostino, H., Semeraro, S., and Gamini, A., Creative Commons Attribution (CC BY) license, Mar Drugs. 2016 May; 14(5): 99. https://doi.org/10.3390/md14050099. Reproduced from the Ref. [136]

137, 139]. Bottom-up techniques are the typical physical crosslinking techniques used to prepare chitosan nanoparticles [128]. These bottom-up techniques include ionic gelation, polyelectrolyte complex (self-assemble), and complex coacervation.

Other bottom-up techniques, including reverse micellar, microemulsion, and emulsification solvent diffusion, are used to synthesize chitosan nanoparticles [2, 135, 140]. Ionic gelation and reverse micellar are the most common methods to synthesize nanoparticles [133, 141]. However, ionic gelation is the most commonly used method to prepare polysaccharide-based nanoparticles as drug carriers for pharmaceutical use [142, 143]. The reverse micellar method can synthesize crosslinked nanoparticles with a narrow size distribution. However, some notable disadvantages of the reverse micellar process are organic solvent, time-consuming preparation, and complexity in the washing step [127, 135].

The ionic gelation procedure is the spontaneous electrostatic interaction between protonated amino groups of chitosan and negatively charged ions in solution at room temperature. The formation of intra and intercrosslinkages within/between polymer chains causes ionic gelation. It does not involve any organic solvent, high temperature, or chemical crosslinking [132]. Bodner et al. [144] described the synthesis of chitosan nanoparticles by natural carboxylic acids (e.g., tartaric, malic, succinic, and citric acid). The surface morphology of the nanoparticles was characterized using transmission electron microscopy (TEM). The particle size of the nanoparticles was identified by the dynamic laser light scattering (DLS) technique. They reported that the particle size of chitosan nanoparticles depends on the solution pH [144].

Several groups have investigated the effect of preparation conditions on the dimensions of chitosan nanoparticles obtained by the ionic gelation process. Sodium tripolyphosphate (TPP) was used as a crosslinking agent for this ionic gelation

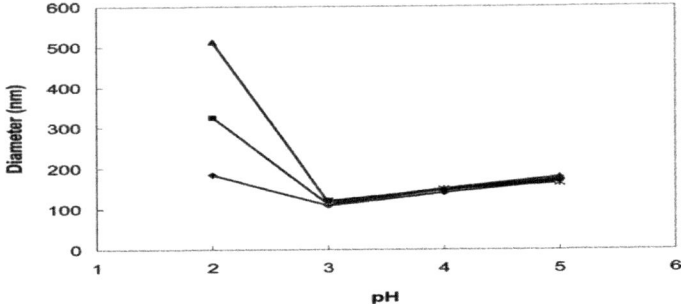

**Fig. 5.15** Typical example of chitosan particle size changes as a function of pH at various temperatures: (♦) 25 °C, (■) 32 °C, and (▲) 37 °C at a chitosan/TPP weight ratio of 5:1. Copyright © 2004, American Chemical Society, Biomacromolecules, 5, 2004, 2461–2468, 2004. Reproduced from Ref. [148] with permission

process [124, 145–148]. In the preparation of chitosan nanoparticles using the ionic gelation process, chitosan molecules become gel when crosslinked with TPP molecules via electrostatic interaction. The process is termed ionotropic gelation, which started with the work initially carried out by Bodmeier to prepare chitosan-TPP complex by dropwise addition of chitosan solution into sodium tripolyphosphate solution [145]. The particle sizes of the chitosan nanoparticles are greatly dependent on the method of preparation, molecular weight, and degree of deacetylation of chitosan [131, 149]. Zhang et al. [148] reported that the concentrations and the ratio of chitosan (CS) and TPP in an aqueous solution play a critical role in controlling the size and the size distribution of the nanoparticles. They also observed that the dimensions of the chitosan nanoparticles were dependent on the pH and the ratio of CS and TPP in the solution. The chitosan nanoparticle size was independent of temperature in the pH range from 3 to 5 (Fig. 5.15). Fan et al. [147] reported that at pH below 4.5, the particle size distribution of chitosan nanoparticles is not unimodal. The formation of microparticles of chitosan may occur in the suspension pH above 5.2.

Fan et al. [147] also reported the effect of chitosan/TPP mass ratio on nanoparticle size (Fig. 5.16). In their experiment, they added different volumes of TPP to 10.0 mL of chitosan solution. They observed that the particle size of the nanoparticles gradually decreased with the increase of TPP volume and then dramatically increased. The zeta potential value also decreased almost linearly with the increase of TPP volume. The decrease of zeta potential value occurs due to the neutralization of protonated amino groups by TPP anions (Fig. 5.16). Jonassen et al. [124] reported the preparation of chitosan nanoparticles by the ionic gelation process. It was concluded that the preparation of nanoparticles should be carried out in the dilute polymer concentration regime to avoid the formation of much larger particles. Liu and Gao [138] provided a process to prepare chitosan nanoparticles, where chitosan and TPP were dissolved in acetic acid and triple-distilled water to obtain solutions of 1.0–5.0 and 0.25–2.0 mg/ml, respectively [138]. The chitosan nanoparticles were

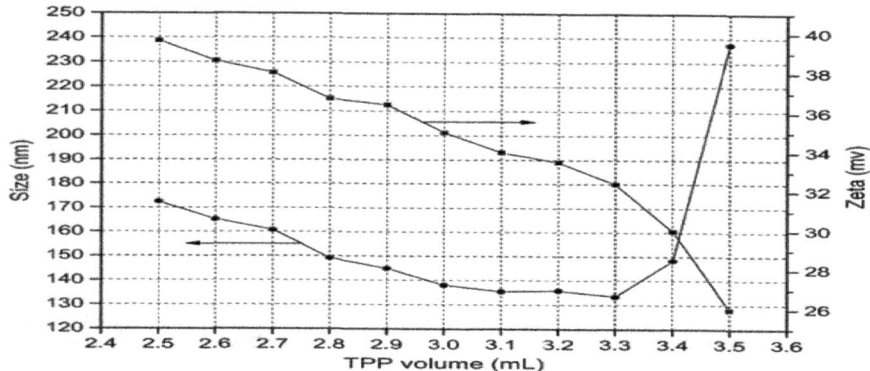

**Fig. 5.16** Effect of the mass ratio of chitosan to TPP on the particle size and zeta potential. Copyright © 2011 Elsevier B.V., Colloids and Surfaces B: Biointerfaces 90:21–27 2012. Reproduced from Ref. [147] with permission

formed upon the addition of 14 ml TPP solution into 35 ml chitosan solution through intra and intermolecular linkage between chitosan amino groups and TPP under mild mechanical stirring (550 rpm) at room temperature. Figure 5.17 shows typical chitosan nanoparticle prepared using TPP as a crosslinking agent.

Koukaras et al. [150] have prepared chitosan/TPP nanoparticles with a range of sizes from 340 to 615 nm. The smallest nanoparticle size was obtained at a CS/TPP w/w ratio near 4/1 [150]. The TPP/chitosan nanoparticle usually displays size polydispersity [151] and low mechanical strength. Hence, their usage in drug delivery is limited [145]. In the process, TPP crosslinks randomly oriented chitosan molecules, which, in turn, are connected to other similarly crosslinked moieties. Such intra and intermolecular crosslinking is relatively uncontrolled and leads to polydispersity in the synthesis process [152]. Grenha [128] reported that larger nanoparticle in the

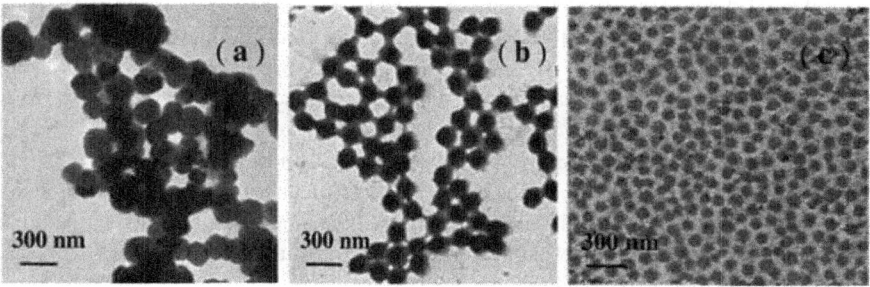

**Fig. 5.17** Transmission electron microscopy (TEM) images of typical chitosan nanoparticles using ionic gelation process: **a** commercial, **b** post-deacetylated, and fractionated chitosan. Copyright © 2004, American Chemical Society, Biomacromolecules, 5, 2004, 2461–2468, 2004. Reproduced from Ref. [148] with permission

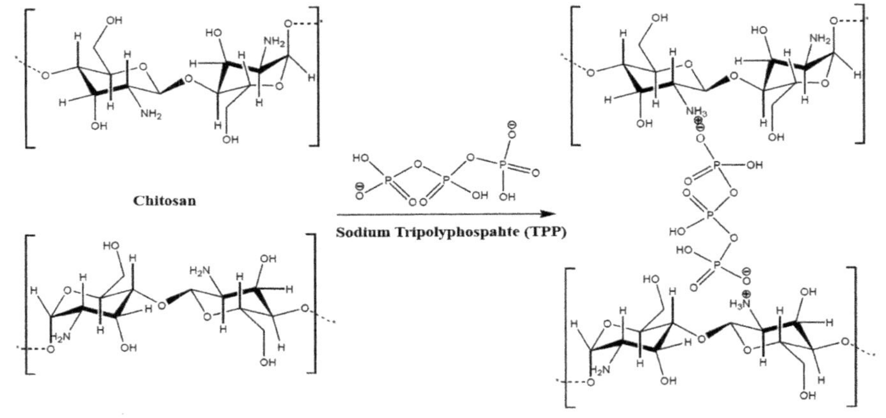

**Fig. 5.18** Schematic of ionic crosslinking of chitosan with TPP. Creative Commons license (CC BY). Fiona C Rodrigues et al. *IOP Conf. Ser.: Mater. Sci. Eng.* 872 (2020) 012109, https://doi.org/10.1088/1757-899X/872/1/012109. Reproduced from the Ref. [153]

polydispersity system might have higher drug loading capacity. In contrast, smaller nanoparticles are expected to have higher efficiency at delivering drugs to tissues or cells. Figure 5.18 shows a typical chemical structure of the chitosan nanoparticle using the ionic gelation process.

In another attempt, Calvo et al. [146] explored the ionic gelation process for the preparation of chitosan–polyethylene oxide nanoparticles using bovine serum albumin (BSA) as a model protein. They observed that these new nanoparticles have high protein loading capacity (entrapment efficiency up to 80% of the protein). They provide a continuous release of the entrapped protein for up to one week.

The physical and chemical properties of nanoparticles play essential roles in the efficacy of nanomedicines. Several studies reviewed the drug release mechanism from drug-loaded chitosan nanoparticles [132, 136, 154]. Chitosan nanoparticles exhibit a pH-dependent drug release because of the solubility of chitosan [132, 154]. The hydrophilicity, swelling rate, and density of the polymer chains also affect the drug release capability [154]. Bellich et al. [136] reported that diffusion, swelling, and erosion are the three main mechanisms to release the drug from the drug-loaded nanoparticles. Moreover, it is strictly dependent on the type and degree of the crosslink [136], crosslink density, and pH of the surrounding medium. Figure 5.19a depicts the possible scenarios of the drug release mechanism. These mechanisms include diffusion, swelling, and erosion of the chitosan nanoparticle and drug matrix. The kinetic of swelling and erosion will affect the initial part of the release curve, determining a characteristic lag phase, shown in Fig. 5.19b.

Wang et al. [151] focused on the effect of nanoparticles' size, shape, modulus, surface charge, and surface chemistry on their interactions with the cells in vitro and their pharmacokinetics and biodistribution in vivo. They concluded that particles in the range of 10–100 nm could avoid renal filtration and are small enough to efficiently

## 5.7 Chitosan Nanofibers

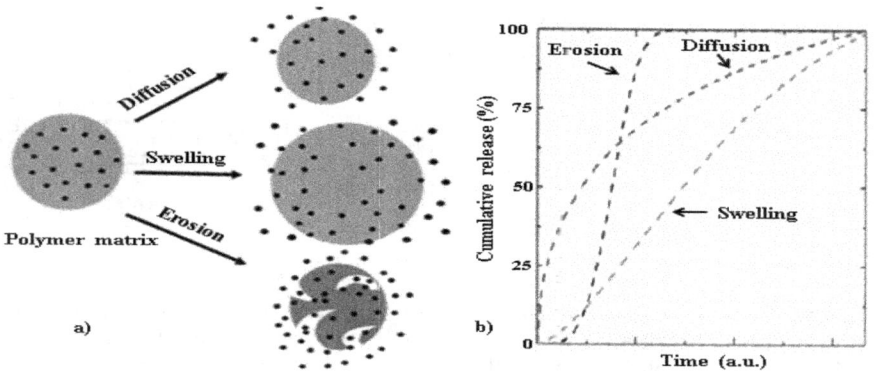

**Fig. 5.19** Possible drug release mechanisms: **a** chitosan nanoparticle and drug matrix and **b** drug release profile. Copyright © 2016 Bellich, B., D'Agostino, H., Semeraro, S., and Gamini, A., Creative Commons Attribution (CC BY) license, Mar Drugs. 2016 May; 14(5): 99, https://doi.org/10.3390/md14050099; Copyright © 2017 Mohammed, M.A., Syeda, J.T.M., Wasan, K.M., Wasan, E.K. Creative Commons Attribution (CC BY) license, Pharmaceutics 2017, 9(4), 53; https://doi.org/10.3390/pharmaceutics9040053. Reproduced from Refs. [136, 154]

accumulate in tumors through enhanced permeability and retention (EPR) effect. It is reported that small particles may facilitate the cellular entry process; there seems to be no size limit up to 5 μm to gain cellular internalization [154]. The adhesive nature of chitosan nanoparticles also causes agglomeration if the nanoparticles are stored in an aqueous medium [127]. Long-term storage of the chitosan nanoparticles in an aqueous medium may cause physical instability, which would restrict their use. Ultrasonication is a more efficient method for obtaining finer particles from nanoparticle clusters. However, there are limited studies regarding the effect of ultrasonication on the particle size and size distribution of chitosan nanoparticles [127].

## 5.7 Chitosan Nanofibers

Nanofibers, in general, have a diameter of less than 100 nm and the diameter to length (aspect) ratio of more than 100 [155]. The nanofibers possess a unique combination of high surface-to-volume ratio, optical properties, and high mechanical strength than the traditional fibers [156, 157]. These properties of nanofibers have increased their interest in the areas of separation technology, textile, and biomedical research. Chitin occurs in nature as a fine crystalline fibrillary structure [158]. This fibrillary structure of chitin is called microfibrils. Microfibrils have alternating crystalline and amorphous domains [159]. Fabritus et al. [160] reported that the chitin nanofibrils consist of 18–25 chitin molecules, which are usually embedded in the protein matrix [160, 161]. Figure 5.20 shows the schematic of the chitin-protein fiber matrix.

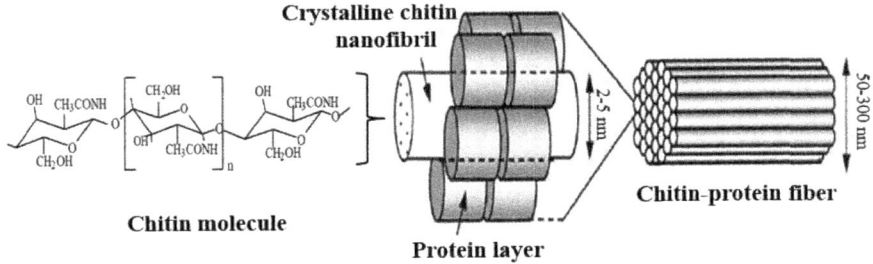

**Fig. 5.20** Schematic illustration of the exoskeleton structure of crab shells. Copyright © 2014 Ifuku, S. Licensee MDPI, Basel, Switzerland. Creative Commons Attribution license, Molecules 2014, 19, 18, 367–18, 380; https://doi.org/10.3390/molecules191118367, Ref. [157]; Copyright © 2009 WILEY-VCH Verlag GmbH & Co. KGaA, Adv. Mater. 21, 2009, 391–400. Reproduced from Ref. [160] with permission

The polymeric linear chain structure of chitin increases its potential for fiber formation abilities similar to those of cellulose [162]. The microfibrils in chitin have lateral dimensions ranging from 2.5 to 25 nm, depending on their biological origin [163]. Chitin and chitosan nanofibers are prepared using several methods. These methods include acid hydrolysis, ultrasonication under acidic conditions, tempo-mediated oxidation, self-assembly, electrospinning, and grinding [155, 164]. Self-assembly is a standard route for the synthesis of many bio-inspired nanofibers; however, water insolubility of chitin and time-consuming processing of continuous nanofibers impedes the use of this approach [164, 165]. In 2,2,6,6-tetramethylpiperidine-1-oxyl (tempo) mediated oxidation, chitin can be transformed into negatively charged tempo-oxidized chitin nanofibers (TOChN), where the oxidation of the $C_6$–OH moieties to $C_6$–COO– on the surface of the crystalline occurs when the chitin was suspended in water containing 2,2,6,6-tetramethylpiperidine-1-oxyl (tempo), sodium bromide, and sodium hypochlorite [156]. This process tends to convert highly crystalline residue, which is insoluble in water, into a stable colloidal suspension by subsequent intense mechanical shearing action [166]. Ifuku and Saimoto [167] prepared chitin nanofibers from the exoskeletons of crabs and prawns by a simple mechanical treatment after removing proteins and minerals by a series of extraction treatments [167]. Figure 5.21 shows the scanning electron micrograph of the woven network structure of the chitin nanofiber from crab shells.

Whiskers or crystalline nanofibrils are the purest crystal form of chitin. The amorphous domains of chitin can be hydrolyzed by acid to release the crystalline nanofibrils. These nanofibrils are known as chitin nanowhiskers [168]. Zhang et al. [156] reported that chitin nanofibers or nanowhiskers could be prepared by acid hydrolysis followed by mechanical disintegration of the hydrolyzed solid in water. The hydrolyzed solids with low aspect ratios are different from the natural chitin fibers in crab shells [155]. Mincea et al. [159] described methods of extraction and preparation of chitin nanowhiskers. The chemical modification and applications of chitin nanowhiskers were also discussed. They reported that partially acetylated

## 5.7 Chitosan Nanofibers

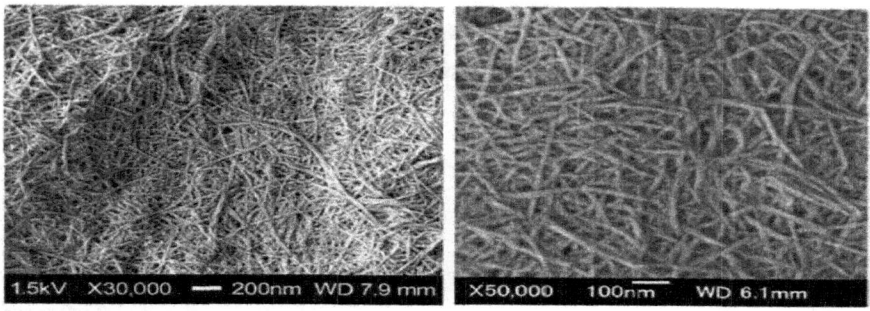

**Fig. 5.21** Scanning electron micrograph of chitin nanofibers from crab shells after grinding treatment. Copyright © 2014 Ifuku, S. Creative Commons Attribution (CC BY) license, Molecules. 2014 Nov; 19(11): 18, 367–18, 380, https://doi.org/10.3390/molecules191118367. Reproduced from the Ref. [157]

whiskers are formed through chain cleavage at random locations along the microfibrils. In a review paper, Muzzarelli [169] discussed chitin nanofibers isolated from the cell walls of five types of mushrooms. The nanofibers were prepared by removing glucans, minerals, and proteins, followed by a simple grinding treatment under acidic conditions. After grinding treatment, the chitin slurry is usually formed as a gel. All these methods for the preparation of chitin nanofibers have their limitations [164]. For instance, harsh conditions of processing (highly acidic or basic) result in depolymerization and deacetylation of chitin.

Polymeric nanofiber using the electrospinning process has proven to be a relatively simple and versatile method for forming non-woven fibrous mats. The electrospinning process is a well-known fabrication technique that can create artificial submicron-sized nanofibers from a wide range of polymers [170]. The polymeric nanofibers prepared by electrospinning have unique properties and, hence, created significant interest in them. High surface-area-to-volume and aspect ratios, increased flexibility in surface functionalities, improved mechanical performances, and smaller pores than fibers produced using traditional methods are a few of the properties [171]. These properties make the non-woven mats composed of electrospun fibers excellent candidates for various applications, e.g., as wound dressings and drug delivery systems [172]. The main advantages of this process include (1) low cost, (2) uniform nanofibers and inexpensive purification, and (3) continuous nanofibers preparation technology [161].

A polymer solution can form continuous jets and ultra-fine fibers rather than obtaining droplets or beads during electrospinning [173]. The electrospun setup mainly consists of a voltage power supply, a capillary tube with a needle, and a metal collector. The process utilizes a strong electrostatic field to obtain ultra-fine fibers from a polymer solution that accelerated toward the grounded collector [174]. The polymer nanofibers are formed between the two electrodes that carry electrical charges of opposite polarity. One electrode is placed into the solution and the other onto a grounded collector to create an electrostatic field [165]. Therefore, the process

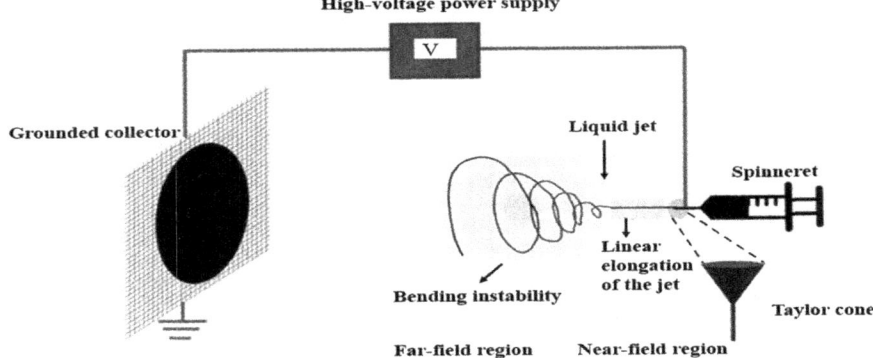

**Fig. 5.22** Schematic of typical electrospinning setup. Copyright © 2021 Xue, C and Wilson, L.D. Creative Commons Attribution (CC BY) license, J. Compos. Sci. 2021, 5, 160. https://doi.org/10.3390/jcs5060160. Reproduced from the Ref. [176]

involves interactions between fluid dynamics, electrically charged surfaces, and electrically charged liquids to prepare nanofiber [157]. Figure 5.22 shows the schematic of a horizontal setup of the electrospinning apparatus. The electrospinning process for polymer nanofibers preparation starts with applying a high voltage to the polymer solution [175]. At a critical electric field, the repulsive electrical forces can overcome the surface tension of the polymer solution. The electrically unstable charged jet of polymer is then ejected through a needle toward a grounded collector with the opposite charge. As the polymer jet travels to the grounded collector, the solvent evaporates, and the stable polymer nanofibers are deposited on the collector [174, 175].

It is reported that the potential voltage difference between the two electrodes depends on the properties of the spinning solution, such as polymer molecular weight and viscosity [165]. The morphology of polymer nanofibers and the ability to electrospun a solution are affected by polymer solution properties, process parameters, and environmental parameters [174, 177, 178]. Table 5.4 summarizes the parameters that affect the characteristics of the obtained nanofibrous structures. These parameters include the polymer type, the conformation of the polymer chain, viscosity (concentration) of the solution, conductivity, polarity, and surface tension of the solvent. Applied voltage, flow rate, the distance between tip and collector, and environmental conditions are adjusted as needed for the final desired product. These process parameters are considered critical for the successful preparation of nanofibers by electrospinning [172, 173, 179]. Electrospinning of biopolymers is difficult because biopolymers (polysaccharides and proteins) often have a distribution of molecular weights or have complex chemical structures [180]. It is reported that the capability to electrospun a polymer is dependent upon finding the optimal solvent and optimizing many other parameters [177]. Chitosan intrinsically has a much wider choice of solvents for electrospinning than chitin since chitosan is soluble in more solvents. Processing of chitin is always challenging as the strong bond renders the material

## 5.7 Chitosan Nanofibers

insoluble in water and non-reactive [164]. Chitin is reported to be soluble only in specific type of solvents such as *N*,*N*-dimethylacetamide (DMAC), lithium chloride (LiCl), hexafluoroacetone, 1,1,1,3,3,3-hexafluoro-2-propanol (HFIP) and saturated calcium-based solvent [177, 181, 182].

**Table 5.4** Effect of process parameters on the structure of nanofibers

| | | |
|---|---|---|
| Solution parameters | Viscosity | – Low viscosity generates beads<br>– High viscosity increases fiber diameter and disappearance of the bead<br>– If viscosity is high, drying of the solution at the tip of the needle occurs, hence preventing electrospinning |
| | Concentration | – Fiber diameter increases with the increase of polymer concentration |
| | The molecular weight of the polymer | – The number of beads decreases with increasing of molecular weight |
| | Conductivity | – Fiber diameter decreases with the increase in conductivity |
| | Surface tension | – Jet instability appears with high surface tension<br>– Beaded fibers are obtained by the collector due to jet instability |
| Processing parameters | Applied voltage | – Fiber diameter decreases with the increase in applied voltage |
| | Distance between tip and collector | – Minimum distance is required for uniform fibers<br>– Fiber diameter decreases with an increase in gap distance<br>– If the distance is too short, beaded fiber and the spun fibers tend to stick to the collecting device and each other. This is due to incomplete solvent evaporation |
| | Feed rate/flow rate of chitosan solution | – Fiber diameter increases with an increase in the flow rate<br>– If the flow rate is too high, beaded fibers are formed |
| Ambient parameters | Humidity | – High humidity results in circular pores on the fibers |
| | Temperatures | – Increase in temperature results in a decrease in fiber diameters |

Copyright © 2008 Taylor & Francis Group, LLC, Polymer Reviews, 48:317–352, 2008, Intechopen, http://dx.doi.org/10.5772/61300; MS Thesis UNC, Reproduced from the Refs. [174, 177, 178] with permission

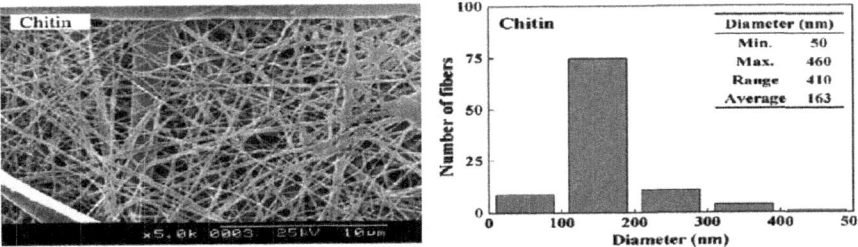

**Fig. 5.23** SEM micrograph of chitin nanofiber and their corresponding diameter distribution. Copyright © 2004 Elsevier Ltd., Polymer, 45, 21, 2004, 7137–7142; Copyright © 2012 Elsevier B.V., Materials Science and Engineering: C32, 2012, 1711–1726. Reproduced from the Refs. [183, 184] with permission

Chitin is soluble in 1,1,1,3,3,3-hexafluoro-2-propanol (HFIP) and can be fabricated into nanofibers using the electrospinning process. In a review article, Schiffman and Schauer [177] discussed the process parameters for the electrospinning of chitin and chitin-containing solutions, the solvent used, special polymer processing, a separation distance between needle and collector (cm), and the advancement speed of solution (mL/h). They discussed the preparation of electrospun practical-grade chitin nanofibers using HFIP as a solvent. They also discussed the characterization of the nanofibers using a field emission scanning electron microscope (FESEM). Min et al. [183] used an electrospinning process to prepare chitin nanofibers with HFIP as a spinning solvent. They reported that the morphology of as-spun and deacetylated chitin (chitosan) nanofibers had a broad fiber diameter distribution, and most of the fiber diameters were less than 100 nm. Figure 5.23 shows a typical example of an SEM micrograph of chitin nanofiber produced using HFIP as a spinning solvent.

Chitosan nanofiber preparation using electrospinning is a challenging task. Chitosan is soluble in most acids. The polycationic nature of chitosan affects the surface tension and rheological properties of the chitosan solution. In the electrospinning process, the repulsive forces between ionic groups within polymer structure arise due to a high electric field. These repulsive forces can restrict the formation of continuous fibers and often produce beads [185]. Chitosan nanofibers can be prepared by electrospinning of chitosan and trifluoroacetic acid (TFA) solution. The mixture of chitosan, TFA, and dichloromethane (DCM) in a solution can also be used for the electrospinning of chitosan nanofibers [174, 177, 186]. The amino groups of the chitosan form salts with TFA, and hence, TFA is very suitable as a spinning solvent [187]. TFA can effectively destroy the intermolecular interactions between the chitosan molecules and thus facilitate electrospinning. The addition of dichloromethane (DCM) to the chitosan-TFA solution improves the homogeneity of the electrospun chitosan fibers. The addition of DCM prevents the formation of interconnected fibrous networks [185]. The use of ionic liquid such as 1,1,1,3,3,3-hexafluoro-2-propanol (HFIP) is also used to prepare chitosan solution.

## 5.7 Chitosan Nanofibers

Optimization of the chitosan solution viscosity is one of the essential tasks to electrospun nanofibers successfully. Singh and Dutta [179] reported that the viscosity of the chitosan solution controls the polymer chain entanglement concentration. The chain entanglement concentration plays a crucial role in the electrospinning of polymeric solutions. In the electrospinning process, the higher or lower threshold of chain entanglement depends on the viscosity of the chitosan solution [178]. At the lower viscosity of the solution, the polymer chain does not entangle; therefore, fiber formation is not possible, and polymer beads are often created. Ohkawa et al. [187] studied the relationship between chitosan concentration and solution viscosity using TFA as a spinning solvent. They reported that homogeneous fiber networks were formed in the viscosity range of 800–1000 cP. They also observed a linear increase of fiber diameter as the concentration of chitosan in the solution decreased. Thus, fiber diameter and polymer concentration had an inverse relationship. Elsabee et al. [184] reviewed the various approaches for successful electrospinning of chitin, chitosan, derivatives, and blends with several other polymers. The authors reported that acetic acid concentration in water strongly influences the surface tension of the chitosan solution. A uniform nanofibrous mat of average fiber diameter of 130 nm without any bead formation can be obtained by electrospinning 7% chitosan solution in aqueous 90% acetic acid in a 4 kV/cm electric field [184]. It is reported that the mechanical properties of electrospun nanofiber scaffolds are not satisfactory due to the random fiber alignment, lack of cohesion between the fibers, and high porosity, which becomes a hurdle in expanding their potential applications in tissue engineering [181]. Sangsanoh and Supaphol [188] discussed the preparation of chitosan nanofibrous membranes using the electrospinning process with chitosan solutions in trifluoroacetic acid (TFA) with or without dichloromethane (DCM) addition. They observed that the electrospun chitosan nanofibrous membranes produce TFA-chitosan salt residues in neutral and weak basic aqueous solutions. They reported that the neutralization of the electrospun nanofiber membrane in $Na_2CO_3$ aqueous

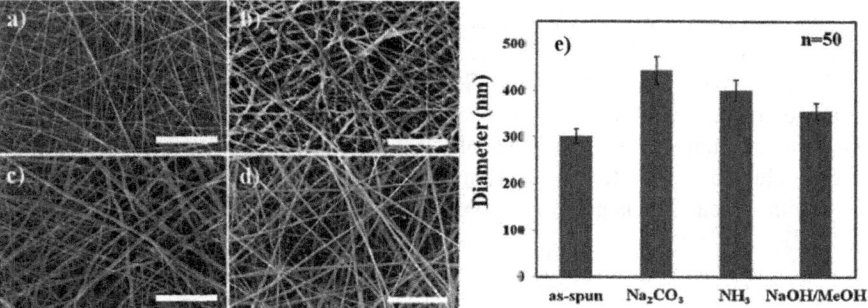

Fig. 5.24 FE-SEM micrographs of the chitosan nanofibers: a non-neutralized as-spun chitosan nanofibers mat and after neutralization with b 5 M $Na_2CO_3$, c 29% $NH_3$ and d 3 M NaOH in MeOH alkali aqueous solutions, e distribution of nanofiber diameters (scale bar: 10 μm). Copyright © 2013 Gu, B.K., Park, S.J., Kim, M.S., et al., Published by Elsevier Ltd, Carbohydrate Polymers, 97, 2013, 65–75. Reproduced from Ref. [189] with permission

solution maintains its fibrous structure even after continuous submersion in phosphate buffer solution (pH = 7.4) or distilled water for 12 weeks. Figure 5.24 shows the neutralization of electrospun chitosan nanofibers with different solutions.

Several authors reviewed the possibility of blending chitosan with other polymers to overcome mechanical and spinnability factors. These factors are significant in limiting the use of chitosan in electrospinning [162, 171, 174]. Blending of chitosan with other polymeric materials such as polyvinyl alcohol (PVA) [173], polyethylene oxide (PEO) [190], alginate [170], and polylactic-co-glycolic acid (PLGA) [191] facilitates its processing and makes electrospinning of chitosan possible. The blending material should have excellent fiber-forming characteristics to create entanglements and physical bonds with chitosan and act as a carrier in the electrospinning process [179]. It is reported that the chitosan to polymer ratio changed the hydrophilic/hydrophobic balance which influenced degradation behavior. The mechanical properties of the composite membrane change during this degradation [173]. For instance, Bhattarai et al. [190] used a chitosan/PEO ratio of 90/10, and they observed that the obtained nanofibers retained excellent integrity of the fibrous structure in water.

Zhang et al. [173] reported nanofiber preparation from a solution consisting of PVA and chitosan (90% DDA) in 2% (v/v) aqueous acetic acid using electrospinning process. The chitosan/PVA nanofiber with an average diameter of $99 \pm 21$ nm was prepared from a 7% chitosan/PVA solution in 40:60 mass ratio. Wang et al. [192] used weight ratio of 50:50 of chitosan and PVA in acetic acid solution to prepare optimal structure (most similar to natural tissue) of composite nanofiber as a potential matrix for tissue engineering as shown in Fig. 5.25. Nokhasteh et al. [192a] prepared nanofiber from acetic acid solution containing 70:30 chitosan/PVA (volume ratio) using electrospinning process at a voltage of 21 kV to the nozzle and distance of 15 cm (nozzle tip to collector). The average diameter of the nanofibers is 56.9 nm (Fig. 5.25d). In both cases, beadles and more efficient fiber was obtained.

Cui et al. [192c] investigated the strain–stress curves of crosslinked PVA/chitosan composite nanofibers with different glutaraldehyde (GA) concentration. It is reported that chitosan/PVA composite nanofibers shows the Young's modulus of 36.5 MPa and tensile strength of 2.2 MPa with an elongation of 30.8%. The Young's modulus and tensile strength of crosslinked PVA/CS composite nanofibers were found to increase with the increasing GA concentration. PCA/CS-0.5% GA composite nanofibers exhibit a maximum Young's modulus of 198.1 MPa and a tensile strength of 4.6 MPa. A list of chitosan-synthetic blend polymer nanofibers and their potential applications in the biomedical area is given in Table 5.5.

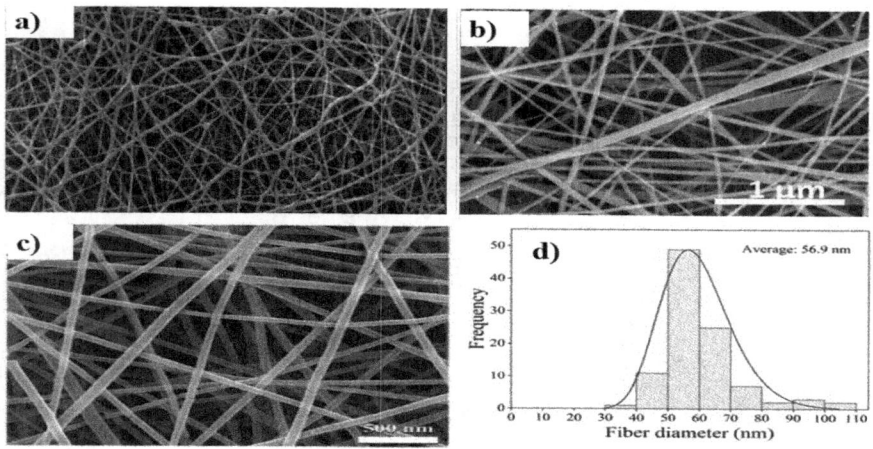

**Fig. 5.25** SEM image of **a** chitosan nanofiber, **b** chitosan/PVA (50:50) nanofibers ($V = 17$ kV, $D = 12$ cm), **c** chitosan/PVA (70:30) nanofibers ($V = 21$ kV, $D = 15$ cm), and **d** normal distribution curve of fiber diameter. **a** Copyright © 2006, American Chemical Society, Biomacromolecules, 7, 2006, 2710–2714. Reproduced from Ref. [188] with permission.; **b** Copyright © 2017Wang, Roy and Webster. Creative Commons Attribution License (CC BY). Front. Physiol. 7:683. https://doi.org/10.3389/fphys.2016.00683; Reproduced from Ref. [192a]; **c** and **d** Copyright © 2019 Nokhasteh, S., et al. Published by IOP Publishing Ltd. Commons Attribution License (CC BY); Mater. Res. Express 7 (2020) 015,401, https://doi.org/10.1088/2053-1591/ab572c. Reproduced from Ref. [192b]

## 5.8 Chitosan Coated on to Inorganic Materials for the Preparation of Adsorbent Bead

Chitosan is very effective in adsorbing metal ions because of its ability for complexation and high content of amino functional groups. In their natural form, these materials are soft and tend to agglomerate or form gels. The specific binding sites of the biosorbents in their natural form are not readily available for sorption. Hence, it is necessary to increase the accessibility of the binding sites for process applications. Several attempts have been made to modify chitosan to increase its capacity for metal ions by coating chitosan onto materials such as alumina, activated carbon, clay minerals, perlite, and Fuller's earth [194–196]. A typical process for coating chitosan onto perlite is described below.

Chitosan was coated on perlite, an inorganic porous aluminosilicate, and formed into beads. Perlite is the generic name of naturally occurring gray to brown, dense glassy volcanic rock with an amorphous structure. It consists essentially of fused sodium potassium aluminum silicate and 3–5% water. Perlite expands four to twenty times from its original volume when heated to 1600 °F (871 °C). This expansion is due to the water that is combined with various compounds in the crude rock. The expanded perlite is white. It can be manufactured to a density of 2 lb/ft$^3$ (32 kg/m$^3$), making it adaptable for numerous industrial uses. Perlite is classified as chemically

**Table 5.5** Electrospinning of nanofibers and their applications in a biomedical area

| Polymer | Molecular weight | Composition ratio | Solvent | Application |
|---|---|---|---|---|
| Chitin | 910 kD | – | HFIP | – |
| Chitin | 920 kD | – | HFIP | – |
| Chitosan | – | – | TFA/MC (70/30) | – |
| Chitosan/PEO (w/wo metal ions) | 50 kDa | 30/70 | Aq. AA | Protein resistance |
| Chitosan/PEO | 190 kDa/900 kDa | 90/10 | Aq. AA, DMSO, Triton X-100™ | Bone tissue engineering |
| Chitosan/PCL | | 80/20 | TFA/TFE | Nerve therapy |
| Chitosan/LA | 120–1600 HD | | TFA/MC (80/20) | Tissue engineering scaffold |
| Chitosan/PVA | | 10/90, 20/80, 25/75 | Aq. AA | – |
| Chitosan/PLA | | | TFA/TCM | Antimicrobial application |
| Chitosan/PET | | | TFA | Wound dressing |
| Chitosan/silk fibroin | 220 kDa | | FA | – |
| Chitosan/HAp/UHMWPEO | | 70/30 | HAc/DMSO (10:1) | Bone tissue engineering |
| Chitosan/collagen | 1000 kDa | 80/20, 50/50, 20/80 | HFIP/TFA | Wound dressing |
| Chitosan-g-PEG/PLGA | | | DMF/THF | Wound dressing |
| Q-chitosan/PVA | 400 kDa | | Aq. AA | – |

(continued)

**Table 5.5** (continued)

| Polymer | Molecular weight | Composition ratio | Solvent | Application |
|---|---|---|---|---|
| CECS/PVA | 120 kDa | 100/0,80/20,60/40, 50/50,30/70,20/80,10/90,0/100 | Water | Skin regeneration |
| CECS/PVA | 390 kDa | 75/25 | Aq. AcrA | Tissue scaffold |
| CMCS/PVA | 405 kDa | 50/50 | Water | Antimicrobial application |
| Chitosan/gelatin w/silver nanoparticles | 51 kDa | – | Aq. AA | Wound dressing |
| Chitosan-HoBt/PVA | 110 kDa | 10/90 to 90/10 | Distilled water | Drug delivery |

Copyright © 2012 Mahoney C, et al. Creative Commons Attribution License, J Nanomedic Biotherapeu Discover 2:102. https://doi.org/10.4172/2155-983X.1000102. Reproduced from the Ref. [193] with permission

*UHMWPEO* Ultra-high molecular weight poly(ethylene oxide); *PET* Poly(ethylene terephthalate); *HAp* Hydroxyapatite; *PLA* Polylactic acid; *PCL* Polycaprole acetone; *CECS* Carboxyethyl chitosan; *CMCS* Carboxymethyl chitosan; *PVA* Polyvinyl alcohol; Q-chitosan: Quaternized chitosan; *PVP* Poly(vinyl Pyrrolidone); *DMF* Dimethyl formamide; *THF*: Tetrahydrofuran; *TFA* Tetra fluoro acetic acid; *Aq AA* Aqueous acetic acid; *MC* Methylene chloride; *Aq AcrA* Aqueous acrylic acid solution; *TCM* Trichloromethane; *DMSO* Dimethyl sulfoxide; *HFIP* 1,1,1,3,3,3-Hexafluoro-2-propanol

**Table 5.6** Physical and chemical properties of perlite

| Physical properties | |
|---|---|
| Color | White |
| Free moisture (maximum) | 0.5% |
| pH | 6.5–8.0 |
| Bulk density | 2–25 lb/ft$^3$ |
| Solubility | Soluble in hot concentrated alkali and in HF<br>Moderately soluble (<10%) in 1 N NaOH<br>Slightly soluble (<3%) in mineral acids (1 N)<br>Very slightly soluble (<1%) in water or weak acid |
| *Chemical properties* | |
| Silicon | 33.8% |
| Aluminum | 7.2% |
| Potassium | 3.5% |
| Sodium | 3.4% |
| Iron | 0.6% |
| Calcium | 0.6% |
| Magnesium | 0.2% |
| Trace minerals | 0.2% |
| Bound water | 3.0% |

Unpublished work by authors

inert. Typical chemical and physical properties of perlite are given in Table 5.6. They are broadly applicable to all expanded perlite.

Perlite uses are classified into three general categories: construction applications, horticultural applications, and industrial applications. It is used as an anticake agent, filter aid, or pressing aid in processing foods and feed ingredients. Occasionally, it is also used, to a limited extent, as an adsorbent for methylene blue, trivalent chromium, and as solid support in chromatography. It was expected that more active sites of chitosan would be available due to the coating, thus, enhancing the adsorption capacity. Chitosan beads were prepared by adding dropwise an acidic solution of chitosan and perlites into a sodium hydroxide precipitation bath (Fig. 5.26). The mixture of oxalic acid, chitosan, and perlite was placed in the beaker and was stirred and heated. Once a homogeneous mixture was formed, the mixture was pumped through the nozzle into the precipitation bath. The steps followed in the preparation of the chitosan-coated perlite beads are discussed in detail below:

1. Preparation of the perlite substrate: The perlite was first crushed in a ball mill and sieved through a 35-mesh screen. About 100 g of screened perlite was then soaked with 1 L of 0.2 M oxalic acid solution and stirred for 4 h at room

## 5.8 Chitosan Coated on to Inorganic Materials for the Preparation ...

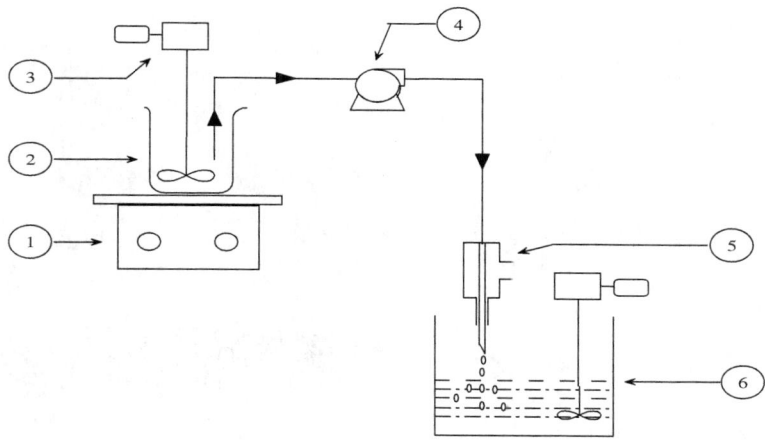

**Fig. 5.26** Apparatus for preparing chitosan beads. Unpublished work by authors. (1) Heater; (2) chitosan solution; (3) stirrer; (4) peristaltic pump; (5) nozzle; (6) NaOH precipitation bath

temperature. The acid from the mixture was filtered with a Whatman 41 filter. The filtered perlite was washed with deionized water until the filtrate pH was between 6 and 6.9. After washing, the perlite was dried overnight at 70 °C and again sieved through a 35-mesh sieve. The perlite powder was stored in a desiccator for later use.

2. Preparation of chitosan gel: Oxalic acid was used to dissolve chitosan. About 30 g of medium molecular weight chitosan was slowly added to 1 L of 0.2 M oxalic acid solution under continuous stirring. The solution was heated to 40–50 °C to facilitate acylation. Finally, a viscous gel, whitish in color, was formed.
3. Mixing of perlite with chitosan: About 60 g of treated perlite powder was mixed with 300 mL deionized water, slowly added to the gel, and stirred for 4 h at 40–50 °C to obtain a homogeneous mixture.
4. Bead preparation: The spherical beads of chitosan, coated on perlite, were prepared by the dropwise addition of perlite gel mixture into a 0.7 M NaOH precipitation bath. Maintaining this concentration of NaOH was critical in forming the beads and in subsequent washing of beads. The objective of adding the acidic perlite-chitosan mixture to the NaOH solution was to rapidly neutralize oxalic acid so that the spherical shape could be obtained. If the concentration is lower than 0.7 M, beads tend to disintegrate and cannot retain the spherical shape. Although it did not change the spherical shape, a higher concentration of NaOH took a long time to wash off the residual NaOH from the beads. The higher concentration of NaOH did not change the spherical shape. The viscosity of the acidic chitosan perlite mixture was found to be critical in forming the beads.
5. Drying: The beads were separated from the NaOH bath and washed several times with deionized water to a neutral pH. Three methods—freeze, oven, and

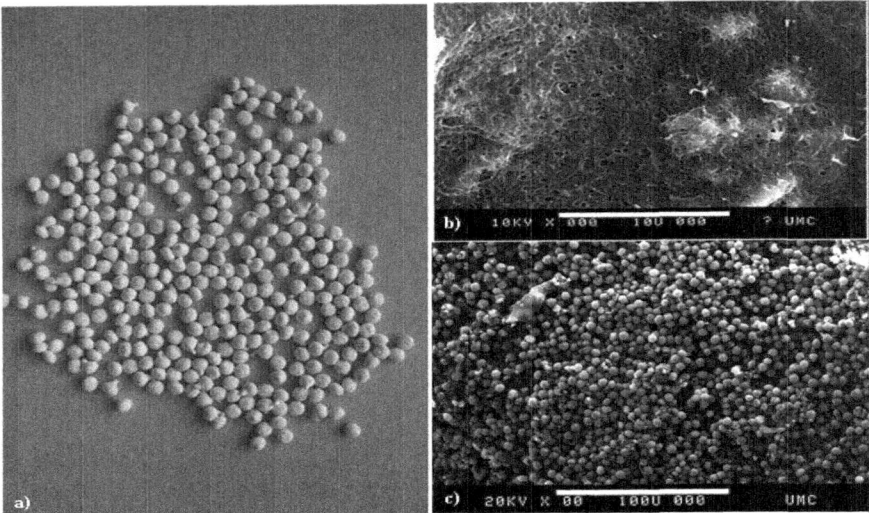

**Fig. 5.27** **a** Chitosan-coated perlite bead, **b** scanning electron microscopy (SEM) micrograph for the surface morphology of chitosan perlite bead, and **c** SEM micrograph of chitosan-coated zeolite bead. Unpublished work by authors

air drying—were used for drying the beads. The freeze-drying method appears to maintain the spherical shape better than the other two methods. The bulk density of freeze-dried beads was the highest. In the oven-drying method, beads are also dried in a vacuum oven at a temperature of 40–500 °C and at a vacuum of about 29″ mercury. In the air-drying method, beads were dried in room air at a temperature of about 25 °C. The beads were spread in a thin layer in the room and dried with air blowing from a fan.

Figure 5.27 shows chitosan-coated perlite beads and scanning electron microscopy (SEM) micrograph of the surface of chitosan-coated perlite beads and also chitosan-coated zeolite beads. As can be seen from the SEM micrographs, the surface morphology of chitosan-coated perlite bead and chitosan-coated zeolite bead retain the spherical shape. The porous nature of the outer surface is evident from this micrograph. Hasan et al. [22] conducted an X-ray photoelectron spectroscopic analysis of the chitosan-coated perlite bead. The analysis was to investigate the chemical interaction of the components in both chitosan and perlite during the crosslinking process. They observed that oxygen-containing functional groups of chitosan might not actively participate during the crosslinking of the chitosan perlite bead formation process. The amino functional groups of chitosan may have formed complexes through crosslinking with the constituents of perlite, such as alumina and silica, from the coating process. The possible reaction mechanism for chitosan perlite bead formation is given in Fig. 5.28.

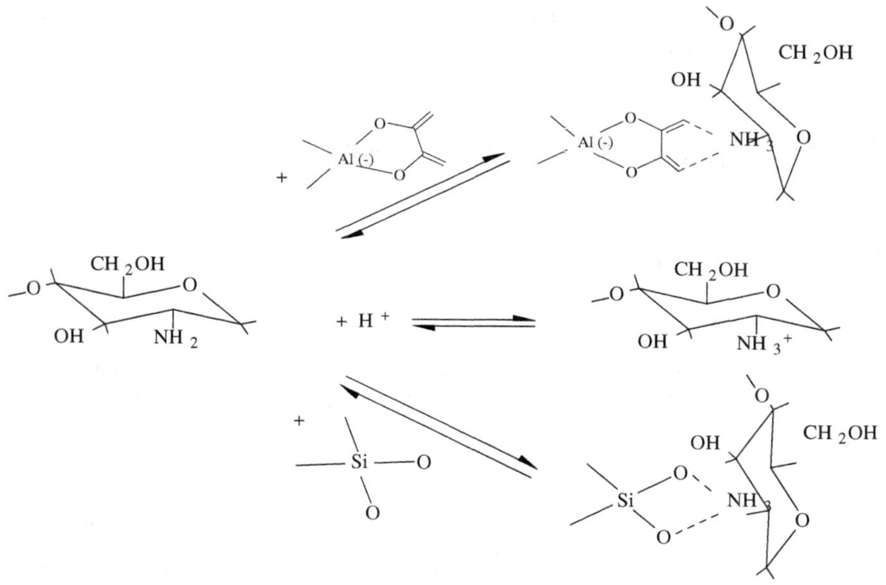

**Fig. 5.28** Possible reaction mechanisms for chitosan perlite bead formation. Copyright © 2006, American Chemical Society, Ind Eng Chem Res 45:5066. Reproduced from Ref. [22] with permission

## 5.9 Summary

Chitosan can be modified through crosslinking to improve its physical and chemical properties for various applications. Crosslinked chitosan becomes more resistant to shear, high temperature, and low pH compared to pure chitosan. Chitosan is insoluble in water but soluble in acidic solutions. This solubility property of chitosan limits its specific application in biomedical and pharmaceutical fields.

Chitosan contains abundant amino and hydroxyl groups. Chemical modification of these functional groups increases the solubility of chitosan in a wide pH range of aqueous solutions and organic solvents. Chitosan-based hydrogel, nanoparticles, and nanofibers are a few examples of crosslinked derivatives of chitosan. Chitosan-based hydrogels are crosslinked hydrophilic materials capable of retaining a large amount of water and biological fluids.

The formulation of chitosan-based nanoparticle through the physical and chemical crosslinking process is widely investigated. Chitosan nanoparticles have a high affinity for macromolecules. The physical and chemical properties of nanoparticles play an essential role in the efficacy of nanomedicine. Chitosan-based nanoparticles have the potential as a carrier for the therapeutic drug delivery system. The polymeric linear chain structure of chitin increases its potential for fiber formation abilities similar to those of cellulose. Because of the bioactivity, biocompatibility, and biodegradability properties, the application of chitosan-based nanofiber has been investigated in the biomedical area. The potential applications of chitosan-based hydrogel, nanoparticles, and nanofibers in biomedical, food, pharmaceutical, biological fields, and environmental fields are evolving. Numerous derivatives of chitin and chitosan have been developed, but most of them are still on the laboratory scale, and large-scale commercial use is yet to be determined.

# References

1. Walker KA, Markoski LJ, Deeter GA, Spilman GE, Martin DC, Moors JS (1994) Crosslinking chemistry for high performance polymer networks. Polymer 35(23):5012–5017
2. Berger J, Reist M, Mayer JM, Felt O, Peppas NA, Gurny R (2004) Structure and interactions in covalently and ionically crosslinked chitosan hydrogels for biomedical applications. Eur J Pharm Biopharm 57:19–34
3. Crini G (2005) Recent developments in polysaccharide-based materials used as adsorbents in wastewater treatment. Prog Polym Sci 30:38–70
4. Ahmed TA, Aljaeid BM (2016) Preparation and characterization, and potential application of chitosan, chitosan derivatives, and chitosan metal nanoparticles in pharmaceutical drug delivery. Drug Des Dev Therapy 10:483–507
5. Szymanska E, Winnicka K (2015) Stability of chitosan—a challenge for pharmaceutical and biomedical applications. Mar Drugs 13:1819–1846
6. Hamman JH (2010) Chitosan based polyelectrolyte complexes as potential carrier materials in drug delivery systems. Mar Drugs 8:1305–1322
7. Terao K (2014) Poly (acrylic acid) (PAA). In: Encyclopedia of polymeric nanomaterials. Springer Berlin Heidelberg. https://doi.org/10.1007/987-3-642-36199-9
8. Berger J, Reist M, Mayer JM, Felt O, Gurny R (2004) Structure and interactions in chitosan hydrogels formed by complexation or aggregation for biomedical applications. Eur J Pharm Biopharm 57:35–52
9. Varshosaz J, Alinagari R (2005) Effect of citric acid as cross-linking agent on insulin loaded chitosan microspheres. Iran Polym J 14(7):647–656
10. Sionkowska, A, Kaczmarek B, Lewandowska K (2014) Characterization of chitosan after cross-linking by tannic acid. Progr Chem Appl Chitin Its Derivatives XIX:135
11. Mane S, Ponrathnam S, Chavan N (2015) Effect of chemical cross-linking on properties of polymer microbeads: a review. Can Chem Trans 3(4):473–485
12. Kapoor M (1996) How to cross-linked proteins, pp 1–6. www.fgsc.net/neurosporaprotocols/How%20to%20cross-link%20proteins.pdf
13. Zhu A, Zhang M, Wu J, Shen J (2002) Covalent immobilization of chitosan/heparin complex with a photosensitive hetero-bifunctional crosslinking reagent on PLA surface. Biomaterials 23:4657–4665
14. Kitagawa T, Shimozono T, Aikawa T, Yoshida T, Nishimura H (1981) Preparation and characterization of hetero-bifunctional cross-linking reagents for protein modification. Chem Pharm Bull 29(4):1130–1135
15. Gonçalves VL, Laranjeira MCM, Fávere VT (2005) Effect of crosslinking agents on chitosan microspheres in controlled release of diclofenac sodium. Polímeros: Ciência e Tecnologia 15(1):5–12 (2005)
16. Farris S, Song J, Huang Q (2010) Alternative reaction mechanism for the cross-linking of gelatin with glutaraldehyde. J Agric Food Chem 58:998–1003
17. Migneault I, Dartiguenave C, Bertrand MJ, Waldron KC (2004) Glutaraldehyde: behavior in aqueous solution, reaction with proteins, and application to enzyme crosslinking. Biotechniques 37:790–802

# References

18. Rithe SS, Kadam PG, Mhaske ST (2014) Preparation and analysis of novel hydrogels prepared from the blend of guar gum and chitosan: cross-linked with glutaraldehyde. Adv Mater Sci Eng Int J (MSEJ) 1(2)
19. Kil'deeva NR, Perminov PA, Vladimirov LV, Novikov VV, Mikhailov SN (2009) On the mechanism of the reaction of glutaraldehyde with chitosan. Russ J Bioorganic Chem 35(3):360–369
20. Hasan S (2014) Preparation of chitosan based microporous composite material and its applications. US Patents, US 8,911695 B2, Dec. 16
21. Silva RM, Silva GA, Coutinho OP, Mano JF, Reis RL (2004) Preparation and characterization in simulated body conditions of glutaraldehyde crosslinked chitosan membranes. J Mater Sci—Mater Med 15:1105–1112
22. Hasan S, Krishnaiah A, Ghosh TK, Viswanath DS, Boddu VM, Smith ED (2006) Adsorption of divalent cadmium (Cd (II)) from aqueous solutions onto chitosan-coated perlite beads. Ind Eng Chem Res 45:5066–5077
23. Anita A, Rejinold NS, Bumgardner JD, Nair SV, Jayakumar R (2012) Approaches for functional modification or cross-linking of chitosan. In: Sarmento B, Neves J (eds) Chitosan-based systems for biopharmaceuticals: delivery, targeting and polymer therapeutics, 1st edn. Wiley
24. Fang Y, Hu D-D (1999) Cross-linking of chitosan with glutaraldehyde in the presence of citric acid—a new gelling systems. Chin J Polym Sci 17(6):551–556
25. Wan Ngah WS, Endud CS, Mayanar R (2002) Removal of copper (II) ions from aqueous solution onto chitosan and cross-linked chitosan beads. React Funct Polym 50:181–190
26. Hudson SM, Wei YC, Hudson SM, Mayer JM, Kaplan DL (1992) Crosslinking with epichlorohydrin maintains the cationic amino function and improves the mechanical properties of the material. J Polym Sci 30(1):2187–2193
27. Torres MA, Vieira RS, Beppu MM (2007) Production of chemically modified chitosan microsphere by a spraying and coagulation method. Mater Res 10(4):347–352
28. IAEA-TECDOC 1420 (2004) Advances in radiation chemistry of polymers. In: Proceedings of a technical meeting held in Notre Dame, Indiana, USA, 13–17 Sept 2003. International Atomic Energy Agency
29. Sabharwal S, Varshney L, Chaudhary AD, Ramani SP (2004) Radiation processing of natural polymers: achievements & trends. In: Radiation processing of polysaccharides. IAEA, pp 29–37
30. Kume T, Takehisa M (1982) Effect of gamma-irradiation on sodium alginate and carrageenan powder. Agric Biol Chem 47:889–890
31. Ulanski P, Rosiak JM (1992) Preliminary studies on radiation induced changes in chitosan. Radiat Phys Chem 39(1):53–57
32. Mishra S, Bajpai R, Katare R, Bajpai AK (2007) Radiation induced crosslinking effect on semi-interpenetrating polymer networks of poly (vinyl alcohol). eXPRESS Polym Lett 1(7):407–415
33. IAEA-TECDOC-1422 (2004) Radiation processing of polysaccharides. International Atomic Energy Agency
34. Pengfei L, Maolin Z, Jilan W (2001) Study on radiation-induced grafting of styrene onto chitin and chitosan. Radiat Phys Chem 61:149–153
35. Yoksan R, Akashi M, Miyata M, Chirachanchai S (2004) Optimal $\gamma$-ray dose and irradiation conditions for producing low-molecular weight chitosan that retains its chemical structure. Radiat Res 161:471–480
36. Shweta A, Sonia P (2013) Pharmaceutical relevance of crosslinked chitosan in microparticulate drug delivery. Int Res J Pharm 4(2):45–50
37. Muzzarelli RAA (1977) Chitin. Pergamon Press, New York
38. (a) Xue C, Wilson LD (2021) An overview of the design of chitosan-based fiber composite materials. J Compos Sci 5:160. https://doi.org/10.3390/jcs5060160. (b) Qin Y (1990) The production of fibers from chitosan. Ph.D. thesis, University of Leeds, UK. https://etheses.whiterose.ac.uk/11299/1/531880.pdf

39. Riva R, Ragelle H, Rieux A, Duhem N, Jérôme C, Préat V (2011) Chitosan and chitosan derivatives in drug delivery and tissue engineering. Adv Polym Sci 244:19–44
40. Brasselet C, Pierre G, Dubessay P et al (2019) Modification of chitosan for the generation of functional derivatives. Appl Sci MDPI 9(7):33. https://doi.org/10.3390/app9071321
41. Zhao D, Yu S, Sun B et al (2018) Biomedical applications of chitosan and its derivative nanoparticles. Polymers 10:462. https://doi.org/10.3390/polym10040462
42. Tien CL, Lacroix M, Ispas-Szabo P, Mateescu M-A (2003) N-acylated chitosan: hydrophobic matrices for controlled drug release. J Control Release 93:1–13
43. Hafdani FN, Sadeghini N (2011) A review on application of chitosan as a natural antimicrobial. Int J Pharmacol Pharmaceut Sci 5(2):46–50
44. Kubota N, Eguchi Y (1997) Facile preparation of water-soluble N-acylated chitosan and molecular weight dependence of its water solubility. Polym J 29:12–127
45. Sashiwa H, Shigemasa Y, Roy R (2000) Novel N-alkylation of chitosan via Michael Type reaction. Chem Lett, 862–863
46. Chen JL (2012) Investigation of film forming properties of β-chitosan from jumbo squid pens (*Dosidius gigas*) and improvement of water solubility of β-chitosan. M.Sc. thesis, Oregon State University, 27 April 2012
47. Qin C, Du Y, Xiao L, Li Z, Gao X (2002) Enzymatic preparation of water-soluble chitosan and their antitumor activities. Int J Biol Macromol 31:111–117
48. Wang W, Meng Q, Li Q et al (2020) Chitosan derivatives and their application in biomedicine. Int J Mol Sci 21:487. https://doi.org/10.3390/ijms21020487
49. Buschmann MD, Merzouki A, Lavertu M, Thibault M, Jean M, Darras V (2013) Chitosans for delivery of nucleic acids. Adv Drug Deliv Rev. https://doi.org/10.1016/j.addr.2013.07.005
50. Bobu E, Nicu R, Lupei M, Ciolacu Fl, Desbrieres J (2011) Synthesis and characterization of N-alkyl chitosan for paper making applications. Cellulose Chem Technol 45(9–10):619–625
51. Jayakumara R, Selvamurugana N, Tokurab S, Tamurab H (2008) Preparative methods of phosphorylated chitin and chitosan—an overview. Int J Biol Macromol 43:221–225
52. Aranaz I, Harris R, Heras A (2010) Chitosan amphiphilic derivatives. Chemistry and applications. Curr Organ Chem 14:308–330
53. Chen M-C, Mi F-L, Liao Z-X, Sung H-W (2011) Chitosan: its applications in drug eluting devices. Adv Polym Sci 243:185–230
54. Strnad S, Sauperl O, Fras-Zemljic L (2010) Cellulose fibers functionalized by chitosan: characterization and applications. In: Elnashar M (ed) Biopolymers. InTech. ISBN 978-953-307-109-1. Available from: http://www.intechopen.com/books/biopolymers/cellulose-fibres-functionalised-by-chitosancharacterization-and-application
55. Ying G-Q, Xiong W-Y, Wang H, Sun Y, Liu H-Z (2011) Preparation, water solubility and antioxidant activity of branched-chain chitosan derivatives. Carbohyd Polym 83:1787–1796
56. Nikmawahda HT, Sugita P, Arifin B (2015) Synthesis and characterization of N-alkyl chitosan as well as its potency as a paper coating material. Adv Appl Sci Res 6(2):141–149
57. Thatte MR (2004) Synthesis and antibacterial assessment of water-soluble hydrophobic chitosan derivatives bearing quaternary ammonium functionality. Ph.D. dissertation, Louisiana State University, Agricultural & Mechanical College, December 2004
58. Champagne LM (2008) The synthesis of water-soluble N-acyl chitosan derivatives for characterization as antibacterial agents. Ph.D. dissertation, Louisiana State University, May 2008
59. Dutta PK, Dutta J, Tripathi VS (2004) Chitin and chitosan: chemistry, properties 688 and applications. J Sci Ind Res 63:20–31
60. Rajshree R, Rahate KP (2013) An overview on various modifications of chitosan and its applications. IJPSR 4(11):4175–4193
61. Mourya VK, Inamdar NN (2008) Chitosan-modifications and applications: opportunities galore. React Funct Polym 68:1013–1051
62. Philloppova OE, Korchagina EV (2012) Chitosan and its hydrophobic derivatives: preparation and aggregation in dilute aqueous solutions. Polym Sci, Ser A 54(7):552–572

63. Kyzas GZ, Bikiaris DN (2015) Recent modification of chitosan for adsorption applications: a critical and systematic review. Mar Drugs 13:312–337
64. Xu C, Pan H, Jiang H et al (2008) Biocompatibility evaluation of N, O-hexanoyl chitosan as a biodegradable hydrophobic polycation for controlled drug release. J Mater Sci Mater Med 19:2525–2532
65. Kim S-K, Nghiep ND, Rajapakse N (2006) Therapeutic prospective of chitin, chitosan and their derivatives. J. Chitin Chitosan 11(1):1–10
66. Vonchang P, Sajomsang W, Kasinrerk W, Damrus S, Kongtawalert P (2003) Anticoagulant activities of the chitosan polysulfate synthesized from marine crab shell by semi-heterogeneous conditions. Sci Asia 29:115–120
67. Pires NR, Cunha PLR, Maciel JS et al (2013) Sulfated chitosan as tear substitute with no antimicrobial activity. Carbohyd Polym 91:92–99
68. Tachaboonyakiat W, Netswasdi N, Srakaew V, Opaprakasit M (2010) Elimination of inter- and intramolecular crosslinks of phosphorylated chitosan by sodium salt formation. Polym J 42:148–156
69. Amaral IF, Granja PL, Barbosa MA (2005) Chemical modification of chitosan by phosphorylation: an XPS, FT-IR and SEM study. J Biomater Sci Polymer Edn 16(12):1575–1593
70. Li B, Huang L, Wang X, Mab J, Xie F (2011) Biodegradation and compressive strength of phosphorylated chitosan/chitosan/hydroxyapatite bio-composites. Mater Des 32:4543–4547
71. Wang K, Liu Q (2014) Chemical structure analyses of phosphorylated chitosan. Carbohydrate Res 386:48–56
72. Martins AF, Facchi SP, Follmann HDM, Pereira AGB, Rubira AF, Muniz EC (2014) Antimicrobial activity of chitosan derivatives containing N-quaternized moieties in its backbone: a review. Int J Mol Sci 15:20800–20832. https://doi.org/10.3390/ijms151120800
73. Gruškienė R, Deveikytė R, Makuška R (2013) Quaternization of chitosan and partial destruction of the quaternized derivatives making them suitable for electrospinning. Chemija 24(4):325–334
74. Goya RC, Moraisb STB, Assis OBG (2016) Evaluation of the antimicrobial activity of chitosan and its quaternized derivative on E. Coli and S Aureus growth. Revista Brasileira de Farmacognosia 26:122–127
75. Britto D, Goya RC, Filho SPC, Assis OBG (2011) Quaternary salts of chitosan: antimicrobial features, and prospects. Int J Carbohydrate Chem 2011, Article ID 312539, 12 p
76. Gao TT, Kong M, Cheng X-J et al (2014) A thermosensitive chitosan-based hydrogel for controlled release of insulin. Front Mater Sci 8(2):142–149
77. Gun'ko VM, Savina IN, Mikhalovsky SV (2017) Properties of water bound in hydrogels. Gels 3:37 (2017). https://doi.org/10.3390/gels3040037. www.mdpi.com/journal/gels
78. Kalhapure A, Kumar R, Singh VP, Pandey DS (2016) Hydrogels: a boon for increasing agricultural productivity in water-stressed environment. Curr Sci 111(11):1773–1779
79. Ahmadi F, Oveisi Z, Samani SM, Amoozgar Z (2015) Chitosan based hydrogels: characteristics and pharmaceutical applications. Res Pharm Sci 10(1):1–16
80. Pasqui D, De Cagna M, Barbucci R (2012) Polymers 4:1517–1534. https://doi.org/10.3390/polym4031517
81. Salvador M, Martin JS, Panicker AB, Josheph A (2017) Preparation of chitosan hydrogels and its pharmaceutical drug delivery. World J Pharm Sci 6(4):703–714
82. Ullah F, Othman MBH, Javed F, Ahmad Z, Akil HM (2015) Classification, processing and application of hydrogels: a review. Mater Sci Eng, C 57:414–433
83. Calo E, Khutoryanskiy VV (2015) Biomedical applications of hydrogels: a review of patents and commercial products. Eur Polymer J 65:252–267
84. Gulrez SKH, Al-Assaf S, Phillips G (2011) Hydrogels: methods of preparation, characterization and applications. In: Carpi A (ed) Progress in molecular and environmental bioengineering—from analysis and modelling to technology applications. In Tech. ISBN 978-953-307-268-5. Available from: http://www.intechopen.com/books/progress in molecular and environmental bioengineering from analysis and modelling to technology applications/hydrogels methods of preparation characterization and applications

85. Moura MJ, Figueiredo MM, Gil MH (2007) Rheological study of genipin cross-linked chitosan hydrogels. Biomacromol 8:3823–3829
86. Buwalda SJ, Vermonden T, Hennink WE (2017) Hydrogels for therapeutic delivery: current developments and future directions. Biomacromol 18(2):316–330
87. Ebara M (2011) Carbohydrate derived hydrogels and microgels. In: Narain R (ed) Engineered carbohydrate-based materials for biomedical applications: polymers, surfaces, dendrimers. Wiley
88. Chai Q, Jiao Y, Yu X (2017) Hydrogels for biomedical applications: their characteristics and the mechanisms behind them. Gels 3:6. https://doi.org/10.3390/gels3010006www.mdpi.com/journal/gels
89. Vashist A, Ahmad S (2013) Hydrogels: smart materials for drug delivery. Orient J Chem 29(3):861–870
90. Garg S, Garg A (2016) Hydrogel: classification, properties, preparation and technical features. Asian J Biomater Res 2(6):163–170
91. Peppas NA, Mikos AG (1986). In: Peppas NA (ed) Hydrogels in medicine and pharmacy—fundamentals, vol I. CRC Press, Florida, pp 1–25
92. Wang J, Gao S, Tian J, Cui F, Shi W (2020) Recent developments and future challenges of hydrogels as draw solutes in forward osmosis process. Water 12:692. https://doi.org/10.3390/w12030692
93. Koetting MC, Peters JT, Steichen SD, Peppas NA (2015) Stimulus-responsive hydrogels: theory, modern advances, and applications. Mater Sci Eng R 93:1–49
94. Moura MJ, Faneca H, Lima MP et al (2011) The potential of a thermosensitive chitosan hydrogel cross-linked in situ with different loads of genipin. In: Fernandes PR et al (eds) ECCOMAS—International conference on tissue engineering, Lisbon, Portugal, 2–4 June 2011
95. Hu X, Gao C (2008) Photo initiating polymerization to prepare biocompatible chitosan hydrogels. J Appl Polym Sci 110:1059–1067
96. Sa-Lima H, Caridade SG, Mano JF, Reis RL (2010) Stimuli-responsive chitosan-starch injectable hydrogels combined with encapsulated adipose-derived stromal cells for articular cartilage regeneration. Soft Matter 6:5184–5195
97. Nie J, Wang Z, Hu Q (2016) Difference between chitosan hydrogels via alkaline and acidic solvent systems. Sci Rep 6:36053. https://doi.org/10.1038/srep36053
98. Ahmad EM (2015) Hydrogel: preparation, characterization, and applications: a review. J Adv Res 6:105–121
99. Jalani G, Rosenzweig DH, Makhoul G et al (2015) Tough, in-situ thermogelling, injectable hydrogels for biomedical applications. Macromol Biosci 15(4):473–480
100. Kim S, Tsao H, Kang Y et al (2011) In vitro evaluation of an injectable chitosan gel for sustained local delivery of BMP-2 for osteoblastic differentiation. J Biomed Mater Res Part B 99B:380–390
101. Yan J, Yang L, Wang G et al (2010) Biocompatibility evaluation of chitosan-based injectable hydrogels for the culturing mice mesenchymal stem cells in vitro. J Biomater Appl 24:625–637
102. Ganji F, Abdekhodaie MJ, Ramazani SAA (2007) Gelation time and degradation rate of chitosan based injectable hydrogel. J Sol-Gel Sci Technol 42:47–53
103. Han HD, Nam DE, Seo DH et al (2004) Preparation and biodegradation of thermosensitive chitosan hydrogel as a function of pH and temperature. Macromol Res 12(5):507–511
104. Moura MJ, Figueiredo MM, Gil MH (2008) Rheology of chitosan and genipin solutions. Mater Sci Forum 587–588:27–31
105. Kim GO, Kim N, Kim DY et al (2012) An electrostatically cross-linked chitosan hydrogel as a drug carrier. Molecules 17:13704–13711
106. Muzzarelli RAA, El Mehtedi M, Bottegoni C, Aquili A, Gigante A (2015) Genipin-crosslinked chitosan gels and scaffolds for tissue engineering and regeneration of cartilage and bone. Mar Drugs 13:7314–7338. https://doi.org/10.3390/md13127068
107. Dimida S, Demitri C, De Benedictis VM, Scalera F, Gervaso F, Sannino A (2015) Genipin-cross-linked chitosan-based hydrogels: reaction kinetics and structure-related characteristics. J Appl Polym Sci 132(28):42256–42263

108. Lai J-Y (2012) Biocompatibility of genipin and glutaraldehyde cross-linked chitosan materials in the anterior chamber of the eye. Int J Mol Sci 13:10970–10985
109. Verestiuc L, Ivanov C, Barbu E, Tsibouklis J (2004) Dual-stimuli-responsive hydrogels based on poly(N-isopropylacrylamide)/chitosan semi-interpenetrating networks. Int J Pharm 269(1):185–194
110. Alvarado AG, Ortega A, Perez-Carrillo LA et al (2017) Synthesis, characterization, and drug delivery from pH and thermoresponsive poly(N-isopropylacrylamide)/chitosan core/shell nanocomposites made by semicontinuous heterophase polymerization. J Nanomater 2017, Article ID 6796412, p 7. https://doi.org/10.1155/2017/67964112
111. Temtem M, Barroso T, Casimiro T et al (2012) Dual stimuli responsive poly(N-isopropylacrylamide) coated chitosan scaffolds for controlled release prepared from a non-residue technology. J Supercritical Fluids 66:398–404
112. Laszlo K, Kosik K, Geissler E (2004) High-sensitivity isothermal and scanning microcalorimetry in PNIPA hydrogels around the volume phase transition. Macromolecules 37:10067–10072
113. Ziminska M, Wilson JJ, McErlean E et al (2020) Synthesis and evaluation of a thermoresponsive degradable chitosan-grafted PNIPAAm hydrogel as a "smart" gene delivery system. Materials 13:2530. https://doi.org/10.3390/ma13112530. www.mdpi.com/journal/materials
114. Lanzalaco S, Armelin E (2017) Poly(N-isopropylacrylamide) and copolymers: a review on recent progresses in biomedical applications, Gels 3:36. https://doi.org/10.3390/gels3040036. www.mdpi.com/journal/gels
115. Bao H, Li L, Leong WC, Gan LH (2010) Thermo-responsive association of chitosan-graft poly(N-isopropylacrylamide) in aqueous solutions. J Phys Chem B 114:10666–10673
116. Yang Y, Tsai Y-T, Ba L, Deng J (2016) Study on the thermal stability of thermosensitive hydrogel. Procedia Eng 135:501–509
117. Xu X, Liu Y, Fu W et al (2020) Poly(N-isopropylacrylamide)-based thermoresponsive composite hydrogels for biomedical applications. Polymers 12:580. https://doi.org/10.3390/polym12030580
118. Alvarez-Lorenzo C, Concheiro A, Dubovik AS, Grinberg NV, Burova TV, Grinberg VY (2005) Temperature-sensitive chitosan-poly(N-isopropylacrylamide) interpreted networks with enhanced loading capacity and controlled release properties. J Control Release 102(3):629–641
119. Zhang W, Gaberman I, Ciszkowska M (2002) Diffusion and concentration of molecular probes in thermoresponsive poly(N-isopropylacrylamide) hydrogels: effect of the volume phase transition. Anal Chem 74:1343–1348
120. Zang X, Yan Y, Wang F, Chung T (2002) Thermosensitive poly (N-isopropylacrylamide-co-acrylic acid) hydrogels with expanded network structures and improved oscillating swelling-deswelling properties. J Am Chem Soc 18:2013–2018
121. Hasan S, Iasir ARM, Ghosh TK et al (2019) Characterization and adsorption behavior of strontium from aqueous solutions onto chitosan-fuller's earth beads. Healthcare 7(1):52. https://doi.org/10.3390/healthcare7010052
122. Hasan S, Ghosh TK (2013) Synthesis of silica-coated uranium oxide nanoparticles by surfactant-templated sol-gel process for use as catalysts. Nucl Technol 181:371–379
123. Zhao L-M, Shi L-E, Zhang Z-L et al (2011) Preparation and application of chitosan nanoparticles and nanofibers. Braz J Chem Eng 28(30):353–362
124. Jonassen H, Kjøniksen A-L, Hiorth M (2012) Effects of ionic strength on the size and compactness of chitosan nanoparticles. Colloid Polym Sci 290:919–929
125. Shard P, Bhatia A, Sharma D (2014) Optimization and effects of physico-chemical parameters on synthesis of chitosan nanoparticles by ionic gelation technique. Int J Drug Delivery 6:58–63
126. Biró E, Németh AS, Sisak C et al (2008) Preparation of chitosan particles suitable for enzyme immobilization. J Biochem Biophys Methods 70:1240–1246
127. Gokce Y, Cengiz B, Yildiz N et al (2014) Ultrasonication of chitosan nanoparticle suspension: influence on particle size. Colloids Surf A Physicochem Eng Aspects 462:75–81

128. Grenha A (2012) Chitosan nanoparticles: a survey of preparation methods. J Drug Target 20(4):291–300
129. Al-Remawi MMA (2012) Properties of chitosan nanoparticles formed using sulfate anions as crosslinking bridges. Am J Appl Sci 9(7):1091–1100
130. Wang JJ, Zeng ZW, Xiao RZ et al (2011) recent advances of chitosan nanoparticles as drug carriers. Int J Nanomed 6:765–774
131. Ahmed TA, Aljaeid BM (2016) Preparation, characterization, and potential application of chitosan, chitosan derivatives, and chitosan metal nanoparticles in pharmaceutical drug delivery. Drug Des Dev Therapy 10:483–507
132. Elgadir MA, Uddin MS, Ferdosh S et al (2015) Impact of chitosan composites and chitosan nanoparticle composites on various drug delivery systems: a review. J Food Drug Anal 23:619–629
133. Wang Y, Li P, Tran TT-D et al (2016) Manufacturing techniques and surface engineering of polymer based nanoparticles for targeted drug delivery to cancer. Nanomaterials 6:26. https://doi.org/10.3390/nano6020026. www.mdpi.com/journal/nano-materials
134. Najafi S, Pazhouhnia Z, Ahmadi O et al (2014) Chitosan nanoparticles and their applications in drug delivery: a review. Curr Res Drug Discovery 1(1):17–25
135. Sailaja AK, Mareshwar P, Chakravarty P (2011) Different techniques used for the preparation of nanoparticles using natural polymers and their application. Int J Pharm Sci 3(2):45–50
136. Bellich B, D'Agostino H, Semeraro S, Gamini A (2016) "The good and bad and ugly" of chitosans. Mar Drugs 14:99. https://doi.org/10.3390/md14050099. www.mdpi.com/journal/marinedrugs
137. Silva VJDD, Preparation and characterization of chitosan nanoparticles for gene delivery. Masters degree thesis, NABL (Nucleic Acids Bioengineering Laboratory) of BERG/CEBQ/IBB, at Instituto Superior Técnico-Alameda
138. Liu H, Gao C (2009) Preparation and properties of ionically cross-linked chitosan nanoparticles. Polym Adv Technol 20:613–619
139. López-León T, Carvalho ELS, Seijo B, Ortega-Vinuesa JL, Bastos-González D (2005) Physicochemical characterization of chitosan nanoparticles: electrokinetic and stability behavior. J Colloid Interface Sci 283:344–351
140. Divya L, Mittapally VK (2016) Preparation and applications of chitosan nanoparticles: a brief review. Res Rev J Mater Sci 04. https://doi.org/10.4172/2321-6212.1000r001
141. Kafshgari MH, Khorram M, Mansouri M, Osfouri ASS (2012) Preparation of alginate and chitosan nanoparticles using a new reverse micellar system. Iran Polym J 21(2):99–107
142. Nogueira DR, Tavano L, Mitjans M et al (2013) In vitro antitumor activity of methotrexate via pH-sensitive chitosan nanoparticles. Biomaterials 34(11):2758–2772
143. Barros APH, Morantes MTA, Hoyos MIC, Ospino LJM (2016) Preparation of chitosan nanoparticles modified with sodium alginate with potential for controlled drug release. Revista EIA 12(3):E75–E83. ISSN 1794-1237
144. Bodnar M, Hartmann JF, Borbelly J (2005) Preparation and characterization of chitosan-based nanoparticles. Biomacromol 6:2521–2527
145. Rampino A (2010) Polysaccharide-based nanoparticles for drug delivery. Doctoral Thesis in Nanotechnologies, Universitas Degli Studi Di Trieste, 161 p
146. Calvo P, Remuna C, Pez N-Lo, Vila-Jato JL, Alonso MJ (1997) Novel hydrophilic chitosan-polyethylene oxide nanoparticles as protein carriers. J Appl Polym Sci 63:1997, 15–132
147. Fan W, Yan W, Xu Z, Nia H (2012) Formation mechanism of monodisperse, low molecular weight chitosan nanoparticles by ionic gelation technique. Colloids Surf, B 90:21–27
148. Zhang H, Oh M, Allen C, Kumacheva E (2004) Monodisperse chitosan nanoparticles for mucosal drug delivery. Biomacromol 5:2461–2468
149. Vandana M, Sahoo SK (2009) Optimization of physicochemical parameters influencing the fabrication of protein-loaded chitosan nanoparticles. Nanomedicine 4(2):773–785
150. Koukaras EN, Papadimitriou SA, Bikiaris DN, Froudakis GE (2012) Insight on the formation of chitosan nanoparticles through ionotropic gelation with tripolyphosphate. Mol Pharmaceutics 9:2856–2862

151. Wang J, Byrne JD, Napier ME, DeSimone JM (2011) More effective nanomedicines through particle design. Small 7(4):1919–1131
152. Kamat V et al (2016) Chitosan nanoparticles synthesis caught in action using microdroplet reactions. Sci Rep 6:22260. https://doi.org/10.1038/srep22260
153. Rodrigues FC et al (2020) IOP Conf Ser: Mater Sci Eng 872:012109. https://doi.org/10.1088/1757-899X/872/1/012109
154. Mohammed MA, Syeda JTM, Wasan KM, Wasan EK (2017) An overview of chitosan nanoparticles and its application in non-parental drug delivery. Pharmaceutics 9:53. https://doi.org/10.3390/pharmaceutics9040053. www.mdpi.com/journal/pharmaceutics
155. Azuma K, Ifuku S, Osaki T et al (2014) Preparation and biomedical applications of chitin and chitosan nanofibers. J Biomed Nanotechnol 10:2891–2920
156. Zhang Y, Jiang J, Liu L et al (2015) Preparation, assessment, and comparison of α-chitin nano-fiber films with different surface charges. Nanoscale Res Lett 10:226
157. Ifuku S (2014) Chitin and chitosan nanofibers: preparation and chemical modifications. Molecules 19:18367–18380. https://doi.org/10.3390/molecules191118367
158. Araki J, Yamanaka Y, Ohkawa K (2012) Chitin-chitosan nanocomposite gels: reinforcement of chitosan hydrogels with rod-like chitin nanowhiskers. Polym J 44:713–717
159. Mincea M, Negrulescu A, Ostafe V (2012) Preparation, modification, and applications of chitin nanowhiskers: a review. Rev Adv Mater Sci 30:225–242
160. Fabritius H-O, Sachs C, Triguero PR, Raabe D (2009) Influence of structural principles on the mechanics of a biological fiber-based composite material with hierarchical organization: the exoskeleton of the lobster *Homarus americanus*. Adv Mater 21:391–400
161. Ding F, Deng H, Du Y et al (2014) Emerging chitin and chitosan nanofibrous materials for biomedical applications. Nanoscale 6:9477–9493
162. Pillai CKS, Paul W, Sharma P (2009) Chitin and chitosan polymers: chemistry, solubility and fiber formation. Prog Polym Sci 34:641–678
163. Dufresne A (2007) Polymer nanocomposites from biological sources. Encycl Nanosci Nanotechnol 21:59–83
164. Hassanzadeh P (2015) Self-assembled chitin nanofibers: properties and applications. Ph.D. dissertation, Material Science and Engineering, University of Washington
165. Huang Z-M, Zhang Y-Z, Kotaki M, Ramakrishna S (2003) A review on polymer nanofibers by electrospinning and their applications in nanocomposites. Compos Sci Technol 63:2223–2253
166. Nair KG, Dufresne A (2003) Crab shell chitin whisker reinforced natural rubber nanocomposites. 1. Processing and swelling behavior. Biomacromolecules 4(3):657–665
167. Ifuku S, Saimoto H (2012) Chitin nanofibers: preparations, modification, and applications. Nanoscale 4:3308–3318
168. Heath L, Zhu L, Thielemans W (2013) Chitin nanowhisker aerogels. Chemsuschem 6:537–544
169. Muzzarelli RAA, El Mehtedi M, Mattioli-Belmonte M (2014) Emerging biomedical applications of nano-chitins and nano-chitosans obtained via advanced eco-friendly technologies from marine resources. Mar Drugs 12:5468–5502. https://doi.org/10.3390/md12115468
170. Sill TJ, von Recum HA (2008) Electrospinning: applications in drug delivery and tissue engineering. Biomaterials 29:1989–2006
171. Pillai CKS, Sharma P (2009) Electrospinning of chitin and chitosan nanofibers. Trends Biomater Artif Organs 22(3):179–201
172. Ignatova M, Manolova N, Rashkov I (2007) Novel antibacterial fibers of quaternized chitosan and poly (vinyl pyrrolidone) prepared by electrospinning. Eur Polymer J 43:1112–1122
173. Zhang Y, Huang X, Duan B, Wu L, Li S, Yuan X (2007) Preparation of electrospun chitosan/poly (vinyl alcohol) membranes. Colloid Polym, Sci 285:855–863
174. Ibrahim HM, El-Zairy EMR (2015) Chitosan as a biomaterial- structure, properties, and electrospun nanofibers. In: Concepts, compounds and the alternatives of antibacterial. https://doi.org/10.5772/61300.http://www.intechopen.com/books/concepts-compounds-and-thealternatives-of-antibacterials
175. Jeong S, Krebs MD, Bonino CA et al (2010) Electrospun alginate nanofibers with controlled cell adhesion for tissue engineering. Macromol Biosci 10:934–943

176. Xue C, Wilson LD (2021) An overview of the design of chitosan-based fiber composite materials. J. Compos. Sci. 5:160. https://doi.org/10.3390/jcs5060160
177. Schiffman JD, Schauer CL (2008) A review: electrospinning of biopolymer nanofibers and their applications. Polym Rev 48:317–352
178. Nandgainkar AG (2011) Electrospinning of chitosan and its correlation with degree of deacetylation and rheological property. M.Sc. thesis, North Carolina State University
179. Singh BK, Dutta PK (2016) Chitin, chitosan, and silk fibroin electrospun nanofibrous scaffolds: a prospective approach for regenerative medicine. In: Chitin and chitosan for regenerative medicine. Springer series on polymer and composite materials, pp 151–189. https://doi.org/10.1007/978-81-322-2511-9_7
180. Zhao L-M, Shi L-E, Zhang Z-L et al (2011) Preparation and application of chitosan nanoparticles and nanofibers. Braz J Chem Eng 28(03):353–362
181. Jayakumar R, Nair SV, Furuike T, Tamura H (2010) Perspective of chitin and chitosan nanofibrous scaffolds in tissue engineering, Chap 10, pp 205–223. http://www.intechopen.com/books/tissue-engineering
182. Schiffman JD, Stulga LA, Schauer CL (2009) Chitin and chitosan: transformations due to the electrospinning process. Polym Eng Sci, 1918–1928. https://doi.org/10.1002/pen.21434
183. Min B-M, Lee SW, Lin JN et al (2004) Chitin and chitosan nanofibers: electrospinning of chitin and deacetylation of chitin nanofibers. Polymer 45(21):7137–7142
184. Elsabee MZ, Nagub HF, Morsi RE (2012) Chitosan based nanofibers, review. Mater Sci Eng, C 32:1711–1726
185. Sun K, Li ZH (2011) Preparations, properties and applications of chitosan-based nanofibers fabricated by electrospinning. eXPRESS Polym Lett 5(4):342–361
186. Lemma SM, Bossard F, Rinaudo M (2016) Preparation of pure and stable chitosan nanofibers by electrospinning in the presence of Poly (ethylene oxide). Int J Mol Sci 17:1790. https://doi.org/10.3390/ijms17111790
187. Ohkawa K, Minato K-I, Kumagai G et al (2006) Chitosan nanofibers. Biomacromolecules 7:3291–3294
188. Sangsanoh P, Supaphol P (2006) Stability improvement of electrospun chitosan nanofibrous membranes in neural or weak basic aqueous solutions. Biomacromolecules 7:2710–2714
189. Gu BK, Park SJ, Kim MS et al (2013) Fabrication of sonicated chitosan nanofiber mat with enlarged porosity for use as hemostatic materials. Carbohyd Polym 97:65–75
190. Bhattarai N, Edmondson D, Veiseh O, Matsen FA, Zhang M (2005) Electrospun chitosan-based nanofibers and their cellular compatibility. Biomaterials 26:6167–6184
191. Ajalloueian F, Tavanai H, Hilborn J et al (2014) Emulsion electrospinning as an approach to fabricate PLGA/Chitosan nanofibers for biomedical applications. BioMed Res Int 2014, Article ID 475280, 13 p. https://doi.org/10.1155/2014/475280
192. (a) Wang M, Roy AK, Webster TJ (2017) Development of chitosan/poly(vinyl alcohol) electrospun nanofibers for infection related wound healing. Front Physiol 7:683. https://doi.org/10.3389/fphys.2016.00683; (b) Nokhasteh S et al (2020) Preparation of PVA/chitosan samples by electrospinning and film casting methods and evaluating the effect of surface morphology on their antibacterial behavior. Mater Res Express 7:015401. https://doi.org/10.1088/2053-1591/ab572c; (c) Cui Z, Zheng Z, Lin L et al (2017) Electrospinning and crosslinking of polyvinyl alcohol/chitosan composite nanofiber for transdermal drug delivery. Adv Polym Technol 00:1–12. https://doi.org/10.1002/adv.21850
193. Mahoney C, Mccullough MB, Sankar J, Bhattarai N (2012) Nanofibrous structure of chitosan for biomedical applications. J Nanomed Biotherapeutic Discovery 2(1):2–9. https://doi.org/10.4172/2155-983X.1000102
194. Boddu VM, Abburi K, Randolph AJ, Smith ED (2008) Removal of copper (II) and nickel (II) ions from aqueous solutions by a composite chitosan biosorbent. Sep Sci Technol 43:1356–1381

195. Hasan S, Krishnaiah A, Ghosh TK, Viswanath DS, Boddu VM, Smith ED (2003) Adsorption of chromium (VI) on chitosan-coated perlite. Sep Sci Technol 38(15):3775–3793
196. Hasan S, Ghosh TK et al (2007) Preparation and evaluation of fullers earth beads for removal of cesium from waste streams. Sep Sci Technol 42(4):717–738

# Chapter 6
# Adsorption—Heavy Metals Removal

**Abstract** Chitosan is very effective in adsorbing metal ions from aqueous solutions. Chitosan is soft and tends to agglomerate or form gel in acidic solutions. Chitosan contains reactive hydroxyl (OH) and amine ($NH_2$) functional groups on its structure. Although the amino and hydroxyl groups in chitosan are mainly responsible for the adsorption of metal ions, these active binding sites are not readily available for sorption. Various methods are used to modify chitosan to avoid this problem. The adsorption theory, adsorption processes, and the adsorption mechanisms of chitosan and its derivative for metal ions from aqueous solutions are discussed in this chapter.

## 6.1 Introduction

Environmental pollution by the process wastewater from mining operations, power generation facilities, tanneries, and metal finishing industries has received attention due to their toxicological effect on ecosystems, agriculture, and human health. The wastewater from these sources typically contains significant toxic heavy metals such as copper, cadmium, nickel, lead, zinc, chromium, and other trace metals and organic chemical wastes. For instance, the metal finishing industry uses various physical, chemical, and electrochemical processes, which produce a significant amount of wastewater. Table 6.1 shows the potential source of process wastewater containing heavy metals generated from different industrial applications and the toxicity effects of heavy metals on the ecosystem and human health. Apart from this, the non-radioactive wastes produced from a nuclear fuel cycle can also be grouped as organic and inorganic wastes. Metals and chemicals are generally present in the wastewater as soluble ionic form. Several techniques, such as coagulation, flocculation, ion exchange, adsorption, electrodeposition, and membrane separation, are used to remove metal contaminants from the effluent wastewater. However, these methods have their inherent advantages and limitations.

Chitosan has been investigated to remove heavy metals from water and wastewater due to its non-toxicity, chelating ability with metals, and biodegradability. Chitosan contains a reactive hydroxyl (OH) and an amine ($NH_2$) group, available for characteristic coordination bonding with metallic ions. Several researchers studied chitosan

**Table 6.1** Sources of typical heavy metals in process wastewaters from industrial applications [1, 1a, 2, 2a].

| Heavy metal | Properties | Sources of wastewater from industrial applications | Toxicity | MCL (mg/L) |
|---|---|---|---|---|
| Cd | Cadmium (Cd) is a transition metal of Group IIB in the periodic table. The atomic number and atomic weight of cadmium are 48 and 112.4. It is white metal with a bluish tinge and has considerable ductility. Cd (II) is the main hydrolyzed species in the aqueous phase. Cadmium has a good thermal conductivity that makes it suitable for rolling, forging, and polishing | Cadmium is widely used in the plating industry due to its excellent corrosion resistance for several chemicals at high temperatures. It is also extensively used in paints, enamel, plastic, batteries, alloys, coatings, stabilizers, and fertilizers | Cadmium accumulates in the food chains, which is toxic to humans, aquatic organisms and animals. It is also considered a human carcinogen and causes kidney damage, renal disorder | 0.01 |
| Cu | Copper (Cu) is the first element of Group IB in the periodic table. The atomic number and atomic weight of copper are 29 and 63.546. It is reddish with a bright metallic luster. Copper has good heat conductivity, high ductility, malleability, and low corrosion properties. Copper can exist in oxidation states from 0 to +3, but the only important valence state in an aqueous solution is cupric ion Cu (I) or Cu(II). The Cu(I) state, such as CuCl, is highly insoluble in water | Copper has high plating efficiency. The high electrical conductivity of copper makes it an excellent and inexpensive coating for electrical cables and electronic products. Copper compounds are also used in fungicides and anti-fouling paint industries | The waste streams containing copper and other complexing agents pose a significant threat to aquatic life and render natural water unsuitable for public use. Liver damage, Wilson disease, insomnia | 0.25 |

(continued)

## 6.1 Introduction

**Table 6.1** (continued)

| Heavy metal | Properties | Sources of wastewater from industrial applications | Toxicity | MCL (mg/L) |
|---|---|---|---|---|
| Cr | The transition metal chromium (Cr) is the first element of Group 6 in the periodic table. The atomic number and atomic weight of chromium are 24 and 52. It is a blue–white metal that is hard, brittle and has high corrosion resistance. Chromium exists mainly in two oxidation states, $+3$ to $+6$ | Chromium is released into the environment primarily in the form of hexavalent [Cr (VI)]. The wastewater containing chromium is mainly generated from chrome plating and processing units, tanneries, electronic device manufacturing facilities, etc. | Allergic dermatitis, Headache, diarrhea, others. Nausea, vomiting, carcinogenic | 0.05 |
| Ni | Nickel (Ni) is a transitional metal with an atomic number 28 and atomic weight 58.69. Nickel is a white silvery lustrous metal with a slight golden tinge. The common oxidation state of nickel is 2$+$ | Wastewater containing nickel is mainly generated from nickel electroplating and metal finishing industries. Nickel contaminated wastewater is generated from plastics manufacturing, nickel–cadmium batteries, fertilizers, pigments, mining, and metallurgical operations | Nickel is a known carcinogen to humans. It may cause kidney failure, damage to the lungs, and gastrointestinal distress. Skin dermatitis, chronic asthma, and nausea may result from the ingestion of nickel in higher concentrations through the water | 0.20 |

(continued)

Table 6.1 (continued)

| Heavy metal | Properties | Sources of wastewater from industrial applications | Toxicity | MCL (mg/L) |
|---|---|---|---|---|
| As | Arsenic (As) has *atomic number* 33, and atomic weight is 74.921<br>Inorganic arsenic species are generally found in groundwater and surface waters as trivalent (arsenite) or pentavalent (arsenate) form | Arsenic enters the groundwater by the dissolution of minerals and ores depending on the pH and soil conditions of the geological area. Many industrial processes and mining activities further contribute to the total amount of arsenic if the waste streams from these activities discharged directly into the water bodies<br>The As (III) (arsenite) compounds are 25–60 times more toxic and tend to be bound weakly to soil minerals than As (V) (arsenate) compounds | Consumption of arsenic for a significant period through drinking water causes cancer of the skin, lungs, urinary bladder, and kidney<br>It also changes the pigmentation of the skin, and subsequently, the skin thickens | 0.01 |
| Pb | Lead (Pb) element is in Group 14 of the periodic table with atomic number 82 and atomic weight 207.2<br>It is soft, malleable, and has a relatively low melting point. Lead shows two primary oxidation states: +4 and +2 | Lead containing wastewater may be generated from the activities of construction, plumbing, lead-acid batteries, *bullets, pewters, fusable alloys, paints, leaded gasoline, and radiation shielding-related activities* | Lead can accumulate in human bones and soft tissues. It can interfere with the function of biological enzymes<br>Lead is considered a neurotoxin as it can damage nervous systems that cause neurological disorders and behavioral problems | 0.006 |

(continued)

Table 6.1 (continued)

| Heavy metal | Properties | Sources of wastewater from industrial applications | Toxicity | MCL (mg/L) |
|---|---|---|---|---|
| Hg | Elemental mercury (Hg) is a silvery-white liquid from Group 12 of the periodic table. Its atomic number is 80, and its atomic weight is 200.592. Mercury is considered a fair conductor for electricity but a poor conductor of heat. Mercury has two oxidation states, I and II Hg (II) is the most common oxidation state | Mercury-containing wastewater may be generated from dental practice wastes, fertilizers, batteries, landfill leachate, paints, domestic waste inputs, and chloro-alkali industrial waste | Rheumatoid arthritis and disease of the kidneys Circulatory and nervous system | 0.00003 |

*MCL* maximum contamination level: The highest contaminant level that is allowed in drinking water

as a coagulating/flocculating agent to address the colloidal particles for wastewater treatment [3–7]. Most colloidal particles in water and wastewater carry a negative charge. Compared to alum or ferric chloride, chitosan can act as a coagulation agent that neutralizes or reduces these colloidal particles [4]. The main mechanisms by which chitosan may serve as a coagulant and flocculant to suspended particles in wastewater treatment are (i) electrostatic attraction, (ii) sorption, and (iii) bridging [5]. The physiochemical principle behind coagulation is reducing the repulsive electrical potential among electronegative colloidal particles in water, followed by aggregation of colloidal and fine suspended particles to form microflocs through van der Waal's force of attraction [3, 4]. Chitosan is a linear polysaccharide, and it possesses a positive ionic charge in acidic media. Therefore, it can bridge flocs together by electrostatic binding of several flocs with the opposite charge along different loci of the biopolymer chains [6] that eventually settle out of the system. Gidas et al. [7] reported that chitosan as a primary coagulant successfully removed than 88% of the zinc, copper, aluminum, chromium, and iron in sanitary wastewater. Garcia et al. [8] studied chitosan acetate and chitosan lactate salts as coagulants/flocculants for fish processing industry wastewater. They reported that chitosan lactate salt is more effective than chitosan acetate salt to remove total suspended solids (TSS) and chemical oxygen demand (COD) from the wastewater. Krishna and Sahu [4] reported that chitosan use as a coagulant agent reduces 83% of COD and 90% more color of the soap and detergent-related wastewaters. Chitosan and bentonite (30:70) mixture as a coagulant can remove turbidity from the wastewater with an efficiency of 97% [9]. Abebe et al. [3] reported chitosan coagulation to improve microbial and turbidity removal by ceramic water filtration for household drinking water treatment.

The solubility of chitosan in low pH conditions imposes some limitations to its use if acid is used for regeneration in a column. At low pH, swelling of chitosan occurs that involves protonation of the amino groups. This protonation leads to chain repulsion and diffusion of protons and counterions and water inside the chain. However, in its natural form, chitosan is soft and tends to agglomerate or form a gel. Although the amino and hydroxyl groups in chitosan are mainly responsible for the adsorption of metal ions, these active binding sites are not readily available for sorption in their natural form. Several studies have investigated pure chitosan beads as an adsorbent to remove heavy metals from aqueous streams [10–13]. According to Rorrer et al. [10], chitosan flake or powder swells and crumbles, making it unsuitable for use in an adsorption column. Chemical modification of chitosan (also known as crosslinking) can improve its physical and mechanical properties, making it suitable for a column. Although *the crosslinked chitosan* has good mechanical properties, its adsorption capacity for metal ions may decrease compared to pure chitosan. The reduction in metal uptake capacity of crosslinked chitosan is due to ligand mobility [11]. Guibal et al. [12] noted that the maximum uptake of chitosan flakes was approximately half of that obtained with chitosan beads for molybdate. Bodmeier et al. [13] noted that the freeze drying of chitosan gel produced particles with a high internal surface area, which boosted the metal binding capacity.

The capacity of chitosan for metal ions can be enhanced through chemical crosslinking of chitosan or modification using physical supports that can increase the accessibility of the binding sites [14]. Su et al. [15] studied the adsorption characteristics of heavy metal ions from wastewater streams using chitosan-chelated beads. The adsorption of nickel, zinc, and copper compared well with the values for commercial adsorbents. A similar study on the adsorption of chromium (III) and cadmium was carried out by Luisa and coworkers [16]. The authors found a 99% retention of cadmium and 83% retention of chromium. Hastuti et al. [17] reported that chitosan crosslinked with epichlorohydrin (ECH) improved adsorption capacity for chromium from 74% (pure chitosan) to 89% with a contact time of 30 min at pH 3. In another study, Bhatt et al. [18] reported that the chitosan crosslinked with diethylentriamine-pentaacetic acid (DTPA) has an adsorption capacity of 192.3 mg chromium/gram of crosslinked chitosan. The adsorption of chromium on to DTPA crosslinked chitosan reported to have occurred in three steps: (1) the binding of anionic $Cr^{6+}$ to protonated positively charged amino groups on the adsorbent surface, (2) the reduction of $Cr^{6+}$ to $Cr^{3+}$ by adjacent electron–donor groups, and (3) the formation of clusters of $Cr^{3+}$ on the adsorbent. Genc et al. [19] evaluated the Precion Green H-4G immobilized poly (hydroxyethyl methacrylate/chitosan) composite membranes for the removal of cadmium (II), lead (II), and mercury (II) from aqueous media. They were able to achieve maximum retention of 44, 68, and 48 mg/g for Cd (II), Pb (II), and Hg (II), respectively. Kawamura et al. [20] prepared a porous polyaminated chitosan chelating resin by introducing poly (ethylene amine) onto the crosslinked chitosan beads. The resultant beads showed a high capacity and high selectivity for metal ions' adsorption, including cadmium. Hsien and Rorrer [21] investigated the adsorption of cadmium on porous magnetic chitosan beads. They found that the adsorption capacities for 1 and 3-mm beads were 518 and 188 mg Cd/g of adsorbent, respectively. They also investigated the effects of acylation and crosslinking of chitosan using glutaraldehyde [22]. They later noted that the crosslinked chitosan beads became more resistant to acid, and the capacity of crosslinked chitosan for cadmium decreased significantly compared to non-crosslinked beads. Chitosan crosslinked with crown ethers [23] showed improved adsorption capacity for cadmium and high selectivity for Ag(I) or Pd (II) in the presence of Pb (II) and Cr (III). Becker et al. [24] studied cadmium's adsorption on dialdehyde, or tetracarboxylic acid crosslinked chitosan.

In yet another study, Seo et al. [25] used water-soluble chitosans in a study to remove heavy metals. The removal efficiency of water-soluble chitosans improved when they were crosslinked with Na N, N-diethyldithiocarbamate trihydrate, or Na salicylate. Reviews of chitosan application for the treatment of wastewaters containing solids and heavy metals, and other contaminants have been presented by No and Myers [26] and Anastopolous et al. [27]. Based on batch experiments, Evans et al. [28] have shown that intraparticle diffusion to be the rate-controlling step in the adsorption of cadmium using chitosan. The data were fitted using the Freundlich adsorption isotherm model. The work of Bassi et al. [9] shows that the order of metal ion adsorption by chitosan decreased from Cu (II) to Zn (II) as copper > lead > cadmium > zinc. The efficiency of adsorption appears to have increased with the amount of adsorbent. The authors find the heavy metal adsorption to be independent of pH with a maximum at pH 6.0 and 7.0. In another study, Bassi et al. [29] carried out equilibrium and kinetic studies for the adsorption of heavy metals

Zn, Cu, Cd, and Pb. In general, a 50% reduction of metal ions was observed. Ngah et al. [30] investigated the adsorption of Cu (II) ions onto chitosan and crosslinked chitosan beads. The beads were crosslinked with glutaraldehyde (GLA), epichlorohydrin (ECH), and ethylene glycol diglycidyl ether (EGDE), respectively, to make them insoluble in water. The authors found a pH of 6.0 to be optimum and Cu (II) uptake was 80.7 mg/g for GLA, 60 mg/g for ECH, and 46 mg/g for EGDE beads. Gotoh et al. [31] used alginate-chitosan hybrid gel beads for the adsorption of Cu (II), Co (II), and Cd (II). The beads prepared were crosslinked with glutaraldehyde to exchange the amine groups with aldehyde groups. Rae and Gibb [32] compared four naturally occurring chitinous materials such as chitin, chitosan, cryogenically milled carapace, and mechanically milled carapace for their effectiveness in removing transition state and heavy metals. They found chitosan to be the most effective among all four chitinous material to remove heavy metals from aqueous solutions. Chitin removes only 35%, chitosan more than 99%, with the other two in the range of 90%. Working on adsorption of Cr (VI) and Cu (II) on chitin, chitosan, and Rhizopus arrihizus, Sag and Aktay [33] examined the sorption mechanism and rate-controlling steps. They found that pseudo-second-order rate expression fitted the data better than any other model. Gyliene et al. [34] use fly larva shells as adsorbents to treat organic wastes containing heavy metals. The sorbents were characterized using techniques such as FTIR, pH-potentiometric titration, and porosimetry. The sorption ability for free metal ions of chitin decreased in the order Fe (III) > Cu (II), Pb (II) > Zn (II) > Ni (II) > Mn (II), and that of chitosan decreased in the order Cu (II) > Mn (II) > Ni (II) > Zn (II) > Pb (II) > Fe (III). Fly larva shells adsorbed complexed metal ions up to 0.2–0.4 mmol/g. Evans et al. [28] used chitosan-based crab shells for their studies of adsorption of cadmium. Three different particle sizes with average diameters of 0.65, 1.43, and 3.38 mm with an average pore diameter of 300–540 Å and surface area of approximately 30 m$^2$/g were used in their study. The authors found that the adsorption equilibrium data followed the Freundlich relation and were not dependent on particle size. Kopecky et al. [35] described Cu (II) adsorption from copper sulfate solution on chitosan with a degree of deacetylation of 64%. The maximum sorption capacity was 200 mg Cu/g chitosan. El-Sawy et al. [36] studied the adsorption of heavy metals on chitosan as a part of their studies on corrosion. The studies were conducted using poly (DEAEMA)-chitosan-graft-copolymer, poly (COOH)-chitosan-graft-copolymer, poly (V–OH)-chitosan-graft-copolymer, and carboxymethyl chitosan. Wang [37] reviewed low-cost adsorbents other than chitin and chitosan for the adsorption of three heavy metals ions Cd, Hg, and Pb. Sag [38] reviews the biosorption capacities of various fungi (free or immobilized or subjected to physical and chemical treatments), chitin, and chitosan. This review stresses the need to consider multicomponent adsorption as most industrial wastewaters contain several metal species. The effects of various metal ions on the biosorption capacity of various fungi are discussed. The actions of the metal ion combinations synergistic or antagonistic are identified. Cardenas et al. [39] prepared chitosan mercaptans using mercaptoacetic acid and 1-chloro-2, 3-epoxy propane propionic acid along with crustacean shells and used it for sorption of Cu and Hg. The adsorbent was characterized using FTIR, TGA, and elemental analysis. A wide

## 6.1 Introduction

range of concentrations from 10 to 104 ppm was used in their studies. Bayramoglu et al. [40] prepared hydroxyethyl methacrylate membranes coated with chitosan for their studies on the adsorption of Pb (II), Hg (II), and Cd (II). The authors found the maximum adsorption capacities of heavy metal ions onto the composite membrane to be 64.3 mmol/m$^2$ for Pb (II), 52.7 mmol/m$^2$ for Hg (II), and 39.6 mmol/m$^2$ for Cd (II), and the affinity order to be Pb (II) > Hg (II) > Cd (II). Su et al. [41] used chitin from mycelium biomass fermentation industries to study Ni (II) ions' adsorption and compared the results with some cation exchange adsorbents. In a patent, Zeng et al. [42] describe the synthesis of silica gel-crosslinked chitosan adsorbents for heavy metals. Shao et al. [43] studied the adsorption of Cu (II), Cd (II), Pb (II), Zn (II), and Ni (II) using chitin modified by L-cysteine. The adsorption capacities of L-cysteine-chitin for Cd (II), Pb (II), and Zn (II) were 214.6, 351.3, and 107.0 mg/g, respectively. Taboada et al. [44] used chitosan modified with 3,5-diaminobenzoic acid to obtain N-(3,5-diaminobenzoate) chitosan (Ch–N–DAB) in their study relating to the adsorption of nickel and zinc. Matejka et al. [45] used chitosan crosslinked by ethylene glycol diglycidyl ether, both in bead and fiber forms, for the adsorption of Mo, W, and V oxoanions as binuclear complexes. They found that the steric configuration of OH groups was not suitable for uptake of As, B, or Ge oxoanions as a mononuclear complex. The sorption efficiency of chitosan fibers was found to be superior to beads. Juang and Shao [46] used chitosan for their work on the adsorption of single and binary metal ion systems. They measured the sorption of Cu (II), Ni (II), and Zn (II) from the water on crosslinked chitosan. A model proposed by the authors predicts the amounts of sorption well in the total metal concentration of 0.77–1.77 mol/m$^3$ and a pH range of 2–5. A graft copolymer of N-isopropyl-acrylamide and chitosan was prepared and tested for Cu (II) sorption by Uehara et al. [47]. The collection efficiency of Cu (II) ion with the grafted copolymer chitosan at pH 6.5 was about 50% similar to the untreated chitosan.

Water-soluble chitosan crosslinked with sodium N, N-diethyldithiocarbamate trihydrate, or sodium salicylate was evaluated for their metal ion sorption capacity by Seo et al. [25]. These researchers found that the crosslinked chitosan was better than non-crosslinked water-soluble chitosan. Navarro and Tatsumi [48] modified chitosan with polyethyleneimine and poly (glycidyl methacrylate) to enhance metal chelating properties of the adsorbents. Batch adsorption studies showed an improved capacity for Cu (II), Zn (II), and Pb (II) ions. The studies were conducted at different pH, and a pH of 4.0 was found to be optimum. Yang and Yang [49] prepared and characterized two chitosan azacrown ethers bearing hydroxyl groups (CTS-DH and CTS-DO) by the reaction of 3-hydroxy-1, 5-diazacycloheptane and 3-hydroxy-1, 5-diazacyclooctane with epoxy-activated chitosan. They investigated the adsorption and selectivity properties of the hydroxyl azacrown ethers chitosan derivatives for Ag+, Cr (III), Cd (II), and Pb (II). Zhuang et al. [50] synthesized a biomaterial using 4-(5-chloro-2-pyridylazo)-1,3-phenylenediamine (5-Cl-PADAB), chitosan, and EDTA as a metal indicator, biosorbent, and crosslinker and chelating agent. The modified chitosan bead was a selective probe for $Co^{2+}$ with a remarkable color change from white to pink. Compared with chitosan beads, the equilibrium uptake ($q_e$) of the modified one increased from 2.00 to 7.97 mg/g. Yin et al. [51] prepared a novel biosorbent by immobilizing *Saccharomyces cerevisiae* in magnetic chitosan microspheres. The maximal adsorption capacity was 81.96 mg strontium/g chitosan microsphere by the Langmuir model. In another study, Zhuang et al. [52] modified chitosan by gamma radiation-induced grafting with maleic acid and then used it to remove cobalt ions from aqueous solutions. The results showed that the optimal dose for grafting was 2 kGy, and the equilibrium adsorption capacity of cobalt ions increased from 2.00 mg/g to 2.78 mg/g after chitosan modification. Deng et al. [53] prepared engineered biochar using chitosan and pyromellitic dianhydride (PMDA), and its performance was evaluated to remove heavy metal ions from a single metal and mixed-metal solutions (Cd, Cu, and Pb). They reported that the engineered biochar had selective adsorption capacity for copper due to the N-containing functional groups. It was suggested that the carbonyl groups of biochar also participated in removing copper and may reduce Cu(II) to Cu(I). Najafabadi et al. [54] prepared chitosan/graphene oxide nanofibrous adsorbent using the electrospinning process. The maximum monolayer adsorption capacity of $Pb^{2+}$, $Cu^{2+}$, and $Cr^{6+}$ metal ions using chitosan/graphene nanofibers was 461.3, 423.8, and 310.4 mg/g at an equilibrium time of 30 min and temperature of 45 °C. Thermodynamic data showed that the nature of the metal ions sorption by chitosan/graphene nanofibers was endothermic and spontaneous.

Schmuhl et al. [55] investigated the adsorption capacity of chitosan for Cu (II) and Cr (V). Studies were conducted with crosslinked and non-crosslinked chitosan for Cu (II) and Cr (VI) from aqueous solutions for different concentrations. The authors concluded that pH did not have a significant influence on Cu adsorption, where Cr showed maximum adsorption at a pH of 5.0. Wan et al. [56] studied chitosan-coated sand containing 5% chitosan for copper adsorption. The authors used only 5% chitosan to make the adsorbent economical for practical applications. Cheung et al. [57] found the equilibrium adsorption of copper to be 1.26 and 1.12 mmol/g at pH 3.5 and 4.5, respectively. They used chitosan in the particle range of 355–500 μm and analyzed their data using four different models. In yet another study, Sag and Aktay [58] used Langmuir and Redlilch–Peterson models to fit their sorption data. In addition, single-resistance models were used to determine external film mass transfer and the intraparticle diffusion data. Tan and Pend [59] used Penicillium chrysogenum mycelium for the adsorption of Cr (III), Zn (II), and Ni (II) from tannery wastes. The adsorption capacities were for Cr (III) from 18.6 to 27.2 mg/g, Ni (II) from 13.2 to 19.2 mg/g, and Zn (II) from 6.8 to 24.5 mg/g. Wu et al. [60] studied the adsorption of three commercial dyes and Cu (II) using chitosan with or without complexing agents such as EDTA. The authors modeled their data using pseudo-first-order, pseudo-second-order, and intraparticle diffusion models. Bayramoglu et al. [61], as a continuation of their earlier work [40], have looked at the lysozyme adsorption capacities of the dye-ligand, dye-ligand-Fe (III), and dye-ligand-Cu (II) immobilized IPNs membranes. da Rocha Filhoa et al. [62] investigated certain graft copolymers of N-vinyl-2-pyrrolidone with chitin for heavy metal removal from wastewater effluents. The authors studied the effect of temperature, pH, and solute concentration on the chelating properties of the grafted and ungrafted chitin. Dhakal et al. [63] used O, O′-decanoyl chitosan and O, O′-decanoyl

N, N-didecanoyl chitosan, the lipophilic chitosans, as well as O, O'-decanoyl chitin, the lipophilic chitin, for the extraction of Cu(II), Ni(II), Co(II), Zn(II), Fe(III), and Pb(II) from aqueous solution with some success. Hasan [64] used chitosan-coated perlite beads, chitosan-Fuller's earth beads, and chitosan–iron composite beads for the removal of Cd (II), Cr (V), Cs (II), Sr (II), As (III), and As (V) from wastewater, ground, and surface water, respectively. Nomanthay and Palanisamy [65] have studied the removal of heavy metals from industrial wastewater using chitosan coated with shell charcoal. This composite bio-adsorbent showed an absorption capacity of 154 mg Cr/g. Pontoni and Fabbricino [66] provided an excellent review of the use of chitosan and chitosan-based adsorbents for the removal of arsenic from water. Hasan et al. [67] explored the use of FeOOH with chitosan for the removal of anions from aqueous solutions. Later, Vasireddy [68a] used this method and achieved average adsorption of 70–75% for both As (III) and As (V). Chen and Chung used chitosan beads to remove As (III) and As (V) from water under both batch and continuous operations [68]. They concluded that chitosan beads favored the adsorption of As (V), but not As (III). A similar observation was reported by Gerente et al. [69] that chitosan presents a positive charge at pH 5 that leads to favorable electrostatic attractions with arsenate As (V) ions over As (III).

## 6.2 Adsorption Methods

Adsorption involves transferring a substance from one phase to a surface where intermolecular forces bound it. Although adsorption is usually associated with the transfer of a substance from gas or liquid to a solid surface, the transfer from a gas phase to a liquid phase also occurs. The substance, which is concentrated on the surface, is defined as the adsorbate, and the solid surface on which the adsorbate accumulates is defined as the adsorbent.

An effective adsorption process requires information about the amount of adsorbate that is adsorbed onto the adsorbent as a function of adsorbate concentration in the gas or liquid phase at different temperatures. This type of information is generally referred to as an adsorption isotherm or equilibrium adsorption data. It is a function of the surface characteristics of the adsorbent and the physical and chemical properties of the adsorbate. Various shapes of curves representing the isotherm data can be obtained. The great majority of the isotherms observed to date can be classified into the five types shown in Fig. 6.1. The different shapes indicate the actual adsorption mechanism, which is related to the adsorbate and adsorbent properties.

Adsorption isotherm data are frequently obtained using a static system. The amount of adsorbate adsorbed on the solid surface is measured as a function of pressure or concentration when the solid adsorbent is exposed to an atmosphere containing the adsorbate at a constant temperature. Experimental data can also be obtained by using a flow system. In a flow (dynamic) system, a stream containing the adsorbate (contaminants) is passed through a fixed bed of solid adsorbent. As the bed becomes saturated, the adsorbate concentration in the outlet stream becomes

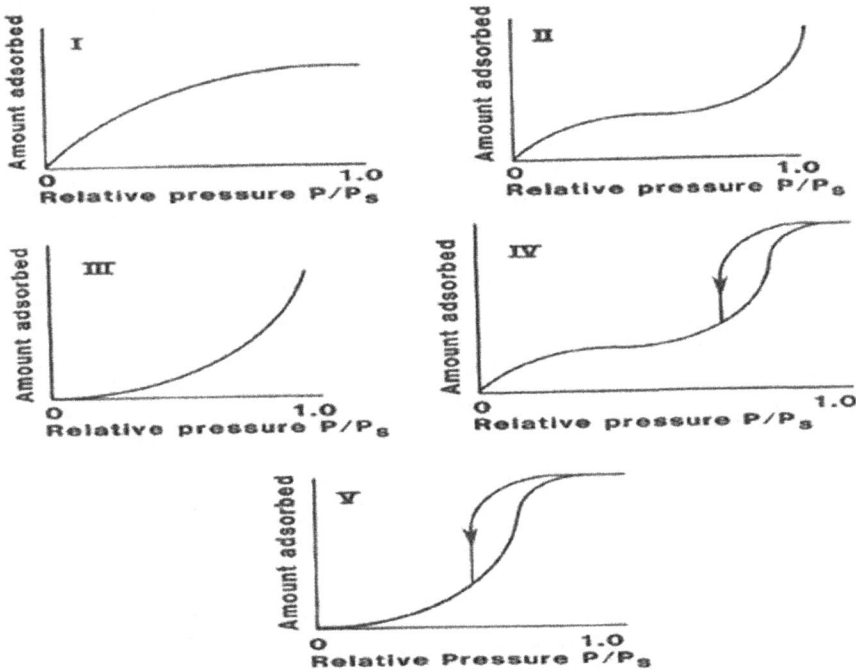

**Fig. 6.1** Brunauer's classification of adsorption isotherms, showing amount adsorbed versus final concentration in the fluid. Copyright © 2015 IUPAC & De Gruyter, M. Thommes et al.: Physisorption of gases, with special reference to the evaluation of surface area and pore size distribution (IUPAC Technical Report), Pure Appl. Chem. 2015; aop, https://doi.org/10.1515/pac-2014-1117, Adapted from open access Ref. [70a]

equal to that of the inlet stream. The data obtained from such systems are expressed in terms of the concentration profile of the adsorbate in the effluent gas/liquid stream as a function of time and are typically referred to as a breakthrough curve. A typical example of the properties of several adsorbents is given in Table 6.2.

## 6.2.1 Experimental Procedure for Equilibrium Uptake of Metal Ions on Adsorbent from Aqueous Solution

Equilibrium batch adsorption studies are carried out by exposing the beads to aqueous metal ion solutions of different concentrations in 125 mL Erlenmeyer flasks to a predetermined temperature. A typical example of a batch adsorption process is as follows [64]. About 0.25 g beads are added to 100 mL of solution. Such an amount of beads and solution assured that an equilibrium condition is reached, i.e., the entire metal ion is not adsorbed by the beads, which would make it difficult to determine

## 6.2 Adsorption Methods

**Table 6.2** Adsorbent properties[a]

| Adsorbents | Types | Pore volume (cm³/g) | Particle density (g/cm³) | Bulk density (g/cm³) | Surface area (m²/g) | Physical form |
|---|---|---|---|---|---|---|
| Silica gel | Regular density [b] | 0.43 | 1.13 | 0.72–0.77 | 750–800 | Granular |
|  | Intermediate density[b] | 1.15 | 0.62 | 0.40 | 340 | Granular |
| Molecular sieve | 5A[b] | 0.32–0.33 | 1.14–1.16 | 0.71 | – | Pellets |
|  | 13X[b] | 0.41 | 1.13 | 0.72 | 395 | Spherical |
|  | 5A[c] | 0.32 | 1.07 | 0.72(bead) | – | Pellets |
| Activated alumina | F1[d] | 0.20 | 1.40 | – | 250 | Granular |
|  | A-201[e] | 0.50 | 1.60 | 0.77 | 325 | Spherical |
| Activated carbon | Type BPL (coal-based)[f] | 0.80 | 0.80 | 0.48 | 1050–1150 | Granular |
|  | Type PCB (coconut-based)[f] | 0.72 | 0.85 | 0 44 | 1150–1250 | Granular |

[a]Manufacturer's data unless stated otherwise; [b]Davison Chemical Division of W.R. Grace & Co.; [c]Linde Division of Union Carbide Co.; [d]Aluminum Co. of America (Alcoa); [e]LaRoche Chemicals; [f]Calgon carbon Corporation

the equilibrium point. The solutions' pH is adjusted by adding either 0.1 mol/L hydrochloric acid or 0.1 mol/L sodium hydroxide. The flasks are placed in a constant temperature shaker bath for a specific time period. Following the exposure of beads to the metal ion, the samples are collected at predetermined time intervals. The solutions are filtered, and the filtrates are analyzed for metal ion by analytical instruments, for example, atomic absorption spectrometer (AAS) or inductively coupled plasma mass spectrometry (ICP-MS). The adsorption isotherm at a particular temperature can be obtained by varying the initial concentration of metal ions. The amount of metal ion adsorbed per unit mass of adsorbent ($Q_e$) can be calculated using the following equation:

$$Q_e = \frac{(C_i - C_e)V}{M}, \qquad (6.1)$$

where

$C_i$ = the initial concentration of a metal ion in the solution (mg/L),
$C_e$ = the equilibrium concentration of metal ion (mg/L), V is the volume of the solution, and
$M$ = the amount of sorbent.

## 6.2.2 Adsorption Controlling Parameters

Guibal [71] reported several experimental parameters, such as the crystallinity of the sorbent, diffusion properties, and speciation of metal ions that control adsorption properties. Hasan et al. [72] reported that adsorbing metal ions on an adsorbent from a solution, it should form an ion in the solution. The types of ions formed in the solution and the degree of ionization are dependent on the solution pH. The variation of adsorption of a metal ion at various pH can be explained based on metal chemistry in solution and the surface chemistry of the adsorbent. The effect of pH on the adsorption of metal ions on adsorbent may be explained based on an aqua complex formation owing to the oxide present in the adsorbent. The protonation and dissociations of the surface functional groups are as follows[73]:

$$SOH + H^+ \rightleftharpoons SOH_2^+, \tag{6.2}$$

$$SOH \rightleftharpoons SO^- + H^+. \tag{6.3}$$

On chitosan, the leading functional group responsible for metal ion adsorption is the amine group ($-NH_2$). Depending on the solution pH, these amine groups can undergo protonation [$NH_3^+$ or $(NH_3 - H_3O)^+$], and the extent of protonation will be dependent on the solution pH. Therefore, the surface charge on the chitosan bead will determine the type of bond formed between the metal ion and the adsorbent surface. The $pH_{max}$ where maximum adsorption of metal takes place seems to be related to the pK or the first hydrolysis product of the metal ions. The exact nature and distribution of hydroxo-complexes depend on the concentration of ligands, i.e., solution pH and soluble metal concentration. For instance, the effect of pH on Cd (II) adsorption by chitosan-coated perlite beads was studied by varying the pH of the solution (Fig. 6.2). The pH of the cadmium solution was first adjusted over a range

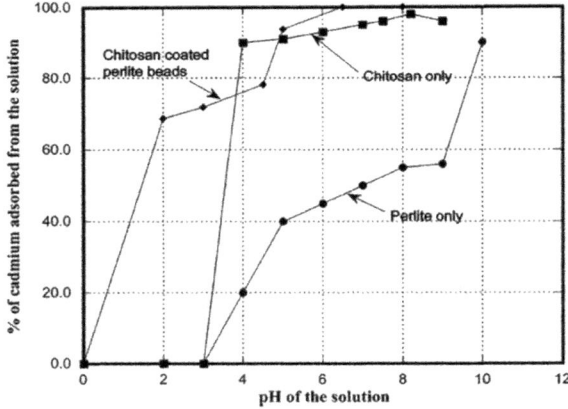

**Fig. 6.2** Effect of pH on cadmium uptake by pure perlite, pure chitosan, and chitosan-coated perlite beads. Copyright © 2006, American Chemical Society, Ind Eng Chem Res 45:5066, Reproduced from Ref. [72] with permission

of 2–8 using either 0.1 N $H_2SO_4$ or 0.1 M NaOH, and then chitosan-coated perlite beads were added. As the adsorption progressed, the pH of the solution increased slowly. No attempt was reported to be made to maintain a constant pH of the solution during the experiment. The amount of cadmium uptake at the equilibrium solution concentration is shown for different initial pHs of the solution in Table 6.3, along with the final pH of the solution. The uptake of Cd (II) by chitosan beads increased as the pH increased from 2 to 8.

Although a maximum uptake was noted at a pH of 8, as the pH of the solution increased to >7, cadmium started to precipitate out from the solution. Therefore, experiments were not conducted at pH >8.0. The increased capacity at pH >7 may be a combination of both adsorption and precipitation on the surface. It is concluded that the beads had a maximum adsorption capacity at a pH of 6 if the precipitated amount is not considered in the calculation.

Table 6.3 Cadmium uptake at equilibrium at various solution pH at 298 K

| Concentration at the equilibrium of the liquid phase (mmol/L) | Uptake by the solid phase (mmol/g) | Final pH of the solution |
|---|---|---|
| *Initial pH of the solution: 2* | | |
| 0.303 | 0.0285 | 2.1 |
| 0.589 | 0.0607 | 2.1 |
| 1.607 | 0.1250 | 2.4 |
| 3.482 | 0.1964 | 2.5 |
| 6.964 | 0.3214 | 2.6 |
| *Initial pH of the solution: 4.5* | | |
| 0.089 | 0.036 | 4.5 |
| 0.402 | 0.098 | 4.6 |
| 1.250 | 0.196 | 4.6 |
| 3.036 | 0.286 | 5.0 |
| 6.696 | 0.424 | 4.9 |
| *Initial pH of the solution: 6.0* | | |
| 0.225 | 0.080 | 6.2 |
| 0.402 | 0.143 | 6.2 |
| 1.160 | 0.268 | 6.3 |
| 2.679 | 0.375 | 6.5 |
| 6.160 | 0.536 | 6.6 |
| *Initial pH of the solution: 8.0* | | |
| 0.089 | 0.089 | 8.0 |
| 0.320 | 0.161 | 8.1 |
| 1.160 | 0.321 | 8.2 |
| 2.770 | 0.446 | 8.3 |
| 5.940 | 0.580 | 8.6 |

Copyright © 2006, American Chemical Society, Ind Eng Chem Res 45:5066, Reproduced from Ref. [72] with permission

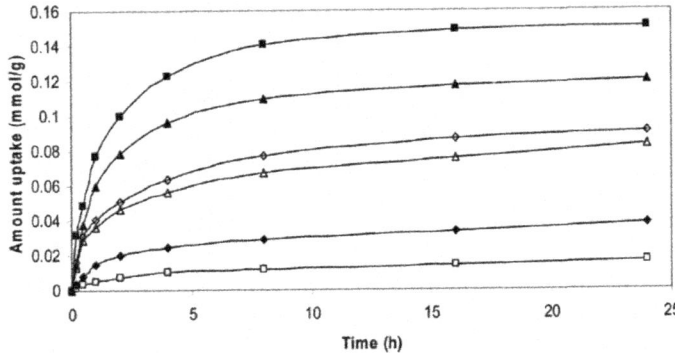

**Fig. 6.3** Effect of time and concentration on cesium removal by chitosan-Fuller's earth bead. The initial concentrations of the solutions were (■) 7.52 mmol/L, (▲) 3.76 mmol/L, (◊) 1.5 mmol/L, (△) 0.752 mmol/L (♦) 0.376 mmol/L, and (□) 0.075 mmol/L, respectively. Copyright © 2007, Taylor & Francis, Sep. Sci. and Technol. 42, 717–738, 2007. Reproduced from Ref. [74] with permission

### 6.2.3 Effect of Contact Time and Sorbate Concentration in Solution on Adsorption Process

Adsorption is a slow process, and adequate contact time is necessary to allow the system to approach equilibrium. This delay provides information about the time required for the considerable amount of metal ion adsorbed on the solid adsorbent and kinetic control mechanism. Figure 6.3 shows an example of a metal ion such as cesium adsorption on chitosan-Fuller's earth beads at pH 6.5 and 298 K [74].

The concentration profiles during cesium uptake by the beads from various concentrations of the solution are shown in Fig. 6.3. The removal of cesium from the aqueous solution by adsorption increased with time. Almost 60% of cesium was adsorbed onto beads during the first 240 min of a run, and then the equilibrium was attained monotonically at 360 min in most of the runs. As shown in Fig. 6.3, the concentration versus time plots increased monotonically to saturation at various concentrations of cesium ion in solution. This indicates the possibility of monolayer adsorption of cesium on the adsorbent.

### 6.2.4 Equilibrium Considerations

The selection of the adsorbent includes consideration of surface area as well as the type of adsorbate involved in the adsorption process since these relate to the types of bonds formed. Depending on the types of bonds, adsorption is described as either physical adsorption or chemical adsorption (chemisorption). Physical adsorption results when the adsorbate adheres to the surface by van der Waals forces (by dispersion and Coulombic forces). Although displacement of electrons may occur, electrons are not shared between the adsorbent and adsorbate. During adsorption, a quantity of heat, described as the heat of adsorption, is released. The quantity of heat released during physical adsorption is approximately equal to the heat of condensation, resulting in adsorption being frequently described as a condensation process. As expected, the quantity of material physically absorbed increases as the adsorption temperature decreases. The nature of the forces for the physical adsorption is such that multiple layers of the adsorbate can accumulate on the surface of the adsorbent.

The primary difference between physical adsorption and chemisorption is the bond between the adsorbed molecule and the adsorbent surface. In the chemisorption process, the sharing of electrons occurs between the adsorbent and the adsorbate, which results in the liberation of a quantity of heat that is approximately equal to the heat of the reaction. As a result of electron sharing with the surface, chemisorbed materials are restricted to forming a monolayer. Interaction forces responsible for adsorptive bonds are given in Table 6.4

Although different thermal effects characterize chemical and physical adsorption, a clear distinction between the two adsorption mechanisms does not exist. It is observed that the amount of adsorbate chemisorbed on a surface increases with increased temperature. The main differences between chemisorption and physisorption are the heat of adsorption, reversibility, and layer of adsorption, as given in Table 6.5.

Figure 6.1 shows the main form of equilibrium isotherm. Brunauer [70] classified the equilibrium isotherm into five principal forms based on the adsorption mechanism. Type I is classified as the Langmuir type and is characterized by a monotonic approach to a limiting adsorption capacity that corresponds to the formation

**Table 6.4** Typical adsorptive bond

| Adsorptive bond | Types of interaction | Strength of interaction (E*) |
|---|---|---|
| Weak | van der Waals force | E <1 to 4 kJ/mole |
|  | Hydrogen bond | E ranges from 4 to 30 kJ/mole |
|  | π-interaction | E ranges from 5 to 40 kJ/mole |
| Strong | Ionic | E > 100 kJ/mole |
|  | Covalent | E ranges from 150 to 1100 kJ/mole |
|  | Coordination (metallic) | E ranges from 60 to 300 kJ/mole |

Copyright © 2013 Dhotel, A.; Chen, Z.; Delbreilh, L et al. licensee MDPI, Basel, Switzerland. Creative Commons Attribution CC BY license, Int. J. Mol. Sci. 2013, 14, 2303–2333. https://doi.org/10.3390/ijms14022303. Reproduced from Ref. [75]

E* = Typical binding energy

**Table 6.5** Physisorption and chemisorption [Unpublished work by authors]

| Parameters | Chemisorption | Physisorption |
|---|---|---|
| Force | Covalent bond (ionic or covalent or a mixture of two) | Physical bond (involves the balancing of weak attractive forces such as van der Waals, H-bond) |
| Layers | One monolayer (No incorporation of adsorbate into the subsurface region) | Monolayers and multilayers (as condensation) |
| Reactivity | Can cause reactivity changes in the adsorbate | Little change |
| Specificity | Highly specific (chemical bonding between adsorbate and adsorbent) | Non-specific (easy to adsorb, needs low temperature to get substantial amounts) |
| Kinetic of activation | Can be activated | Non-activated |
| Rate | Slow at low T, fast at high T | High (second) |
| Reversibility | Not reversible | Reversible |
| $\Delta H_{ads}$ (adsorption heat) | ~40–1000 kJ/mole ($\approx$ reaction heat) | ~10–40 kJ/mole ($\approx$ condensation heat) |

of a complete monolayer. This type is found for systems in which the adsorbate is chemisorbed. Type I isotherms have been observed for microporous adsorbents such as charcoal, silica gel, and molecular sieves in which the capillaries have a width of only a few molecular diameters. Type II isotherm is characteristic of forming multiple layers of the adsorbate molecule on the solid surface. Type II isotherm, which is known as the BET after Brunauer, Emmett, and Teller (1938), has been found to exist for non-porous solids [76]. Type III isotherms, although similar to Type II because they have been observed for non-porous solids, are relatively rare. The shape of Type III isotherms also suggests the formation of multilayers. Type IV and Type V isotherms reflect capillary condensation since they level off when the saturation pressure of the adsorbate vapor is reached. Both types of isotherms exhibit a hysteresis loop during desorption; porous adsorbents provide isotherms of this shape.

## *6.2.5 Single Component Monolayer Models*

Adsorption equilibrium studies represent net mass transfer between phases. Equilibrium data are used to determine the distribution of an adsorbate between the bulk fluid phase and the phase adsorbed on a solid adsorbent surface. The equilibrium distribution (equilibrium isotherm) is generally measured at a constant temperature. Several mathematical models have been developed to describe the adsorption process. In

## 6.2 Adsorption Methods

addition to monolayer and multilayer adsorption equations, models have been developed to describe situations in which the adsorbate occurs either locally on specific sites or is mobile over the surface of the adsorbent. Consideration has also been given to cases where the adsorbed molecules interact with both the surface and each other. Rather than attempting to discuss all the isotherm equations, only the isotherms that have been used to describe metal ions uptake onto chitosan and chitosan derivatives from the aqueous solution will be presented in this section. Table 6.6 shows the adsorption capacity and experimental conditions of chitosan and its derivatives for various metal ions from aqueous solutions. For a comprehensive discussion of the various isotherms and the mechanism of adsorption, the reader is referred to the monographs of Young and Crowell, Ruthven, Yang [77–79].

### 6.2.5.1 Langmuir Isotherm

Langmuir isotherm is one of the oldest and most frequently used isotherms equations [100]. The Langmuir isotherm models the monolayer coverage of the adsorption surface. The model assumes that each adsorption site accepts only one molecule, and the adsorbed molecules are organized as a monolayer. The Langmuir equation applies to cases in which there is no interaction among molecules on the surface, and the surface is energetically homogeneous. It provides a reasonable description of Type I systems and is often justified based on its ability to fit equilibrium data. The Langmuir equation can be expressed as

$$q_e = \frac{Q_0 b C_e}{1 + b C_e}, \qquad (6.4)$$

where

$q_e$ = equilibrium uptake of adsorbate by the adsorbent corresponding to concentration $C_e$,
$Q_0$ = weight of adsorbate contained in the monolayer on the surface,
$C_e$ = concentration of adsorbate in the fluid phase in equilibrium with the concentration $q_e$ on the solid,
$B$ = constant.

The Langmuir equation can be reduced to a linear form:

$$\frac{C_e}{q_e} = \frac{1}{Q_0 b} + \frac{C_e}{Q_0}. \qquad (6.5)$$

If the equilibrium provides a good fit to the data, a plot of $C_e/q_e$ versus $C_e$ should give a straight line, and the two constants $Q_0$ and b may be evaluated from the slope and intercept, respectively.

Table 6.6 Adsorption capacities and experimental conditions of chitosan and its derivatives for various metal ions from wastewater

| Adsorbent | Adsorbate | Uptake (mg/g) | pH | Kinetic model | Isotherm | References |
|---|---|---|---|---|---|---|
| Chitosan-coated diatomite | Hg (II) | 116.2 | 5 | Pseudo-second-order | Langmuir, Freundlich, D-R model | Caner et al. [80] |
| Chitosan bead | Zn (II) | 109.18 | 6 | Pseudo-second-order | Langmuir, Freundlich | Salih and Ghosh [81] |
| Chitosan bead | Cr (VI) | 79.56 | 5 | Pseudo-second-order | Langmuir, Freundlich | Salih and Ghosh [81] |
| Chitosan microsphere (MA-DETA) | Pb (II) | 239.2 | 4 | Pseudo-second-order | Langmuir, Freundlich | Zhang et al. [82] |
| Chitosan microsphere (MA-DETA) | Cd (II) | 201.6 | 5 | Pseudo-second-order | Langmuir, Freundlich | Zhang et al. [82] |
| Magnetic chitosan bead | Hg (II) | 145 | 5 | Pseudo-first-order, second-order, Elovich | Langmuir, Freundlich | Kyzas and Deliyanni [83] |
| Chitosan-ECH-TPP | Cu (II) | 130.72 | 6 | – | Langmuir, Freundlich | Laus and de Favere [84] |
| Chitosan/PVA | Cu (II) | 47.85 | 6 | Second-order | Langmuir | Wan Ngah et al. [85] |
| Chitosan-ECH-TPP | Cd (II) | 83.75 | 7 | – | Langmuir, Freundlich | Laus and de Favere [84] |
| Chitosan/zeolite film | Cr (VI) | 66.8 | 4 | – | Langmuir, Freundlich Redlich–Peterson | Batista et al. [85] |
| Chitosan/perlite | Cr (VI) | 153.8 | 4 | – | Langmuir | Hasan et al. [17] |
| Chitosan/perlite | Cd (II) | 178.6 | 6 | – | Langmuir two site | Hasan et al. [71] |
| Chitosan/perlite | Cu (II) | 104.0 | 4.5 | – | Langmuir | Hasan et al. [86] |
| Chitosan/ceramic alumina | Ni (II) | 78.1 | 4 | – | Langmuir, Freundlich Redlich–Peterson | Veera et al. [87] |
| Chitosan/alginate | Cu (II) | 67.66 | 4.5 | Pseudo-second-order | Langmuir | Wan Ngah and Fatinathan [88] |

(continued)

6.2 Adsorption Methods

Table 6.6 (continued)

| Adsorbent | Adsorbate | Uptake (mg/g) | pH | Kinetic model | Isotherm | References |
|---|---|---|---|---|---|---|
| Magnetic chitosan-TU | Hg (II) | 135 ± 3 | 5 | Pseudo-first-order, second-order | Langmuir, Freundlich | Monier and Abdel-Latif [89] |
| Magnetic chitosan-TU | Cd (II) | 120 ± 1 | 5 | Pseudo-first-order, second-order | Langmuir, Freundlich | Monier and Abdel-Latif [89] |
| Magnetic chitosan-TU | Zn (II) | 52 ± 1 | 5 | Pseudo-first-order, second-order | Langmuir, Freundlich | Monier and Abdel-Latif [89] |
| Diatomite chitosan composite | Hg (II) | 195.7 | 3 | Pseudo-first-order, second-order | Langmuir, Freundlich | Fu et al. [90] |
| Magnetic carboxymethyl chitosan | Pb (II) | 114.7 | 4–5 | Pseudo-first-order, second-order, Elovich | Langmuir, Freundlich | Wang et al. [91] |
| Chitosan nanofiber | As (V) | 11.2 | 7.2 | Pseudo-first-order, second-order, intraparticle | Langmuir, Freundlich | Min et al. [92] |
| Chitosan–iron composite flake | As (V) | 22.47 ± 0.56 | 7.0 | – | Langmuir | Gupta et al. [93] |
| Chitosan–iron composite flake | As (III) | 16.15 ± 0.32 | 7.0 | – | Langmuir | Gupta et al. [93] |
| Chitosan-FeOOH composite bead | As (V) | 5.4 | 6.5 | Pseudo-first-order | Langmuir | Hasan et al. [94] |
| Chitosan-FeOOH composite bead | As (III) | 7.2 | 6.5 | Pseudo-first-order | Langmuir | Hasan et al. [94] |
| Magnetic chitosan nanoparticle | As (III) | 60.2 | 6.8 | – | Langmuir, Freundlich | Liu et al. [95] |
| Magnetic chitosan nanoparticle | As (V) | 65.5 | 6.8 | – | Langmuir, Freundlich | Liu et al. [95] |

(continued)

Table 6.6 (continued)

| Adsorbent | Adsorbate | Uptake (mg/g) | pH | Kinetic model | Isotherm | References |
|---|---|---|---|---|---|---|
| Composite chitosan | Cd (II) | 108.7 | 6.0 | Pseudo-first-order, second-order, Elovich | Langmuir, Freundlich, D-R, Temkin | Madala et al. [96] |
| Chitosan bead | Cu (II) | 7.02 | 4.0 | Pseudo-first-order, second-order, Elovich | Langmuir, Freundlich, Temkin | Patrulea et al. [97] |
| Chitosan-glutaraldehyde composite | Cu (II) | 8.67 | 6.0 | Pseudo-first-order, second-order, Elovich | Langmuir, Freundlich, Temkin | Patrulea et al. [97] |
| Chitosan-eggshell composite | Cu (II) | 48.3 | 6.0 | Pseudo-first-order, second-order, intraparticle | Langmuir, Freundlich, Temkin | Anantha and Kota [98] |
| EDTA functionalized magnetic chitosan-graphene oxide composite | Pb (II) Cu (II) As (III) | 206.52 207.26 42.75 | 5.0 5.5 8.0 | Pseudo-first-order, second-order, | Langmuir, Freundlich | Shahzad et al. [99] |
| Chitosan with graphene oxide nanofiber | Pb (II) Cu (II) Cr (VI) | 461.3 423.8 310.4 | 6.0 6.0 3.0 | Pseudo-first-order, second order, and double exponential kinetic model | Langmuir, Freundlich, Redlich–Peterson | Najafabadi et al. [54] |
| Chitosan crosslinked with PADAB | Co (II) detection Co (II) | Color change 7.97 | 5–10 | Pseudo-second-order | Langmuir, Freundlich | Zhuang et al. [50] |
| Chitosan grafted with maleic acid by gamma radiation | Sr (II) | 2.78 | – | Pseudo-second-order | Langmuir, Freundlich, Temkin | Zhuang et al. [52] |
| Magnetic chitosan bead with *Saccharomyces cerevisiae* | Sr (II) | 81.96 | 8 | – | Langmuir, Freundlich | Yin et al. [51] |
| Biochar modified with chitosan and pyromellitic anhydride (CPMB) | Pb (II) Cd (II) Cu (II) | 11.91 33.89 92.19 | 5 | Pseudo-second-order | Langmuir, Freundlich at 303 K | Deng et al. [53] |

## 6.2 Adsorption Methods

The Langmuir isotherm can be correlated with a dimensionless separation factor $R$, which is defined by Equation 6.6 [101]:

$$R = \frac{1}{1 + bC_0}, \tag{6.6}$$

where

$b$ = the Langmuir constant (L/mmol), and
$C_0$ = the initial concentration (mmol/L) of the adsorbate in the solution.

The value of R indicates the shape of the isotherm to be unfavorable ($R > 1$), linear ($R = 1$). The adsorption is considered irreversible when $R = 0$, favorable when $0 < R < 1$ [102]. It is important to note that a good fit of the data by the model is a necessary but not sufficient condition to verify the accuracy of the assumptions imposed on the equation [103].

***Example 6.1*** The adsorption of cesium on chitosan-Fuller's earth bead was studied by Hasan et al. [74] at 303 K. Using the data presented below, determine if the Langmuir equation can be used to model the data.

| Initial concentration of cesium (mmol/L) | Liquid-phase concentration of cesium at equilibrium ($C_e$, mmol/L) | Uptake by the adsorbent at equilibrium ($q_e$, mmol/g solid) |
|---|---|---|
| 0.414 | 0.323 | 0.0361 |
| 0.752 | 0.602 | 0.0602 |
| 1.654 | 1.391 | 0.1053 |
| 3.834 | 3.421 | 0.1504 |
| 7.368 | 6.955 | 0.1654 |

Copyright © 2007, Taylor & Francis, Sep. Sci. and Technol. 42, 717–738, 2007. Above table is reproduced from Ref. [74] with permission

If the Langmuir equation can be used, then a plot of $C_e/q_e$ versus $C_e$ should give a straight line.

| $C_e$ (mmol/L) | $C_e/q_e$ (g solid/L) |
|---|---|
| 0.323 | 8.95 |
| 0.602 | 10 |
| 1.391 | 13.21 |
| 3.421 | 22.746 |
| 6.955 | 42.05 |

From Fig. 6.4, the intercept is $I = \frac{1}{Q_0 b} = 6.587$, and the slope is $\frac{1}{Q_0} = 5.005$.

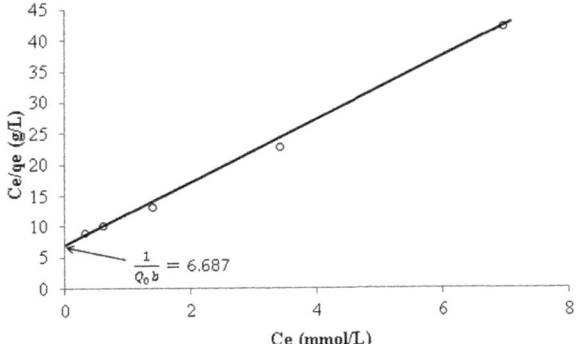

**Fig. 6.4** Solution to Example 6.1 (obtaining Langmuir parameters)

Thus, $Q_0 = 1.998$ mmol/g solid.

$$\frac{1}{Q_0 b} = 6.687.$$

$$b = \frac{1}{6.687 Q_0} = 7.48 \times 10^{-2} (\text{L/mmol}).$$

Ghosh [104] reported that the projected area of a molecule on the solid surface could be calculated from the following expression:

$$\alpha = 1.091 \left( \frac{M}{N_0 \rho} \right)^{2/3}, \tag{6.7}$$

where

$M$ = the molecular weight,
$\rho$ = the density of the adsorbed molecules, and
$N_0$ = Avogadro's number.

Once the projected area $\alpha$ is known, the surface area can be calculated from the following equation.

$$S = \frac{V_m N_0 \alpha}{V_s}, \tag{6.8}$$

where

$S$ = the surface area,
$V_m$ = the monolayer coverage of solids, and
$V_s$ = the standard volume at STP.

## 6.2 Adsorption Methods

Chiu and Wang [105] reported a similar equation to estimate the surface area of the adsorbent for cobalt uptake from an aqueous solution:

$$S = \frac{q_m N_0 \alpha}{M}, \qquad (6.9)$$

where

$S$ = the specific surface area,
$q_m$ = the monolayer adsorption of cobalt on the adsorbent,
$N_0$ = the Avogadro's number,
$\alpha$ = the cross-sectional area of cobalt ions, and
$M$ = the molecular weight of cobalt ions.

### 6.2.5.2 Freundlich Isotherm

Another frequently used isotherm is the semiempirical Freundlich equation as

$$q_e = K(C_e)^{1/n}. \qquad (6.10)$$

The equation can be rearranged to linear form:

$$\log q_e = \log K + \frac{1}{n} \log C_e, \qquad (6.11)$$

where $q_e$ = equilibrium uptake of adsorbate by the adsorbent corresponding to concentration $C_e$; $C_e$ = concentration of adsorbate in the fluid phase that is in equilibrium with the concentration $q_e$ on the solid:

$$K, n = \text{constants}.$$

A plot of $\log q_e$ versus $\log C_e$ will yield a straight line with a slope of $1/n$ and an intercept equal to $\log K$, provided that the data point follows the isotherm. This isotherm is related to the heterogeneous surface and multilayer adsorption and is frequently used for the adsorption data at low to intermediate solution concentration range. The factor $1/n$ indicates if the adsorption process is favorable for the adsorbate ion. If the values of $1/n$ are smaller than 0.5, it shows that the adsorbate is easily adsorbed, and as the value gets closer to zero reflects more surface heterogeneity. However, if the values of $1/n$ are larger than 2, it indicates that the adsorbate is hardly adsorbed [106]. It is reported that the value of $1/n$ below unity implies the chemisorption process or the most common and corresponds to a normal L-type Langmuir isotherm [107, 108]. The L-type isotherm indicates that the proportion of sorbate adsorbed increases more slowly as the amount of material adsorbed increases, where $1/n$ above one indicates cooperative adsorption, which involves strong interactions between the adsorbate molecules [107]. It is important to note that the Langmuir equation reduces to a linear form, frequently identified as Henry's law equation, as the fluid phase concentration approaches zero; whereas, the Freundlich equation does not. Henry's law equation is given as

$$q_e = K'C_e. \tag{6.12}$$

### 6.2.5.3 Dubinin–Radushkevich Isotherm Model

The Dubinin–Radushkevich (D–R) isotherm is an empirical model that is generally used to express the temperature-dependent adsorption mechanism with a Gaussian energy distribution to estimate the heterogeneity of the surface site energy. The D–R isotherm predicts the porous structure of the sorbent and the apparent energy of adsorption [109]. Dubinin–Radushkevich equation (D–R) is very widely used to describe the adsorption isotherm of subcritical vapors in microporous solids such as activated carbon and zeolite [110]. One of the assumptions of this model is that the adsorption has a multilayer character with no constant potential energies [111]. The linearized form of the D–R equation is

$$\ln q_e = \ln q_{max} - K'\varepsilon^2, \tag{6.13}$$

where

$q_e$ = the adsorbate concentration in the solid at equilibrium (mol/g),
$q_{max}$ = the maximum adsorption capacity (mol/g),
$\varepsilon$ = the Polanyi potential (kJ/mol), and it is equal to:

$$\varepsilon = RT \ln\left(1 + \frac{1}{C_e}\right), \tag{6.14}$$

where

$C_e$ = the adsorbate ion concentration in the solution at equilibrium (mol/L),
$R$ = the gas constant ($8.314 \times 10^{-3}$ kJ/mol K), and
$T$ = the absolute temperature ($K$).

$K'$ = the constant related to the mean free energy of the adsorption per mole of adsorbate (mol$^2$/kJ$^2$). The constant $K'$ can be used to estimate adsorption energy ($E$). The adsorption energy ($E$) is related to the mean free energy of sorption per molecule of the adsorbate. It is transferred to the surface of the solid from infinity in the solution. The adsorption energy can be estimated from the following equation:

$$E = (2K')^{-1/2}. \tag{6.15}$$

## 6.2 Adsorption Methods

The value of $E$ provides valuable information about the adsorption mechanism. A value of E between 8 and 16 kJ/mol indicates ion exchange as the main adsorption process in the system. If the value is less than 8 kJ/mol, physisorption is the main adsorption mechanism [112]. The D–R isotherm model is more general than the Langmuir isotherm model because it does not assume a homogeneous surface or constant adsorption model [113]. It has often successfully fitted high solute activities and the intermediate range of concentration data well. It is reported that this isotherm has unsatisfactory asymptotic properties and does not predict Henry's law at low pressure, which is a necessity for a thermodynamically consistent isotherm [114].

### 6.2.5.4 Temkin Isotherm

Temkin equation is originally used to describe hydrogen adsorption on platinum electrodes in acidic solutions that describe chemisorption systems. The isotherm suggests that the heat of adsorption of all molecules in the layer should decrease linearly with the surface coverage because of adsorbate–adsorbent interactions. The derivation of the Temkin equation implies a homogeneous distribution of binding energies to all adsorption sites. The linear form of the Temkin isotherm model is as follows:

$$q_e = \frac{RT}{b} \ln K_T + \frac{RT}{b} \ln C_e. \tag{6.16}$$

The equation can be written assuming $K_1 = RT/b$:

$$q_e = K_1 \ln K_T + K_1 \ln C_e, \tag{6.17}$$

where

$R$ is gas constant (J mol$^{-1}$ K$^{-1}$),
$T$ = the temperature in Kelvin,
$K_T$ = dimensionless Temkin isotherm constant, and
$K_1$ = Temkin constant related heat of adsorption (L/g).

The constants can be determined from a plot of $q_e$ versus $\ln C_e$. This isotherm has Henry's law limit, and the liquid-phase adsorption isotherm shows an inadequate fit than in the case of the Langmuir [110, 115].

### 6.2.5.5 Redlich–Peterson Isotherm

Redlich–Peterson [116] developed an empirical isotherm by incorporating the features of both Langmuir and Freundlich isotherm. This isotherm applies to heterogeneous adsorbents and can function by more than one mechanism [115]. The isotherm model is as follows:

$$q_e = \frac{K_R C_e}{1 + a_R C_e^\beta}, \qquad (6.18)$$

where

$K_R$ = Redlich–Peterson isotherm constant (L/g),
$a_R$ = constant (Lmg$^{-1}$).

The exponent $\beta$ lies between 0 and 1. When $\beta = 1$, the Redlich–Peterson equation reduces to the Langmuir equation. When $\beta = 0$, Henry's law form results considering $(1/(1+a_R))$ as a constant. At low concentrations, the Redlich–Peterson isotherm approximates Henry's law, and at high concentrations, its behavior approaches that of the Freundlich isotherm [117]. Ayawei et al. [118] reported that this isotherm model has a linear dependence on concentration in the numerator and an exponential function in the denomination, which altogether represent adsorption equilibrium over a wide range of concentration of adsorbate, which is applicable in either homogeneous or heterogeneous systems because of its versatility.

### 6.2.5.6 Sips Isotherm

Sips (1948) reported an isotherm, which is a combination of Langmuir and Freundlich isotherm [119]. This isotherm is derived from the limiting behavior of the Langmuir and Freundlich isotherm that predicts heterogeneous adsorption systems. The Sips isotherm is used for the evaluation of isosteric heat of gas adsorption [120]. At low adsorbate concentration, the model reduces to the Freundlich isotherm model and does not provide the correct Henry's law limit. At high adsorbate concentration, it predicts monolayer adsorption characteristics of the Langmuir model. The isotherm for the liquid phase is as follows:

$$q_e = \frac{q_m (K_s C_e)^{1/n}}{1 + (K_s C_e)^{1/n}}. \qquad (6.19)$$

The linearized form of the Sips isotherm equation is as follows [121]:

$$\frac{1}{q_e} = \frac{1}{q_m K_s}\left(\frac{1}{C_e}\right)^{1/n} + \frac{1}{q_m}, \qquad (6.20)$$

where

$K_s$ (L/mg) = Sips equilibrium constant, and
$q_m$ (mg/g) = maximum adsorption capacity values.

The maximum adsorption capacity value can be obtained from the slope and the intercept of the plot of $1/q_e$ and $(1/C_e)^{1/n}$. The Sips isotherm equation is characterized by the dimensionless heterogeneity factor, $n$, which can also be employed to

## 6.2 Adsorption Methods

describe the system's heterogeneity when $n$ is between 0 and 1. When $n = 1$, the Sips equation reduces to the Langmuir equation, and it implies a homogeneous adsorption process [121]. Do [110] reported that the behavior of this isotherm is the same as the Freundlich equation except that the Sips equation possesses a finite saturation limit when the pressure is sufficiently high. Table 6.7 shows adsorption isotherms that have been used to express heavy metal ion uptake on the chitosan-based adsorbent.

Jovanovic [122] proposed another simple equation that applies to Type I isotherms, which can be expressed in terms of the volume adsorbed as

$$V = V_m(1 - \exp(-\alpha P/P_s)), \tag{6.21}$$

where

$V_m$ = the volume adsorbed in the monolayer,
$P_s$ = the saturation pressure, and
$\alpha$ = constant that describes the adsorption in the monolayer.

The term $\alpha$ is expressed as

$$\alpha = \sigma \tau P_s / \sqrt{2\pi mkT}, \tag{6.22}$$

where

$\sigma$ = the area occupied by one molecule on the surface,
$\tau$ = the average settling time of a molecule adsorbed in the first layer,
$m$ = the mass of one molecule,
$T$ = the absolute temperature, and
$k$ = the Boltzman constant.

The Jovanovic equation has been used extensively to develop several more complicated and accurate isotherm models that apply to heterogeneous surfaces. Ayawei et al. [118] reported the Jovanovic isotherm model that is predicted on the assumptions contained in the Langmuir model for liquid-phase adsorption. The linear form of Jovanovic isotherm is given as follows [118]:

$$\ln q_e = \ln q_{\max} - K_j C_e, \tag{6.23}$$

where

$q_e$ = the amount of adsorbate at equilibrium (mg/g),
$q_{\max}$ = the maximum uptake of adsorbate obtained from the plot of $\ln q_e$ versus $C_e$, and
$K_j$ = Jovanovic constant.

All the above models apply to the Type I isotherms but frequently fail to fit the equilibrium data with suitable accuracy. This is often attributed to the heterogeneous

**Table 6.7** List of adsorption isotherms

| Isotherm | Nonlinear form | Linear form | Plot | Remark |
|---|---|---|---|---|
| Langmuir | $q_e = \frac{Q_0 b C_e}{1 + b C_e}$ | $\frac{C_e}{q_e} = \frac{1}{Q_0 b} + \frac{C_e}{Q_0}$ | $\frac{C_e}{q_e}$ versus $C_e$ | – Chemisorption and physical adsorption<br>– Monolayer coverage of the adsorption surface<br>– Surface is energetically homogeneously<br>– Henry's law equation as fluid phase concentration approaches to zero |
| Freundlich | $q_e = K(C_e)^{1/n}$ | $\log q_e = \log K + \frac{1}{n} \log C_e$ | $\log q_e$ versus $\log C_e$ | – Chemisorption and physical adsorption at low coverages<br>– The isotherm is related to heterogeneous surface<br>– Multilayer adsorption<br>– Use adsorption data at low to intermediate solution concentration<br>– Does not follow Henry's law equation even the fluid phase concentration approaches to zero |

(continued)

**Table 6.7** (continued)

| Isotherm | Nonlinear form | Linear form | Plot | Remark |
|---|---|---|---|---|
| Dubinin–Radushkevich | $q_e = q_{max} \exp(-K'\varepsilon^2)$ | $\ln q_e = \ln q_{max} - K'\varepsilon^2$ | $\ln q_e$ versus $\varepsilon^2$ | – Describe temperature-dependent adsorption mechanism<br>– Estimate heterogeneity of surface site energy<br>– It fitted well high and intermediate-Range of concentration data<br>– Isotherm does not predict Henry's law at low pressure<br>– Multilayer formation in microporous solids |
| Temkin | $q_e = \dfrac{RT}{b} \ln K_T C_e$ | $q_e = \dfrac{RT}{b} \ln K_T + \dfrac{RT}{b} \ln C_e$ | $q_e$ versus $\ln C_e$ | – Chemisorption<br>– Assume homogeneous distribution of binding energies to all adsorption sites<br>– Isotherm has Henry's law limit<br>– Liquid-phase adsorption isotherm shows a poor fit |

(continued)

**Table 6.7** (continued)

| Isotherm | Nonlinear form | Linear form | Plot | Remark |
|---|---|---|---|---|
| Redlich–Peterson | $q_e = \dfrac{K_R C_e}{1 + a_R C_e^{\beta}}$ | $\ln\left(K_R \dfrac{C_e}{q_e} - 1\right) = \beta \ln C_e + \ln a_r$ | $\ln\left(K_R \dfrac{C_e}{q_e} - 1\right)$ versus $\beta \ln C_e$ | – Applicable to heterogeneous surface<br>– At low concentration, the isotherm approximates Henry's law<br>– At high concentration, its behavior approaches Freundlich isotherm |
| Sips | $q_e = \dfrac{q_m (K_s C_e)^{1/n}}{1 + (K_s C_e)^{1/n}}$ | $\dfrac{1}{q_e} = \dfrac{1}{q_m K_s}\left(\dfrac{1}{C_e}\right)^{1/n} + \dfrac{1}{q_m}$ | $\dfrac{1}{q_e}$ versus $\left(\dfrac{1}{C_e}\right)^{1/n}$ | – Predicts heterogeneous adsorption systems<br>– At low concentration, it reduces to Freundlich isotherm and does not provide the correct Henry's law limit |

Copyright © 2009 Elsevier B.V. Chemical Engineering Journal, 156, 2–10, 2010. Reproduced from Ref. [108] with permission

nature of the adsorbent surface. As a result, other models have been developed for Type I systems considering the energy of an adsorption site. It is well known that surface heterogeneity plays a vital role in determining adsorption characteristics for a large class of solid–vapor systems, and its effects require adequate treatment. The general approach used to describe adsorbent heterogeneity is to postulate that the heterogeneous surface exhibits a distribution of adsorptive potentials grouped in patches or distributed randomly on the surface. Even small variations in the adsorption potential have been shown to influence the adsorption behavior.

Ross and Olivier [123] reported that the overall isotherm could be obtained by integrating the contribution of each patch over the energy distribution range. The adsorption isotherm is thus given as

$$Q(P, T) = \int_0^\infty Q_1(P, T, e) E(e) de, \qquad (6.24)$$

where

$E(e)$ = probability distribution function, and
$Q(P, T)$ = the overall adsorption isotherm on the heterogeneous adsorbent.

The term $Q_1(P, T, e)$ describes a specific adsorption isotherm for homotactic sites of adsorption energy $e$. Both $Q$ and $Q_1$ are the amounts adsorbed per unit mass of adsorbent. Ross and Olivier [123] used a number of probability distribution functions for $E(e)$, including a Gaussian function. They employed the two-dimensional van der Waals equation of state for the adsorption isotherm $Q(P, T)$ [123], and a detailed review of the studies that employ the approach is given by Jaroniec et al. [124].

### 6.2.6 Single Component Multilayer Models

The formation of multiple adsorbed layers has been observed for a large number of adsorbate–adsorbent systems. Although multilayer adsorption is a physical process, only a limited number of models have contributed significantly to its understanding. These models often fail to fit equilibrium adsorption data over a wide range of pressures.

The earliest multilayer model was proposed by Brunauer et al. [75] and included the basic assumptions of the Langmuir equation, except that multilayers are formed, and the heat of adsorption for these layers is different from the first layer. Their model, which is typically referred to as the BET equation, is given as

$$\frac{P}{V(P_s - P)} = \frac{1}{V_m C} + \frac{(C-1)P}{(CV_m P_s)}, \qquad (6.25)$$

where

$V$ = the volume adsorbed at pressure $P$,
$V_m$ = the volume occupied in a monolayer,
$P_s$ = the saturation pressure, and
$C$ = constant that is related to the energy of adsorption.

Although the BET equation typically is valid over a relative pressure range, $P/P_s$, from 0.05 to 0.35, and it would consequently not be used to fit adsorption data, it is widely used as a method for determining surface areas of adsorbents. A plot of $P/(V(P_s - P))$ versus $P/P_s$ should give a straight line with a slope:

$$S = \frac{C-1}{V_m C}, \qquad (6.26)$$

and an intercept,

$$I = \frac{1}{V_m C}. \qquad (6.27)$$

The volume of the adsorbed gas that corresponds to a monolayer can be obtained by solving the equations. Thus,

$$V_m = \frac{1}{S+I}. \qquad (6.28)$$

A more recent multilayer adsorption model proposed by Jovanovic provides a method of correlating data over a wider relative pressure range, $0.2 < P/P_s < 0.7$, than does the BET equation [122]. This model is given as

$$V = V_m(1 - \exp(-\alpha\, P/P_s))\exp(b\, P/P_s), \qquad (6.29)$$

where $\alpha$ was defined earlier for the Jovanovic monolayer model and b is related to uptake in the second and higher layers.

## 6.3 Error Analysis for Isotherm Studies

In isotherm modeling of experimental data, the error function equation can evaluate the fit of the isotherm model with the experimental equilibrium data. Various error functions equation has been used to evaluate the isotherm fit to the experimental data, and they are as follows [115]:

1. The sum of the square errors (ERRSQ).
2. The hybrid functional error function (HYBRID).

## 6.3 Error Analysis for Isotherm Studies

3. The average relative error (ARE).
4. Marquardt's percent standard deviation (MPSD).
5. The sum of absolute error (EABS).
6. The coefficient of determination.
7. Nonlinear chi-square test.

Table 6.8 shows five nonlinear common error functions considered useful for metal ion–chitosan.

As shown in Table 6.8, the equations use $q_{e,\exp}$ value that is the metal ion uptake on solid adsorbent from the experimental data. The $q_{e,\text{calc}}$ value is the theoretical uptake of metal ions on solid adsorbent that is calculated from the isotherm model. The most commonly used error function is the sum of the squares error (SSE).

The SSE error function's major drawback is its inability to provide better adjustment of the data in the final portion of the isotherm due to the magnitude of the errors, which causes an increase in squared errors as the adsorbate concentration increases [125, 126]. The HYBRID error function is an extension of SSE. This function was developed to improve fit the SSE at low concentration by dividing the value of SSE by the experimental value [125]. Chan et al. [127] reported that the error function Marquardt's percent standard deviation (MPSD) is similar to some respects of a modified geometric mean error distribution according to the number of degrees of freedom in the system. The sum of the absolute error (EABS) provides better adjustment for higher concentrations. This phenomenon occurs because of an increase in the concentration range, which causes an increase in error. The average relative error (ARE) function attempts to minimize the fractional error distribution across the entire concentration range.

The traditional approach of determining the kinetic parameters by linear regression of kinetic equations is the correlation coefficients (r), which are in the range of

**Table 6.8** List of error functions

| Error function | Expression |
|---|---|
| Sum squares errors (ERRSQ/SSE) | $\sum_{i=1}^{P} (q_{ei,\exp} - q_{ei,\text{calc}})_i^2$ |
| Composite fractional error function (HYBRID) | $\sum_{i=1}^{P} \left[ \frac{(q_{ei,\exp} - q_{ei,\text{calc}})^2}{q_{ei,\exp}} \right]_i$ |
| Average relative error (ARE) | $\sum_i^P \left\lvert \frac{q_{ei,\exp} - q_{ei,\text{calc}}}{q_{ei,\exp}} \right\rvert_i$ |
| Marquardt's percent standard deviation (MPSD) | $\sum_i^P \left( \frac{q_{ei,\exp} - q_{ei,\text{calc}}}{q_{ei,\exp}} \right)_i^2$ |
| Sum of absolute error (EABS) | $\sum_i^P \lvert q_{ei,\exp} - q_{ei,\text{calc}} \rvert_i$ |

Copyright © 2017 Nimibofa Ayawei et al. Creative Commons Attribution CC BY License, Journal of Chemistry, Volume 2017, Article ID 3,039,817, 11 pages, https://doi.org/10.1155/2017/3039817. Reproduced from Ref. [118]; Copyright © 2002 Elsevier Science (USA). All rights reserved. Journal of Colloid and Interface Science, Volume 255, Issue 1, 1 November 2002, p 64–74. Reproduced from Ref. [118a] with permission

zero to unity. The coefficient of determination represents that the percentage of variability in the dependent variable is employed to analyze the fitting degree of isotherm and kinetic models with the experimental data [108]. The coefficient of determination ($R^2$) is generally used to test the best fit kinetic model to the experimental data can be expressed as follows [128]:

$$R^2 = \sum \frac{(q_m - \overline{q}_t)^2}{(q_m - \overline{q}_t)^2 + (q_m - q_t)^2}, \qquad (6.30)$$

where

$q_m$ = the amount of metal ion uptake on solid obtained from the kinetic model (mg/g), and
$q_t$ = the amount of metal ion uptake on solid at any time $t$ (mg/g) obtained from the experiment,
and $\overline{q}_t$ = the average of $q_t$ (mg/g).

The use of $R^2$ is limited to solving linear forms of the kinetic equation. The $R^2$ value measures the difference between the experimental and theoretical data in linearized plots only, but not the errors in the nonlinear form of kinetic curves [129]. Another important statistical tool in the best fit of an adsorption system is the nonlinear chi-square ($\chi^2$) test. It can be obtained by the difference between experimental and calculated data, with each square difference divided by its corresponding values calculated from the model [108, 118]. A small $\chi^2$ value indicates its similarities, while a larger number represents the variation of the experimental data [108]. The $\chi 2$ value can be obtained using the following equation:

$$\chi^2 = \sum_{i=1}^{n} \frac{(q_{e,\text{cal}} - q_{e,\text{exp}})^2}{q_{e,\text{exp}}}. \qquad (6.31)$$

Ho and McKay [109] reported that each error criteria is likely to produce a different set of isotherm parameters. Several studies reported a calculation method for the sum of normalized error [108, 115, 118] that is provided by Ho and McKay [109] is as follows:

(a) Selection of one isotherm model and error function and determine the adjustable parameters that minimize the error function.
(b) Determine the values for all the other error functions for that isotherm parameter set.
(c) Calculate all other parameter sets and all their associated error function values (initiation of the procedure by minimizing the error function) for that isotherm.
(d) Normalization and selection of the maximum parameter sets with respect to the largest error measurement.
(e) Summation of each parameter set, which generates the minimum normalization error.

## 6.4 Thermodynamic Parameters

### 6.4.1 Pseudo-First-Order Kinetics and Equilibrium Adsorption Isotherm

Adsorption kinetics data explain the rate of adsorption of metal ions on the solid phase. The adsorption rate could be described by the Lagergren equation, which is a special case of the general Langmuir rate equation [130]. For a batch contact process, where the rate of sorption of adsorbate onto the adsorbent surface is proportional to the amount of adsorbate uptake from the solution phase, a pseudo-first-order equation may be used. The equation is given by

$$\frac{dq_t}{dt} = k_1(q_e - q_t), \quad (6.32)$$

where

$k_1$ = the rate constant of pseudo-first-order adsorption, and
$q_e$ = the amount adsorbed at equilibrium.

The integration of Eq. 6.32 with the following initial conditions provides Eq. 6.34:

$$q_t = 0 \text{ at } t = 0, \quad (6.33a)$$

$$q_t = q_t \text{ at } t = t. \quad (6.33b)$$

Equation 6.10 becomes

$$\ln(q_e - q_t) = \ln q_e = k_1 t. \quad (6.34)$$

The constant $k_1$ is obtained by plotting $\ln(q_e - q_t)$ versus $t$. The slope of the straight line provides the value of $k_1$. The logarithmic form of the equation is as follows:

$$\log(q_e - q_t) = \log q_e - k_1 t/2.303. \quad (6.35)$$

Gerente et al. [115] reported that the pseudo-first-order equation of Lagergren does not give theoretical $q_e$ values that agree with experimental $q_e$ values, and the plots are only linear over the initial 20–30 min of the sorption process [115].

## 6.4.2 Pseudo-Second-Order Kinetic Model

The pseudo-second-order equation is greatly influenced by the sorption capacity on the solid phase. The rate is directly proportional to the number of active surface sites. The differential equation is as follows [131]:

$$\frac{dq_1}{dt} = k(q_e - q_t)^2. \tag{6.36}$$

Integrating equation for the boundary conditions $t = 0$ to $t = t$ and $q_t = 0$ to $q_t = q_t$ gives

$$\frac{1}{(q_e - q_t)} = \frac{1}{q_e} + kt. \tag{6.37}$$

The equation can be rearranged to obtain a linear form

$$\frac{t}{q_t} = \frac{1}{k_2 q_e^2} + \frac{1}{q_e} t. \tag{6.38}$$

The product $k_2 q_e^2$ is the initial sorption rate "$h$"

$$h = k_2 q_e^2, \tag{6.39}$$

where

$q_e$ = the amount of strontium uptake on adsorbent at equilibrium (mg/g),
$q_t$ = the uptake at any time t.

The pseudo-second-order rate constant $k_2$ (g. mg$^{-1}$. min) can be obtained by plotting $\frac{t}{q_t}$ versus t at different adsorbate concentrations. The slope and intercept of the straight line of second-order kinetic plots provide the value of $k_2$ and $q_e$, respectively.

## 6.4.3 The Elovich Equation

The equation is first proposed by Roginsky and Zeldovich but now generally known as the Elovich equation has been extensively applied to chemisorption data [132]. The general expression of the Elovich equation is as follows:

$$\frac{dq_t}{dt} = \alpha \exp(-\beta q_t). \tag{6.40}$$

## 6.4 Thermodynamic Parameters

Integrating equation for the boundary conditions $t = 0$ to $t = t$ and $q_t = 0$ to $q_t = q_t$ gives

$$q_t = \beta \ln(\alpha\beta) + \ln(t), \tag{6.41}$$

where

$q_t =$ the sorption capacity at any time t (mg/g),
$\alpha =$ the initial sorption rate (mg/g. min), and
$\beta =$ the desorption constant (g/mg) of the Elovich model during any one experiment.

The constants can be obtained from the slope and intercepts of the plot of $q_t$ versus $\ln(t)$. This equation can be tested for the applicability of the Elovich equation.

### 6.4.4 Intraparticle Diffusion

The adsorption mechanism can also be identified by investigating the contribution of intraparticle diffusion in the adsorption process. Sundaram et al. [128] reported that the solute transfer is usually characterized either by particle diffusion or pore diffusion control for a solid–liquid sorption process [128]. A simple equation for particle diffusion-controlled sorption process is as follows:

$$\ln\left(1 - \frac{C_t}{C_e}\right) = -K_p t, \tag{6.42}$$

where

$K_p =$ the particle rate constant (min$^{-1}$).

The value of the particle rate constant is obtained by the slope of $\ln\left(1 - \frac{C_t}{C_e}\right)$ against $t$. Fierro et al. [133] reported that the overall adsorption might be controlled either by one or more processes [133], for example, film or external diffusion, pore diffusion, surface diffusion, and adsorption on the pore surface, or a combination of more than one step. In general, a process is diffusion controlled if its rate depends on the rate at which components diffuse toward one another. In this regard, the Weber–Morris model is the most commonly used technique to investigate the kinetics of adsorbate uptake onto adsorbent [134]:

$$q_t = k_{id} t^{1/2} + \theta. \tag{6.43}$$

In Eq. 6.43,

$q_t =$ the amount of adsorbate uptake (mg/g) at any time $t$,

$K_{id}$ = the rate constants of intraparticle transport (mg/g. min$^{1/2}$),
$\theta$ (mg/g) = constant related to the thickness of the boundary layer diffusion.

If the intraparticle diffusion is the rate-controlling step in the adsorption process, then the linear plot of $t^{1/2}$ versus the amount of adsorbate uptake, according to the Weber–Morris model, should pass through the origin.

In order to understand the rate-controlling mechanism of intraparticle diffusion, the metal ion sorption kinetics was further analyzed for the adsorbent particle with radius r at any time t that can be expressed following the equation given by Crank [135]:

$$F = 1 - \frac{6}{\pi^2} \sum_{m=1}^{\infty} \frac{1}{m^2} \exp\left(-\frac{Dm^2\pi^2 t}{r^2}\right), \tag{6.44}$$

where $F = \frac{Q_t}{Q_e}$ is the fractional attainment of the metal ion. The ability of Eq. 6.44 to describe the sorption kinetics is limited due to the complex pore size, pore size distribution, particle size, and physicochemical changes such as swelling that may occur in adsorbent beads [136] To investigate the function of diffusion in controlling the rate factor for the sorption of metal ion on an adsorbent bead from aqueous solution, Eq. 6.44 can be approximated to

$$(1 - F) = \frac{6}{\pi^2} \exp(-D\pi^2 t/r^2), \tag{6.45}$$

wherein $(1 - F)$ is the fraction of metal ion uptake from solution on the adsorbent bead at any time ($t$). Therefore, in terms of the fraction of metal ion remaining in the bead, Eq. 6.45 can be rewritten as

$$\ln(1 - F) = \ln\left(\frac{6}{\pi^2}\right) - \frac{D\pi^2 t}{r^2}. \tag{6.46}$$

Assuming $\ln\left(\frac{6}{\pi^2}\right)$ as a constant $\varphi_r$ so that it includes any anomalies arising out of such factors as swelling, and replacing $D\pi^2/r^2$ by a constant $K$, we can rewrite Eq. 6.46 for a particle of radius $r$ as

$$\ln(1 - F) = -Kt^n + \varphi_r, \tag{6.47}$$

where $n = 0.5$ represents Fickian diffusion and $1 > n > 0.5$ represents anomalous diffusion.

**Example 6.2** Typical data for the fractional uptake of strontium on a chitosan-based bead from a 220 mg/L solution at different temperatures are given below [137]. (Conditions: the amount of bead 0.25 g; the amount of solution 100 mL and the pH of the solution is 6.5). a) Estimate the diffusion kinetics (Table 6.9).

## 6.4 Thermodynamic Parameters

**Table 6.9** Typical example of fractional uptake of strontium on the chitosan-based bead

| Time (h) | Fractional uptake F at different temperature | | |
|---|---|---|---|
| | 293 K | 298 K | 308 K |
| 0.17 | 0.1486 | 0.1596 | 0.1443 |
| 0.5 | 0.272 | 0.2702 | 0.2615 |
| 1 | 0.402 | 0.4006 | 0.4139 |
| 2 | 0.556 | 0.5453 | 0.5414 |
| 4 | 0.7313 | 0.7391 | 0.7619 |
| 8 | 0.9096 | 0.9031 | 0.9377 |
| 16 | 0.9596 | 0.9689 | 0.9817 |
| 24 | 1 | 1 | 1 |

Copyright © 2019 Hasan et al. Licensee MDPI, Basel, Switzerland. Creative Commons Attribution (CC BY) license. Healthcare, 2019, 7, 52, https://doi.org/10.3390/healthcare7010052, Reproduced from Ref. [137]

**Solution**

According to equation $\ln(1 - F) = -Kt^n + \varphi_r$, the values of the constant $K$ and $\varphi_r$ are obtained from the slope and intercept of the plot of $\ln(1 - F)$ versus $t^n$. The resulting plot is shown in Fig. 6.5.

For each temperature, the data were treated for "$n$" values 0.5 and 0.65, and it was observed that strontium uptake from the solution on the bead showed a straight line when the n value was 0.65 (Fig. 6.5). The plots are linear and passed through close to the origin (Fig. 6.5) for all the temperatures. This indicates that the adsorption of strontium onto chitosan-Fuller's earth bead may follow anomalous diffusion [137], but further experimental work is needed to quantify these data. It is reported that the polymer whose glass transition temperature is higher than the experimental temperature may follow anomalous diffusion [137a].

Diffusion parameters obtained from Fig. 6.5 are as follows (Table 6.10).

The diffusion coefficient was estimated from the slope ($K$) and the mean bead radius of 0.11 cm using the equation:

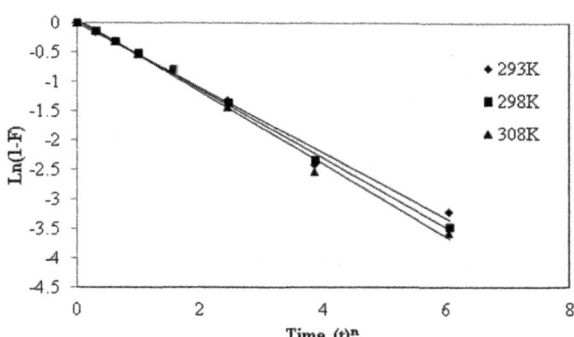

**Fig. 6.5** Effect of temperature on strontium diffusion on CF beads. Copyright © 2019 Hasan et al. Licensee MDPI, Basel, Switzerland. Creative Commons Attribution (CC BY) license., Healthcare, 2019, 7, 52, https://doi.org/10.3390/healthcare7010052, Reproduced from Ref. [137]

**Table 6.10** Diffusion parameters

| Temperature (K) | $K$ | $\varphi_r$ | $R^2$ | D (cm$^2$/sec) | Absolute Error (%) |
| --- | --- | --- | --- | --- | --- |
| 293 | 0.554 | 0.0101 | 98.93 | $1.56 \times 10^{-7}$ | $-1.86$ |
| 298 | 0.5854 | 0.0469 | 99.72 | $1.65 \times 10^{-7}$ | 0.7634 |
| 308 | 0.687 | 0.118 | 99.09 | $1.94 \times 10^{-7}$ | 1.1622 |

Copyright © 2019 Hasan et al. Licensee MDPI, Basel, Switzerland. Creative Commons Attribution (CC BY) license., Healthcare, 2019, 7, 52, https://doi.org/10.3390/healthcare7010052, Reproduced from Ref. [138]

For 293 K temperature, the value for $D = \frac{0.554 \times (0.11)^2}{(3.14)^2} = 1.56 \times 10^{-7} \text{cm}^2/\text{s}$.

The function yield values range from $1.56 \times 10^{-7}$ to $1.94 \times 10^{-7}$ cm$^2$/s for the particle diffusion coefficient for the temperature range of 293–308 K [137].

## 6.5 Thermodynamic Parameters

Gibbs free energy $\Delta S$ change equation can be used to gain insight into the thermodynamic nature of the sorbate sorption process on to adsorbent. The Gibbs free energy change, $\Delta G$, can be used to identify whether the chemical reaction that occurs during the adsorption process is a spontaneous reaction or not. Therefore, it is considered an essential criterion for spontaneity. The Gibbs free energy equation for an adsorption process can be expressed as follows:

$$\Delta G = -RT \ln k_c, \qquad (6.48)$$

where $k_c$ is the sorption equilibrium constant, $R$ is the gas universal constant, and $T$ is the absolute temperature (K). The sorption equilibrium constant ($k_c$) can be calculated from:

$$k_c = \frac{F}{(1-F)}, \qquad (6.49)$$

where $F$ is the fraction attainment of metal ion sorbate at equilibrium.

Both enthalpy and entropy factors can be estimated from the Gibbs free energy of the process. The Gibbs free change can be represented by the following equation:

$$\Delta G = \Delta H - T\Delta S. \qquad (6.50)$$

The values of enthalpy change ($\Delta H$) and entropy change ($\Delta S$) can be calculated from the intercept and slope of the plot of ($\Delta G$) versus $T$.

## 6.5 Thermodynamic Parameters

**Table 6.11** Equilibrium constant and thermodynamic parameters at different temperatures

| Temperature (K) | $K_c$ | $-\Delta G$ (kJ/mol) | $\Delta H$ (kJ/mol) | $\Delta S$ (kJ/mol. K) |
|---|---|---|---|---|
| 293 | 2.722 | 2.44 | 8.2831 | 0.0365 |
| 298 | 2.833 | 2.5803 | | |
| 308 | 3.2 | 2.98 | | |

Copyright © 2019 Hasan et al. Licensee MDPI, Basel, Switzerland. Creative Commons Attribution (CC BY) license. Healthcare, 2019, 7, 52, https://doi.org/10.3390/healthcare7010052, Reproduced from Ref. [138]

***Example 6.3*** Kc values are obtained from Table 6.9 for fractional uptake of strontium uptake on the chitosan-based bead at different temperatures. Calculate $\Delta H$ and $\Delta S$ values from Eq. 6.50 [137] (Table 6.11).

**Solution**

(a) The $\Delta G$ values at different temperatures are calculated according to the Gibbs free energy equation:

$$\Delta G = -RT \ln k_c,$$

$$R = 8.314 \frac{J}{mol.K},$$

$$T = 293 K,$$

$$k_c = 2.722.$$

$$\Delta G = -8.314 \times 293 \times \ln(2.722),$$

$$\Delta G = -2.4393 \frac{kJ}{mol}.$$

Similarly, $\Delta G$ values are also calculated for 298 and 313 K and are given as the example in Table 6.11. The obtained negative values of the $\Delta G$ confirmed the feasibility of the process and also the spontaneous nature of the adsorption process. On the contrary, the positive value of $\Delta G$ indicates that the adsorption reaction requires energy to continue the reaction process.

(b) The values of enthalpy change ($\Delta H$) and entropy change ($\Delta S$) can be calculated from the intercept and slope of the plot of ($\Delta G$) versus $T$. The plot is shown in Fig. 6.6.

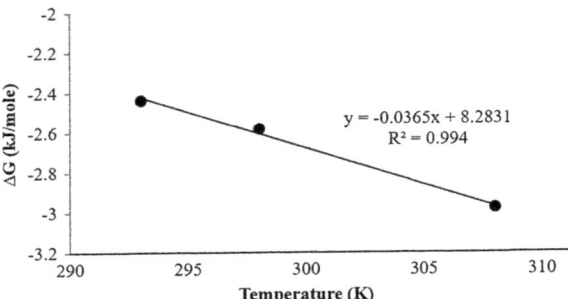

**Fig. 6.6** The relationship between the changes in Gibbs free energy and temperature of the adsorption of strontium on chitosan bead. Copyright © 2019 Hasan et al. Licensee MDPI, Basel, Switzerland. Creative Commons Attribution (CC BY) license., Healthcare, 2019, 7, 52, https://doi.org/10.3390/healthcare7010052, Reproduced from Ref. [138]

The equation obtains from the plot (Fig. 6.6) is as follows:

$$y = -0.0365x + 8.2831.$$

($\Delta S$) value from the slope = 0.0365.
($\Delta H$) value from the intercepts = 8.2831.

The positive enthalpy $\Delta H$ indicates that the energy is adsorbed as the adsorption proceeds, and the reaction is said to be endothermic. The positive value of entropy change ($\Delta S$) reveals that the freedom of metal ions is not too restricted in the adsorbent. This phenomenon also suggests that the randomness at the solid/solution interface may increase with structural changes in the adsorbate and adsorbent. It is suggested that the positive entropy change may be due to the dissociation reaction or the release of the water molecule as the reaction proceeds. Such adsorption phenomena are not favorable at high temperatures. A negative value of ($\Delta S$) suggests that the adsorption process involves an associative mechanism. Also, a negative value of ($\Delta S$) reflects that no significant change occurs in the internal structures of the adsorbent during the adsorption process. Saha and Chowdhury [138] suggested that the negative entropy change ($\Delta S$) would indicate faster interaction with the active surface sites of adsorbent (forward reaction).

In the case of a gas or liquid molecule adsorption on a solid surface, a quantity of kinetic energy is released from the molecule. This kinetic energy is also termed as the heat of adsorption. The magnitude of the heat of adsorption and its variation with coverage can provide useful information about the nature of the surface and the adsorbed phase. The isosteric heat can be defined as the ratio of the change in the adsorbate enthalpy to the change in the amount adsorbed [110]. In the adsorption process, the isosteric heat of adsorption measures the interaction between the adsorbate molecules and the adsorbent surface atoms. The isosteric heat of adsorption depends on the extent of the surface coverage. An estimation of the variation in isosteric heats of adsorption with coverage was estimated from the Clausius–Clapeyron equation

## 6.5 Thermodynamic Parameters

$$\Delta H_{iso} = -R\left(\frac{\partial \ln C_e}{\partial (1/T)}\right)_q, \qquad (6.51)$$

where

$\Delta H_{iso}$ = the enthalpy change for the sorption process (cal/mole),
$R$ = the gas constant (cal/(mole K)),
$C_e$ = the equilibrium concentration of sorbate in solution (mmol/L), and
$T$ = the temperature in K.

The equation can be rearranged as

$$\ln C_e = -\left(\frac{\Delta H_{iso}}{R}\right)\frac{1}{T} + K, \qquad (6.52)$$

where $K$ is a constant and the heat of adsorption is obtained from the slope of the plot $\ln C_e$ versus $1/T$.

### 6.5.1 Polanyi's Potential Theory

The Polanyi potential theory was modified to correlate the adsorption data at various temperatures and to check the consistency of the data. The Polanyi theory assumes that the adsorbent exerts long-range attractive forces on the gas or vapor surrounding it. These attractive forces generate a potential field, which decreases as the distance between the gas and adsorbent surface increases. This theory was initially developed for multilayer gas adsorption and has been used widely to correlate gas phase adsorption data. The potential theory of Polanyi can be expressed as

$$\varepsilon = RT \ln(P_s/P). \qquad (6.53)$$

Polanyi further assumed that the adsorption potential given by Eq. (6.53) is independent of temperature. Therefore, a plot of adsorption potential ($\varepsilon$) versus the volume adsorbed on the solid surface should yield a single characteristic curve. Several researchers have extended the theory to the adsorption of solutes from the aqueous phase to solid by expressing the potential in terms of saturation and equilibrium concentration [139–141]. However, no attempts were made to correlate metal ions' adsorption data from an aqueous solution on a solid adsorbent. Hasany et al. [142] used the Dubinin–Radushkevich (D–R) equation to correlate cadmium (II) adsorption data on CdS. In order to apply the D–R equation, they calculated the potential using Polanyi's theory. A good fit to the data was reported. The adsorption potential ($\varepsilon$) for the sorbate ion uptake may be calculated from the following equation.

$$\varepsilon = RT \ln\left(\frac{a_s}{a}\right), \qquad (6.54)$$

where

$\varepsilon$ = adsorption potential (kcal.mol$^{-1}$),
$a_s$ = activity of sorbate ion at equilibrium concentration (mmol/L),
$a$ = activity of sorbate at equilibrium at any concentration (mmol/L),
$R$ = gas constant (kcal/mol), and
$T$ = temperature (K).

The activity is expressed in terms of mole fraction and activity coefficients ($a_s = x_s \gamma_s$). The activity coefficient of sorbate ion in the solution was calculated using the Davis equation [71].

**Example 6.4** Using the data for the adsorption of strontium on a chitosan-based bead, (a) show the data can be described by a single characteristic curve as required by the potential theory and calculate the (b) isosteric heat of adsorption (Table 6.12).

According to Polanyi's potential theory, a plot of metal ion uptake ($Q_e$) versus $RT \ln\left(\frac{a_s}{a}\right)$ is independent of temperature. The equilibrium concentrations of strontium in the solution are given as 10.594 mmol/L, 10.625 mmol/L, and 10.739 mmol/L at 293K, 298K, and 308K, respectively. The activity coefficient value $\gamma$ for 0.01 M strontium solution is 0.744 [143]. The following table is generated using the value of the activity coefficient $\gamma$ for strontium and equilibrium strontium concentration in the solution (Table 6.13).

The resulting potential curve is shown in Fig. 6.7.

As shown in Fig. 6.7, a single characteristic curve was obtained for three temperatures, as suggested by the Polanyi theory. The molar volume of strontium is assumed to be independent of temperature. Therefore, the equilibrium data can be calculated at various temperatures from the single characteristic equation. The heat of adsorption can be obtained from the Eq. 6.52 and given in Fig. 6.8.

**Table 6.12** Strontium uptake onto chitosan-based beads at different temperature

| 293 K | | 298 K | | 308 K | |
|---|---|---|---|---|---|
| $C_e$ (mmol/L) | $Q_e$ (mmol/g) | $C_e$ (mmol/L) | $Q_e$ (mmol/g) | $C_e$ (mmol/L) | $Q_e$ (mmol/g) |
| 0.0199 | 0.0193 | 0.0227 | 0.0182 | 0.0268 | 0.0166 |
| 0.0583 | 0.0222 | 0.0568 | 0.0227 | 0.0611 | 0.0210 |
| 0.1079 | 0.0477 | 0.1136 | 0.0455 | 0.1222 | 0.0420 |
| 0.412 | 0.0852 | 0.4403 | 0.0738 | 0.4602 | 0.0648 |
| 0.8369 | 0.1198 | 0.8523 | 0.1136 | 0.8778 | 0.1034 |
| 2.0222 | 0.1911 | 2.043 | 0.1829 | 2.0869 | 0.1665 |
| 2.6295 | 0.2210 | 2.6704 | 0.2045 | 2.6989 | 0.1932 |

Copyright © 2019 Hasan et al. Licensee MDPI, Basel, Switzerland. Creative Commons Attribution (CC BY) license., Healthcare, 2019, 7, 52, https://doi.org/10.3390/healthcare7010052, Reproduced from Ref. [138]

## 6.5 Thermodynamic Parameters

**Table 6.13** Strontium uptake ($Q_e$) versus $RT \ln\left(\frac{a_s}{a}\right)$

| 293 K | | 298 K | | 308 K | |
|---|---|---|---|---|---|
| $Q_e$ (mmol/g) | $RT \ln\left(\frac{a_s}{a}\right)$ | $Q_e$ (mmol/g) | $RT \ln\left(\frac{a_s}{a}\right)$ | $Q_e$ (mmol/g) | $RT \ln\left(\frac{a_s}{a}\right)$ |
| 0.0193 | 3.483 | 0.0182 | 3.465 | 0.0166 | 3.486 |
| 0.0222 | 2.857 | 0.0227 | 2.922 | 0.0210 | 2.983 |
| 0.0477 | 2.498 | 0.0455 | 2.512 | 0.0420 | 2.559 |
| 0.0852 | 1.72 | 0.0738 | 1.71 | 0.0648 | 1.747 |
| 0.1198 | 1.306 | 0.1136 | 1.32 | 0.1034 | 1.352 |
| 0.1911 | 0.7915 | 0.1829 | 0.8005 | 0.1665 | 1.822 |
| 0.2210 | 0.6391 | 0.2045 | 0.6426 | 0.1932 | 0.6642 |

Copyright © 2019 Hasan et al. Licensee MDPI, Basel, Switzerland. Creative Commons Attribution (CC BY) license., Healthcare, 2019, 7, 52, https://doi.org/10.3390/healthcare7010052, Reproduced from Ref. [138]

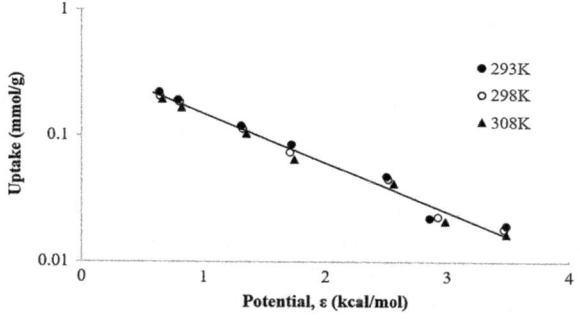

**Fig. 6.7** Plot of adsorption data according to Polanyi's potential theory for strontium uptake on chitosan-Fuller's earth bead. Copyright © 2019 Hasan et al. Licensee MDPI, Basel, Switzerland. Creative Commons Attribution (CC BY) license., Healthcare, 2019, 7, 52, https://doi.org/10.3390/healthcare7010052, Reproduced from Ref. [137]

**Fig. 6.8** Isosteric heat of adsorption from the loading of (■) 0.08 mmol/g, (□) 0.04 mmol/g, and (◆) 0.022 mmol/g at different temperatures. Unpublished work by authors

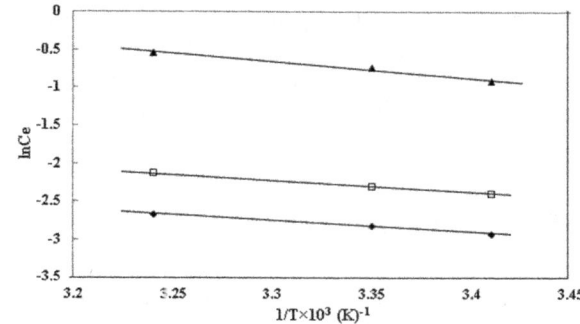

**Fig. 6.9** Isosteric heat of adsorption of strontium on chitosan-Fuller's earth bead. Unpublished work by authors

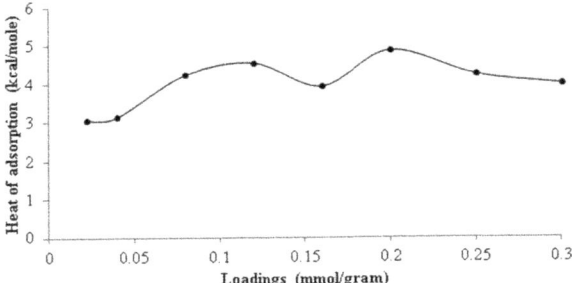

The slopes obtained from the plot of $\ln C_e$ versus $1/T$ for different loadings are as follows:

(♦) Loading 0.022 mmol/g = −1.536,
(□) Loading 0.04 mmol/g = −1.5784,
(■) Loading 0.08 mmol/g = −2.1283.

| Loading (mmol/g) | 0.022 | 0.04 | 0.08 |
|---|---|---|---|
| $H_{st}$ (kcal/mole) | 3.052 | 3.136 | 4.23 |

The heat of adsorption at different loadings is shown in Fig. 6.9. The heat of adsorption describes the binding strength between strontium molecules and the surface-active sites of the beads. The heat of adsorption of strontium increased as the loading increased up to a particular loading, and then it started to decrease. However, the isosteric heat of adsorption seems to have been approaching its integral heat of adsorption as the saturation capacity was approached. The initial increase in the heat of adsorption may be attributed to lateral interactions between the adsorbed strontium ions, which are known to form complex molecules on a solid surface. This phenomenon suggests that the surface became saturated with strontium, and the heat of adsorption was approaching its equilibrium value. The subsequent decrease in the values of heat of adsorption can be attributed to the heterogeneity of the surface and multilayer coverage [144]. It is expected that amine groups in chitosan will be homogeneous energetically, and, therefore, a constant heat of adsorption should be obtained. However, the substrate Fuller's earth is not expected to be homogeneous energetically. Although Fuller's earth was coated with chitosan, individual amine groups seem to become heterogeneous because of the presence of Fuller's earth.

## 6.5.2 Activation Energy

In a chemical reaction, the activation energy plays a vital role in the interaction between the solute and the sorbent. It is the minimum energy required for a reactant to overcome before a chemical reaction can occur. In general, the Arrhenius equation

## 6.6 Dynamic Adsorption

**Table 6.14** Typical thermodynamic parameters of adsorption processes [112a, b, 138]

| Types of sorption | Heat of adsorption (kJ/mol) | Adsorption energy (kJ/mol) | Activation energy (kJ/mol) | Refs |
|---|---|---|---|---|
| Physiosorption | 80 | 2.1–20.9 | No more than 4.2 | [112b, 138] |
|  | <80 | <8 | <40 | [112a] |
| Chemisorption | 80–400 | 80–200 | 8.4–83.7 | [138] |
|  | >80 | >16 | >40 | [112a] |
| Ion-exchange | <40 | 8–16 | 24–40 | [112a] |

is used to determine the activation energy of sorbate uptake on adsorbent is given by the equation [131]:

$$k = k_o \exp\left(\frac{-E}{RT}\right), \qquad (6.55)$$

where

$E$ = activation energy,
$k$ = the rate constant of sorption, g/mg. min,
$k_o$ = the temperature-independent factor, g/mg. min,
$R$ = the gas constant, 8.314 J/mol. K, and
$T$ = solution temperature, K.

The $k$ values in Eq. 6.55 can be obtained from both pseudo-first-order and second-order kinetic models. The k values are plotted as a function of the reciprocal of the Kelvin temperature. The slope of the Arrhenius plot is used to correlate the rate constant at different temperatures. The positive activation energy indicated the minimum energy required to facilitate the forward ion-exchange process. Table 6.14 shows thermodynamic parameters for the adsorption process as follows [112].

## 6.6 Dynamic Adsorption

In dynamic adsorption of metal ion on adsorbent solid, the sorbate (metal ion) solution is passed through the column containing adsorbent. For instance, under dynamic conditions, an all-glass column was used to study heavy metal ion adsorption, such as Cd (II). Approximately, 35 g of chitosan-based beads were used to make a 10-in. column. The bed volume is calculated from the expression:

$$\text{Bed volume} = \pi r^2 h,$$

where $R$ is the radius of the column and h is the bed height. The influent solution is allowed to pass through the bed at a predetermined constant flow rate during a run. Each run was continued for approximately 180 min, and samples at the bed outlet were collected at a regular time interval. The bed became saturated during this time, as indicated by the outlet Cd-(II) concentration. The breakthrough adsorption capacity of adsorbate was obtained in the column at different cycles using the equation [145]

$$q_e = \left[\left(\frac{C_i - C_e}{m}\right)\right] \times \text{bv}, \tag{6.56}$$

where

$C_i$ = initial metal ions concentration (mg/L),
$C_e$ = the equilibrium (at breakthrough) of heavy metal ions concentration (mg/L),
bv = the breakthrough volume of the heavy metal ions solution in liters,
$m$ = the mass of the adsorbent used (g).

In the case of Cd (II) solution passing through the chitosan-based adsorbent column, the inlet concentration was 100 mg/L (0.893 mmol/L), and the flow rate was 3.5 mL/min through the bed, cadmium broke through the column after 20-bed volumes (see Fig. 6.10). Complete saturation of the bed occurred after 70-bed volumes.

It can also be noted that, under this condition, after cadmium broke through the column, it reached 90% of its inlet concentration reasonably quickly by another 20-bed volumes. Then, another 30-bed volumes were required to saturate the column. After the column was exhausted, the column drained off the remaining aqueous solution by pumping air. The breakthrough time can be delayed through the use of a larger quantity of the adsorbent.

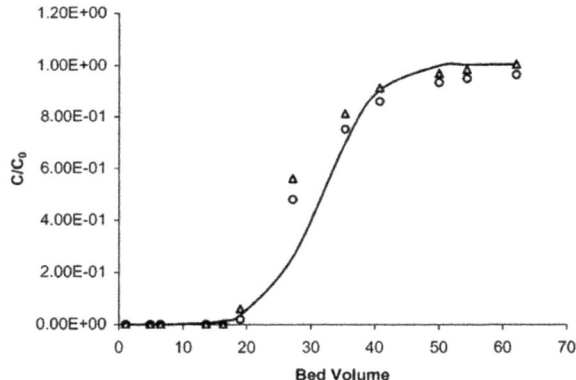

**Fig. 6.10** Breakthrough curve for adsorption of cadmium on chitosan-coated perlite bead. Copyright © 2006, American Chemical Society, Ind Eng Chem Res 45:5066, Reproduced from Ref. [72] with permission

## 6.6.1 Fundamentals of Dynamic Adsorption

The analysis of adsorption in packed beds is based on the development of effluent concentration–time curves, which are a function of adsorber geometry and operating conditions and equilibrium adsorption data. These breakthrough curves are obtained by flowing a fluid that contains an adsorbable solute with an initial concentration $C_0$ through a packed bed that contains a clean or regenerated adsorbent. As the fluid flow continues, the bed becomes saturated with the adsorbate at a given position, and a concentration distribution is established within the bed, as shown in Fig. 6.11.

At time $t_i$, the solute first appears in the effluent stream. Time $t_b$ is defined as the time required to reach the breakpoint concentration, indicated as $C_b$. This corresponds to the maximum concentration allowable in the effluent stream. Time $t_e$ is the time at which the bed becomes saturated with the adsorbate. At this time, the bed is exhausted and must be regenerated prior to reuse. The time for $t_i$ to $t_e$ corresponds to the thickness of the adsorption or mass transfer zone in the bed and is related to the mechanism of the adsorption process. It is readily seen that the area behind the breakthrough curve represents the quantity of adsorbate retained by the adsorbent contained in the column. This corresponds to a point on the equilibrium isotherm. If the isotherm can be represented by the Langmuir equation, which is expressed as

$$q^\infty = \frac{Q K C_0}{1 + K C_0} \quad \text{or} \quad C_s^\infty = \frac{Q' K C_0}{1 + K C_0}, \tag{6.57}$$

where $C_0$ is the concentration of the solute in the influent solution and $q^\infty$ and $C_s^\infty$ are the saturation capacities of the adsorbate in the adsorbent bed. Consistent concentration units must be used in these equations.

To discuss dynamic adsorption, the equilibrium isotherms are classified as (a) favorable, (b) linear, or (c) unfavorable. These are shown in Fig. 6.12. If the isotherm is concave in the direction of the fluid phase concentration, as shown by curve (a),

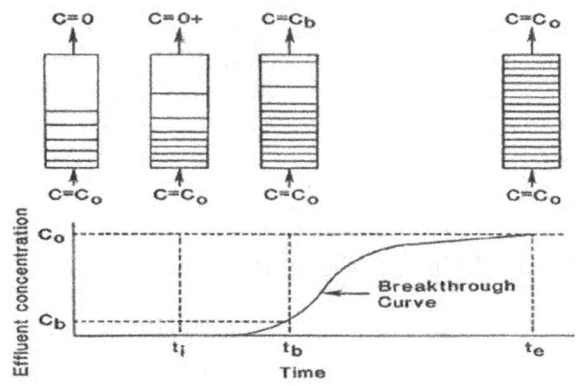

**Fig. 6.11** Adsorption wavefront. Hines and Maddox (1985); Hines et al. (1993) Reproduced from Ref. [102] with permission

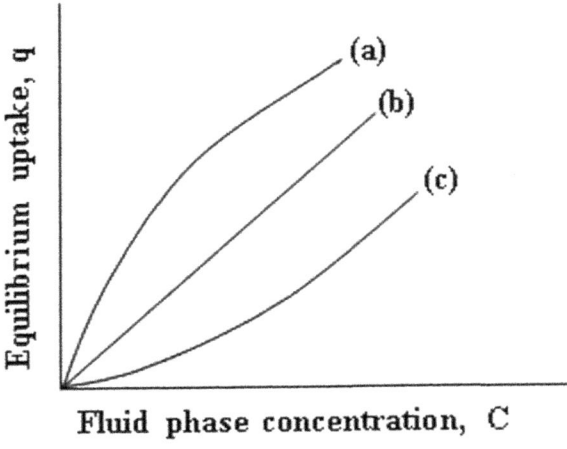

**Fig. 6.12** Shape of equilibrium isotherms. Copyright © 1999 Elsevier Science Ltd. Bioresource Technology, 72 (2000) 153–158, Reproduced from Ref. [104] with permission

layers of high concentration in the bed move faster than layers of lower concentration. This results in the adsorption zone becoming thinner as the wavefront moves through the bed and gives a breakthrough curve that is self-sharpening. For a favorable isotherm, the breakthrough curve develops and moves through the packed column in a constant pattern. The unfavorable isotherm results in a breakthrough curve that becomes more diffuse as it traverses the bed length. For non-equilibrium adsorption, a favorable isotherm will yield a constant pattern breakthrough curve after a period of time. In most industrial adsorption systems, the equilibrium is not reached between the adsorbate and adsorbent. The velocity of the gas/liquid stream flowing through the bed also changes the shape of the breakthrough curve and influences the adsorption rate. Low flow rates facilitate equilibrium conditions, but the adsorbate's axial dispersion can be significant and must be considered in the bed design. For higher flow rates, axial dispersion is typically insignificant, relative to its impact on the mass transfer rate and its influence on the shape of the breakthrough curve, but adsorbate–adsorbent equilibrium is not attained. The overall rate of adsorption is controlled by several factors, including external mass transfer to the surface, internal mass through the fluid that fills the pores of the adsorbent, internal mass transfer across the solid surface, and the actual rate of adsorbate uptake. Various mechanisms that impact the rate of adsorption in a bed are discussed by Yang, and Hines and Maddox [79, 102].

## *6.6.2 LUB Equilibrium Method*

The bed region over which the adsorbate concentration changes from the specified breakthrough concentration, $C_b$, to the inlet or fully loaded, $C_0$, is then defined as the height or thickness of the mass transfer zone. The thickness of the zone is a function

## 6.6 Dynamic Adsorption

of the adsorbate–adsorbent system. It depends specifically on the mass transfer rate in the solid, the transfer rate to the solid surface, and the concentration difference. Ideally, an effluent concentration curve, in addition to equilibrium data, is required before calculating the amount of adsorbent necessary to make a specific separation. However, it may be more expedient to obtain a breakthrough curve than to measure equilibrium isotherm data if neither is available. A breakthrough curve obtained at the expected design temperature and pressure provides uptake data under dynamic conditions and the height of the mass transfer zone, which gives the length of the unused bed (LUB). The LUB equilibrium method is frequently used in adsorber design. In this method, the pack-bed adsorber is viewed as consisting of two sections: the equilibrium section and the length of unused bed (LUB) section. The size of the equilibrium section is found from equilibrium adsorption data at the bed design temperature. The length of the equilibrium section represents the shortest bed length possible. It can be described as the stoichiometric length since the adsorbent in the bed is assumed to be in equilibrium with the adsorbate in the fluid. The stoichiometric wavefront moves through the bed as a step function.

Because of the presence of the mass transfer zone, all the adsorbent behind the actual wavefront will not be at its maximum capacity. Therefore, we must add an additional quantity of adsorbent to the bed to compensate for the mass transfer zone, described as the LUB. The stoichiometric wavefront, relative to the actual stable wavefront, is shown in Fig. 6.13. In this Fig. 6.13, $t_b$ is defined as the time when the leading edge of the breakthrough curve leaves the bed, $t_e$ is the time when the trailing

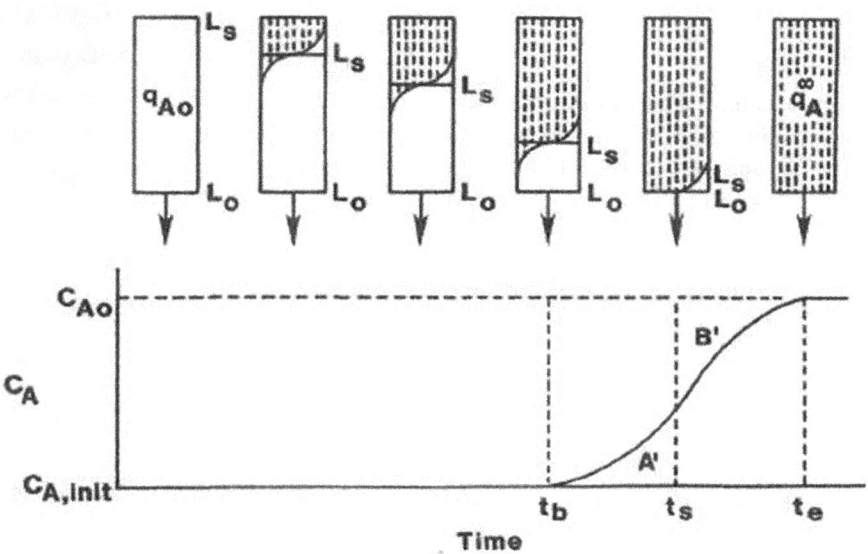

**Fig. 6.13** Stoichiometric front relative to the mass transfer zone. Hines and Maddox (1985), Hines et al. (1993). Reproduced from Ref. [102] with permission

edge of the wavefront leaves the bed, and $t_s$ is the time when the stoichiometric wavefront would leave. The stoichiometric time is found by adjusting $t_s$ until the areas indicated by $A'$ and $B'$ are equal. At breakthrough, the leading edge of the actual wavefront, relative to the stoichiometric front, is

$$\text{LUB} = L_0 - L_s, \qquad (6.58)$$

where

$L_0 =$ the total bed length, and
$L_s =$ the distance the stoichiometric front has moved through the bed. The important assumptions on which the LUB equilibrium section concept is based have been summarized by Collins [146].

From Fig. 6.13, the length of the unused bed can be shown to be

$$\text{LUB} = L_0 \frac{(t_s - t_b)}{t_s}. \qquad (6.59)$$

As shown by Collins [146], the length of the stoichiometric wavefront can be expressed as

$$L_s = \frac{U(C_{A0} - C^*_{A,\text{init}}) t_b M_A}{\rho_b (q_A^\infty - q_{A0})}, \qquad (6.60)$$

where

$q_A^\infty =$ saturation capacity of the adsorbent, g solute/g solid,
$q_{A0} =$ initial concentration of solute on the adsorbent, g solute/g solid,
$C_{A0} =$ concentration of solute in the influent, g mol/cm$^3$,
$C^*_{A,\text{init}} =$ concentration of solute in equilibrium with $q_{A0}$, g mol/cm$^3$,
$M_A =$ molecular weight of the adsorbate,
$U =$ superficial velocity, and
$\rho_b =$ density of the adsorbent in the bed, g solid/cm$^3$.

From this equation, the velocity of the stoichiometric wavefront is

$$u' = \frac{L_s}{t_b} = \frac{U(C_{A0} - C^*_{A,\text{init}}) t_b M_A}{\rho_b (q_A^\infty - q_{A0})}. \qquad (6.61)$$

Examination of the material balance used by Collins [146] shows that the maximum driving force for mass transfer was used for both phases and the driving force is constant. The approximate nature of the material balance above should be recognized since the concentration differences vary with time and position. The equation does provide a method of obtaining the length of an adsorber bed if the adsorption cycle time is specified.

The complete design of an adsorption system requires equilibrium adsorption data, along with an effluent concentration–time curve, or the ability to predict one using mass transfer data and an appropriate model. The amount of adsorbent necessary to make a certain separation can be estimated using isotherm data. In reality, this method underestimates the amount of adsorbent needed for a specific separation because when the maximum breakthrough concentration, given as the breakthrough point $t_b$ in Fig. 6.13, is reached, the operation of the bed must be discontinued. This results in the amount of adsorbent in the bed between points $t_b$ and $t_e$ unavailable for adsorption. To account for this loss of capacity, a design based on equilibrium data only must be increased by some value, say 20–25%.

## 6.7 Sorption Mechanisms

Metal ions sorption performance and mechanisms by chitosan depend on the pH, the types of metal ligands present in the solution, and chitosan's structural parameters such as degree of acetylation and crystallinity [70]. In an aqueous solution, crystallinity controls chitosan's swelling and diffusion properties. The amine group of the chitosan has a lone pair of electrons from nitrogen, which primarily acts as an active site for forming the chitosan metal ion complex. As mentioned previously, at lower pH values, the amine group of chitosan undergoes protonation, forming $NH_3^+$, which increases electrostatic attraction between $NH_3^+$ and the sorbate anion. Chitosan can form chelates with metal ions with the release of hydrogen ions. A chelate formation may require the involvement of two or more complexing groups from the molecule. As the adsorption progress, the pH of the solution increases suggesting that metal forms a covalent bond with $NH_2$ groups. The $NH_2$ group could come from two different glucosamine residues of the same molecule or two different molecules of chitosan. The protonation of $NH_2$ occurs at low pH. As the pH increases, the deprotonation of amine groups may occur, resulting in a decrease of competition between proton and metal species for surface sites of chitosan. At a low pH range, chitosan is positively charged due to the protonation of amino groups, leading to increased electrostatic interaction with negatively charged entities and inducing covalent bonds. This type of bond is stronger than the electrostatic interactions formed by the anionic molecules [7]. In addition to amino groups, hydroxyl groups on the chitosan structure can also react with ionic molecules. The detailed mechanism by which transition metals are adsorbed by chitosan is not known because of the complexity of the gel network and the flexibility of the polymer chains of chitosan. Mitani et al. [147] suggested bridge models and pendant models to address the binding mechanism of transition metals to chitosan. In the bridge model, inter and intramolecular crosslinking between chitosan is assumed to occur with the intervention of transition metal ions such as copper ions. In this case, both amino and hydroxyl groups in the chitosan molecule may chelate with metal ions. In the pendant model, a copper ion is assumed to be linked to an amino group. Monteiro and Airoldi [148] conducted a calorimetric titration study with chitin and chitosan to demonstrate their ability to interact with cations. They observed the negative entropic values of chitosan to be the opposite of chitin. These negative entropic values suggest that the presence of cation bonding amino groups in one or different polymeric chains may order the random coil disposition of the chains. It is well known that the acetyl group is bonded to nitrogen in the chitin structure. A disorder in solvent behavior causes an increase in entropy; however, the deacetylated biopolymers can order the solvent molecules of the systems as the complexation process progresses. Zhou et al. [149] conducted an FTIR study with chitosan and the chitosan-copper complex. They found that the copper-chitosan complex showed a decrease in intensity of the amino $\nu$-CN at 1153.4 cm$^{-1}$ and the disappearance of the $\delta$-NH of the amino at 1600.8 cm$^{-1}$, indicating the formation of a complex band N $\rightarrow$ Cu. They also observed a shift in the $\nu$-CO band, from 1074.3 to 1064.6 cm$^{-1}$, and in the $\nu$-OH band, from 3435.0 to 3400.3 cm$^{-1}$. These changes led the authors to conclude the possibility of a complex reaction of copper with chitosan at the amino and the secondary hydroxyl group sites. Sorption mechanisms reported by different studies showed that amino groups are the main metal-binding sites. The presence of both amine and hydroxyl groups in the chitosan structure may also form coordination bonds with transitional metal ions, which vary with pH of the solution, metal concentration, molecular weight, and degree of acetylation of chitosan [147, 149–151].

In an aqueous solution, chromium (VI) forms stable complexes, such as $Cr_2O_7^{2-}$, $HCrO_4^-$, $CrO_4^{2-}$, and $HCr_2O_7^-$, depending on the pH of the solution. The fraction of any particular species depends on the chromium concentration and pH of the solution [152, 153]. Speciation studies of Cr (VI) in aqueous solution based on spectrophotometry, electrochemistry, freezing point depression, and NMR indicated the existence of the following equilibria:

$$H_2CrO_4 = H^+ + HCrO_4^-, \tag{6.62}$$

$$HCrO_4^- = H^+ + CrO_4^{2-}, \tag{6.63}$$

$$2HCrO_4^- = Cr_2O_7^{2-}. \tag{6.64}$$

The equilibrium between the species in the solution is dependent on pH, with $Cr_2O_7^{2-}$ and $HCrO_4^-$ existing primarily in acidic media and $CrO_4^{2-}$ being the only species of Cr (VI) above pH 7.0. Ramsey et al. [154], using Raman spectroscopic analysis of dilute chromate solutions, concluded that at pH 4.0 and below 0.007 mol of Cr (VI), the predominant species was $HCrO_4^-$ rather than $Cr_2O_7^{2-}$. Thus, the chelation is expected to be dependent on pH. Kaminski and Modrzejewska [155] suggested that chitosan can form chelates with metal ions with hydrogen ion release, which suggests that the adsorption of a metal ion on chitosan should depend strongly on the pH of the solution. The extent of Cr (VI) adsorption increased with a decreasing

## 6.7 Sorption Mechanisms

pH of the solution. At lower pH values, the amine groups of chitosan undergo protonation, leading to the increased electrostatic attraction between $NH_3^+$ and sorbate anion. As the pH increases, the amino group's deprotonation occurs, resulting in decreased adsorption. Gao et al. [156] reported that chitosan adsorbed some metals quantitatively as oxyanions or anionic chloro complexes in sample solutions by an ion-exchange mechanism. This implies that the interaction occurs between $NH_3^+$ functional groups in chitosan and an anion such as $Cr_2O_7^{2-}$, and the interaction is mainly electrostatic in nature. Dambis et al. [157] conducted XPS studies, which identified the sorption sites involved and the forms of species adsorbed and found that Cr (VI) sorption occurred on amine functional groups of chitosan.

Cadmium in an aqueous solution can be hydrolyzed with the formation of various species, depending on the solution pH. Moreover, $Cd^{2+}$, which is the main hydrolyzed cadmium species in the pH range of 5–7, appears in the form of $Cd(OH)^+$, $Cd(OH)_2^0$, and $Cd(OH)_3^-$ 3. Among them, $Cd^{2+}$ is the predominant species in the solution within this pH range. Therefore, the fraction of negatively charged hydrolysis products in the solution increases as pH increases. Various hydrolysis reactions are given by Reed and Matsumoto [158] and Baes and Mesmer [159].

Hasan et al. [71] prepared chitosan-coated perlite beads in the laboratory via the phase inversion of a liquid slurry of chitosan dissolved in oxalic acid and perlite to an alkaline bath for better exposure of amine groups ($NH_2$). The surface morphology of beads was observed using scanning electron microscopy (SEM), and microanalysis was obtained using energy-dispersive X-ray spectroscopy (EDS), which also revealed their porous nature. The chitosan content of the beads was 32%, as determined using a thermogravimetric method. Under both equilibrium and dynamic conditions, the adsorption of $Cd^{2+}$ from an aqueous solution on chitosan-coated perlite beads was studied. The adsorption of $Cd^{2+}$ on chitosan was pH dependent. The maximum adsorption capacity of chitosan-coated perlite beads was determined to be 178.6 mg/g at 25 °C, and the pH of the solution was 6.0. Chitosan forms chelates with cadmium ions ($Cd^{2+}$) with the release of $H^+$ ions. A chelate formation may require the involvement of two or more complexing groups from the molecule. The Cd ion may seek two or more amine groups from chitosan to form the complex. This should typically reduce the pH of the solution. The increase in the pH may be due to the exchange of $H^+$ ions between the surface of the bead and the solution. In the case of chitosan, the protonation of $NH_2$ groups occurs at a relatively low pH range. The pH of the solution increased as the adsorption progressed suggests that $Cd^{2+}$ formed a covalent bond with the $NH_2$ group. The two $NH_2$ groups could come from two different glucosamine residues of the same molecule or two different molecules of chitosan. Jha et al. [160] compared the stability constants for ammonia and amino complexes with those for cadmium chloro complexes. They noted that the formation of a covalent bond with amine nitrogen is the more-preferred reaction. The XPS data suggest that cadmium was mainly adsorbed as $Cd^{2+}$ and attached to the $NH_2$ and OH groups in the chitosan structure, as shown in Fig. 6.14. The adsorption data were fitted to a two-site Langmuir adsorption isotherm. According to a modified Polanyi's potential theory, the data obtained at various temperatures provided a single characteristic curve when correlated.

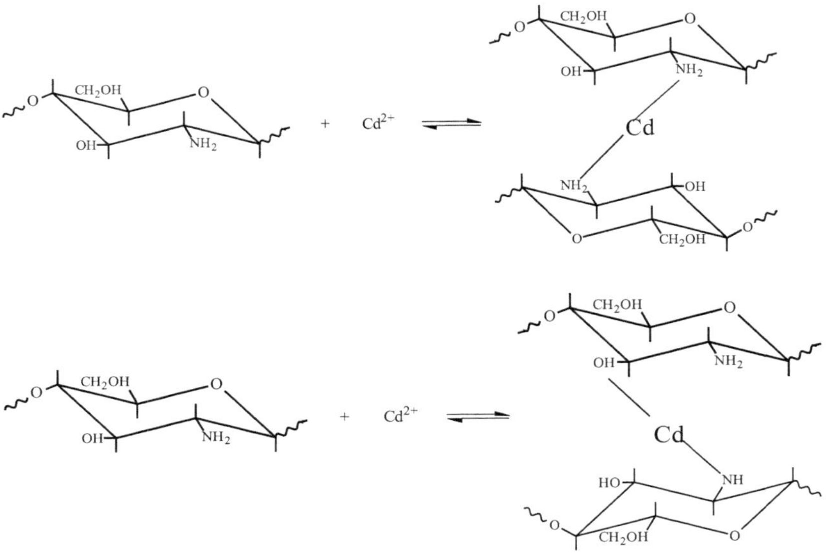

**Fig. 6.14** Possible reaction mechanism of cadmium uptake onto chitosan perlite bead. Copyright © 2006, American Chemical Society, Ind Eng Chem Res 45:5066, Reproduced from Ref. [72] with permission

In the case of Cu (II), the uptake of Cu (II) by chitosan beads increased as pH increased from 1 to 4.5 [86]. It then started to decrease as the pH was further increased. At a pH of 7.0, Cu (II) started to precipitate out from the solution. Verbych et al. [161] also reported an increase in copper adsorption on chitosan with the increase of solution pH in the range of 1–5, having a maximum capacity within the pH range of 4–5. The species formed by hydrolysis of copper salt are discussed by Chu and Hashim [162] and Baes and Mesmer [159]. The main hydrolyzed copper species in the pH range of 3–6 appear to be Cu (II) (unhydrolyzed species), $Cu(OH)^+$, and $Cu(OH)_2^{0-}$. Cu (II) is the predominant species in the solution within this pH range. Amine groups in chitosan are generally considered as the main active sites for the adsorption of metal ions. In the pH range of 2.0–4.5, both Cu (II) and H$^+$ ions were present. However, Cu (II) was able to compete better for active sites. Based on the molecular weight, the theoretical amount of –NH$_2$ for metal binding sites on chitosan is estimated to be about 6.9 mmol/g. However, it is reported that the amount of Cu (II) adsorbed on chitosan is about 5.12 mmol/g of chitosan [86]. Therefore, it is unlikely that two chitosan molecules were involved in chelating or binding with Cu (II). As noted by Monteiro and Airoldi [148], the monomer of chitosan can participate in copper adsorption. Also, it appears that some copper may have been adsorbed as Cu (OH)$^+$. Sakkayawong et al. [163] conclude the mechanism of adsorption of certain dyes from wastewater by chitosan to be a chemical process based on their studies. This observation is based on their experimental studies using the reflectance FTIR technique and evaluating free energy and entropy changes due to adsorption.

Various theories have been put forward to describe and interpret metal ion interaction at the solid-solution interface. According to Mc Naughton and James [164], the metal ion can be removed from an aqueous solution by the following mechanisms:

(a) Ion-exchange reaction.
(b) Metal ion adsorption at hydrated oxides of the surface.
(c) Metal hydroxyl species adsorption at the hydrated oxide surface.

The ion-exchange reaction differs from adsorption. Since adsorption and desorption accompany nearly every ion-exchange process, adsorption sometimes is indistinguishable from ion exchange.

## 6.8 Case Study I

### 6.8.1 Sorption Mechanism of Arsenic on to Chitosan–Iron Bead

Hasan et al. [94] also studied surface charge, XPS, FTIR, and RAMAN spectra analysis of chitosan–iron composite beads exposed to both As (III) and As (V) solution, respectively. Chitosan was crosslinked with hydrous iron oxide, and the crosslinked adsorbent was prepared as spherical beads using a simple and innovative method. In this study, the bead is known as a chitosan–iron oxyhydroxide (CFOH) bead. Most of the adsorbents reported in the literature are ineffective for As (III) removal from aqueous solution without the pre-oxidation process. In this work, the CFOH bead was evaluated to remove both As (V) and As (III) without pre-oxidation.

CFOH bead was prepared using chitosan, oxalic acid, and ferric nitrate. Approximately, 1.5 g chitosan was first dissolved in 100 mL of 0.2 M oxalic acid solution. About 2.3 g ferric nitrate was added to the chitosan oxalic acid solution. The mixture was heated at 323–343 K (50–70 °C) to form a gel. The bead was prepared by dropwise addition of the gel into NaOH solution using the procedure mentioned in Chap. 5. The beads prepared using ferric nitrate or chloride were separated from the solution and washed with deionized water until the washed solution's pH was 7. The beads were dried in a vacuum oven for subsequent use. The pH plays a significant role in arsenic adsorption from an aqueous solution usually explained in terms of ionization of both adsorbates and adsorbents. The solution pH of the solution affects the degree of ionization, the surface charge, and the speciation of arsenic, all of which can impact the adsorption mechanism and the uptake capacity. The effect of pH on arsenic removal from aqueous solutions was determined over the pH range of 5–11. The effect of pH on As (V) and As (III) uptake onto CFOH beads is shown in Fig. 6.15a. The concentration of both As (III) and As (V) was maintained at 1000 µg/L in all the runs. The results indicated that pH had no significant effect on As (III) removal from solution at pH 5.8 to 7.3, with removal efficiency greater than 88% (Fig. 6.15a).

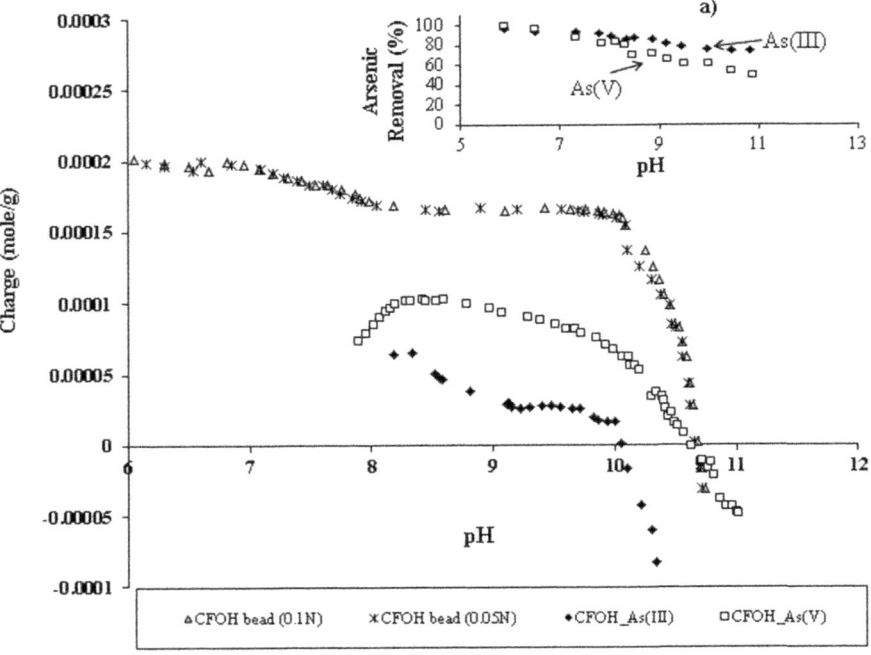

**Fig. 6.15** Surface charge of CFOH bead in the presence of (△) 0.1 N NaNO$_3$ and (∗) 0.05 N NaNO$_3$ solution. Surface charge of CFOH bead in the presence of 1.76 mol/L (♦) As (III) and (□) As (V) in the solution, respectively. **a** Effect of pH on the As (III) and As (V) uptake by CFOH beads (Condition: Initial concentration 1000 μg/L beads (Condition: Initial concentration 1000 μg/L; Temp: 298 K). Copyright © 2014, Taylor & Francis, Sep, Sci, and Technol. 49, 2863–2877, 2014. Reproduced from Ref. [94] with permission

Both As (III) and As (V) in an aqueous solution are hydrolyzed with the formation of various species, depending on the solution pH. For As (III), H$_3$AsO$_3$, which is the predominant species in the pH range of 2–9, appears neutral in the aqueous solution, whereas H$_2$AsO$_3^-$, HAsO$_3^{2-}$, AsO$_3^{2-}$, and AsO$_3^{3-}$ are stable species at pH < 9. At pH 6 to 9.0, As (III) occurs at neutral H$_3$AsO$_3$, and the CFOH bead undergoes surface protonation. The CFOH beads showed two distinct regions in the pH curve for As (V). The percentage removal of As (V) decreased by a small amount in the pH range of 5.8–7.3. At pH 7.3, arsenic removal from aqueous solution was about 85%. A significant decrease in As (V) adsorption was not observed in this study until the pH was increased to >9.0. When the pH was further increased to 11, the percentage removal was 72% for As (III) and 50% for As (V). The result suggests that the material would be sufficient for most water supplies, which generally have a pH range of 6.5–8.5.

Depending on the solution pH, the iron oxide and amine groups in the CFOH bead can undergo protonation to Fe(OH)$_2^+$, Fe$^{3+}$, and NH$_3^+$, or (NH$_2$–H$_3$O)$^+$, respectively. The extent of protonation will depend on the solution pH. Therefore, the surface

## 6.8 Case Study I

charge on the bead may provide a better understanding of the type of bond formed between arsenic species and the surface of the CFOH bead. The surface charge of the CFOH bead was determined by a potentiometric titration method in the presence of a symmetric electrolyte (sodium nitrate) [71]. The surface charges were measured for the point of zero charge (PZC). The pH of the PZC at a given surface is dependent on the relatively basic and acidic properties of the solid and allows estimation of the net uptake of $H^+$ and $OH^-$ ions from the solution.

As shown in Fig. 6.15, the PZC value of the CFOH bead was determined to be 10.7, whereas the PZC value of pure chitosan is reported to be in the range of 6.2–6.8. [86] Goldberg and Johnston [165] reported that PZC value of the amorphous iron oxide is 8.5. The PZC value of 10.7 and the behavior of surface charge of the CFOH bead could be due to the modification of chitosan when crosslinked with iron oxide. The pKa value of chitosan was determined to be 6.5. The protonation of the beads increased at the pH range of 10.7–9.2, making the surface positive (Fig. 6.15). At different ionic strengths, the surface charge of the bead was almost identical, and the pH of the bead suspension did not increase when an electrolyte salt was added. The point to note from this Fig. 6.15 is that PZC shifted toward 10.05 in the presence of As (III). The iron oxide and amine group presence in the CFOH bead was assumed to act as active sites for chitosan-arsenic-iron complex formation. It is worth noting that the pure chitosan had an almost-negligible adsorption capacity for As (III) [68].

The CFOH bead can form chelates with As (III) with the release of $H^+$ ions. A chelate formation requires two or more complexing groups from the molecule. The As (III) ion may seek two or more iron oxide or amine groups from the CFOH bead to form the complex. This should typically reduce the pH of the solution. It is reported that the increase in pH may be attributed to the exchange of released $H^+$ ions between the surface of the bead and the solution [71]. In the case of the CFOH bead, the protonation of the surface site occurs at a wide pH range. The pH of the solution decreased as the adsorption progressed suggests that As (III) formed a covalent bond with the iron oxide or amine group. The adsorption yield decrease above pH 9 may increase negatively charged As (III) species and negatively charged surface sites. It was evident that the amount of negatively charged species increased with the increase of pH, while the positively charged surface site decreased up to pHzpc (Fig. 6.15). In this connection, the As (III) can be adsorbed through specific adsorption (Lewis acid–base interaction) between the neutral species and positively charged surface sites at lower pH values. A similar observation was reported by Gou and Chen [166] for arsenic removal from groundwater by cellulose bead loaded with iron oxyhydroxide.

In solution, As (V) species has three pKa values: $pKa_1 = 2.20$, $pKa_2 = 6.98$, and $pKa_3 = 11.6$. In the pH range of 6–9, the $H_2AsO_4^-/H_2AsO_4^{2-}$ of As (V) is the predominant anionic species in the solution. The CFOH bead surfaces exhibited a positive charge in the pH range of 5–9, thus anionic As (V) presumably the major species being adsorbed by Coulombic interactions. As the equilibrium pH increased from lower pH to pHzpc, the decreased percentage removal of As (V) was attributed to the decreasing electrostatic attraction between the surface of CFOH bead and anionic As (V) species. Figure 6.16 shows the XPS spectra of CFOH. The binding

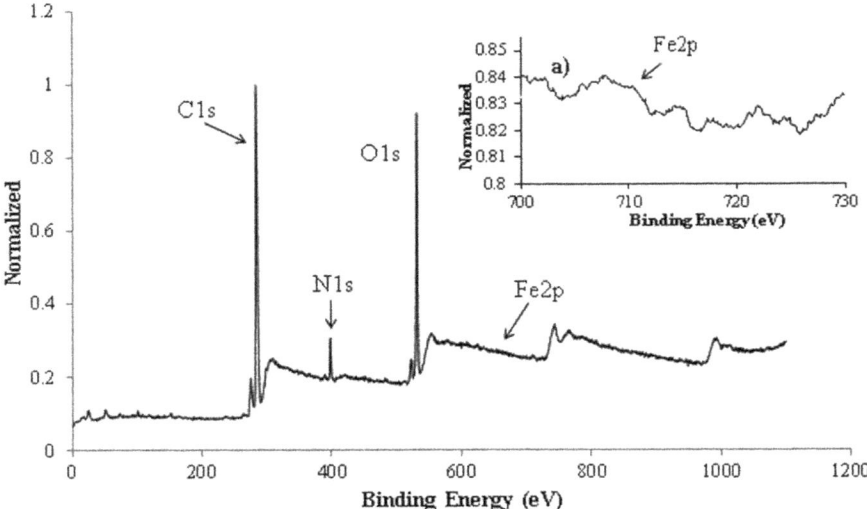

**Fig. 6.16** XPS survey scan spectra of CFOH beads. The binding energy of core level C1s, O1s, N1s, and Fe2p in the composite beads shows at 283.5, 531.0, 398 eV, and 710 eV, respectively. **a** Close-up of the Fe2p XPS peak. Copyright © 2014, Taylor & Francis, Sep, Sci, and Technol. 49, 2863–2877, 2014. Reproduced from Ref. [94] with permission

energy of core level C1s, O1s, N1s, and Fe2p in the composite beads is shown at 283.5, 531.0, 398 eV, and 710 eV, respectively.

Figures 6.17 and 6.18 show the peak positions of carbon, oxygen, and nitrogen obtained by the XPS for CFOH bead exposed to As (V) and As (III) solution. The magnitude of total C-1s, O-1s, peaks binding energy changed when the CFOH beads was exposed to As (III) solution (Table 6.9), while no change was observed for the beads exposed to As (V) solution. The magnitude of the binding energy shift, depending on the concentration of different atoms, in particular on the surface of materials, indicates the strength of the surface oxyanion interaction without giving explicit information of the nature of this interaction [167]. Compared with the XPS (Fig. 6.17a–c), the N-1s and O-1s peaks of CFOH beads did not shift before and after exposure to As (V) solution. There was also no shift of PZC observed in the surface charge experiment (Fig. 6.15), indicating that the chemical state of N and O atoms was not changed after adsorption of As (V) from the solution.

The N1s peak for CFOH beads exposed to As (III) was 398.1 eV, while for the CFOH beads, two peaks were observed at 397.6 and 398.3 eV (Fig. 6.18c). The N1s peak is around 397.6 eV of the CFOH bead, which was assigned to the amino nitrogen (–NH$_2$) in chitosan. After exposure to As (III), the peak was observed at 398.1 eV and assigned to the C–NH$_2$–Fe–As complex in the composite bead's structure. Figure 6.19 shows the XPS spectra of CFOH beads exposed to As (V) and As (III), respectively. The binding energy value of arsenic in the spectra was seen at 44 keV.

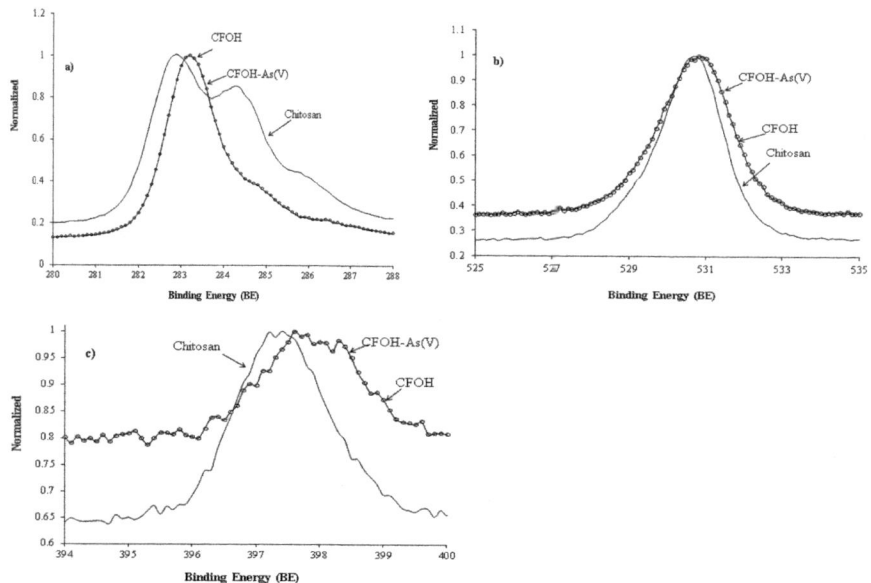

**Fig. 6.17** **a** C1s spectra of CFOH beads before (dot line overlap by dark line) and after (dark line) exposed to As (V) solution. The gray color line represents pure chitosan flake. **b** O1s spectra of CFOH beads before (dot line overlap by dark line) and after (dark line) exposed to As (V) solution. The gray color line represents pure chitosan flake. **c** N1s spectra of CFOH beads before (dot line overlap by dark line) and after (dark line) exposed to As (V) solution. The gray color line represents pure chitosan flake. Copyright © 2014, Taylor & Francis, Sep, Sci, and Technol. 49, 2863–2877, 2014. Reproduced from Ref. [94] with permission

Ding et al. [167] studied the adsorption mechanisms of arsenic on iron oxyhydroxide. They suggested that the surface-bound Fe (III) acts as a Lewis acid, which interacts with As (V) admolecules. A simple formula connecting O1s and Fe–2p chemical shift and charge can be evaluated by the following equations [168]:

$$Q_o = -4.372 + [385.023 - 8.976 \times (545.509 - O1sBE)]/4.488, \quad (6.65)$$

$$QFe(2p) = 0.3233 Fe2pBE - 228.51, r2 = 0.999, \quad (6.66)$$

where $Q_o$ and $Q_{Fe}$ were the actual oxygen and iron charge in the material at their respected binding energy. The iron charge can be calculated after the oxygen charge of the compounds has been determined from the O1s XPS shift. These iron charges can be correlated with the observed iron core level binding energies and iron charge. The values are given in Table 6.15. In the case of CFOH beads exposed to As (V) as shown in Fig. 6.19, the oxygen charge of the Fe-As surface complexes was similar to the CFOH beads, and there was no shift on PZC observed in the surface charge

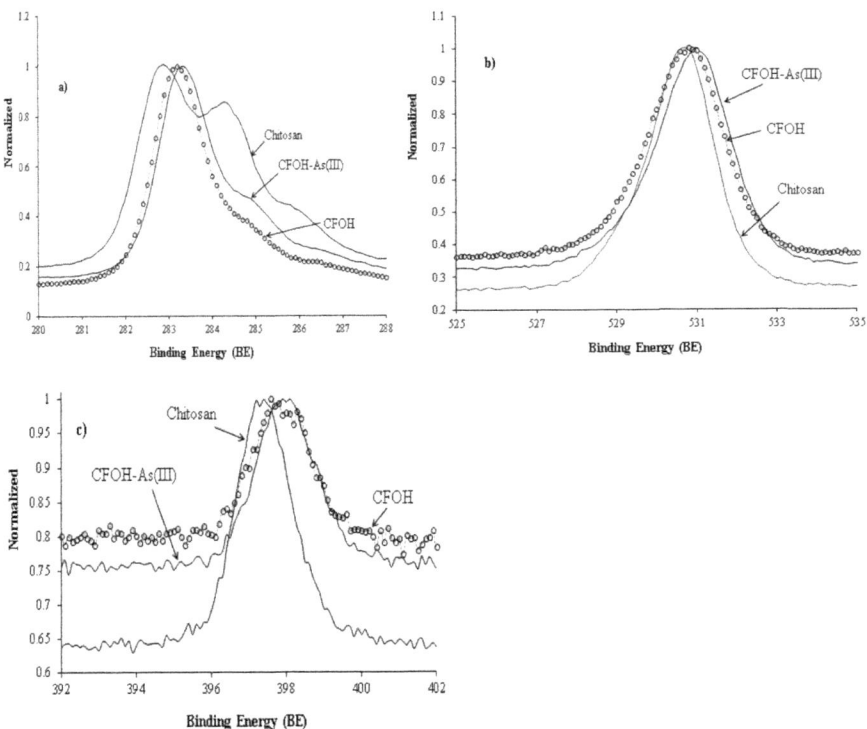

**Fig. 6.18 a** C1s spectra of CFOH bead before (dot line) and after (dark line) exposed to As (III)solution. The gray color line represents pure chitosan flake. **b** O1s spectra of CFOH bead before (dot line) and after (dark line) exposed to As (III) solution. The gray color line represents pure chitosan flake. **c** N1s spectra of CFOH beads before (dot line) and after (dark line) exposed to As (III) solution. The gray color line represents pure chitosan flake. Copyright © 2014, Taylor & Francis, Sep, Sci, and Technol. 49, 2863–2877, 2014. Reproduced from Ref. [94] with permission

experiment (Fig. 6.15). This indicates that the adsorption of As (V) on the iron surface did not involve noticeable electron transfer between the surface and adsorbate and suggests that As (V) may undergo Coulombic as well as Lewis acid–base interaction with the surface of the beads (Table 6.15).

In the case of CFOH beads exposed to As (III) solution, it was estimated that iron became more positive and oxygen-less negative charges compared to iron and oxygen charges in the CFOH beads (Table 6.15). Moreover, PZC of CFOH beads was found to be shifted toward positive when it was exposed to As (III) solution (Fig. 6.15), suggesting that surface oxygen atoms and iron atoms act as Lewis bases on adsorption of As (III) onto the CFOH beads. A similar observation was reported by Peak et al. [169]. It is also reported that outer-sphere surface complexes of neutral As $(OH)_3$ molecules have been observed to form complexes with iron oxide at pH below PZC of the metal oxide [170].

6.8 Case Study I

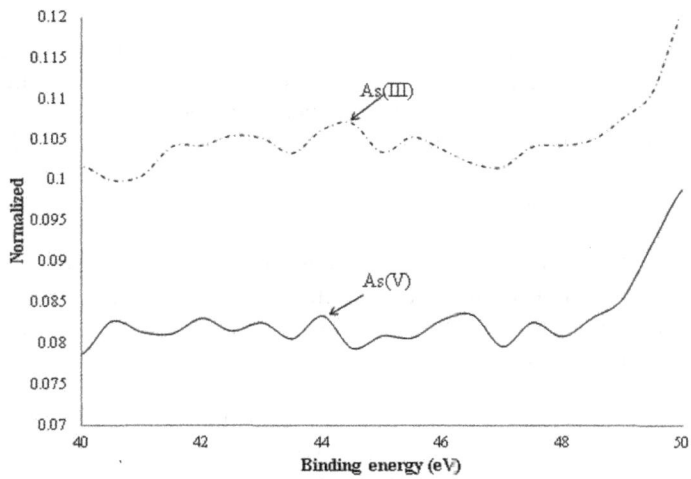

**Fig. 6.19** XPS of beads exposed to As (III) (dot line) and As (V) (dark line). The As (V) peak shows at 44 eV, whereas the As (III) peak shows a broader peak at around 44 eV. Copyright © 2014, Taylor & Francis, Sep, Sci, and Technol. 49, 2863–2877, 2014. Reproduced from Ref. [94] with permission

**Table 6.15** XPS binding energy of C-1s, O-1s, N-1s, and Fe of CFOH beads before and after exposure to As (III) and As (V) solution

| Element | C-1s | N-1s | O-1s | | Fe-2p | | Sorption interaction |
|---|---|---|---|---|---|---|---|
| | BE (eV) | BE (eV) | BE (eV) | Calculated charge $Q_o$ | BE (eV) | Calculated charge $Q_{Fe}$ | |
| CFOH | 283.5 | 397.6/398.3 | 531.0 | −0.8154 | 710.0 | 1.033 | |
| CFOH–As (V) | 283.5 | 397.6/398.3 | 531.0 | −0.8154 | 710.0 | 1.033 | Coulombic as well as Lewis acid–base interaction |
| CFOH–As (III) | 283.2 | 398.1 | 531.2 | −0.8029 | 727.0 | 6.5291 | Lewis acid–base interaction |

Copyright © 2014, Taylor & Francis, Sep, Sci, and Technol. 49, 2863–2877, 2014. Reproduced from Ref. [94] with permission

It was assumed that surface oxygen in (–C–O–Fe) moieties of CFOH beads accepts or donates electrons by a difference in electron negativity between adsorbate and iron (III). Therefore, it was hypothesized that both As (III) and As (V) adsorbed primarily on the iron oxide through complex formation with (–C–O–Fe) moieties. In the case of As (III), the shift of binding energy of C1s, O1s, and N1s suggests that As (III) formed a strong complex compound with the surface (–C–O–Fe) functional groups.

Furthermore, it was evident from Fig. 6.18c that the N1s spectra of CFOH beads shifted comparatively in the presence of As (III). Therefore, the $NH_3^+$ functional groups in the beads might also participate in the complex formation, thus further increasing the uptake of arsenic (III) from aqueous solution.

The intermolecular interaction between chitosan and Fe (III) is reflected by changes in the characteristics of IR peaks. Figure 6.20 shows the FTIR spectra of chitosan and CFOH bead. The phase composition of solid hydrolytic products of $Fe^{3+}$ ions depends on the concentration of Fe (III)-salt, type of anion, temperature, and hydrolysis time. The $Fe^{3+}$ hydrolysis results in the release of $H^+$ ions, and consequently, a decrease in pH occurs. FTIR spectra at 886 and 798 $cm^{-1}$ are typical for $\alpha$–FeOOH [171]. Figure 6.20 shows that the FeOH band occurs at 852 and 786 $cm^{-1}$ in the CFOH sample spectrum.

The peaks of 1510 $cm^{-1}$(–$NH_2$), 1612 $cm^{-1}$ (amide I band), 1019 $cm^{-1}$ (C–N), and 1091 $cm^{-1}$ (C–N) represent the chitosan peaks. In the region of 2900 $cm^{-1}$ to 3500 $cm^{-1}$ of the spectrum, both chitosan and iron composite beads exhibit peaks at 3450 $cm^{-1}$ corresponding to the stretching O–H and N–H groups. In both cases, the OH band tends to overlap the stretching band of N–H (near 3300 $cm^{-1}$). In the CFOH bead, the spectra at 2912 $cm^{-1}$ corresponding to CH stretching vibration in –CH, and –$CH_2$ are distinguishable compared to the spectra observed at 2912 $cm^{-1}$ in chitosan flake. In the region of 1000–1200 $cm^{-1}$, chitosan shows two peaks 1095 and 1015 $cm^{-1}$, corresponding to the stretching of the C–O bond of C3 from chitosan

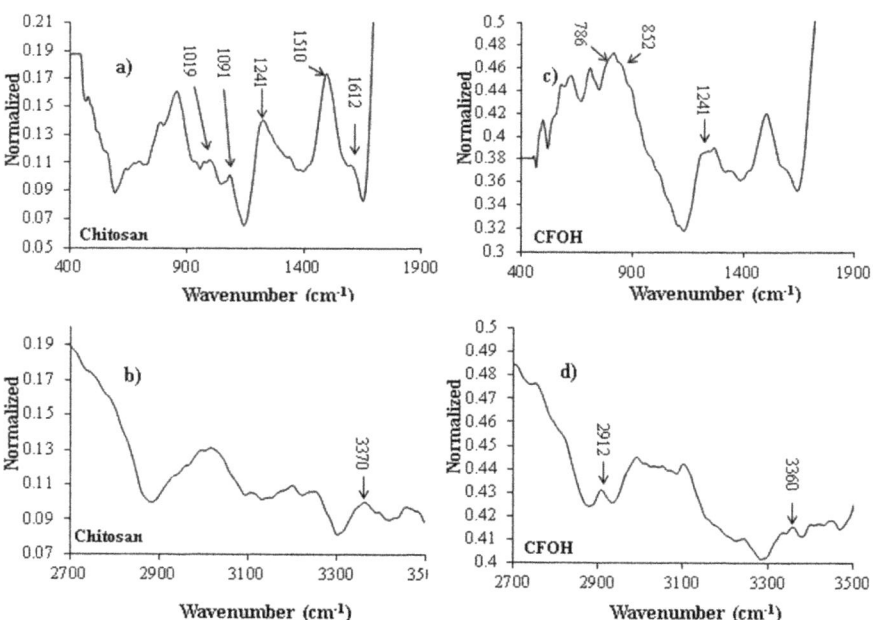

**Fig. 6.20** DRIFT-FTIR of chitosan flake (**a**, **b**) and CFOH beads (**c**, **d**). Unpublished work by authors

(secondary OH) and C–O stretching of C6 of chitosan (primary OH) [172]. Compared with the C–O spectrum of chitosan obtained at 1015 cm$^{-1}$, the absorption peaks of the secondary hydroxyl group of CFOH bead become folded and in the O–H band reduced and shifted from 3370.0 to 3363.0 cm$^{-1}$. This evidence of chemical bond constant of C–O and O–H band decrease supports the existence of a complexing reaction of iron (III) hydroxide species [FeOH$^{2+}$, Fe (OH)$_2^+$] and iron(III) aqua complexes may take place with the participation of surface oxygen functionalities and, thus, with chitosan at the secondary hydroxyl group [149].

The complicated nature of the adsorption bands in the 1650–1500 cm$^{-1}$ region suggests that aromatic ring bands and double-bond (C=C) vibrations overlap the C=O stretching vibration bands and OH binding vibration bands. The peaks expected in the region of FTIR spectra of chitosan include in ammonium (–NH$_3^+$), amine (–NH$_2$), and carbonyl (–CONHR) band [173, 174]. Figure 6.20 (i and ii) shows peaks at 1510 cm$^{-1}$ and at 1612 cm$^{-1}$, which could be assigned to NH$_3^+$ band and NH$_2$ (or CONH$_2$) groups of chitosan, respectively [175]. For primary aromatic amine, C–N stretching vibration falls between 1350 and 1250 cm$^{-1}$ [176]. There are peaks observed for both chitosan and CFOH bead at 1241 cm$^{-1}$. The peaks at 1241 cm$^{-1}$ for CFOH beads are found to be weakened. However, the presence of these peaks in CFOH spectra at 1510 cm$^{-1}$ (NH$_3^+$), 1612 cm$^{-1}$ (NH$_2$), and 1241 cm$^{-1}$ (C–N) is depicted in Fig. 6.20a, b, suggesting that there was no interaction between the NH$_3^+$ and iron (III) nitrate salt and the change in the peak at 1241 cm$^{-1}$ for chitosan–iron composite may be related to the chemical bonds with nitrogen, indicating that amine groups present in the chitosan may react with iron and form a complex as R–N–Fe$^+$ or changed to be protonated amine.

FTIR spectra of CFOH adsorbent exposed to As (V) solution containing 1.75 mmol/L of As (V) at a pH of 6.5 for 24 h are shown in Fig. 6.21. The pH effluent pH increased from 6.5 to 6.8–7.3 following adsorption for 24 h (data not shown). It has appeared from Fig. 6.21 that a new peak appeared at 1087, 1029, and 3181 cm$^{-1}$ following exposure of the beads to As (V) solution, which are assigned to H–OH stretching and bending vibration of a hydroxyl group of Fe-oxide and water.

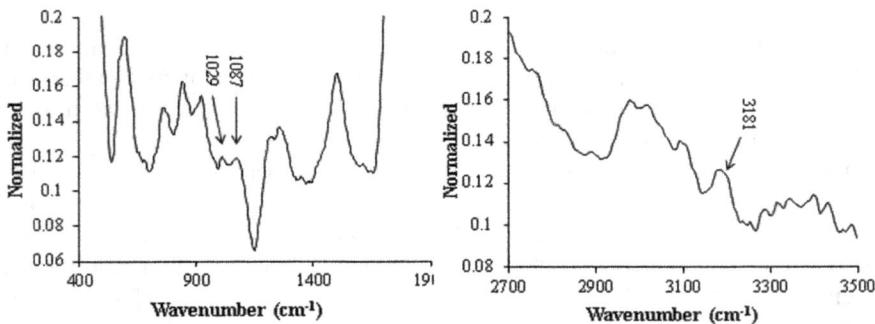

**Fig. 6.21** FTIR spectra (DRIFT Mode) of CFOH beads exposed to As (V) solution. Unpublished work by authors

**Fig. 6.22** DRIFT difference FTIR spectra of As (V). Unpublished work by authors

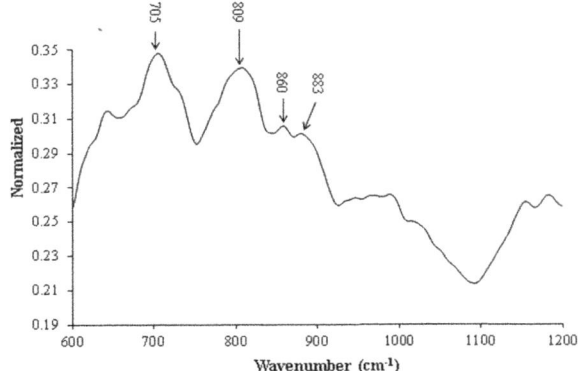

The DRIFT mode IR spectra of the CFOH bead were deducted from the IR spectra of the CFOH bead that was exposed to As (V) solution. Therefore, the DRIFT difference spectrum of As (V) on beads showed peaks at 705, 809 cm$^{-1}$, and 860 cm$^{-1}$, with a shoulder at 883 cm$^{-1}$ (Fig. 6.22).

Goldberg and Johnson [170] reported that IR spectra for arsenate sorbed onto Fe-oxide at 817 cm$^{-1}$ and 854 cm$^{-1}$ at pH 9.0 and 824 cm$^{-1}$ 861 cm$^{-1}$ at pH 5.0, respectively. They assigned band at 817–824 cm$^{-1}$ to the Fe–O–As groups, and the 854–861 cm$^{-1}$ band corresponds to non-surface-complexed As–O bonds of the adsorbed As (V) species. Suarez et al. [177] reported vibrational spectra for the solution of $Na_2HAsO_4$ at pH 8, where $HAsO_4^{2-}$ is the dominant aqueous As species, has a large broad peak at 862 cm$^{-1}$. This band is assigned to the IR active $\upsilon_1$ and $\upsilon_3$ modes and suggests that $HAsO_4^{2-}$ has a similar spectrum to $AsO_4^{3-}$. At pH 5, where $H_2AsO_4^{-}$ is dominant, it has two peaks at 909 and 878 cm$^{-1}$, corresponding to the splitting of the $\upsilon_3$ modes.

They speculated that protonation to $HAsO_4^{2-}$ should lower the symmetry to $C_{3\upsilon}$ with further protonation to $H_2AsO_4^{-}$ lowering the symmetry to $C_{2\upsilon}$. Arai and sparks [178] studied the phosphate adsorption mechanism onto the ferrihydride–water interface. They found that bidentate binuclear ($C_{2\upsilon}$) and monodentate mononuclear phosphate complexes ($C_1$) predominantly form at the ferrihydride–water interface at pH $\geq$ 7.5 and pH < 7, respectively. Using the DRIFT IR spectrum of As (V) that was adsorbed on to amorphous Fe (OH)$_3$, Suarez et al. [177] reported a broad peak at 834 cm$^{-1}$ with a shoulder at 880 cm$^{-1}$ as $\upsilon_3$ mode of $HAsO_4^{2-}$ species.

The spectra obtained for As (V) sorbed onto CFOH beads at 809 cm$^{-1}$ and at 860 cm$^{-1}$ cannot be assigned to $\upsilon_3$ mode as the split is much more than as suggested by Suarez et al. [177]. The position of Raman spectra for As (V) sorbed to beads showed three strong peaks at 905, 873, and 762 cm$^{-1}$ (See Fig. 6.23). The Raman spectra for As (V) at 765 and 875 cm$^{-1}$ were assigned by Tossel [179] to an asymmetric stretching vibration of As–OH and symmetric stretching vibration of As–O bond, respectively. The Raman spectra of As (V) sorbed to CFOH beads at 762 cm$^{-1}$ are assigned to $\upsilon$(As–OH), and 905 cm$^{-1}$ and 873 cm$^{-1}$ are to $\upsilon$(As–O) vibration,

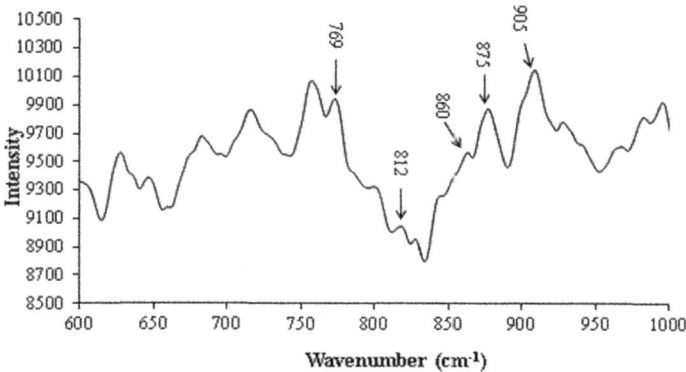

**Fig. 6.23** Raman spectra of chitosan–iron composite beads exposed to As (V) solution. Unpublished work by authors

respectively [170, 178]. The band at 860 cm$^{-1}$ and 809 cm$^{-1}$ may correspond to symmetric and asymmetric stretching modes of two equivalent As–O–Fe bonds. It is suggested that both $HAsO_4^{2-}$ and $H_2AsO_4^-$ species might participate in surface complexation reaction with the beads, and the spectra can be assigned to As–OH and As–O–Fe stretching vibration. The spectra assume that the substitution of Fe-OH groups plays a major role in adsorption mechanisms [180, 181]. Based on the DRIFT IR spectra, the following reactions are dominant when As (V) species are adsorbed on to beads surface.

The DRIFT difference spectrum of As (III) sorbed onto CFOH beads at pH 6.5 shows two main peaks at 794 cm$^{-1}$ and 620 cm$^{-1}$ with a minor peak at 458 cm$^{-1}$ (See Fig. 6.24). The $\upsilon_3$ asymmetric stretching modes and $\upsilon_1$ symmetric stretching modes reported for $H_3AsO_3$ in Raman spectra are 655 and 710 cm$^{-1}$, respectively [177]. The As (III) Raman spectra show strong peak at 667 and 754 cm$^{-1}$ (See Fig. 6.25). It is also observed that the IR spectra at 794 cm$^{-1}$ have two shoulders at 779 cm$^{-1}$ and 809 cm$^{-1}$, respectively. This could have resulted from bidentate binuclear ($C_{2\upsilon}$)

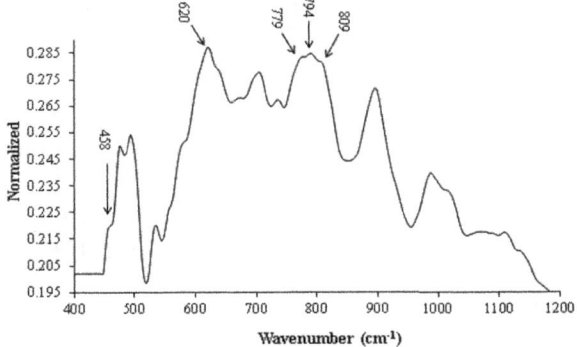

**Fig. 6.24** DRIFT difference FTIR spectra of CFOH beads that exposed to As (III) solution. Unpublished work by authors

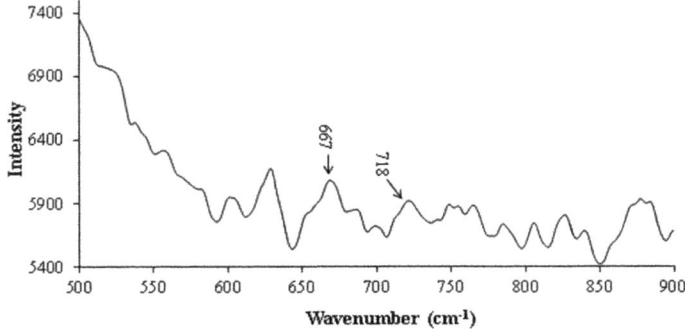

**Fig. 6.25** Raman spectra of chitosan–iron composite (CFOH) beads exposed to As (III) solution. Unpublished work by authors

and monodentate mononuclear arsenic complexes ($C_1$) form at the iron-water interface. The spectrum at 794 cm$^{-1}$ resembles the $\upsilon_1$ As–O mode for $H_2AsO_3^-$ and at 630 cm$^{-1}$ for $\upsilon_5$ As–OH asymmetric stretching modes [177]. The shoulder position at 809 cm$^{-1}$ can be assigned to As–O asymmetric mode and 779 cm$^{-1}$ for $\upsilon_1$ As–O stretching mode for $HAsO_3^{2-}$ species. It is suggested from the IR and Raman spectra that when the bead was exposed to As (III) solution at pH 6.5, both $H_2AsO_3^-$ and $HAsO_3^{2-}$ undergo inner-sphere surface complex reaction with the Fe-O group in the beads. The following reactions can be responsible for As (III) uptake onto the bead surface.

The pH study concluded that the CFOH beads had a maximum adsorption capacity at a pH of 6.5. Therefore, the equilibrium adsorption capacity of beads for different percentages of As (III) and As (V) in the solution in the presence of 0.05 M NaCl was determined at pH 6.5 and 298 K. The data are shown in Fig. 6.26. As expected, the adsorption capacity was dependent on the arsenic concentration in the solution. It may also be noted that the equilibrium uptake capacity of the CFOH beads for As (III) is comparatively higher than As (V).

CFOH bead contains amino and iron oxide groups available for characteristic coordination bonding with metal ions. The following one-site Langmuir equation may be used to describe pH-dependent metal ion adsorption:

$$-SH - S + H^+; \quad k_h, \tag{6.67}$$

$$-S + M - SM; \quad k_M M : \text{metal ion}, \tag{6.68}$$

$$q = \frac{q_m \alpha K_m [M]}{1 + \alpha K_m [M]}; \tag{6.69}$$

where
$q =$ the adsorption capacity corresponding to metal ion concentration [M],

## 6.8 Case Study I

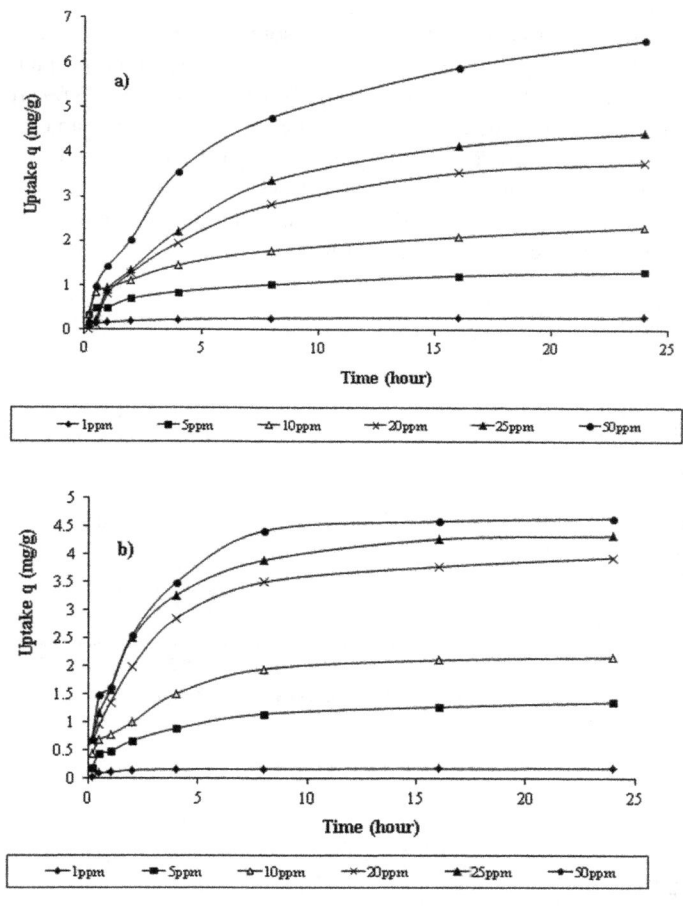

**Fig. 6.26** a Effect of time and concentration on **a** As (III) and **b** As (V) uptake onto CFOH beads (Condition: pH: 6.5; Temp: 298 K). Copyright © 2014, Taylor & Francis, Sep, Sci, and Technol. 49, 2863–2877, 2014. Reproduced from Ref. [94] with permission

$K_m$ is equilibrium constant, and
$q_m$ = maximum adsorption amount of metal ions (mg/g).

The effect of pH was incorporated by introducing a parameter "$\alpha$" that is dependent on pH of the solution.

$$\alpha = \frac{K_H}{(K_H + [H^+])}, \tag{6.70}$$

where
$[H^+]$ is the hydrogen ion concentration,
$K_H$ is the equilibrium constant.

Equation 6.69 was used to correlate the adsorption capacity of the beads. The equilibrium data for As (V) could be correlated with the Langmuir equation within ± 5% of the experimental values (Fig. 6.27b). Constants of the Eq. 6.69 are obtained by nonlinear regression of the experimental data and are given in Table 6.15. It was noted that Eq. 6.69 represented the adsorption behavior of As (V) on the CFOH bead adequately, but it did not fit the data well for As (III) (Fig. 6.27).

To obtain a better fit of the As (III) data, the Langmuir one-site model was extended to the Langmuir two-site model, given below:

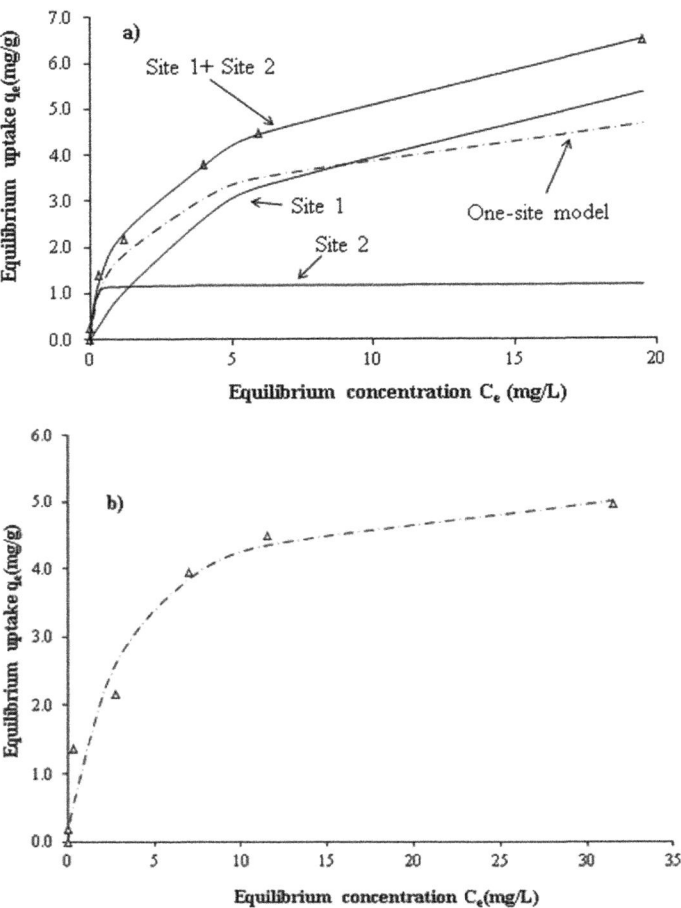

**Fig. 6.27** a Langmuir isotherm of **a** As (III)adsorption onto chitosan composite bead [dotted line is Langmuir one site; (Δ) experimental data and solid line are Langmuir two-site model (Site1 + Site 2). **b** Langmuir isotherm of As (V) adsorption [dotted line is Langmuir one site; (■) experimental data]. Copyright © 2014, Taylor & Francis, Sep, Sci, and Technol. 49, 2863–2877, 2014. Reproduced from Ref. [95] with permission

6.8 Case Study I

**Table 6.16** Estimated parameters for Langmuir model

| Langmuir model | $q_{m1}$ (mg/g) | $q_{m2}$ (mg/g) | $k_{m1}$ | $k_{m2}$ |
|---|---|---|---|---|
| As (III) (one-site model) | 7.24 | – | 1.282 | – |
| As (III) (two-site model) | 7.26 | 1.18 | 0.117 | 24.52 |
| As (V) (one-site model) | 5.486 | – | 0.2785 | – |
| As (V) (two-site model) | 4.7801 | 1.0857 | 7.219 | 0.1611 |

Copyright © 2014, Taylor & Francis, Sep, Sci, and Technol. 49, 2863–2877, 2014. Reproduced from Ref. [95] with permission

$$q = \frac{q_{m1}\alpha_1 K_{M1}[M]}{1+\alpha_1 K_{M1}[M]} + \frac{q_{m2}\alpha_2 K_{M2}[M]}{1+\alpha_2 K_{M2}[M]}. \tag{6.71}$$

A nonlinear regression method was also used to obtain the constants of Eq. 6.71 and are given in Table 6.16. The two-site Langmuir isotherm plot is shown in Fig. 6.27a. Again, a better fit to the experimental data for As (III) was observed.

## 6.8.2 Summary

Spherical CFOH beads of a diameter of 1 mm were prepared to remove As (III) and As (V) from water. Scanning electron micrograph and energy-dispersive X-ray spectrometry analysis showed that beads were porous, and rod-shaped iron oxide particles were dispersed homogeneously in the bead. The CFOH beads adsorbed both As (III) and As (V) when exposed to arsenic solutions. XPS studies revealed that As (V) uptake on the bead was based on Coulombic and Lewis acid–base interaction, whereas As (III) uptake was a Lewis acid–base interaction. Furthermore, As (III) species may undergo complex formation with $NH_3^+$ group presence in the beads. The results are consistent with the data obtained from pH and surface charge analysis—that As (III) may undergo reaction with iron oxide and an amine group presence in the CFOH beads. The equilibrium adsorption capacity of both As (III) and As (V) was observed to depend on the solution pH. The Langmuir model provided the best fit of the adsorption data for As (V), while the two-site Langmuir model provided the best fit for As (III) adsorption data.

## 6.9 Case Study 2

### 6.9.1 Sorption Mechanism of Uranium on to Chitosan-Coated Perlite (CP) Beads

Hasan et al. [182] also studied the adsorption of uranium on to chitosan-coated perlite beads. Chitosan is a natural biopolymer, hydrophilic, and can form complexes with metals. It is also a non-toxic, biodegradable, and biocompatible material. Chitosan tends to agglomerate or form a gel in aqueous media. Although the amine ($-NH_2$) and hydroxyl groups (OH) in chitosan are mainly responsible for the adsorption of metal ions, these active binding sites are not readily available for sorption on a gel or in its natural form. In this study, a chitosan-coated perlite bead was prepared by dispersing chitosan on an inert substrate (perlite) to enhance its adsorption capacity by exposing more active sites, in other words, by increasing the accessibility of the metal-binding sites for adsorption. Chitosan was coated on perlite, and the coated adsorbent was prepared as a spherical bead following the process mentioned elsewhere. The adsorbent is known as chitosan perlite (CP) bead in this study. The adsorption capacity of the CP bead was evaluated for uranium under batch studies at different pH and uranium concentrations in the solution. The equilibrium adsorption data were correlated using Langmuir isotherm equations. To gain a better understanding of the uranium adsorption process, the surface properties of the CP bead before and after expose to uranium solution were characterized by scanning electron micrograph (SEM), energy-dispersive X-ray spectrometry (EDS) microanalysis, transmission electron microscopy (TEM), and X-ray photoelectron spectroscopy (XPS) analysis.

#### 6.9.1.1 Adsorption Studies

The chitosan coated on perlite (CP) beads was prepared following the process mentioned in Chap. 5 (Sect. 5.7). The adsorption capacity of CP beads for uranium was determined using a batch technique. One gram of CP bead was suspended in 100 mL solution containing U(VI) in the range of 10–1000 mg/L. The pH of the uranium-containing solution was adjusted between 1 and 8 by adding either 0.1 M NaOH or 0.1 N $H_2SO_4$ to the solution. Two types of control experiments were carried out. The first control contained only the pH adjusted uranium solution but no CP beads. The other control contained the beads only in an aqueous solution; there was no uranium ion in the solution. The flasks containing the beads and U(VI) were then placed in a shaker at 298 K and agitated at 200 rpm for 24 h along with the controls. The mixtures were filtered to separate the beads, and the filtrates were analyzed for uranium concentration using a quadrupole ICP-MS using external calibrations. A correlation constant of 0.9999 was achieved for the calibration series. Each sample was calculated at a relative standard deviation of <1% over ten runs. The controls provided a check of any systematic errors introduced during experimentation.

## 6.9.2 Results and Discussion

### 6.9.2.1 Characterization of the Chitosan Beads

After oven drying, the CP beads were characterized for their particle size, pore volume, and pore size. Particle size was determined by measuring the size of 50 beads using Vernier calipers. The average diameter of the beads was 2.2 mm. The surface area, pore volume, and pore size were obtained from the BET surface area measurement using $N_2$ at 77 K. Various physical properties of the beads are given in Table 6.17.

The chitosan content of the CP beads was determined by a thermogravimetric analyzer (TGA). Beads were heated up to 800 °C (1073 K) at a rate of 10 °C (10 K) per min in air. Figure 6.28a shows that the bead started to decompose at 200 °C (473 K). At a temperature of 500 °C (773 K), the weight of the sample became constant, suggesting that all of the chitosan of the beads had been burnt out at this temperature. A 32% weight loss was calculated from the data. The porous structure of beads was investigated using a scanning electron microscope (SEM). Figure 6.28b shows the SEM image of the cross section of the CP bead. As can be observed from this figure, the beads were porous.

The pore size distribution of chitosan flake, pure chitosan, and chitosan-coated perlite beads is shown in Fig. 6.28c. The average pore diameter of CP beads was found to be 0.0068 μm, whereas the average pore diameter of chitosan flake and pure chitosan beads were 0.004 and 0.0047 μm, respectively. The average pore volume of chitosan flake, pure chitosan beads, and CP beads were found to be $1 \times 10^{-3}$, 1.01

**Table 6.17** Physical properties for chitosan-coated perlite (CP) beads

| Properties | Values of the parameters |
|---|---|
| Avg. particle diameter, $d_p$ (m) | $2.2 \times 10^{-3}$ |
| Particle density, $\rho_p$ (kg/m$^3$) | 272 |
| Particle porosity, $\varepsilon_p$ | $2.56 \times 10^{-3}$ |
| Shape | Spherical |
| Chitosan content (wt%) | 32 |
| Surface area (m$^2$/g) | 4.5 |
| Pore volume (m$^3$/kg) | $9.4 \times 10^{-6}$ |
| Pore radius (m) | $4.13 \times 10^{-9}$ |

Copyright © 2007, Taylor & Francis. Nuclear Technology, 2007, 159, 59–71. Reproduced from Ref. [182] with permission

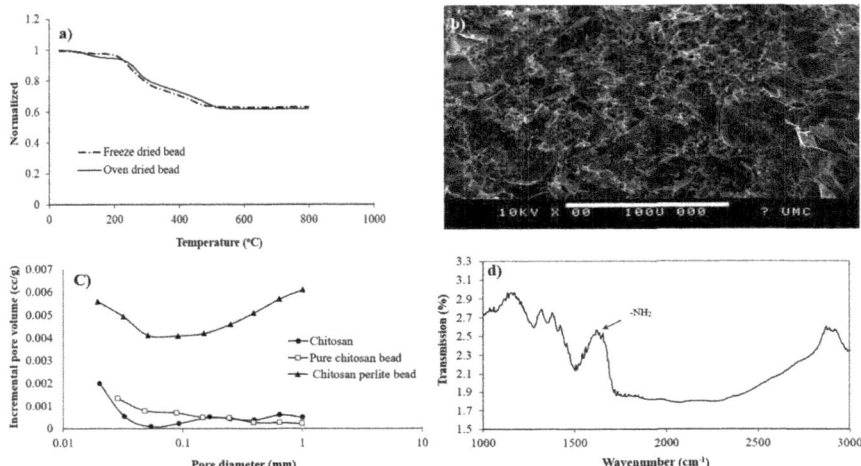

**Fig. 6.28** a Thermogravimetric analysis (TGA), **b** scanning electron microscopy (SEM) micrograph, **c** pore size and pore volume, and **d** Fourier transform infrared spectrum of CP bead. Copyright © 2007, Taylor & Francis. Nuclear Technology, 2007, 159, 59–71. Reproduced from Ref. [182] with permission

$\times 10^{-3}$, and $9.41 \times 10^{-3}$ cm$^3$/g, respectively. The increase in the pore volume of CP beads suggests more intraparticle space than pores within perlite particles. Chitosan film held these particles together and was dispersed on perlite in the process. The CP beads were crushed to powder, and the Fourier transform infrared (FTIR) spectrum was obtained in a diffuse reflectance mode (Fig. 6.28d). The peak at 1595 cm$^{-1}$ is due to an amine (–NH$_2$) group, which is considered to be the main functional group of chitosan for bonding with metal ions.

The energy-dispersive spectroscopic (EDS) X-ray microanalysis was used for elemental analysis of the CP beads. In this attempt, the bead was exposed to uranium solution containing 100 mg/L of uranium ion for 24 h and then dried in a vacuum oven at 60 °C overnight. The SEM micrograph of the cross section of the bead shows uranium uptake occurs at the outer rim of the bead (bright portion), as shown in Fig. 6.29a. The EDS analysis of the same sample shows a strong characteristic peak for uranium at ~3.2 keV (Fig. 6.29b). An EDS X-ray mapping of uranium present in the CP bead was carried out to understand the adsorption mechanism better. Figure 6.29c shows a traditional micrograph generated by the reconstructed images of the particles using the X-ray. Both the surface and the inside cross-section of the bead were analyzed. All of the metal ions present in the bead were targeted. Later, other ions were phased out of the micrograph except for the uranium ion in the bead. This was done by using only 3.2 keV energy during X-ray mapping. This was the only signal recorded due to uranium and was used to reconstruct the image shown in Fig. 6.29d. The data showed that uranium progressively diffused from the outer

## 6.9 Case Study 2

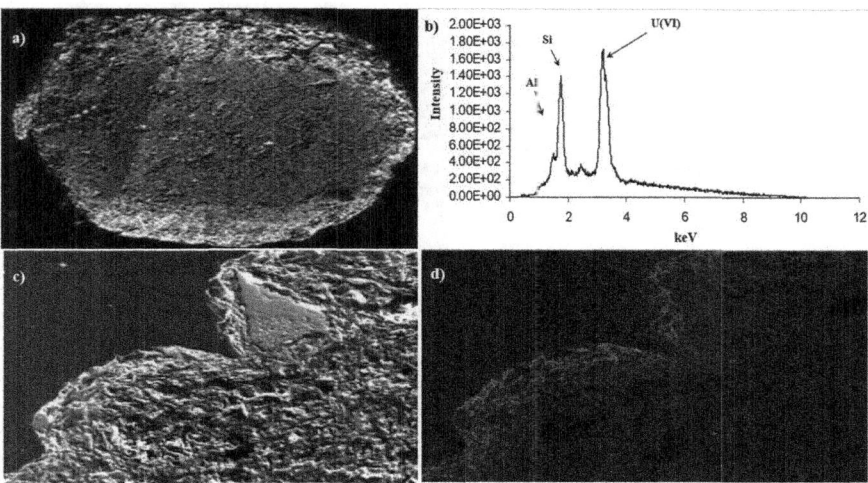

**Fig. 6.29** a Scanning electron microscopy (SEM), b energy-dispersive X-ray spectrometry (EDS) microanalysis of chitosan perlite bead exposed to uranium solution, c EDS X-ray mapping of chitosan perlite beads that were exposed to uranium, and d EDS X-ray map of beads exposed to uranium showing the adsorption of uranium around the outer rim of the beads. The constituents of perlite are blocked out during mapping using 3.2 keV energy only. Copyright © 2007, Taylor & Francis. Nuclear Technology, 2007, 159, 59–71. Reproduced from Ref. [182] with permission

surface into the interior of the bead. This suggests that the mobility of U(VI) was not hindered by perlite substrate, and U(VI) can bind to chitosan.

Transmission electron microscope (TEM) analysis of the bead was performed to investigate the presence of the uranium ions in the bead and the heterogeneity of the sample. The sample was prepared using the microtome technique. TEM utilizes a NORAN pioneer ultra-thin 0.5 μm polymer sheet window detector. TEM of the samples after exposure to uranium ions is shown in Fig. 6.30. It may be noted that metal uranium showed a strong affinity toward chitosan. Uranium is detected only in the vicinity of chitosan, not near the perlite substrate. Also, it can be observed that the beads were highly heterogeneous. The chitosan film seems to surround the perlite particles. Also, a cluster of particles covered with the films could be seen. It is hypothesized that uranium (VI) formed a covalent bond with amine groups in the chitosan and as explained later in this section. The EDS analysis of the same sample is also shown in Fig. 6.30. The EDS system was mounted on the TEM system. Figure 6.30 showed a characteristic peak at 0.3 keV representing carbon that might have come from chitosan, a peak at 0.525 keV due to oxygen $K_\alpha$ peak generated from oxygen atoms present either in perlite or chitosan. Finally, a peak around 3.2 keV confirms the presence of uranium on the bead.

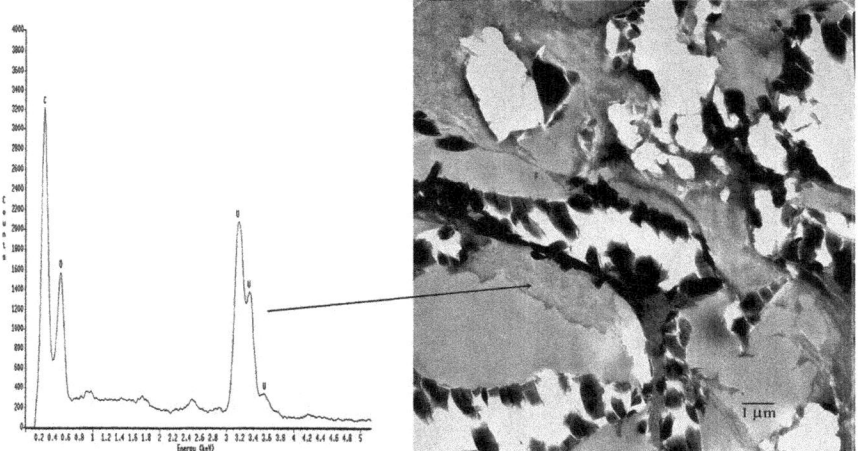

**Fig. 6.30** Transmission electron microscopy (TEM) micrograph of chitosan-coated perlite beads exposed to uranium, and EDS X-ray microanalysis of the same sample shows the presence of uranium on the bead. Copyright © 2007, Taylor & Francis, Nuclear Technology, 2007, 159, 59–71. Reproduced from Ref. [182] with permission

### 6.9.2.2 Effect of pH on Uranium Uptake by Chitosan-Coated Perlite (CP) Beads

The pH of the solution is important during the sorption of metal ions. The pH of the uranium solution before adding chitosan beads was adjusted between 1 and 8 using either 0.1 N $H_2SO_4$ or 0.1 M NaOH. The pH of the solution was found to increase at the end of an adsorption run. No attempt was made to maintain a constant pH of the solution during the experiment. The uptake of U(VI) by CP beads increased as the pH increased from 1 to 5 (Fig. 6.31).

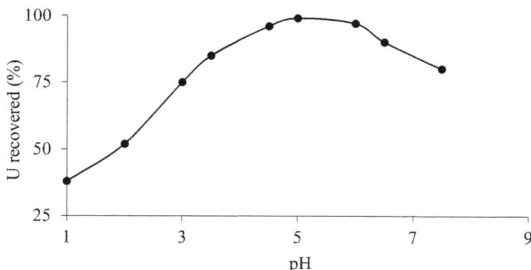

**Fig. 6.31** Effect of pH on U(VI) removal by chitosan-coated perlite beads from aqueous solution. (Solution concentration was 0.42 mmol/L of U(VI) at 298 K). Copyright © 2007, Taylor & Francis, Nuclear Technology, 2007, 159, 59–71. Reproduced from Ref. [182] with permission

## 6.9 Case Study 2

As the pH of the solution further increased, the uptake capacity started to decrease. Amine groups (–NH$_2$) in chitosan are considered the active functional groups for the adsorption of metal ions. For adsorption of U(VI) on chitosan, it must form an ion in the solution. The extent of ionization of uranium in the solution and ionization species depends on the concentration and pH of the solution.

The hydrolysis of uranyl nitrate can proceed in the following manner:

$$UO_2(NO_3)_2 \Leftrightarrow UO_2^{2+} + 2NO_3^-, \tag{6.72}$$

$$UO_2^{2+} + H_2O \Leftrightarrow UO_2OH^+ + H^+, \tag{6.73}$$

$$2UO_2^{2+} + 2H_2O \Leftrightarrow (UO_2)_2(OH)_2^{2+} + 2H^+, \tag{6.74}$$

$$2UO_2^{2+} + H_2O \Leftrightarrow (UO_2)_2(OH)^{3+} + H^+, \tag{6.75}$$

$$3UO_2^{2+} + 4H_2O \Leftrightarrow (UO_2)_3(OH)_4^{2+} + 4H^+, \tag{6.76}$$

$$3UO_2^{2+} + 5H_2O \Leftrightarrow (UO_2)_3(OH)_5^+ + 5H^+. \tag{6.77}$$

The relative distribution of various uranium ions formed in aqueous solutions is further calculated from MINEQL software and is shown in Fig. 6.32. It is observed that uranium in an aqueous solution is mainly present as a cation at a lower pH. The main hydrolyzed uranyl species in the pH range of 4–6 are $UO_2^{2+}$, $UO_2(OH)^+$, $(UO_2)_2(OH)_2^{2+}$, and $(UO_2)_3(OH)^{5+}$. The fraction of negatively charged hydrolysis products in the solution increases as the pH increases [183]. The possible hydrolysis reactions forming negative uranium ions are as follows:

$$UO_2^{2+} + 3H_2O \Leftrightarrow UO_2(OH)_3^- + 3H^+, \tag{6.78}$$

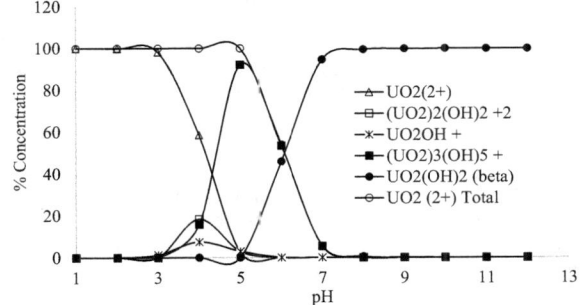

**Fig. 6.32** Distribution of uranium species in aqueous solution at different pH as calculated using MINEQL (The U(VI) concentration used in the calculation was 1000 μg/L). Copyright © 2007, Taylor & Francis, Nuclear Technology, 2007, 159, 59–71. Reproduced from Ref. [182] with permission

$$UO_2^{2+} + 4H_2O \Leftrightarrow UO_2(OH)_4^{2-} + 4H^+, \quad (6.79)$$

$$3UO_2^{2+} 7H_2O \Leftrightarrow (UO_2)_3(OH)_7^- + 7H^+. \quad (6.80)$$

Therefore, it is expected that $UO_2^{2+}$, $UO_2(OH)$-positive ions will attach to $-NH_2$ groups in chitosan during the adsorption process in the pH range of 4–6.

### 6.9.2.3 Equilibrium Studies for Uranium Uptake Onto Beads

The equilibrium adsorption capacity of chitosan beads for uranium was determined in the concentration range of 10–1000 mg/L at 298 K. As mentioned in the previous section, the maximum adsorption capacity for uranium occurred at a pH of 5.

The amount of uranium adsorbed as a function of time from various uranium concentrations of the solution is shown in Fig. 6.33. Almost 60% of uranium was adsorbed during the first 60 min of a run, and then the equilibrium was attained monotonically at 240 min in most of the runs. The adsorption isotherm data were obtained at 298 K and pH 5. The data were correlated according to the Langmuir isotherm equations. The Langmuir equation provided the best fit of the data, as shown in Fig. 6.34. The Langmuir isotherm generally provides a good fit of data when the isotherm is Type I, which is due to the monolayer coverage of a surface containing a finite number of identical sites. The Langmuir equation can be expressed as

$$q_e = \frac{Q_0 b C_e}{1 + b C_e}, \quad (6.81)$$

where

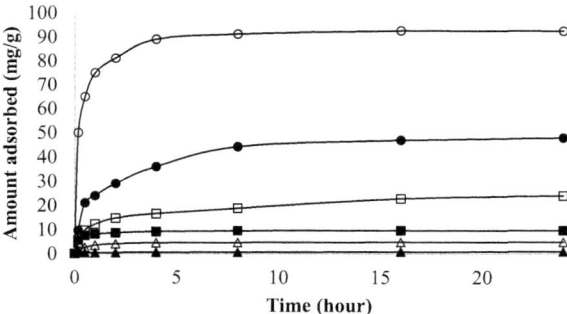

**Fig. 6.33** Effect of time and concentration on uranium adsorption onto CP bead. The initial concentrations of the solution were (○) 1000, (●) 500, (□) 250, (■) 100, (△) 50, and (▲) 10 mg/L, respectively (pH: 5.0 and temperature: 298 K). Copyright © 2007, Taylor & Francis, Nuclear Technology, 2007, 159, 59–71. Reproduced from Ref. [182] with permission

## 6.9 Case Study 2

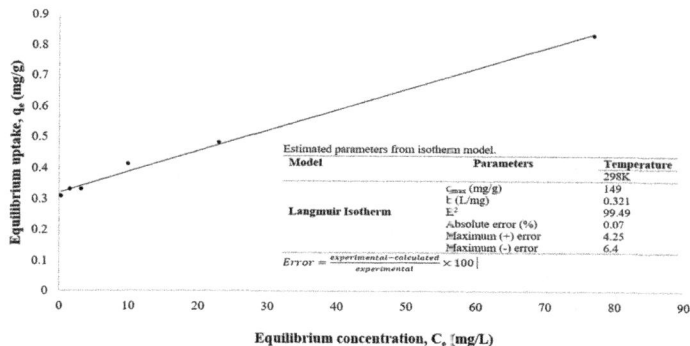

**Fig. 6.34** Langmuir plot at 298 K temperature. The $q_{max}$ and b are the Langmuir constants. Copyright © 2007, Taylor & Francis, Nuclear Technology, 2007, 159, 59–71. Reproduced from Ref. [182] with permission

$q_e$ = equilibrium uptake of adsorbate by the adsorbent corresponding to concentration $C_e$,
$Q_0$ = weight of adsorbate contained in the monolayer on the surface,
$C_e$ = concentration of adsorbate in the fluid phase in equilibrium with the concentration $q_e$ on the solid, and
$b$ = constant.

The values of adsorption constants for Langmuir isotherm are listed on Fig. 6.34. As can be seen, the Langmuir equation provided a good fit of the data with a maximum absolute error of less than 1%.

### 6.9.2.4 Sorption Mechanism of Uranium Onto CP Beads

The XPS analysis of the beads before and after the adsorption of uranium was used to better understand the adsorption sites onto which uranium was adsorbed. To understand the binding of uranium to the active sites on CP bead, it was exposed to 100 mL of a 100 mg/L uranium solution at pH 5.0. After 24 h of exposure, the beads were removed from the solution and oven dried at 60 °C overnight. Figure 6.35 (survey scan) shows the binding energies of various components in CP bead and CP bead exposed to uranium, as noted during XPS analysis. The survey scan of XPS analysis of CP beads shows peaks at 283.5, 398, and 530.5 that corresponded to C-1s, N-1s, and O-1s, respectively (Fig. 6.35). The survey scan of XPS analysis of the CP beads exposed to uranium solution showed an additional peak at around 379.5 eV that corresponded to the core level U-4f peak for uranium (Fig. 6.35). U-4f core level spectrum is generally characterized by a strong satellite peak. The uranium spectra in the binding energy range of 370–400 eV show the two most intense peaks at ~379.5 eV and ~389 eV (Fig. 6.35a). Each of the peaks consists of asymmetric lines and broad satellites at higher binding energy. These two peaks at ~379.5 eV for

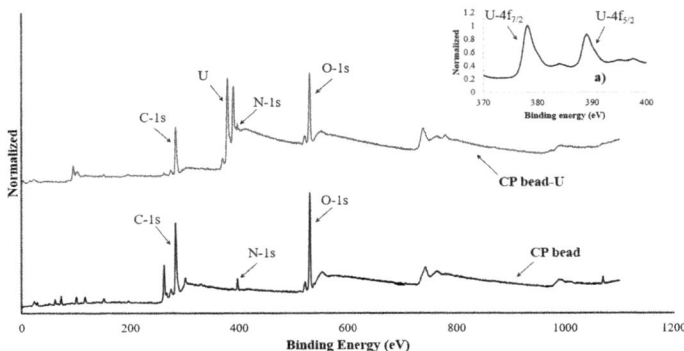

**Fig. 6.35** Survey scan of CP bead and the CP beads exposed to uranium solution. Inset shows a spin–orbit splitting spectrum corresponding to $4f_{7/2}$ and $4f_{5/2}$ of uranium. Copyright © 2007, Taylor & Francis, Nuclear Technology, 2007, 159, 59–71. Reproduced from Ref. [182] with permission

U-$4f_{7/2}$ and ~389 eV for U-$4f_{5/2}$ are the fingerprints of two valences of uranium. The splitting of the U-$4f_{7/2}$ and U-$4f_{5/2}$ peaks is caused by the spin–orbital interaction [184].

Table 6.18 shows the binding energy of various uranium ions reported by various researchers. It is generally accepted that the U-$4f_{7/2}$ and U-$4f_{5/2}$ are the main peaks for $UO_2$ located in the range of 379.2–380.9 eV and 390.0–391.8 eV, respectively [185].

The broad satellite peak intensity, which is located at least ~3.5 eV higher than the main peak, is rather significant (Fig. 6.35a). These two additional satellite bands at 385 eV and 395 eV may represent unsaturated electronic configuration. Table 6.17 shows the position of U-4f spectrum at around 379.5 eV indicates the presence of $UO_2^+$ ion in the bead. It should be noted that the characteristic satellite peaks were observed along with the main peaks (Fig. 6.35a) of uranium, suggesting that the

**Table 6.18** Absolute binding energy of uranium ions to different materials

| Sample type | U ion | Binding energy (BE), eV | | Binding energy of satellite peak, eV | | References |
|---|---|---|---|---|---|---|
| | | $4f_{7/2}$ | $4f_{5/2}$ | $4f_{7/2}$ | $4f_{5/2}$ | |
| U/Si (111) | USi | ~378.2 | 389 | ~384 | ~396 | [186] |
| U (Rh$_{1-x}$ Pd$_x$)$_3$ | URh$_3$ | ~379.3 | 390.1 | ~379 | ~395 | [187] |
| Uranium metal | U | ~376.5 | – | – | – | [188] |
| $UO_2$ | U | 379.2 | – | – | – | [188] |
| $UO_2(NO_3)_2 \cdot 6H_2O$ | U | ~379.5 | ~389 | ~383 | ~395 | [182] |

Copyright © 2007, Taylor & Francis, Nuclear Technology, 2007, 159, 59–71. Reproduced from Ref. [182] with permission

## 6.9 Case Study 2

uranium was adsorbed on the surface of chitosan. The binding energy of different elements present in the beads is given in Table 6.19.

Note from the XPS spectrum, that the magnitude of N-1s binding energy changed slightly only when the CP beads were exposed to uranium solution, whereas no significant change was observed for the C-1s and O-1s peaks (Table 6.18). The chemical shift is considered significant when they exceeded 0.5 eV. The magnitude of the binding energy shift depends on the concentration of different atoms, particularly on the surface of a material. This data indicates that carbon and oxygen-containing functional groups had limited participation in the metal binding reactions. Therefore, the shifts of N-1s peaks could be the results of chemical interactions of uranium with the amine ($-NH_2$) functional groups of chitosan. Furthermore, it is assumed that uranium ion entered into the porous matrix of the bead and formed coordination complexes with the $-NH_2$ groups that present in the CP beads.

When comparing the ratio of U/N with the ratio of U/O, it appears that more uranium ions were attached to nitrogen than to oxygen. It is also observed that the amount of oxygen present in the exposed beads was comparatively higher than that on the unexposed bead. This increased sorption may be due to the adsorption of hydrolyzed products, such as $(UO_2)_2OH$, on hydroxyl groups, $CH_2OH$, or OH, of chitosan. In the pH range of 4–6, the $UO_2^{2+}$, $UO_2(OH)^+$, $(UO_2)_2(OH)_2^{2+}$, and $(UO_2)_3(OH)_5^+$ still dominate in the solution (Fig. 6.32). The XPS spectrum also (Fig. 6.35a) suggests that uranium was mainly adsorbed as $UO_2^{2+}$, but can also be adsorbed as $UO_2(OH)^+$ in this pH range. Based on the pH study and the data from XPS analysis, following mechanism for adsorption of uranium on chitosan was hypothesized:

**Table 6.19** Absolute binding energy of CP bead and the bead exposed to uranium

| Sample | C | | N | | O | | U | | U/N | U/O |
|---|---|---|---|---|---|---|---|---|---|---|
| | BE | Wt% | BE | Wt% | BE | Wt% | BE | Wt% | Wt % | Wt % |
| A | 283.5 | 57.6 | 397.5 | 3.91 | 530.5 | 28.1 | – | – | – | – |
| B | 283.5 | 53.9 | 398 | 3.6 | 530.5 | 30.0 | 379.5 | 9.23 | 2.564 | 0.308 |

Copyright © 2007, Taylor & Francis, Nuclear Technology, 2007, 159, 59–71. Reproduced from Ref. [182] with permission
A = chitosan-coated perlite (CP) beads and
B = chitosan-coated perlite (CP) beads exposed to uranium

$$\text{Chitosan} + UO_2^{2+} \rightleftharpoons \text{Chitosan-}UO_2 \text{ complex}, \quad (6.82)$$

$$\text{Chitosan-}NH_3^+ + (UO_2)_2OH^- \longrightarrow \text{Chitosan-}NH_3^+ \cdots (UO_2)_2OH^-. \quad (6.83)$$

### 6.9.3 Summary

Chitosan-coated perlite particles appear to adsorb uranium ions effectively from an aqueous solution. Scanning electron micrographs showed that the beads are highly porous. The extent of uranium adsorption depended on the pH of the solution, and a maximum capacity was noted at a pH of 5. The adsorption capacity of CP beads for U (VI) at a pH of 5 was found to be 149 mg/g from a solution containing 1000 mg/L of U(VI). The XPS and TEM data suggested that uranium was mainly adsorbed as $UO_2^{2+}$ and was attached to amine groups of chitosan.

## References

1. Agoro MA, Adenji AO, Adefisoye MA et al (2020) Heavy metals in wastewater and sewage sludge from selected municipal treatment plants in Eastern Cape Province, South Africa. Water 12:2746. https://doi.org/10.3390/w12102746; (a) Azimi A, Azari A, Rezakazemi M, Ansarpour M (2017) Removal of heavy metal from industrial wastewaters: a review. Chem Bio Eng Rev 4(1):1–24
2. Barakat M (2011) New trends in removing heavy metals from industrial wastewater. Arabian J Chem 4:361–377; (a) Wang J (2018) Reuse of heavy metal from industrial effluent water. IOP Conf Ser: Earth Environ Sci 199:042002. https://doi.org/10.1088/1755-1315/199/4/042002
3. Abebe LS, Chen X, Sobsey MD (2016) Chitosan coagulation to improve microbial and turbidity removal by ceramic water filtration for household drinking water treatment. Int

J Environ Res Public Health 13:269. https://doi.org/10.3390/ijerph13030269
4. Krishna R, Sahu O (2013) Reduction of COD and color by polymeric coagulant (chitosan). J Polym Biopolymer Phys Chem 1(1):22–25
5. Miranda R, Nicub R, Latoura I et al (2013) Efficiency of chitosan for the treatment of paper making process water by dissolve air flotation. Chem Eng J 231:304–313
6. Bhalkaran S, Wilson LD (2016) Investigation of self-assembly processes for chitosan-based coagulant-flocculant systems: a mini-review. Int J Mol Sci 17:1662. https://doi.org/10.3390/ijms17101662
7. Gidas MB, Gamier O, Gidas NK (1999) Performance of chitosan as a primary coagulant for the wastewater treatment. Trans Ecol Environ 26:47–56. www.witpress.com, ISSN 1743–3541
8. García MA, Montrlongo I, Rivero A et al Treatment of wastewater from fish processing industry using chitosan acid salts. Int Water Wastewater Treat 2(2). https://doi.org/10.16966/2381-5299.121
9. Rozainy MR, Syafalny HM, Pugansshwary P, Afifi A (2014) Combination of chitosan and bentonite as coagulant agents in dissolved air floatation. APCBEE Proc 10:229–234
10. Rorrer GL, Hsien T-Y, Way JD (1993) Synthesis of porous-magnetic chitosan beads for removal of cadmium ions from wastewater. Ind Eng Chem Res 32:2170
11. Bassi R, Prasher SO, Simpson BK (2000) Removal of selected metal ions from aqueous solutions using chitosan flakes. Sep Sci Technol 35(4):547–560
12. Guibal E, Milot C, Tobin JM (1998) Metal-Anion sorption by chitosan beads: equilibrium and kinetic studies. Ind Eng Chem Res 37:1454
13. Bodmeier R, Oh KH, Pramar Y (1989) Preparation and evaluation of drug-containing chitosan beads. Drug Dev Ind Pharm 15:1475
14. Hasan S, Krishnaiah A, Ghosh TK, Viswanath DS, Boddu VM, Smith ED (2003) Adsorption of chromium (VI) on chitosan-coated perlite. Sep Sci Technol 38(15):3775–3793
15. Su H, He X, Tan T (2003) Chromatography, and adsorption properties of heavy metal ions from wastewater with chitosan-chelated bead. Ziran Kexueban 30(2):19–22
16. Luisa CM, Carmen AM, de la Guardia M (2003) Removal of heavy metals by using adsorption on alumina or chitosan. Anal Bioanal Chem 375(6):820–825
17. Hastuti B, Masykur A, Hadi S (2016) Modification of chitosan by swelling and crosslinking using epichlorohydrin as heavy metal Cr (VI) adsorbent in batik industry wastes. IOP Conf Ser: Mater Sci Eng 107:012020. https://doi.org/10.1088/1757-899X/107/1/012020
18. Bhatta R, Sreedharb B, Padmajaa P (2015) Adsorption of chromium from aqueous solutions using crosslinked chitosan–diethylenetriaminepentaacetic. Int J Biol Macromol 74:458–466
19. Genc O, Soysal L, Bayramoglu G, Arica MY, Bektas S (2003) Procion green H-4G immobilized poly(hydroxyethylmethacrylate/chitosan) composite membranes for heavy metal removal. J Hazard Mater 97(1–3):111–125
20. Kawamura Y, Mitsuhashi M, Tanibe H, Yoshida H (1993) Adsorption of metal ions on polyaminated highly porous chitosan chelating agents. Ind Eng Chem Res 32:386
21. Hsien T-Y, Rorrer GL (1995) Effect of acylation and cross-linking on the material properties and cadmium ion adsorption capacity of porous chitosan beads. Sep Sci Technol 30:2455
22. Hsien T-Y, Rorrer GL (1997) Heterogeneous cross-linking of chitosan gel beads: kinetics, modeling, and influence on cadmium ion adsorption capacity. Ind Eng Chem Res 36:3631
23. Peng C, Wang Y, Tang Y (1998) Synthesis of cross-linked chitosan crown ethers and evaluation of these products as adsorbents for metal ions. J Appl Polym Sci 70:501
24. Becker T, Schlaak M, Strasdeit H (2000) Adsorption of Nickel(II), Zinc(II), and Cadmium(II) by new chitosan derivatives. React Funct Polym 44:289
25. Seo S, Kajiuchi T, Kim DI, Lee S, Kim HK (2002) Preparation of water soluble chitosan blendmers and their application to removal of heavy metal ions from wastewater. Macromol Res 10(2):103–107
26. No HK, Meyers SP (2000) Application of chitosan for treatment of wastewaters. Rev Environ Contam Toxicol 163:1–27
27. Anastopoulos I, Bhatnagar A, Bikiaris DN, Kyzas GZ (2017) Int J Mol Sci 18:114. https://doi.org/10.3390/ijms18010114

28. Evans JR, Davids WG, MacRae JD, Amirbahman AA (2002) Kinetics of cadmium uptake by chitosan-based crab shells. Water Res 36(13):3219–3226
29. Bassi R, Prasher SO, Simpson BK (1999) Remediation of metal-contaminated leachate using chitosan flakes. Environ Technol 20(11):1177–1182
30. Ngah WSW, Endu CS, Mayanar R (2002) Removal of copper(II) ions from aqueous solution onto chitosan and cross-linked chitosan beads. React Funct Polym 50(2):181–190
31. Gotoh T, Matsushima K, Kikuchi K-I (2004) Preparation of alginate-chitosan hybrid gel beads and adsorption of divalent metal ions. Chemosphere 55(1):135–140
32. Rae IB, Gibb SW (2003) Removal of metals from aqueous solutions using natural chitinous materials. Water Sci Technol 47(10):189–196
33. Sag Y, Aktay Y (2002) Kinetic studies on sorption of Cr(VI) and Cu(II) ions by chitin, chitosan and Rhizopus arrhizus. Biochem Eng J 12(2):143–153
34. Gyliene O, Rekertas R, Salkauskas M (2002) Removal of free and complexed heavy-metal ions by sorbents produced from fly (Musca domestica) larva shells. Water Res 36(16):4128–4136
35. Kopecky F, Kopecka B, Semjanova O (2002) Properties of chitosan and adsorption of copper(II) ions from copper sulfate solutions on chitosan. Ceska Slov Farm 51(3):134–139
36. El-Sawy SM, Abu-Ayana YM, Abdel-Mohdy FA (2001) Some chitin/chitosan derivatives for corrosion protection and wastewater treatments. Anti-Corros Methods Mater 48227–234.
37. Wang Y-X (2001) Removal of heavy metals from waste waters using low-cost sorbents: applications of biomass and geomaterial to environmental protection. Dixue Qianyuan 8(2):301–307
38. Sag Y (2001) Biosorption of heavy metals by fungal biomass and modeling of fungal biosorption: a review. Sep Purif Methods 30(1):1–48
39. Cardenas G, Orlando P, Edelio T (2001) Synthesis and applications of chitosan mercaptans as heavy metal retention agent. Intl J Biol Macromol 28(2):167–174
40. Bayramoglu G, Yalcin E, Genc O, Arica MY (2002) Dye-ligand immobilized IPNs [interpenetrating polymer networks] membrane for removal [of] heavy metal ions. In: Macromolecular symposia (4th international conference on polymer-solvent complexes and intercalates) 203:219–224
41. Su H, Wang L, Tan T (2002) Study of removal heavy metal ions ($Ni^{2+}$) by mycelium-chitin adsorbent. Huanjing Wuran Zhili Jishu Yu Shebei 3(7):5–8
42. Zeng W, Wan Y, Li J Synthesis of silica gel-crosslinked chitosan adsorbents for heavy metals. Application: CN 2001–127897:20010925
43. Shao J, Yang Y, Shi C (2003) Preparation and adsorption properties for metal ions of chitin modified by L-cysteine. J Appl Polymer Sci 88(11):2575–2579
44. Taboada E, Cabrera G, Cardenas G (2002) Synthesis and application of chitosan derivative to trap heavy metal ions. Adv Chitin Sci 6:227–228
45. Matejka Z, Ruszova P, Parschova H, Jelinek L, Kawamura Y (2002) Selective uptake of (Mo, V, W, As)-oxoanions by crosslinked chitosan; beads versus fibers. Adv Chitin Sci 6:213–216
46. Juang R-S, Shao H-J (2002) A simplified equilibrium model for sorption of heavy metal ions from aqueous solutions on chitosan. Water Res 36(12):2999–3008
47. Uehara N, SawaDa A, Shimizu T (2001) Synthesis of thermo-responsive polymers having chelate functional group and their collection properties of metal ion. Anal Sci 17(Suppl):a371–a374
48. Navarro RR, Tatsumi K (2001) Improved performance of a chitosan-based adsorbent for the sequestration of some transition metals. Water Sci Technol 43 (11, 1st World Water Congress, Part 4: Wastewater Treatment, 2000):9–16
49. Yang Z, Yang Y (2001) Synthesis, characterization, and adsorption properties of chitosan azacrown ethers bearing hydroxyl group. J Appl Polymer Sci 81(7):1793–1798
50. Zhuang ST, Yin YN, Wang JL (2018) Simultaneous detection and removal of cobalt ions from aqueous solution by modified chitosan beads. Int J Environ Sci Technol 15:385–394
51. Yin Y, Wang J, Yan X, Li W (2017) Removal of strontium ons by immobilized *Saccharomyces cerevisiae* in magnetic chitosan microspheres, Nuclear. Eng Technol 49:172–177

52. Zhuang S, Yin Y, Wang J (2018) Removal of cobalt ions from aqueous solution using chitosan grafted with maleic acid by gamma radiation. Nuclear Eng Technol 50:211–215
53. Deng J, Liu Y, Liu S et al (2017) Competitive adsorption of Pb(II), Cd(II) and Cu(II) onto chitosan-pyromellitic dianhydride modified biochar. J Colloid Interface Sci 506:355–364
54. Najafabadi HH, Irani M, Rad LR et al (2015) Removal of Cu2+, Pb2+ and Cr6+ from aqueous solutions using a chitosan/graphene oxide composite nanofibrous adsorbent. RSC Adv 5:16532–16539
55. Schmuhl R, Krieg HM, Keizer K (2001) Adsorption of Cu(II) and Cr(VI) ions by chitosan: kinetics and equilibrium studies. Water SA 27(1):1–7
56. Wan M-W, Petrisor IG, Lai H-T, Kim D, Yen TF (2004) Copper adsorption through chitosan immobilized on sand to demonstrate the feasibility for insitu soil decontamination. Carbohyd Polym 55(3):249–254
57. Cheung WH, Ng JCY, McKay G (2003) Kinetic analysis of the sorption of copper(II) ions on chitosan. J Chem Technol Biotechnol 78(5):562–571
58. Sag Y, Aktay Y (2002) A comparative study for the sorption of Cu(II) ions by chitin and chitosan: application of equilibrium and mass transfer models. Sep Sci Technol 37(12):2801–2822
59. Tan T, Cheng P (2003) Biosorption of metal ions with Penicillium chrysogenum. Appl Biochem Biotechnol 104(2):119–128
60. Wu F-C, Tseng R-L, Juang R-S (2001) Kinetic modeling of liquid-phase adsorption of reactive dyes and metal ions on chitosan. Water Res 35(3):613–618
61. Bayramoglu G, Kaya B, Arica MY (2002) Procion Brown MX-5BR attached and Lewis metals ion-immobilized poly(hydroxyethyl methacrylate)/chitosan IPNs membranes: their lysozyme adsorption equilibria and kinetics characterization. Chem Eng Sci 57(13):2323–2334
62. da Rocha Filhoa JA, Bacha EE, Barrakb ER, de Queirozb AA (2001) Novel material for nickel recuperation. Mat Res 4(2):53–57
63. Dhakal RP, Inoue K, Yoshizuka K, Ohto K, Yamada M, Seki S (2005) Solvent extraction of some metal ions with lipophilic chitin and chitosan. Solvent Extr Ion Exch 23(4):529–543
64. Hasan S (2005) Development of materials for the removal of metal ions from radioactive and non-radioactive waste streams. Ph.D. thesis, The University of Missouri, Columbia
65. Nomanbhay SM, Palanisamy K (2005) Removal of heavy metal from industrial wastewater using chitosan coated oil palm shell charcoal. Electronic J. Biotechnol. 8(1):43–53
66. Pontoni L, Fabbricino M (2012) Use of chitosan and chitosan derivatives to remove arsenic from aqueous solution: a mini review. Carbohyd Res 356:86–92
67. Hasan S, Deng B, Ghosh TK (2006) Synthesis of chitosan-iron composite adsorbent for removal of toxic heavy metals and perchlorate from aqueous streams. Invention disclosure, The University of Missouri-Columbia, MO; (16a) Vasireddy D (2005) Arsenic adsorption onto chitosan-iron composite frfom drinking water. M.S. thesis. The University of Missouri-Columbia, MO
68. Chen C-C, Chung Y-C (2006) Arsenic removal using a biopolymer chitosan sorbent. J Environ Sci Health Part A 41(4):645–658
69. Gerente C, Mckay G, Andres Y, Cloirec P (2005) Interaction of natural aminated polymers with different species of arsenic at low concentrations: application in water treatment. Adsorption 11:859–863
70. Brunauer S (1945) The adsorption of gases and vapors, Princeton University Press, Princeton, N.J; (a) Thommes M et al (2015) Physisorption of gases, with special reference to the evaluation of surface area and pore size distribution (IUPAC Technical Report). Pure Appl Chem aop. https://doi.org/10.1515/pac-2014-1117
71. Guibal E (2004) Interaction of metal ions with chitosan-based sorbents: a review. Sep Purif Technol 38(1):43–74
72. Hasan S, Krishnaiah A, Ghosh TK, Viswanath DS (2006) Adsorption of divalent cadmium (II) from aqueous solutions onto chitosan-coated perlite beads. Ind Eng Chem Res 45:5066–5077
73. Goldberg S (2014) Applications of surface complexation models to anion adsorption by natural materials. Environ Toxicol Chem 33(10):2172–2180

74. Hasan S, Ghosh TK, Viswanath DS et al (2007) Preparation and evaluation of fullers earth bead for removal of cesium from waste streams. Sep Sci Technol 42:717–738
75. Dhotel A, Chen Z, Delbreilh L, Youssef B, Saiter J-M, Tan L (2013) Molecular motions in functional self-assembled nanostructures. Int J Mol Sci 14:2303–2333. https://doi.org/10.3390/ijms14022303
76. Brunauer S, Emmett PH, Teller E (1938) J Am Chem Soc 60:309
77. Young DM, Crowell AD (1962) Physical adsorption of gases. Butterworths, London
78. Ruthven DM (1984) Principles of adsorption and adsorption processes. Wiley, New York
79. Yang RT (1987) Gas separation by adsorption process, Butterworths, Boston, Mass
80. Caner N, Sari A, Tüzen M, (2015) Adsorption characteristics of mercury(II) ions from aqueous solution onto chitosan-coated diatomite. Ind Eng Chem Res 54:7524–7533
81. Salih SS, Ghosh TK (2017) Preparation and characterization of bioadsorbent beads for chromium and zinc ions adsorption. Cogent Environ Sci 3:1401577
82. Zhang H, Dang Q, Liu C et al (2017) Uptake of Pb(II) and Cd(II) on chitosan microsphere surface successively grafted by methyl acrylate and diethylenetriamine, ACS Appl Mater Interfaces 9:11144–11155
83. Kyzas GZ, Deliyanni EA (2013) Mercury(II) removal with modified magnetic chitosan adsorbents. Deliyanni, Mol 18:6193–6214. https://doi.org/10.3390/molecules18066193
84. Laus R, Tadeu de Favere V (2011) Competitive adsorption of Cu(II) and Cd(II) ions by chitosan crosslinked with epichlorohydrin–triphosphate. Bioresour Technol 102:8769–8776
85. Batista ACL, Villanueva ER et al (2011) Chromium (VI) ion adsorption features of chitosan film and its chitosan/zeolite conjugate 13X film. Molecules 16:3569–3579. https://doi.org/10.3390/molecules16053569.
86. Hasan S, Ghosh TK, Viswanath DS, Boddu VM (2008) Dispersion of chitosan on perlite for enhancement of copper(II) adsorption capacity. J Hazard Mater 152(2):826–837
87. Veera MB, Krishnaiah A, Randolph AJ, Edgar DS (2008) Removal of copper (II) and nickel (II) ions from aqueous solutions by a composite chitosan biosorbent. Sep Sci Technol 43:1365–1381
88. Wan Ngah WS, Fatinathan S (2008) Adsorption of Cu(II) ions in aqueous solution using chitosan beads, chitosan–GLA beads and chitosan–alginate beads. Chem Eng J 143:62–72
89. Monier M, Abdel-Latif DA (2012) Preparation of cross-linked magnetic chitosan-phenylthiourea resin for adsorption of Hg(II), Cd(II) and Zn(II) ions from aqueous solutions. J Hazard Mater 209–210:240–249
90. Fu Y, Xu X, Huang Y et al Preparation of new diatomite–chitosan composite materials and their adsorption properties and mechanism of Hg(II). R Soc Open Sci 4:170829. https://doi.org/10.1098/rsos.170829
91. Wang Y, Wu D, Wei Q et al Rapid removal of Pb(II) from aqueous solution using branched polyethylenimine enhanced magnetic carboxymethyl chitosan optimized with response surface methodology. Sci Rep 7:10264. https://doi.org/10.1038/s41598-017-09700-5
92. Min L-L, Zhong L-B, Zheng Y-M et al (2016) Functionalized chitosan electrospun nanofiber for effective removal of trace arsenate from water. Sci Rep 6:32480. https://doi.org/10.1038/srep32480
93. Gupta A, Chauhan VS, Sankaramakrishnan N (2009) Preparation and evaluation of iron–chitosan composites for removal of As (III) and As (V) from arsenic contaminated real life groundwater. Water Res 4(3):3862–3870
94. Hasan S, Ghosh A, Race K et al (2014) Dispersion of FeOOH on chitosan matrix for simultaneous removal of As (III) and As (V) from drinking water. Sep Sci Technol 49:2863–2877
95. Liu C, Wang B, Deng Y et al (2015) Performance of a new magnetic chitosan nanoparticle to remove arsenic and its separation from water. J Nanomater 2015:9, Article ID 191829. https://doi.org/10.1155/2015/191829
96. Madala S, Nadavala SK, Vudagandla S et al (2013) Equilibrium, kinetics and thermodynamics of Cadmium (II) biosorption on to composite chitosan biosorbent. https://doi.org/10.1016/j.arabjc.2013.07.017

97. Patruela V, Negrulescu A, Mincea MM et al (2013) Optimization of the removal of copper (II) ion from aqueous solution of chitosan and chitosan beads. BioResources 8(1):1147–1156
98. Anantha RK, Kota S (2016) An evaluation of the major factors influencing the removal of copper ions using the egg shell (Dromaius novaehollandiae): chitosan (Agaricus bisporus) composite. 3(Biotech 6):83. https://doi.org/10.1007/s13205-016-0381-2
99. Shahzad A, Miran W, Rasool K et al (2017) Heavy metals removal by EDTA-functionallized chitosan graphene oxide nanocomposite. RSC Adv 7:9764
100. Langmuir I (1918) J Am Chem Soc 40:1361
101. Weber TW, Chakravorti RK (1974) Pore and solid diffusion models for fixed-bed adsorbers. AIChE J 20:228–238
102. Hines AL, Maddox RN (1985) Mass transfer: fundamentals and applications, prentice Hall, N.J; a), Hines AL, Ghosh TK, Loyalka SK, Warder RC Jr (1993) Indoor Air: Quality and Control Prentice Hall, N.J., ISBN 0–13–463977–4
103. Hasan S, Hashim MA, Sengupta B (2000) Adsorption of Ni($SO_4$) on Malaysian rubber-wood ash. Biores Technol 72:153–158
104. Ghosh TK Adsorption of Acetaldehyde, Propinaldehyde, and Butaraldehyde on silica gel and Molecular Sieve-13X. Ph.D. thesis, Oklahoma State University
105. Chiu HS, Wang JJ (2009) Adsorption thermodynamics of cobalt ions onto attapulgite. J Environ Prot Sci 3:102–106
106. Kaygun AK, Eral M, Erenturk SA (2017) Removal of cesium and strontium using natural attapulgite: evaluation of adsorption isotherm and thermodynamics data. J Radioanal Nucl Chem 311(2):1459–1464
107. Yu Z, Qi T, Qu J et al (2009) Removal of ca(II) and Mg(II) from potassium chromate solution on Amberlite IRC 748 synthetic resin by ion exchange. J Hazard Mater 167(1–3):406–412
108. Foo KY, Hamid BH (2010) Insights into the modelling of adsorption isotherm systems. Chem Eng J 156:2–10
109. Ho YS, Porter JF, McKay G (2002) Equilibrium isotherm studies for the sorption of divalent metal ions onto peat, copper, nickel, and lead single component systems. Water Air Soil Pollut 141:1
110. Do DD (1998) Adsorption analysis: equilibria and kinetics. Series on chemical engineering, vol 2. Imperial College Press
111. Nguyen C, Do DD (2001) The Dubinin–Radushkevich equation and the underlying microscopic adsorption description. Carbon 39:1327
112. (a) Inglezakis VJ, Zorpas AA (2012) Heat of adsorption, adsorption energy and activation energy in adsorption and ion exchange systems. Desalin Water Treat 39:149; (b) Kumar U (2011) Thermodynamics of the Adsorption of Cd(II) from aqueous solution on NCRH. Int J Environ Sci Develop 2(5)
113. Han F, Zhang G-H, Gu P (2013) Adsorption kinetics and equilibrium modeling of cesium on copper ferrocyanide. J Radioanal Nucl Chem 295(1):369
114. Hutson ND, Yang RT (1997) Theoretical basis for the Dubinin-Radushkevitch (D-R) adsorption isotherm equation. Adsorption 3:189–195
115. Gerente C, Lee VKC, Cloirec PL, McKay G (2007) Application of chitosan for the removal of metals from wastewaters by adsorption-mechanisms and models review. Crit Rev Environ Sci Technol 37:41–127
116. Redlich O, Peterson DL (1959) A useful adsorption isotherm. J Phys Chem 63(1):1024–1026
117. Ho YS, McKay G (1998) Kinetic models for the sorption of dye from aqueous solution by wood. Trans I Chem E 79(Part B):183–191
118. Ayawei N, Ebelegi AN, Wankasi D (2017) Modelling and interpretation of adsorption isotherms. J Chem 2017:11, Article ID 3039817. https://doi.org/10.1155/2017/3039817; (a) Ng JCY, Cheung WH, Mckay G (2002) Equilibrium studies of the sorption of Cu(II) ions onto Chitosan. J Colloid Interface Sci 255(1):64–74
119. Sips R (1948) The structure of a catalyst surface. J Chem Phys 16:490–495
120. Samiey B, Abdollahi Jonaghani S (2015) A new approach for analysis of adsorption from liquid phase: a critical review. J Pollut Eff Cont 3:139. https://doi.org/10.4172/2375-4397.1000139

121. Kumara NTRN, Hamdan N, Petra MI et al (2014) Equilibrium isotherm studies of adsorption of pigments extracted from Kuduk-kudul (Melastoma malamathricum L.) pulp on to $TiO_2$ nanoparticles. J Chem 2014:6, Article ID 468975. https://doi.org/10.1155/2014/468975
122. Jovanovic DS, Kolloid ZZ (1969) Polym 235:1203
123. Ross S, Oliver JP (1964) On physical adsorption. Wiley-Interscience, New York
124. Jaroniec M, Patrykiejew A, Borowko M (1981) Progress in surface and membrane science, vol 14. Academic Press, New York
125. Ng JCY, Cheung WH, McKay G (2003) Equilibrium studies for the sorption of lead from effluents using chitosan. Chemosphere 52:1021–1030
126. Piccin JS, Dotto GL, Pinto LAA (2011) Adsorption isotherms and thermochemical data of FD&C red No 40 biniding by chitosan. Braz J Chem Eng 28(02):295–304
127. Chan LS, Cheung WH, Allen SJ, McKay G (2012) Error analysis of adsorption isotherm models for acid dyes onto Bamboo derived activated carbon. Chin J Chem Eng 20(3):535–542
128. Sundaram CS, Viswanathan N, Meenakshi S (2008) Defluoridation chemistry of synthetic hydroxyapatite at nano scale: equilibrium and kinetic studies. J Hazard Mater 155:206–215
129. Nassar MM, Daifullah AH, Farah JY, Kelany H (2015) Air stirring system for adsorption of hexavalent chromium onto chitosan. Int J Sci Eng Res 6(1):190–197
130. Cestari AR, Vieira EFS (2005) Determination of kinetic parameters of Cu(II) interaction with chemically modified thin chitosan membranes. J Colloid Interface Sci 234(1):204–216
131. Ho YS, McKay GA (1994) Pseudo-second-order model for sorption process. Process Biochem 34:451–465
132. McLintock IS (1967) The Elovich equation in chemisorption kinetics. Nature 1204–1205
133. Fierro V, Torne-Fernandez V, Montane D, Celzard A (2008) Adsorption of phenol onto activated carbons having different textural and surface properties. Microporous Mesoporous Mater 111:276–284
134. Weber WJ, Morris JC (1963) Kinetics of adsorption on carbon from solutions. J Sanit Eng Div 89:31–60
135. Crank J (1957) The mathematics of diffusion. Oxford Press, London, 2nd edn, pp 89–96
136. Chen H-L, Viswanath DSA (1989) generalized model for leaching of chlorine from Illinois coal and the effect of particle size. Fuel 68:1184–1188
137. Hasan S, Iasir ARM, Ghosh TK et al (2019) Characterization and adsorption behavior of strontium from aqueous solutions onto chitosan-fuller's earth beads, Healthcare 7:52. https://doi.org/10.3390/healthcare7010052; (a) Karimi M, Fateme A (2013) Analyzing the diffusion process for polymer solution using FTIR-ATR technique: special consideration. J Text Polym 1(1):1–8
138. Saha P, Chowdhury S (2011) Insight into adsorption thermodynamics. Thermodynamics, Mizutani T (ed), ISBN: 978–953–307–544–0, In Tech. http://www.intechopen.com/books/thermodynamics/insight-into-adsorption-thermodynamics
139. Hansen RS, Fackler WVA (1953) Generalization of the polanyi theory of adsorption from solution. J Phys Chem 57:634
140. Wohleber DA, Manes M (1971) Application of polanyi adsorption potential theory to adsorption from solution on activated carbon. III. Adsorption of miscible organic liquids from water solution. J Phys Chem 75:3720
141. Hasnain MA, Hines AL (1981) Application of the adsorption potential theory to adsorption of carboxylic acids from aqueous solutions onto a macroreticular resin. Ind Eng Chem Process Dev 20:621
142. Hasany SM, Saeed MM, Ahmed M (1998) Uptake of cadmium (II) ions by cadmium sulfide from aqueous solutions. Radiochim Acta 83:205
143. Kielland J (1937) Individual activity coefficients of ions in aqueous solutions. J Am Chem Soc 59(9):1675–1678
144. Ghosh TK, Hines AL (1990) Adsorption of acetaldehyde, propionaldehyde, and buteraldehyde on silica gel. Sep Sci Technol 25:1101

145. Al-Anber MA (2011) Thermodynamics approach in the adsorption of heavy metals. Thermodynamics—interaction studies—solids, liquids and gases. Moreno PirajÃ¡n JC (ed) ISBN: 978-953-307-563-1.InTech, Available from http://www.intechopen.com/books/thermodynamics interaction studies solids liquids and gases/thermodynamics approach in the adsorption of heavy metals
146. Collins JJ (1967) Chemical Eng. Prog Symp Ser 63(74):31
147. Mitani T, Kawakami T, Morishita M, Adachi Y, Ishii H (2003) Effect of copper adsorption on the mechanical properties of chitosan beads. J Appl Polym Sci 88:2988–2991
148. Jr M, OAC, Airoldi C (2005) The influence of chitosans with defined degrees of acetylation on the thermodynamic data for copper coordination. J Colloid Interface Sci 282:32–37
149. Zhou Y-G, Yang Y-D, Guo X-M, Chen G-R (2003) Effect of molecular weight and degree of deacetylation of chitosan on urea adsorption properties of copper chitosan. J Appl Polym Sci 89:1520–1523
150. Navarro R, Guzman J, Saucedo I, Revilla J, Guibal E (2003) recovery of metal ions by chitosan: sorption mechanisms and influence of metal speciation. Macromol Biosci 3:525–561
151. Li N, Bi R (2005) A novel amine-shielded surface cross-linking of chitosan hydrogel beads for enhanced metal adsorption performance. Ind Eng Chem Res 44:6692–6700
152. Udayabhaskar P, Iyengar L, Rao PAVS (1990) Hexavalent chromium interaction with chitosan. J Appl Polym Sci 39:739–747
153. Kawamura Y, Yoshida H, Asai S, Kurahashi I, Tanibe H (1997) Effects of chitosan concentration and precipitation bath concentration on the material properties of porous cross linked chitosan beads. Sep Sci Technol 32:1959–1974
154. Ramsey JD, Xia L, Kendig MW, McCreery RL (2001) Raman spectroscopic analysis of the speciation of dilute chromate solutions. Corrosion Sci 43:1557–1572
155. Kaminski W, Modrzejewska Z (1997) Application of chitosan membranes in separation of heavy metal ions. Sep Sci Technol 32:2659–2668
156. Gao YH, Lee KH, Oshima M, Motomizu S (2000) Adsorption behavior of metal ions on cross-linked chitosan and the determination of oxoanions after pretreatment with a chitosan column. Anal Sci 16:1303–1308
157. Dambies L, Guimon C, Yiacoumi S, Guibal E (2001) Characterization of metal ion interactions with chitosan by X-ray photoelectron spectroscopy. Colloid Surf A Physicochem Eng Asp 177:203–214
158. Reed BE, Matsumoto MR (1993) Modelling cadmium adsorption by activated carbon using the Langmuir and Freundlich isotherm expressions. Sep Sci Technol 28:2179–2195
159. Baes CF, Mesmer RE (1976) The hydrolysis of cations. Wiley, New York
160. Jha IN, Iyenger LR, Prabhakararao AVS (1988) Removal of cadmium using chitosan. J Environ Eng 114:962
161. Verbych S, Byrk M, Chornokur G (2005) Removal of copper(II) from aqueous solutions by chitosan adsorption. Sep Sci Technol 40:1749–1759
162. Chu KH, Hashim AA (2000) Adsorption of copper (II) and EDTA-chelated copper (II) onto granular activated carbons. J Chem Technol Biotechnol 75:1054–1060
163. Sakkayawong N, Thiravetyan P, Nakabanpote W (2005) adsorption mechanism of synthetic reactive dye wastewater by chitosan. Journal of Colloid Interface Science 286:36–42
164. Mc Naughton MG, James RO (1974) Adsorption of aqueous mercury. Complexes at oxide/water interface (II). J Colloid Interface Sci 47:429–431
165. Goldberg S, Johnston CT (2003) Mechanisms of arsenic adsorption on amorphous oxide minerals: implications for arsenic mobility. Environ Sci Technol 37(18):4182–4189
166. Guo X, Chen F (2005) Removal of arsenic by bead cellulose loaded with iron oxyhydroxide from groundwater. Environ Sci Technol 39(17):6808–6818
167. Ding M, de Jong BHWS (2000) XPS studies on the electronic structure of bonding between solid and solutes: adsorption of arsenate, chromate, phosphate, Pb2+, Zn2+ ions on amorphous black ferric oxyhydroxide. Geochemica et Cosmochimica Acta 64(7):1209–1219
168. De Jong BHWS, Ellerbroke D, Spek AL (1994) Low-temperature structure of lithium mesosilicate, $Li_4SiO_4$, and its Li 1s and O1s X-ray photoelectron spectrum. Acta Crystallogr A 50:511–518

169. Peak D, Luther GW (2003) ATR-FTIR spectroscopic studies of boric acid adsorption on hydrous ferric oxide. Geochim Cosmochim Acta 67(14):2551–2560
170. Goldberg S, Johnson CT (2001) Mechanisms of arsenic adsorption on amorphous oxides evaluated using macroscopic measurements, vibrational spectroscopy, and surface complexation modeling. J Colloid Interface Sci 241:317–326
171. Saric A, Music S, Nomura K, Popovic S (1999) FT-IR and 57Fe Mossbauer spectroscopic investigation of oxide phase precipitated from $Fe(NO_3)_3$ solutions. J Mol Struct 480–481:633–636
172. Cardenas G, Cabrera G, Taboada E et al (2006) Synthesis and characterization of chitosan alkyl phosphate. J Chil Chem Soc 51(1):815–820
173. Pakula M, Biniak S, Swiatkowski A (1998) Chemical and electrochemical studies of interactions between iron(III) ions and an activated carbon surface. Langmuir 14:3082–3089
174. Ahmed AL, Sumathi S, Hameed BH (2004) Chitosan: a natural biopolymer for the adsorption of residue oil from oily wastewater. Adsorpt Sci Technol 22(1):75–88
175. Grant J, Cho J, Allen C (2006) Self-assembly and physicochemical and rheological properties of a polysaccharide-surfactant system formed from the cationic biopolymer chitosan and nonionic sorbitan esters. Langmuir 22:4327–4335
176. Smith BC (1999) Infrared spectral interpretation: a systematic approach. CRC Press LLC
177. Suarez DL, Goldberg S, Su C (1998) Evaluation of oxyanion adsorption mechanisms on oxides using FTIR spectroscopy and electrophoretic mobility. Am Chem Soc Symp Ser 715:136–177
178. Arai Y, Sparks DL (2001) ATR-FTIR spectroscopic investigation on phosphate adsorption mechanisms at the ferrihydrate-water interface. J Colloid Interface Sci 241:317–326
179. Tossel JA (1997) Theoretical studies on arsenic oxide and hydroxide species in minerals and in aqueous solution. Geochim Cosmochim Acta 61(8):1613–1623
180. Myneni SCB, Tarina SJ, Waychunas GA, Logan TJ (1998) Experimental and theoretical vibrational spectroscopic evaluation of arsenate coordination in aqueous solutions, solids, and at mineral-water interfaces. Geochim et Cosmochim Acta 62(19/20):3285–3300
181. Zhang Y, Yang M, Dou X-M, He H, Wang D-S (2005) Arsenate adsorption on an Fe-Ce bimetal oxide adsorbent: Role of surface properties. Environ Sci Technol 39:7246–7253
182. Hasan S, Ghosh TK, Prelas MA et al (2007) Adsorption of uranium on a novel bioadsorbent chitosan-coated perlite. Nucl Technol 159:59–71
183. Park GI, Park HS, Woo SI (1999) Influence of pH on the adsorption of uranium ions by oxidized activated carbon and chitosan. Sep Sci Technol 34(5):833
184. Bera S, Sali SK, Sampath S et al (1998) Oxidation state of uranium: an XPS study of alkali and alkaline earth urinates. J Nucl Mater 255:26–33
185. Strehle M (2011) X-ray photoelectron spectroscopy (XPS) study of single crystal UO2 and U3O8 on R-plane sapphire and Yttrium stabilized Zirconium (YSZ) substrates, MS Thesis, University of Illinois at Urbana-Champaign
186. Fujimori S, Saito Y, Yamaki K et al (2000) Photoemission study of U/Si (111) interface. Surf Sci 444:180
187. Fujimori S, Saito Y, Sato N et al (1998) X-ray photoelectron spectroscopy study of $U(Rh_{1-x}Pd_x)_3$ alloys. J Phys Soc Jpn 67(12):4164
188. Howng W-R, Thorn RJ (1979) X-ray photoelectron spectrum of U(4f) in condensates from UF4 and ion bombarded Uf4; spectrum of UF3. Chem Phys Lett 62(1):57

# Chapter 7
# Chitosan-Based Sensors

**Abstract** Chitosan and its derivatives are used to develop biosensors or supporting materials for sensors. Potential of these chitosan-based biosensors in medical applications is reviewed. Chitosan can be crosslinked with a targeted compound to modify its physical, chemical, and electrical properties. The fabrication of various chitosan-based biosensors and their electrochemical reaction mechanisms are briefly discussed.

## 7.1 Introduction

Chitosan uses in biomedical sensors for a variety of applications are increasing. Chitosan has been used as a support material and as a thin film membrane in the sensors. The unique structural flexibility of chitosan makes it an excellent candidate for both applications. Chitosan has a high concentration of primary amines. It is stable at high pH. Chitosan changes its structure based on solution pH and can crosslink with other chemical compounds. These properties of chitosan can be exploited further for biofabrications.

Traditionally, a biological component providing molecular recognition capabilities of nucleic acids, enzymes, and antibodies is integrated with the sensor's signal processor. Coté et al. [1] discussed various promising biomedical sensing technologies including electrochemical, optical, and acoustic wave transducers. Figure 7.1 shows a general diagram of a biomedical sensor system. The size and portability of these sensors were always an issue. However, modern biosensors use micro and nanofabrication techniques with more sophisticated signal processing systems. Miniaturization of the biosensor systems, including lab-on-a-chip type sensors, has occurred. Figure 7.2 shows a schematic diagram of a biochip system.

Chitosan can support developing the bioreceptor part of the sensor chip because chitosan can be fabricated as membranes, thin films, and three-dimensional structures. Furthermore, the fabricated membranes or thin films can be chemically modified to use a sensor or a support material for a sensor. A brief discussion of the fabrication of these chitosan products is given below. Following this, their specific use as a sensor is described.

© The Author(s), under exclusive license to Springer Nature Switzerland AG 2022
S. Hasan et al., *Chitin and Chitosan*, Engineering Materials and Processes, https://doi.org/10.1007/978-3-031-01229-7_7

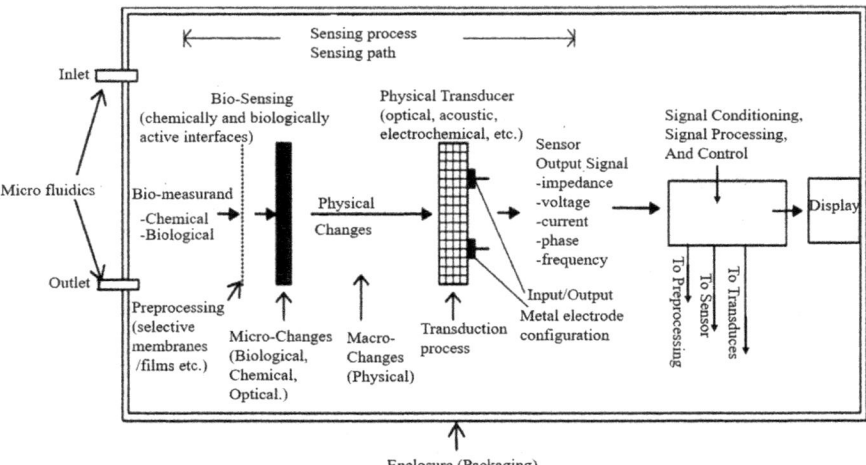

**Fig. 7.1** General diagram of a biomedical sensor system. Reproduced with permission. Copyright © 2003, IEEE. Sensor Journal, IEEE, 3(3), 251–266 (2003). Reproduced from Ref. [1] with permission

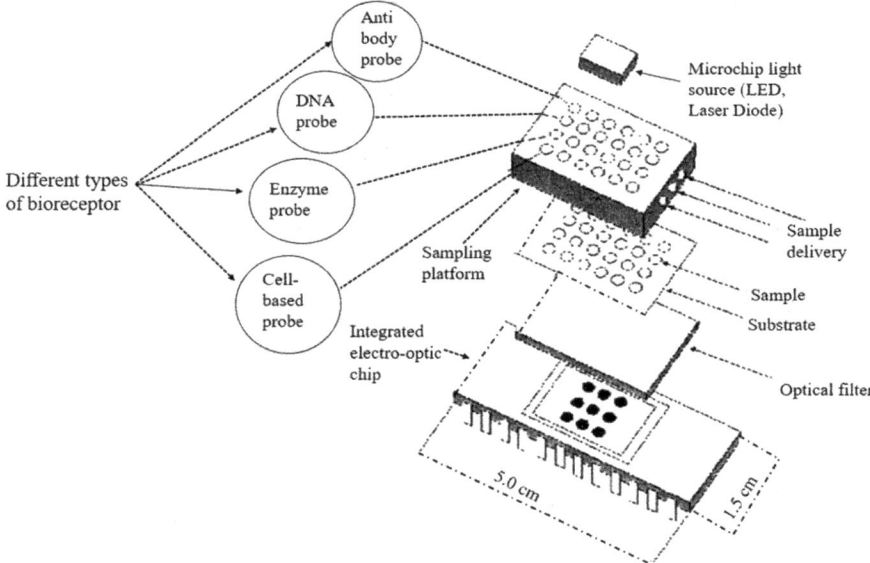

**Fig. 7.2** Schematic diagram of a biochip system (www.ornl.gov/virtual/biomedicalsensors) [1]. Copyright © 2003, IEEE. Sensor Journal, IEEE, 3(3), 251–266 (2003). Reproduced from Ref. [1]. with permission

## 7.2 Fabrication of Chitosan Membranes and Films

The basic approach for preparing chitosan membranes involves the dissolution of chitosan flakes in a dilute organic acid solution to form a homogeneous gel. The gel may be filtered to remove undissolved materials from it. Finally, the chitosan film can be prepared by casting the gel on a glass plate and drying it overnight at room temperature under a humid atmosphere (40 to 60% relative humidity). After drying, the transparent film is peeled off from the glass plate. In this way, about ten–μm-thick films can be produced. However, Kim et al. [2] noted that films made only from chitosan lack water resistance and have low mechanical properties. Also, pure films have less resistance to low pH. To overcome these issues, either the film is crosslinked with various chemicals or mixed with other compounds to make a composite film.

Chitosan is generally crosslinked with glutaraldehyde to make it more resistant to acids. Wan et al. [3] described a process for making a chitosan membrane. The preparation of the membrane involved a two-step process. In the first step, a 1.0 wt% acidic chitosan solution was prepared by dissolving chitosan powder in a 1.0% (by volume) aqueous acetic acid solution. Glutaraldehyde was added to this solution, with vigorous stirring for 3 h at ambient temperature. The amount of glutaraldehyde added depends on the degree of crosslinking desired. Chitosan has been crosslinked with several chemicals for a variety of applications. Jegal and Lee [4] gave another example of crosslinking. They prepared a sulfosuccinic acid crosslinked chitosan membrane of 10 mm thickness by dipping the membrane for 10 h in a methanol solution containing 2 mol % of sulfosuccinic acid. The chitosan membrane crosslinked with sulfosuccinic acid was dried at room temperature. Figure 7.3 shows the reaction mechanism for the preparation of chitosan crosslinked with sulfosuccinic acid [4].

One of the critical elements of developing a chitosan-based sensor (CBAS) is the capability of layer-by-layer formation of chitosan films. The LBL assembly of chitosan allows the construction of multilayer films with different properties of each layer.

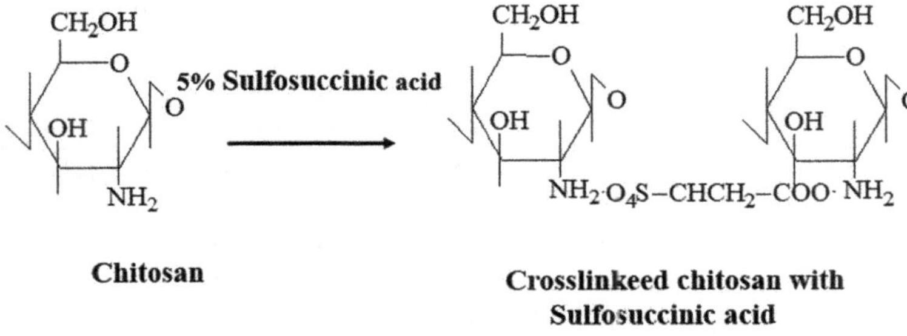

**Fig. 7.3** Reaction of chitosan with sulfosuccinic acid. Copyright © 1999, John Wiley and Sons, Inc., Journal of Applied Polymer Sciences, Vol. 71, Issue, 4, pp. 671–675, 1999. Reproduced from Ref. [4] with permission

Ang et al. [5] fabricated three-dimensional scaffolds of chitosan using rapid prototyping robotic dispensing. About 30 mL solution of chitosan and chitosan–hydroxyapatite (HA) was dissolved in acetic acid. This solution was forced out through a small Teflon-lined nozzle into a mixture of sodium hydroxide ethanol (in the ratio of 7:3). A gel-like layer was formed due to the rapid neutralization of acidic chitosan solution with alkaline NaOH solution. The LBL film of chitosan was fabricated using a preprogrammed pattern. A suitable attachment between layers allows the chitosan matrix to form a fully interconnected channel architecture. Figure 7.4 shows the scaffold fabricated by them. Layer-by-layer assembly to create chitosan-based multilayer films is summarized in Table 7.1.

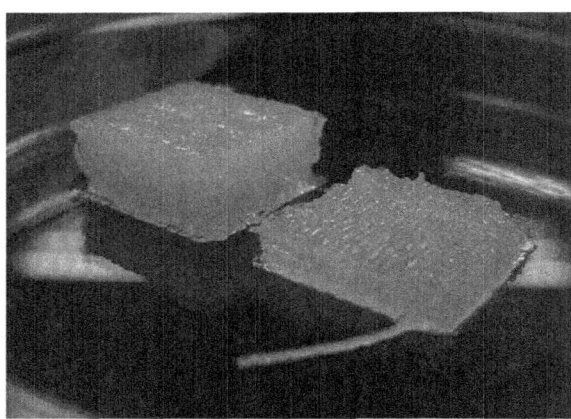

**Fig. 7.4** Chitosan scaffold prepared by LBL deposition. Copyright © 2002 Elsevier Science B.V. Materials Science and Engineering: C, Volume 20, Issues, 1–2, pp. 35–42, 2002; Reproduced from Ref. [5] with permission

**Table 7.1** Typical example of chitosan-based substrate utilized in the layer-by-layer (LBL) self-assembly technique and their applications

| Substrate | Applications | References |
|---|---|---|
| Chitosan and alginate | Scaffold, 3D cell cultivation | [6] |
| Chitosan | Paraoxon detection | [7] |
| Chitosan | Glucose biosensor | [8] |
| Chitosan and chondroitin sulfate | Biomedical applications | [9] |
| Chitosan and SWCNT | Chlorogenic acid | [10] |
| Chitosan gold and Prussian blue | Glucose biosensors | [11] |
| Chitosan and graphene | Glucose biosensor | [12] |
| Chitosan with polydiacetylene | Colorimetric sensing film | [13] |
| Chitosan/poly (styrene sulfonate) | Glucose biosensor | [14] |
| Chitosan CdS | Biosensor | [15] |

## 7.3 Chitosan-Based Sensors

Chitosan is mainly used in the following two manners for the fabrication of sensors.

1. Chitosan serves as a matrix to entrap components within its film and, hence, serves as an electrode for biosensors.
2. The functional groups of chitosan bind directly with the targeted compound, changing various electrical and chemical properties.

### 7.3.1 Glucose Biosensor

Glucose biosensors are used to measure the blood sugar levels of diabetic patients. Glucose biosensors can be categorized based on sensor devices and biological materials used to detect glucose in the sample [16]. The potentiometric, voltammetric (amperometry), conductometric, and field-effect transistor (FET) are some of the electrochemical sensing devices [16]. In electrochemical sensors, the electrical signal results from a chemical process at the transducer/analyte interface.

Several studies reported chitosan and glucose oxidase ($GO_x$)-based glucose biosensors [12, 17–19]. The enzyme $GO_x$ is a type of glycoprotein. $GO_x$ consists of two identical polypeptide chains covalently joined through disulfide bonds. $GO_x$ is widely used to determine glucose in the body fluid. In a chitosan-$GO_x$-based glucose sensor, the enzyme $GO_x$ is immobilized on the chitosan-based porous composite membranes using a covalent bond between $GO_x$ and the chitosan membrane [18]. The enzyme $GO_x$ usually acts as a catalyst for the oxidation of glucose in the presence of oxygen [19]. The formation of $H_2O_2$ occurs due to the enzymatic oxidation reaction of glucose. The electrochemical changes in the reaction and the generated $H_2O_2$ are monitored to determine glucose during sensing. The following describes the overall reaction for the glucose biosensors [17]:

$$\text{Glucose} + O_2 \xrightarrow{\text{Glucose oxidase}} H_2O_2 + \text{gluconic acid}, \quad (7.1)$$

$$H_2O_2 \xrightarrow{\text{electrode}} 2H^+ + O_2 + 2e^- (\sim 0.55 \text{ V}). \quad (7.2)$$

Equation 7.1 shows gluconic acid and $H_2O_2$ produced from the oxidation reaction of glucose. Gluconic acid is the hydrolyzed form of gluconolactone. If a mediator such as ferrocene is used as an electron transfer mediator, the possible reaction could be as follows [18]:

$$\text{Glucose} + (GO)_x \rightarrow \text{Gluconic acid} + (GO)_x (\text{reduction}), \quad (7.3)$$

$$(GO)_x (\text{reduction}) + 2M(\text{oxidation}) \rightarrow (GO)_x (\text{oxidation}) + 2M (\text{reduction}) + 2H^+, \quad (7.4)$$

$$2M \text{ (reduction)} \rightarrow 2M \text{ (oxidation)} + 2e^-. \tag{7.5}$$

M represents the electron transfer mediator. The reduced form of $GO_x$ is reoxidized at the electrode giving a current signal (proportional to the glucose concentration) as the mediator's oxidized form is regenerated.

Gotoh et al. [20] constructed a glucose biosensor using a double-layer membrane and electrodes. One layer of the membrane was chitosan. The other layer was made of glucose oxidase polyvinylchloride. The sensor was found to be more stable and durable in controlling serum and porcine whole blood. Glucose in the physiological buffer was determined in the range of 10–500 mg/dL with a response time of about 2 min.

Amperometric glucose biosensors were constructed by the codeposition of $GO_x$ with chitosan gold (chitosan-AuNP) on gold Prussian blue (Au-PB) nanoparticles modified glassy carbon electrodes (GCE) [21]. The biosensors showed a 95% response time up to 10 s, sensitivity up to 9.5 $\mu A \ M^{-1} \ cm^{-2}$, and linear calibration graph up to 3 mM of glucose. Figure 7.5 shows the glucose diffusion from bulk solution into $GO_x$ film, oxidized with $H_2O_2$ generation. The produced $H_2O_2$ is catalytically reduced with Prussian white (PW), which is the reduced form of Prussian blue (PB). The reduced PB is generated at an electrode potential of $-0.05$ V.

A planar amperometric glucose microsensor with $GO_x$-chitosan film on the PB layer has been developed [22]. It is reported that the optimum detection potential of this amperometric glucose microsensor is 50 mV with an optimum pH of 6.5. The sensor shows an excellent sensitivity of 98 nA/M and a linear range of 0.1–6.0 mM under selective conditions. The apparent Michaelis–Menten constant of the sensor is 21 mM. The response time is less than 60 s. Figure 7.6 shows the calibration of the sensor with a linear response to a glucose concentration of 7 mM. The sensor was used to detect glucose in human blood serum, and its results were compared with that of a standard photometric measurement [22]. As noted in Fig. 7.7, the difference between the two measurements is negligible.

Wang et al. [23] prepared a functional biosensor using chitosan and PB to detect glucose, galactose, glutamate in human blood serum, and fermentation broth solutions. It is reported that ascorbic and uric acids can interfere with the performance of the sensor. Chitosan film has selective permeability for these two compounds, and

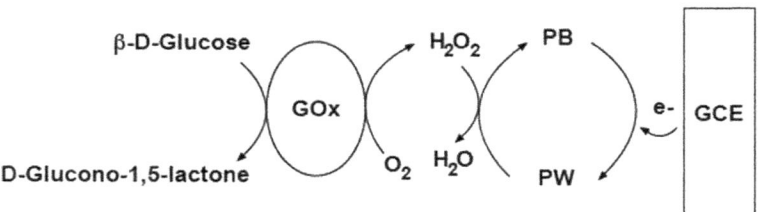

**Fig. 7.5** Schematic of biosensor action. Copyright © 2008 Bentham Open; The Open nanoscience Journal, 2, 34–38, 2008; Reproduced from Ref. [21] with permission

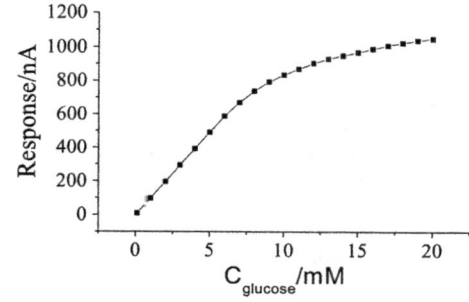

**Fig. 7.6** Calibration curve for glucose microsensor based on $GO_x$ immobilized by chitosan film on PB layer. Copyright © 2002 by MDPI, Creative Common open access license. Sensors 2002, 2(4), 127–136; https://doi.org/10.3390/s20400127. Reproduced from Ref. [22] with permission

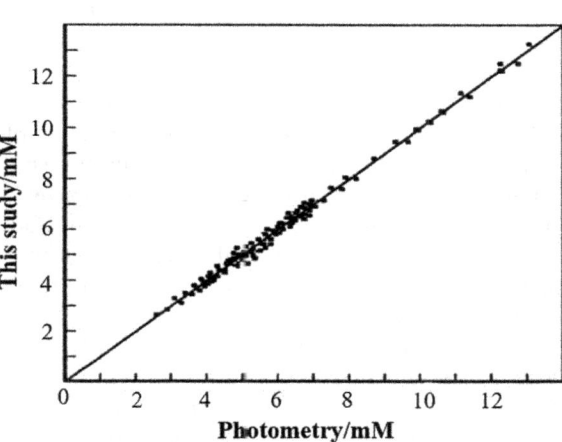

**Fig. 7.7** Comparison of glucose concentration measurement by CBAS and spectrophotometry of 100 human serum samples. Copyright © 2002 by MDPI, Creative Common open access license. Sensors 2002, 2(4), 127–136; https://doi.org/10.3390/s20400127. Reproduced from Ref. [22] with permission

the electrocatalysis of the PB layer can generate $H_2O_2$ [17]. In this way, the interference from ascorbic and uric acids for glucose sensing can be avoided. A composite film of chitosan/silica hybrid was prepared from chitosan and methyltrimethoxysilane (MTOS) solution [24]. The film was deposited on the PB-modified glass carbon electrode to detect glucose. The biosensor fabricated under optimal conditions had a linear response to glucose over the range $5.0 \times 10^{-5}$–$2.6 \times 10^{-2}$ M and a detection limit of $8.0 \times 10^{-6}$ M with the signal-to-noise ratio $(S/N) = 3$. The biosensor had a response time of less than 10 s, high sensitivity of 420 nA $mM^{-1}$, long-term stability of over 60 days, and good selectivity. The sensor was used successfully to determine the glucose concentration in real human blood samples.

A new type of organically modified sol–gel/chitosan composite material was developed and used to construct a glucose biosensor [25]. This material provided good biocompatibility and the stabilizing microenvironment around the enzyme. Ferrocene was immobilized on the surface of the glassy carbon electrode as a mediator. The characteristics of the biosensor were studied by cyclic voltammetry (CV) and chronoamperometry. Enzyme loading, buffer pH, applied potential, and several interferences were investigated to determine their effect on the enzyme

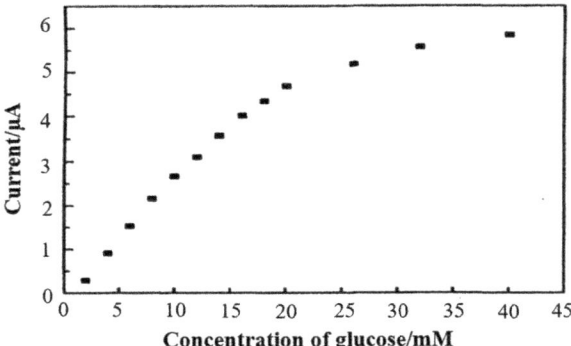

Fig. 7.8 Calibration plot of a chitosan-based sensor for glucose in 50 mM phosphate buffer (with a pH 7.5) and the applied potential of 350 mV. Copyright © 2003 John Wiley & Sons Inc., Electroanalysis, 15 (7), 608–612, 2003; Reproduced from Ref. [25] with permission

electrode's response. The low-cost and straightforward glucose biosensor exhibited high sensitivity and good stability. The calibration of the sensor showed significant improvement in the range where it has a linear response (Fig. 7.8).

Glucose biosensors have also been developed by LBL deposition of chitosan film on electrodes, mainly platinum electrodes [26]. $GO_x$ is immobilized on the film. $GO_x$ was found to be catalytically active, even in multilayer films. The platinum electrode modified with chitosan-glucose oxidase (CS-$GO_x$) multilayer film showed an amperometric response to glucose at the standard and diabetic level [26].

Zare et al. [19] conducted similar studies to immobilize $GO_x$ onto the platinum electrode using chitosan as a substrate. Based on the low value of the Michaelis–Menten constant, they reported that the biosensor has a high affinity to glucose. It was reported that the sensor showed a short response time of 5 s, increased sensitivity of 38.5 $\mu$A mM$^{-1}$ cm$^{-2}$, in the linear range of $2 \times 10^{-4}$–$9.1 \times 10^{-3}$ M and a detection limit of 7.4 $\mu$M for glucose at a signal-to-noise ratio of 3 [19]. The biosensors retained 97% of their initial current response after 30 days.

Ang et al. [16] studied different molecular weights of chitosan as an immobilization matrix for glucose biosensors. They reported that the chitosan sample with medium molecular weight showed a better performance than the chitosan sample with a lower molecular weight in determining glucose. Chen et al. [27] studied the effect of the degree of deacetylation and the thickness of chitosan membranes on the electrochemical response of chitosan membrane-based sensors. They noted that chitosan membranes with various thicknesses and degree of deacetylation did not influence the electrochemical characteristics. The membranes were covalently immobilized with $GO_x$ and deposited onto a platinum electrode. A linear signal was received for glucose concentration of up to 1.0 mM. The sensor was sensitive to 10 $\mu$M glucose concentration, and the signal was reproducible for 50 $\mu$M glucose.

Caseli et al. [28] used the LBL method to fabricate chitosan/$GO_x$ films. They reported that the chitosan/$GO_x$ film detects glucose with a detection limit of 0.2 mmol/L and activity of 40.5 $\mu$A mmol$^{-1}$ L $\mu$g$^{-1}$. Figure 7.9 shows the cathodic response of the chitosan biosensor to glucose concentration in a buffer solution. The

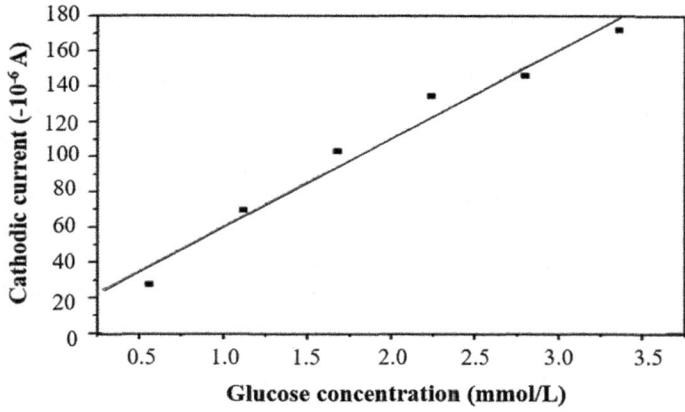

**Fig. 7.9** Cathodic response of the chitosan biosensor to glucose concentration in a buffer solution (pH 7.5) against a potential of 0.43 V. Copyright © 2006 Elsevier Inc., Journal of Colloid and Interface Science, 303(1), 326–331, 2006. Reproduced from Ref. [28] with permission

current values at 0.43 V were plotted against the glucose concentration. Immobilized $GO_x$ in films preserved their catalytic activity toward glucose oxidation. The highest sensitivity for measuring glucose concentration was achieved when only the top layer contained $GO_x$. The diffusion of analyte and electron transport through the deposited layers might have interfered with the sensor [28].

An amperometric glucose detection system was designed based on the self-assembling of a new chitosan derivative, Nafion, and $GO_x$ onto thiolated gold electrodes [29]. First, a layer of chitosan derivative (quaternized or hydrophobic) was deposited onto the gold surface modified with the sodium salt of 3-mercapto-1-propanesulfonic acid, which was followed by a layer of Nafion (as an anti-interference barrier) and alternating deposition of the chitosan derivative and $GO_x$ (as a biocatalytic layer). The optimized sensor with five quaternized chitosan-$GO_x$ bilayers had a sensitivity of $4.9 \pm 0.2 \times 10^2$ nA mM$^{-1}$ and no interference from physiological levels of ascorbic acid and uric acid at their maximum concentration.

Several studies investigated $GO_x$, chitosan, and graphene nanocomposite as a biosensor [30–32] utilizing graphene's high electrical conductivity. Kang et al. [30] reported the direct electrochemistry of a $GO_x$–graphene–chitosan nanocomposites. They reported that direct electron transfer between $GO_x$ and the substrate (two electrons along with two proton reaction) provides an electrochemical response, which can be used to prepare a bio-electrocatalytic sensing device. Flavin adenine dinucleotide (FAD) is a part of the $GO_x$ molecule. FAD undergoes a redox reaction where two protons and two electrons are exchanged as per the following reaction:

$$(GO)_x - FAD + 2e^- + 2H^+ \leftrightarrow (GO)_x - FADH_2. \quad (7.6)$$

This $GO_x$–graphene–chitosan nanocomposite-based biosensor exhibits a broader linearity range from 0.08 mM to 12 mM glucose. It has a detection limit of 0.02 mM and sensitivity of 37.93 $\mu$Am $M^{-1}$ $cm^{-2}$. The large surface-to-volume ratio and high conductivity of graphene promote direct electron transfer between redox enzymes and the surface of the electrodes. Qiu et al. [31] reported homogeneous chitosan–ferrocene/graphene oxide/$GO_x$ (CS–Fc/GO/$GO_x$)-based novel platform to fabricate a glucose biosensor. They observed that the homogeneous distribution of graphene oxide is favorable for the higher loading of $GO_x$ into the chitosan matrix. Besides, the redox mediator ferrocene groups (Fc) improve the electrical conductivity of chitosan and promote the electron transfer between $GO_x$ and electrode as per the following equation:

$$(GO)_x(FADH_2) + 2Fc^+ \rightarrow (GO)_x(FAD) + 2Fc^+ + 2H^+, \qquad (7.7)$$

$$Fc \leftrightarrow Fc^+ + e^-, \qquad (7.8)$$

where

Fc = reduced form of ferrocene,

$Fc^+$ = oxidized form of ferrocene,

$GO_x$ (FAD) = oxidized form of $GO_x$, and

$GO_x$ ($FADH_2$) = reduced form of $GO_x$.

In this process, the mediator takes oxygen in the native enzymatic reaction, and two electrons are transferred from glucose to the FAD of the enzyme. These electrons can then be transferred from $FADH_2$ to the mediator, which is then oxidized at the electrode surface, producing a directly proportional current to the glucose concentration in the solution. Biosensors based on this CS–Fc/GO/$GO_x$ showed a linear response to glucose in the concentration range from 0.02 to 6.78 mM with a detection limit of 7.6 $\mu$M at a signal-to-noise ratio of 3 and exhibited a higher sensitivity of 10 $\mu$A $mM^{-1}$ $cm^{-2}$.

Wu et al. [32] described a bionanocomposite film consisting of glucose oxidase/Pt/functional graphene sheets/chitosan ($GO_x$/Pt/FGS/chitosan) for glucose sensors. The biosensor based on this bionanocomposite film shows good reproducibility, long-term stability, and negligible interfering signals from ascorbic acid and uric acid. This hybrid nanocomposite glucose sensor provides a new opportunity for clinical diagnosis and point-of-care applications.

Figure 7.10a shows the typical amperometric response of $GO_x$/Pt/FGS/chitosan/GCE to successive additions of 0.15 mM glucose with an applied potential at 0.4 V; Fig. 7.10b is the calibration curve. Figure 7.10a shows that the electrochemical response increased as the glucose concentration increased. From the inset in Fig. 7.10a, it can be seen that the sensor is responsive to a low concentration of glucose, such as 0.6 $\mu$M glucose.

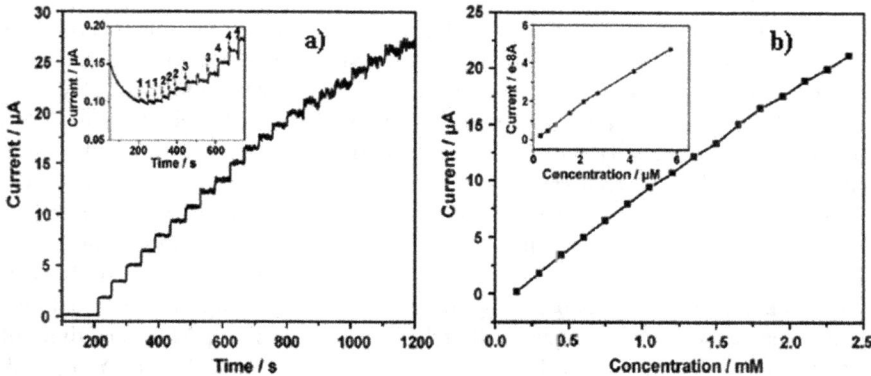

**Fig. 7.10 a** Amperometric response of $GO_x$/Pt/FGS/chitosan/GCE to sequential addition of a series of glucose concentrations from 0.15 mM to 4.2 mM at + 0.4 V. Inset (**a**) amperometric response to the sequential addition of (1) 0.3 μM, (2) 0.6 μM, (3) 1.5 μM, and (4) 3 μM glucose. **b** The calibration curve from (**a**). Inset (**b**) is the lower range of the calibration curve. Copyright © 2009 Published by Elsevier B.V. Talanta, 80, 403–406, 2009. Reproduced from Ref. [32] with permission

Rassas et al. [33] prepared a $GO_x$ encapsulated bionanocomposite film for glucose sensing. The film consists of gold nanoparticles (AuNPs)-doped polyelectrolyte (chitosan/kappa-carrageenan) complex (PEC). The film is assembled on a gold electrode (Au) to fabricate the sensor. Acceptable recovery rates were obtained when PEC/AuNPs/$GO_x$-based biosensor was tested in glucose-spiked saliva samples. Gold nanoparticles (GNp) dispersed in chitosan film were used to immobilize $GO_x$ on the platinum electrode to fabricate a glucose biosensor [34]. The interference from uric acid and ascorbic acid was studied. The oxidation current of uric acid and ascorbic acid at the chitosan-modified electrode was inhibited. Chitosan and nanoparticles systems provided a biological environment for enzymes and electrocatalysis. The linear range for glucose was 0–10 mmol/L; the response time was 8 s. A chitosan film containing $MnO_2$ nanoparticles was deposited electrochemically on electrodes previously modified with $GO_x$ [35]. It was reported that the $MnO_2$ nanoparticles oxidized ascorbic acid and eliminated ascorbate interference on the glucose biosensors based on direct oxidation of hydrogen peroxide [35]. An amperometric glucose biosensor based on ferrocene-doped silica (FcDS) nanoparticles incorporated in the chitosan membrane was developed by Zhang et al. [36]. The ferrocene maintains high electron transfer efficiency and functions as a mediator. The biosensor showed a detection limit of $2.0 \times 10^{-6}$ mol $L^{-1}$ for glucose with a linear range from $5.0 \times 10^{-6}$ to $1.2 \times 10^{-2}$ mol $L^{-1}$. Liu et al. [37] noted that by entrapping $GO_x$ in a composite of CNT/chitosan, the electron transfer rate was significantly enhanced to 7.73 $s^{-1}$ in the $GO_x$/carbon nanotube/chitosan system. The flavin adenine dinucleotide (FAD)/carbon nanotube/chitosan system provides an electron transfer rate of 3.1 $s^{-1}$. The proposed electrode ($GO_x$/carbon nanotube/chitosan) showed a sensitivity of 0.52 μA $mM^{-1}$ for glucose and better stability.

An electrodeposition method of polypyrrole–chitosan–iron oxide (Ppy–CS–$Fe_3O_4$) nanocomposite films (Ppy–CS–$Fe_3O_4$NP/ITO) is developed for nonenzymatic glucose biosensors applications [38]. The fabricated electrode Ppy–CS–$Fe_3O_4$ NP/ITO shows a fast amperometric response with high selectivity to detect glucose non-enzymatically with improved linearity (1–16 mM) and the detection limit of (234 μM) at a signal-to-noise ratio of 3 ($S/N = 3.0$). In another approach, chitosan-functionalized graphene (CG) has been prepared by a one-step ball milling of carboxylic chitosan and graphite [12]. Magnetic $Fe_3O_4$ nanoparticles are introduced into the as-synthesized CG for multifunctional applications beyond biosensors such as magnetic resonance imaging (MRI). The resulting biosensor exhibits a good glucose detection response with a detection limit of 16 μM, a sensitivity of 5.658 mA/cm$^2$/M, and a linear detection range of up to 26 mM glucose. Yang et al. [39] incorporated cobalt hexacyanoferrate nanoparticles (CoNP) and CNT in an aqueous solution of chitosan to form a film. The CoNP-CNT-CS film was deposited on GCE. Glutaric dialdehyde was used to immobilize $GO_x$ onto the electrode surface. The CNT amplified the $H_2O_2$ sensitivity by approximately 70 times compared to a film without CNT. The glucose content of the solution could be determined in pH 6.98 phosphate buffer. The determination of glucose was realized at −0.2 V with a linear range from 0.01 to 10 mM and response time <10 s (Fig. 7.11). The detection limit was 5 μM glucose ($S/N = 3$).

Another approach taken by several researchers to fabricate biosensors is to disperse nanoparticles of a compound into the chitosan matrix to produce a nanocomposite to be deposited on the electrodes. Zhai et al. [40] incorporated PB nanoparticles and multiwalled carbon nanotubes (MWNTs) in chitosan (CS) to prepare CS/PB/MWNT nanocomposite. The CS/MWNTs/PB nanocomposite-modified glassy carbon (GC) electrode amplified the reduction current of hydrogen peroxide by about 35 times compared with that of the CS/MWNTs/GC electrode. It reduced the response time from 60 s for CS/PB/GC to 3 s. The interference from ascorbic acid uric acid and acetaminophen was negligible. The detection limit of the

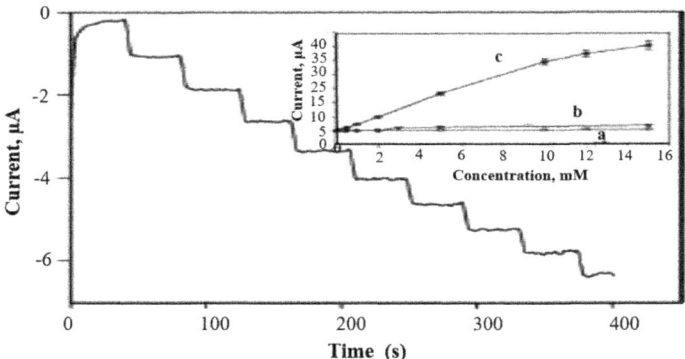

**Fig. 7.11** Glucose determination by chitosan-based biosensor. Copyright © 2005 Elsevier B.V. Biosensors & Bioelectronics (2006), 21(9), 1791–1797. Reproduced from Ref. [39] with permission

biosensor for glucose was 2.5 μM. The response time to detect glucose was less than 5 s, and the linear range was from 4 μM to 2 mM.

Mani et al. [41] reported a robust bionanocomposite for enzymatic biosensing of hydrogen peroxide ($H_2O_2$) and nitrite ($NO_2^-$). This heterostructured nanobiocomposite was prepared using MWCNT, graphene oxide nanoribbons (reduced form) (rGONRs), and chitosan (CS). This nanocomposite-based (MWCNTs@rGONRs/CS) sensor was investigated for direct electron transfer characteristics. These characteristics include redox properties, immobilized active iron, oxygen-binding protein myoglobin, electron transfer efficiency, and durability. It was reported that this nanocomposite sensor (MWCNTs@rGONRs/chitosan) functions as an excellent signal amplifier, which helped in achieving low detection limits (DL) to quantify $H_2O_2$ (1 nM) and $NO_2^-$ (10 nM). The practical feasibility of the biosensor was reported for use in contact lens cleaning solutions and meat samples [41]. Shrestha et al. [17] prepared a bionanohybrid film using polypyrrole (PPy)-Nafion (Nf)-functionalized multiwalled carbon nanotubes (fMWCNTs) nanocomposite [17]. The composite was prepared on the glassy carbon electrode (GCE) as substrate by a facile one-step electrochemical polymerization technique followed by CS-$GO_x$ immobilization on its surface to achieve a high-performance glucose biosensor. The electrochemical behavior of the fabricated biosensor was evaluated using the CV, electrochemical impedance spectroscopy (EIS), and amperometry measurements. The results indicated an excellent catalytic property of bionanohybrid film for glucose detection with improved sensitivity of 2860.3 $\mu A\ mM^{-1}\ cm^{-2}$, the linear range up to 4.7 mM ($R^2 = 0.9992$), and a low detection limit of 5 μM under a signal-to-noise ($S/N$) ratio of 3.

### 7.3.2 Chitosan-Based Sensor by Layer-by-Layer (LBL) Method

Several CBASs have been developed to detect various compounds using the LBL deposition technique. Zhang et al. [42] reported a highly sensitive saliva glucose sensor consisting of a working (sensor) electrode, a counter electrode, and reference electrodes that are integrated on a single chip through several microfabrication processes [42]. The working electrode is functionalized through a LBL assembly of single-walled carbon nanotubes (SWNTs) and multilayer films composed of chitosan (CS), GNp, and $GO_x$, as shown in Fig. 7.12.

They reported that this on-chip saliva glucose sensor has the following features: (1) direct electron transfer between $GO_x$ and the electrode surface; (2) glucose detection down to 0.1 mg/dL (5.6 1 M); (3) good sensing linearity over 0.017–0.81 mM; (4) high sensitivity (61.4 1 A/mM $cm^2$) with a small reactive area (8 $mm^2$); (6) fast response; (7) high reproducibility and repeatability; (8) reliable and accurate saliva glucose detection.

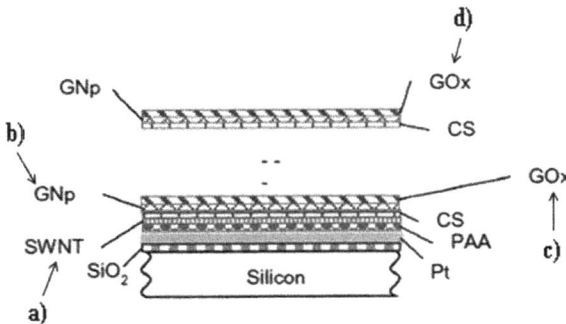

**Fig. 7.12** Cross section of the functional layers on the sensor electrode of **a** one layer of single-wall carbon nanotube (SWNT), **b** one layer of SWNT/gold nanoparticle (GNp), **c** poly(allylamine) (PAA)/SWNT/chitosan/GNp/GO$_x$, and **d** PAA/SWNT/(chitosan/GNp GO$_x$)3 film on the sensor electrode surface. Copyright © 2015 Zhang, W., Du, Y., and Wang, M.L. Published by Elsevier B.V., Sensing and Bio-Sensing Research, 4, 96–102, 2015. Reproduced from Ref. [42] with permission

A chlorogenic acid sensor based on chitosan (CS) and multiwalled carbon nanotubes (MWCNTs) modified glassy carbon electrode (GCE) was fabricated by Cheng et al. [10] via a LBL self-assembly method. A multiple LBL self-assembly film of CS and MWCNTs on a GCE was prepared as follows. The activated GCE was first immersed in CS–acetic acid solution for 5 min. This allows the formation of self-assembly between the negatively charged GCE and positively charged CS through electrostatic interactions. The self-assembly of negatively charged MWCNTs was obtained by further immersing the GCE with a CS layer in a 1 mg dm$^{-3}$ MWCNTs solution for 10 min. After each modification, the resulting GCE was washed alternately with a phosphate buffer solution and distilled water and dried under nitrogen. A multiple self-assembly film of CS and MWCNT-modified GCE was formed by repeating these steps. Compared to the bare GCE, the prepared electrode exhibited excellent catalytic performance for chlorogenic redox reactions. Figure 7.13 shows the reaction of chlorogenic acid occurring on the surface of the electrode.

The sensor shows significant sensitivity in the detection of chlorogenic acid. The detection limit was estimated to be $1.16 \times 10^{-8}$ mol dm$^{-3}$ ($S/N = 3$).

Amemory et al. [43] reported a nitric oxide (NO) sensor fabricated by a combination of LBL assembly and acid-free chitosan dissolving techniques. The following procedure prepared the NO sensor: Mesoporous silica particles were immersed alternately into a 1 mg/mL chitosan (CT) solution containing 25% formic acid and a 1 mg/mL dextran sulfate (DS) solution. The immersion was repeated five times and lasted for 15 min each. The films of chitosan/dextran sulfate were prepared on the silica particles. The particles were soaked in 5 µM aqueous 4,5-diaminofluorescein (DAF-2) solution for 5 days, and then the DAF-2 encapsulated sensor particles were obtained after washing and subsequent freeze drying. Figure 7.14 shows the fabrication of the sensor particle and the sensing reaction mechanism of nitric oxide (NO) [43].

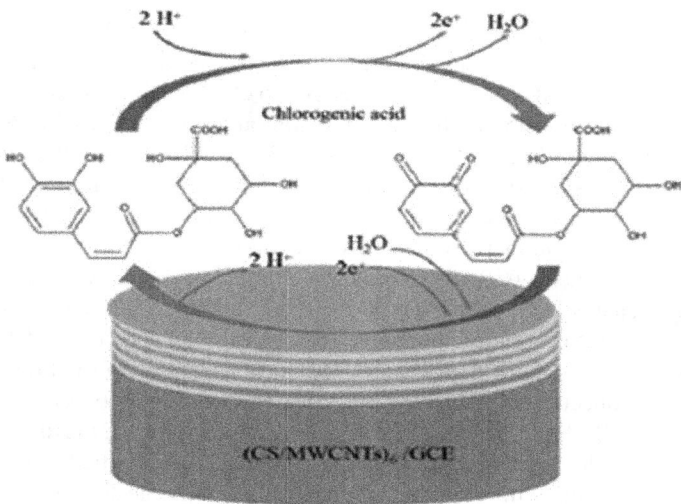

**Fig. 7.13** Mechanism of the electrochemical reaction of chlorogenic acid occurring on the surface of the (CS/MWCNTs)6/GCE. Copyright © The Royal Society of Chemistry 2017, Creative Common open access CC BY license. RSC Adv., 7, 6950, 2017. https://doi.org/10.1039/c6ra26378j. Reproduced from Ref. [10]

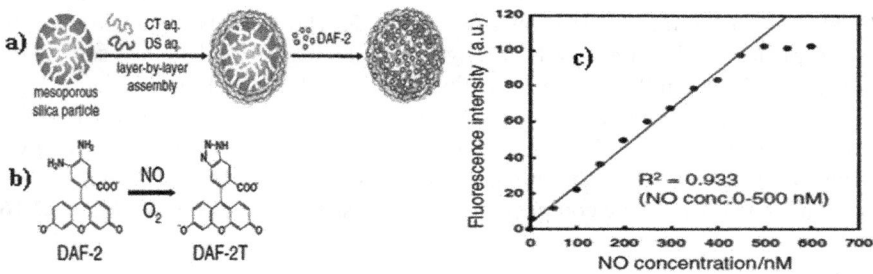

**Fig. 7.14.** a Schematic fabrication process of sensor particle, b possible mechanism in the DAF-2 reaction with NO in the presence of dioxygen, and c fluorescence intensity at 515 nm of 1 mg/mL sensor particle with varied NO concentration in 50 mM Tris-HCl buffer (pH 7.4) at 37 °C. Copyright © 2010 The Chemical Society of Japan; Chem. Lett. 39, 42–43, 2010. Reproduced from Ref. [43] with permission

The sensor particles encapsulating 4,5-diaminofluorescein (DAF-2) can detect 5–500 nM NO and have higher cytocompatibility than DAF-2 molecules (Fig. 7.14c). These intensities were compared to values from the control sample without nitric oxide (NO). These sensor particles will be useful for quantitative and spatial analyses of extracellular NO molecules from living cells.

Chen et al. [44] modified a 100 μm diameter platinum electrode by coating it with a thin layer of chitosan, nickel sulfate, and Nafion successively. The modified electrode was used to detect nitric oxide (NO) in aqueous and biological samples. NO could be oxidized catalytically on the surface of the modified electrodes. The oxidation currents of NO on the electrodes due to electrochemical reactions of NO showed a linear relationship in the concentration range of $4.5 \times 10^{-4}$ to $6.7 \times 10^{-8}$ molL$^{-1}$ NO with the detection limit of $3.0 \times 10^{-8}$ molL$^{-1}$. The electrodes were highly selective against ascorbate (200:1), dopamine (200:1), and nitrite (250:1). The electrodes were suitable for determining NO in vitro or vivo for their higher mechanical strength and increased sensitivity to NO.

Thin films of chitosan (CS), modified by grafting Lucifer yellow VS dye (LYVS) onto chitosan chains and crosslinked with glutaric dialdehyde (GDI), were cast on GCE [45]. The chitosan matrix-supported a fast ion transport as indicated by the apparent diffusion coefficients of Ru (NH$_3$)$_6^{3+}$ and dopamine in the films. Anionic LYVS dye introduced a permselectivity against anions (e.g., Fe(CN)$_6^{4-}$, ascorbate) into the CS-LYVS films. The crosslinking of such films with GDI further increased their permselectivity and their stability. The chemically modified chitosan is an attractive new coating to develop fast, selective, and reversible sensors.

Fartas et al. [46] reported electrochemical biosensors based on immobilization of tyrosinase onto graphene-decorated gold nanoparticle/chitosan (Gr-Au-Chit/Tyr) nanocomposite to detect phenolic compounds. The nanocomposite was deposited onto a screen-printed carbon electrode (SPCE). In the sensing mechanism, the tyrosinase catalyzes the oxidation of phenolic substrates in the presence of molecular oxygen to produce 1,2-dihydroxybenzene (catechol) and the subsequent oxidation of catechol to o-quinone, as shown in Fig. 7.15. The biosensor shows linearity toward phenol in the concentration range from 0.05 to 15 µM with a sensitivity of 0.624 µA/µM and the limit of detection (LOD) of 0.016 µM ($S/N = 3$).

Wang et al. [47] prepared a CBAS for phenol detection. The composite film was prepared by crosslinking chitosan with (3-aminooryloxypropyl) dimethoxymethylsilane. It was followed by the immobilization of tyrosinase [47]. The sensor was fabricated by attaching the composite film to a glassy carbon electrode. Figure 7.16

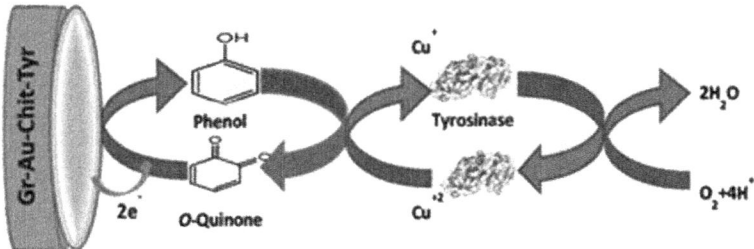

**Fig. 7.15** Schematic of the reaction mechanism of tyrosinase with phenol occurring on the electrodes' surface. Copyright © 2017 Fartas, F.M., Abdullah, J., Yusof, N.A. et al. Creative Commons Attribution (CC BY) license. Sensors, 17, 1132, 2017. https://doi.org/10.3390/s17051132, www.mdpi.com/journal/sensors. Reproduced from the Ref. [46]

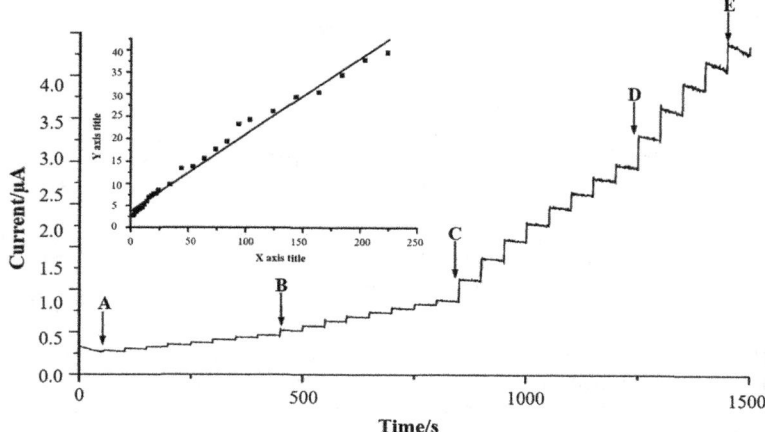

**Fig. 7.16** Hydrodynamic response of the tyrosinase-chitosan biosensor for phenol. (Concentration range A → B, in $1 \times 10^{-10}$ M steps; BC, in $2 \times 10^{-10}$ M steps, CD in $1 \times 10^{-9}$ M steps; DE in $2 \times 10^{-9}$ M steps. Inset: the calibration curve of phenol. Copyright © 2002 Elsevier Science B.V., Bioelectrochemistry (2002), 57(1), 33–38. Reproduced from Ref. [47] with permission

displays a typical current–time response using the enzyme electrode. A high sensitivity (150 nA nM$^{-1}$) for the monitoring of phenol was achieved with the detection limit of $5.0 \times 10^{-11}$ M. The response time of this sensor for phenol was less than 2 s reaching 95% of the steady-state value. It may retain 75% of the activity for at least 70 days.

Constantine et al. [48] fabricated a polyelectrolyte architecture composed of chitosan and organophosphorus hydrolase polycations. Thioglycolic acid-capped CdS quantum dots (QDs) were used as the polyanion. Epifluorescence microscopy was used to detect paraoxon. Chitosan thin films containing pyrene (Py) and β-cyclodextrin (β-CD) units were coated on a quartz plate surface for nitromethane detection [49]. Figure 7.17 shows a schematic representation of the preparation and chemical structure of the chitosan film.

The preparation process included the immobilized pyrene in the cavity of its neighbor β-CD. As per the Stern–Volmer equation, a linear plot of $I_0/I$ versus $Q$ may indicate a homogeneous distribution of pyrene on the substrate surface (Fig. 7.18) [49]. The behavior of the film to nitromethane in methanol and water is similar. However, the quenching efficiency in water is significantly higher than that in ethanol due to the more pronounced effect of the cavity to quencher in the water (Fig. 7.18b).

Figure 7.19 shows that the sensing property of the film is reversible. This unique characteristic could be utilized to prepare a sensor for nitromethane in energetic fuels [49].

Dubas et al. [50] fabricated a polyelectrolyte multilayer (PEM) thin film using a LBL deposition technique. It is reported that acid dye such as (Ph amino)-5-[[4-(3-sulfonatophenyl) azo] -(1-naphthalenyl) azo]-1-naphthalenesulfonic acid disodium

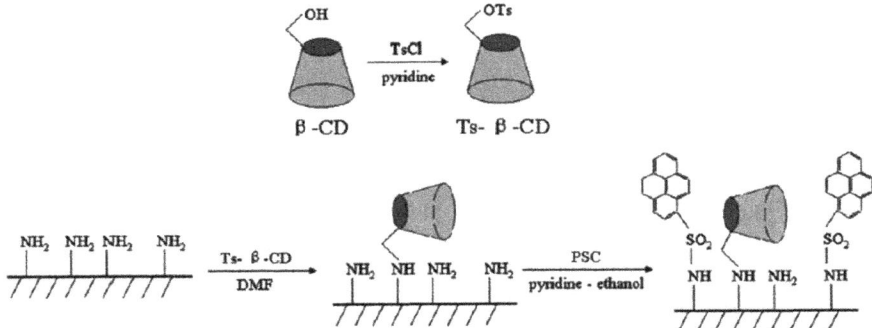

**Fig. 7.17** Schematic representation of the preparation and chemical structure of functional chitosan film. Chitosan is crosslinked with β-CD and 1-pyrenesulfonyl chloride (PSC). Copyright © 2003 Elsevier Science B.V., Thin Solid Films (2003), 440(1, 2), 255–260. Reproduced from Ref. [49] with permission

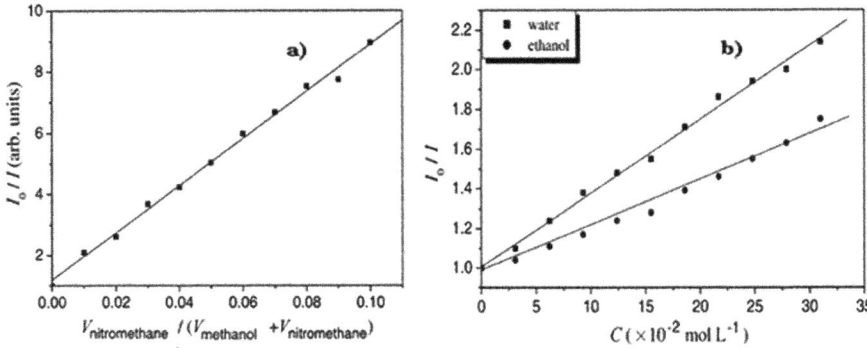

**Fig. 7.18** Stern–Volmer plot for the system of (chitosan + β-CD + PSC) + nitromethane in the medium of **a** methanol and **b** ethanol, or water ($\lambda_{ex}/\lambda_{em} = 344/377$ nm), where $I_0$ and $I$ are the fluorescence intensities in the absence and presence of quencher and $Q$ is the concentration of quencher. Copyright © 2003 Elsevier Science B.V., Thin Solid Films (2003), 440(1, 2), 255–260. Reproduced from Ref. [49] with permission

salt (Nylosan) and chitosan on the glass slide was found to be sensitive to ethanol content in water [50]. In addition, the absorbance ($\lambda_{max}$) spectrum from 540 to 580 nm of PEM thin films increased ethanol concentration in water. The characteristic color shift of the Nylosan dye occurred at an ethanol concentration ranges from 10 to 45% compared to its usual shift in aqueous solution (from 0 to 30%). The absorption response to ethanol content was linear from 10–45% ethanol content (Fig. 7.20).

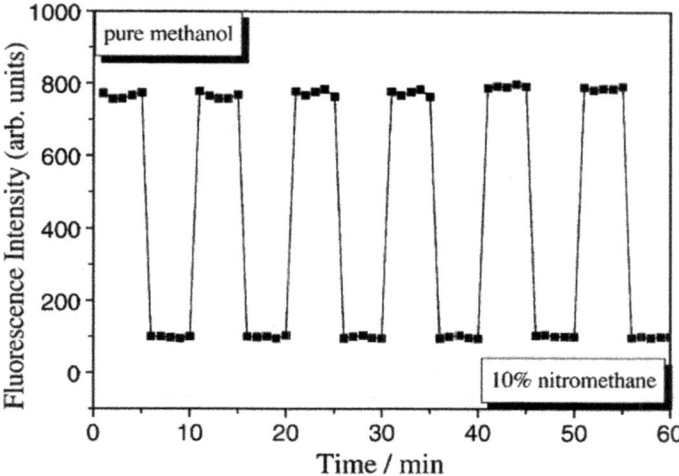

**Fig. 7.19** Reversibility of sensing property of chitosan + β-CD + PSC film to nitromethane. [The medium was changed from pure methanol to nitromethane solution (10% in $V_{nitromethane}/(V_{nitromethane} + V_{methanol})$) and back and forth for several measurements ($\lambda_{340}/\lambda_{377}$ nm)]. Copyright © 2003 Elsevier Science B.V., Thin Solid Films (2003), 440(1,2), 255–260. Reproduced from Ref. [49] with permission

### 7.3.3 Chitosan Nanocomposite-Based Sensor

Incorporating nanoparticles into the chitosan matrix is useful for developing sensors to detect various chemical compounds, including complex biomolecules. CdS nanoparticles were incorporated into chitosan (CS) films during in situ synthesis of CdS nanoparticles. The fluorescence emission of the nanocomposite films is very sensitive to the presence of pyridine, and a small amount of pyridine increases the emission dramatically. The interference from most common ions in water except for copper and iodide was negligible [51]. Zhang et al. [52] prepared a colloidal solution of a chitosan-carbon nanotube (CS-CNT) by dispersing multiwalled CNT in an aqueous solution of chitosan (CS). The colloidal solutions of CS-CNT were placed on the surface of glassy carbon (GC) electrodes to form CS-CNT films, facilitating the electro-oxidation of β-nicotinamide adenine dinucleotide (NADH). The GC/CS-CNT sensor for NADH required approximately 0.3 V less overpotential than the GC electrode. Chitosan film was further modified by crosslinking it with glutaric dialdehyde (GDI). The modified film allowed glucose dehydrogenase (GDH) immobilized in the CS-CNT films. The sensor was both stable and sensitive to glucose detection in the physiological matrix (urine). In pH 7.40 phosphate buffer solution, the detection limit was 3 μM glucose ($S/N = 3$) and had a linear response over the range of 5–300 μM glucose.

Martinez et al. [53] reported sensing properties of a hybrid nanocomposites film prepared using siloxane sols, chitosan, and poly (monomethyl itaconate) (PMMI).

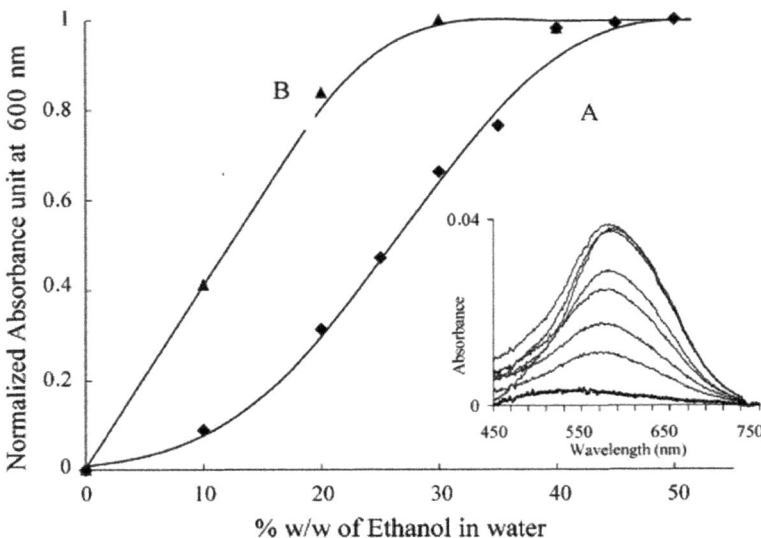

**Fig. 7.20** Increase in absorbance at 600 nm of the Nylosan dye as a function of various environments. **a** Changes in absorbance when the dye is deposited into PEM. **b** Changes in absorbance when the Nylosan dye is dissolved in a water/ethanol mixture. Inset: UV-vis spectrum of the chitosan–Nylosan thin film exposed to increasing concentration of aqueous ethanol (10, 20, 25, 30, 35, 40, 45, 50 wt%). The spectrum of the thin film initially dipped in water was subtracted. It was used as a baseline to improve the clarity of the graph. Copyright © 2005 Elsevier B.V. Sensors and Actuators, B: Chemical (2006), B113(1), 370–375. Reproduced from Ref. [50] with permission

Tetra-ethyl-orthosilicate (TEOS) was used as an inorganic network forming reagent. In nanocomposite preparation, TEOS was used to prepare siloxane sol by acid-catalyzed hydrolysis and condensation reaction. A hybrid film was prepared using siloxane sol, PMMI, and chitosan to construct a solid-state potentiometric electrode. It was reported that the nanocomposite showed good sensing capacity for $NO_3^-$ ion.

$Fe_3O_4$ incorporated chitosan film was used to modify the glassy carbon rotating disk electrode to detect $H_2O_2$ [54]. The amperometric response of $H_2O_2$ was measured at $-0.2$ V (vs. Ag/AgCl). There were several characteristic enhancements by the coated chitosan thin film for the $H_2O_2$ sensor. The calibration curves are linear up to 4.0 and 5.0 mM ($r = 0.999$) in pH 3–7 with the DL of 7.6 and 7.4 $\mu$M L$^{-1}$ (S/N = 3), respectively. The stability was evaluated by the results of half-life time ($t_{50}$%) for 9 months at room temperature and 24 months at 4 °C. In neutral pH, a comparison of the performance for this $Fe_3O_4$/chitosan-modified electrode with other $H_2O_2$ sensors reported in the literature is shown in Table 7.2. Figure 7.21 illustrates the typical calibration plots over the range of 0.025–10 mM $H_2O_2$ at different pHs. The actual response for $Fe_3O_4$/chitosan-modified electrode in pH 3 solution is also shown as the inset. They reported that the $Fe_3O_4$/chitosan-modified electrodes (curve a, b) possess a longer linearity range than the electrodes without chitosan (curve c, d) in corresponding pH [54].

## 7.3 Chitosan-Based Sensors

**Table 7.2** Comparison of the performance of various $H_2O_2$ sensors based on cathodic measurement schemes

| Electrode modifier | Potential applied (vs. Ag/AgCl) (V) | Linear range (mM) | Sensitivity ($\mu A\ mM^{-1}$) | Response time (s) | Detection limit (g) ($\mu M$) | Lifetime (h) (months) |
|---|---|---|---|---|---|---|
| $Fe_3O_4$/chitosan | −0.2 | 0.025–5.0 | 9.6 | 5.2 | 7.4 | 24.0(50%) |
| CoHCF | 0.0 | 0.005–1.1 | 224.0 | 6.5 | 0.063 | 0.5(50%) |
| HRP/sol–gel /hydrogel | −0.05 | 0.1–3.4 | 15.0 | 10.0 | 0.5 | 3.5(92%) |
| HCF/HRF/sol–gel /chitosan | −0.1 | 0.25–3.4 | 14.86 | 10.0 | 3.0 | 1.0(85%) |
| $VZrO_2$/graphite/PE | −0.4 | 0.005–0.4 | 70.0 | – | 2.1 | – |
| Al/MnHCF | 0.0 | 0.0006–7.4 | 194.35 | 4.7 | 0.2 | 3.0(90%) |

Copyright © 2005 Wiley-VCH Verlag GmbH & Co. Electroanalysis (2005), 17(22), 2068–2073. Reproduced from Ref. [55] with permission

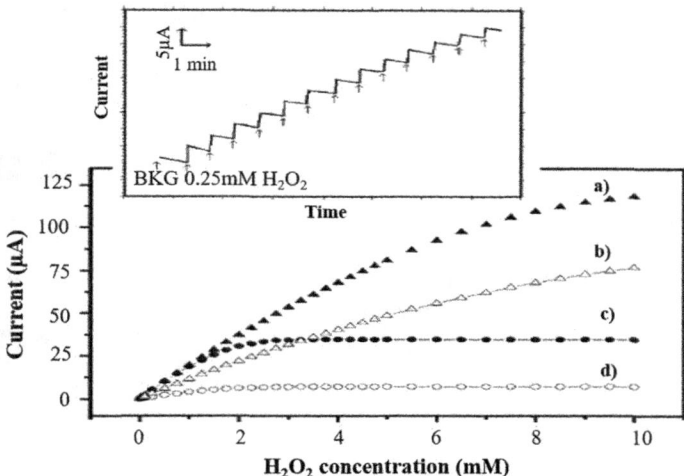

**Fig. 7.21** Calibration curves for the detection of $H_2O_2$ by $Fe_3O_4$ modified electrode with chitosan (**a, b**) or without chitosan (**c, d**). The filled symbol and hollow symbols represent the buffer solution at pH 3 and 7, respectively. The actual response of curve (**a**) was shown as an inset. Other operational conditions: −0.2 V applied potential; 0.05 M pH 3 citrate buffer with 0.1 M KCl; rotating rate 1225 rpm. Copyright © 2005 Wiley-VCH Verlag GmbH & Co. Electroanalysis (2005), 17(22), 2068–2073. Reproduced from Ref. [55] with permission

Darder et al. [55] prepared amperometric biosensors by immobilizing $GO_x$ thin film onto nanoporous $Al_2O_3$ membranes. The active phase of the biosensors was an external coating of chitosan. This coating entrapped enzyme, which remained active for further detection. Darder et al. [56] also developed a nanocomposite based on the intercalation of chitosan in $Na^+$-montmorillonite that provided bidimensional nanostructured materials with anionic exchange sites ($-NH^{+3}X^-$). The method for intercalation of the chitosan in $Na^+$-montmorillonite has been discussed by Drader et al. [56, 57]. A schematic of chitosan-graphite composite biosensor construction is given in Fig. 7.22.

The reaction between the chitosan and montmorillonite clay constituents resulted in the high stability of the intercalated biopolymer. The intercalated biopolymer was stable against desorption or degradation. This prevented the formation of the usual chitosan film. The resulting material was a robust three-dimensional nanocomposite that exhibited structural and functional properties. It was further mixed with graphite particles to provide the active phase and the electronic collector for the developed electrochemical sensor. The resulting sensors were tested for the detection of several

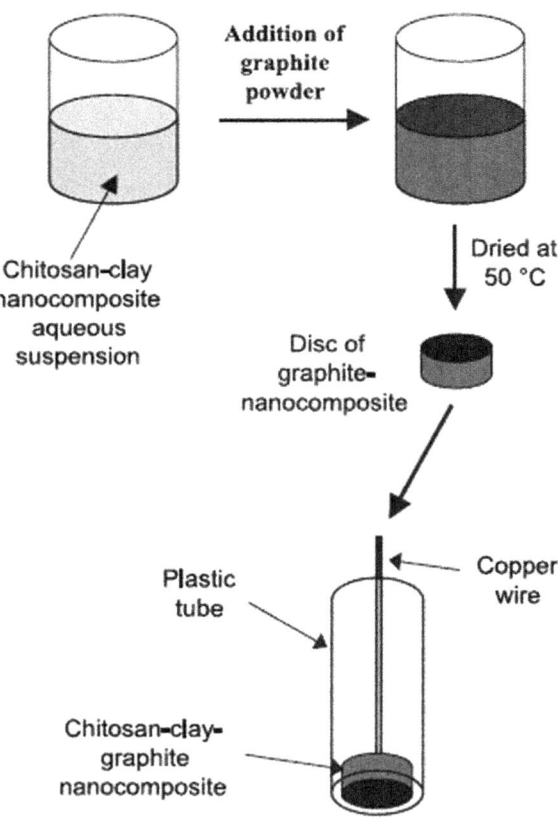

**Fig. 7.22** Chitosan-clay-graphite system preparation and sensor construction. Copyright © 2004 Elsevier B.V. Applied Clay Science (2005), 28(1–4), 199–208. Reproduced from Ref. [56] with permission

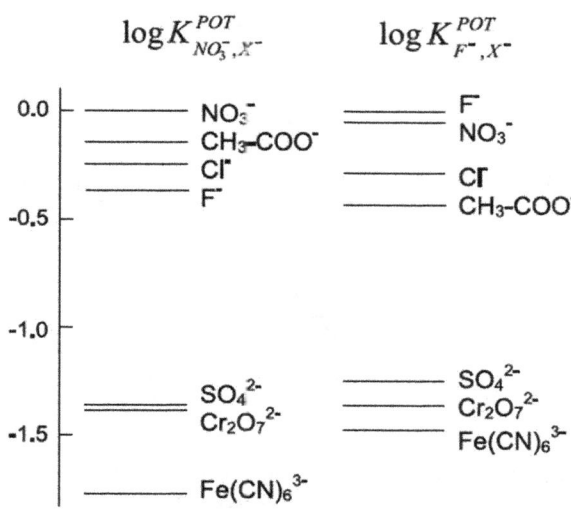

**Fig. 7.23** Selectivity coefficient of the sensor based on the chitosan-montmorillonite nanocomposite with 173.8 mEq/100 g clay of adsorbed chitosan for $NO_3^-$ and $F^-$, against different anions. Copyright © 2003, American Chemical Society, Chemistry of Materials (2003), 15(20), 3774–3780. Reproduced from Ref. [57] with permission

anions. Compared to di- or trivalent anions, a higher selectivity toward monovalent ions was observed. The best potentiometric response was obtained for nitrate ions (Fig. 7.23).

In another work, chitosan (CS) was first crosslinked with a redox mediator Azure dye (AZU) [58]. Then CNT was dispersed into the CS-AZU matrix to form composite films for the amperometric determination of β-NADH [58]. The incorporation of CNT into the CS-AZU matrix facilitated the AZU-mediated electro-oxidation of NADH. The film produced in this manner decreased the potential for the mediated process by an extra 0.30 V. It amplified the NADH current by approximately 35 times (at −0.10 V) while reducing the response time from about 70 s for CS-AZU to approximate 5 s for CS-AZU/CNT films [58]. As compared to CS-TBO, the CS-TBO/CNT films displayed significant amplification of a current due to the TBO-mediated oxidation of NADH at −0.10 V.

The metal ion chelating ability of chitosan has been utilized by Schauer et al. [59] as an optical dipstick sensor to detect aqueous chromium ions. Four different thiolated chitosans were synthesized and spin-coated into colored, thin 110 nm films. These films were shown to be sensitive toward aqueous lead and mercury ions. These thiolated chitosans were also investigated for their ability to bind GNp and were used to make thin biopolymer-gold films. An amperometric immunosensor for human IgG assay based on ZnO/chitosan composite as a sensing platform has been described [60]. A sequential sandwich immunoassay format was performed on the ZnO/chitosan composite supported by a glass carbon electrode (GCE) using a goat antihuman IgG antibody (IgG Ab) and human IgG as a model system. Using hydroquinone as a mediator, amperometric detection at −150 mV (vs. SCE) resulted in a detection range of 2.5–500 ng mL$^{-1}$ and a detection limit of 1.2 ng mL$^{-1}$ (Fig. 7.24).

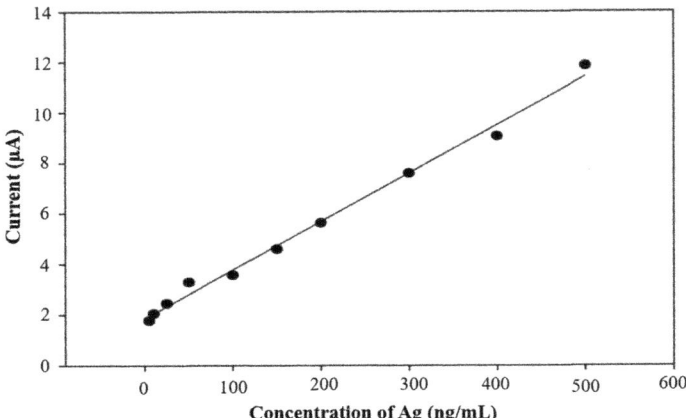

**Fig. 7.24** Current versus concentration plot for ZnO/chitosan-IgG Ag immunosensor after the sandwich immunoassay. The immunosensor was incubated in 0.1 M tris–HCl/1 mM EDTA buffered solution of pH 7.5 containing different amounts of IgG Ag. These experimental conditions are optimal. Copyright © 2005 Elsevier B.V. Talanta (2006), 69(3), 686–690. Reproduced from Ref. [60] with permission

Dopamine is among the essential neurotransmitters ensuring interneuronal communication in the human central nervous system [61]. The electro-active multilayered films assembled onto an indium tin oxide (ITO) electrode were used to detect dopamine. Electro-active nanostructured films of chitosan (CS) and tetra-sulfonated metallophthalocyanines were produced using the LBL technique. The tetra-sulfonated metallophthalocyanines, such as nickel (NiTsPc), copper (CuTsPc), and iron (FeTsPc), are used to crosslink with chitosan to prepare the nanostructures films. These films detected dopamine (DA) concentration ranging from $5.0 \times 10^{-6}$ to $1.5 \times 10^{-4}$ mol $L^{-1}$. The DL of these films is given in Table 7.3. The ITO-(Ch/NiTsPc) n electrodes showed higher electrocatalytic activity for DA oxidation when compared with a bare ITO electrode (Fig. 7.25). The Ch/FeTsPc and Ch/CuTsPc modified electrodes also distinguish between DA and ascorbic acid [62].

Ben Aoun [61] reported a dopamine electrochemical sensor prepared on nitrogen-doped graphene quantum dots–chitosan nanocomposite-modified nanostructured SPCE. Graphene quantum dots were prepared via microwave-assisted hydrothermal

**Table 7.3** Detection limits (DL) exhibited by the three systems employed

| Electrodes | DL (mol $L^{-1}$) |
|---|---|
| Ch/NiTsPc | $4.88 \times 10^{-5}$ |
| Ch/CuTsPc | $9.74 \times 10^{-5}$ |
| Ch//FeTsPc | $3.7 \times 10^{-5}$ |

Copyright © 2006, American Chemical Society, Journal of Physical Chemistry B (2006), 110(45), 22690–22694. Reproduced from Ref. [62] with permission

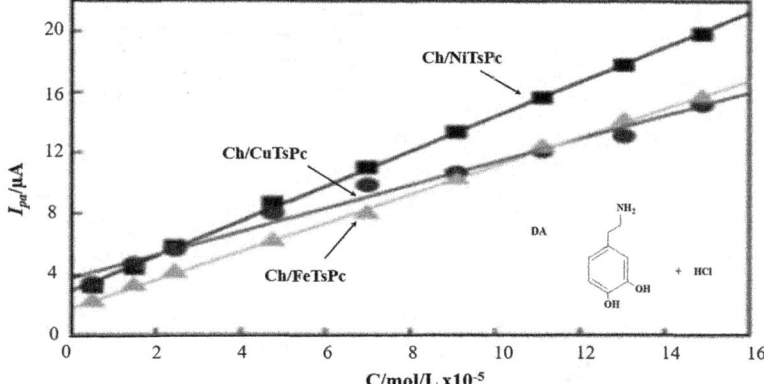

**Fig. 7.25** Linear calibration curves for dopamine (DA) detection for LbL films: (9) Ch/NiTsPc: $I_{pa}$ (*i*A) 2.83 + 1.16–105 $C$, **b** Ch/CuTsPc: $I_{pa}$ ($\mu$A) 3.75 + 0.77–105 $C$ and (2) Ch/FeTsPc: $I_{pa}$ ($\mu$A)) 1.74 + 0.94–105 $C$. Scan Rate: 50 mV s$^{-1}$. Copyright © 2006, American Chemical Society, Journal of Physical Chemistry B (2006), 110(45), 22690–22694. Reproduced from Ref. [62] with permission

reaction of glucose. Nitrogen doping was realized by introducing ammonia in the reaction mixture. It was reported that the incorporation of chitosan played a significant role in the selectivity of the prepared sensor by hindering the ascorbic acid interference and enlarging the peak potential separation between dopamine and uric acid. The sensor as-prepared CS/N, GQDs@SPCE shows high sensitivity (i.e., ca. 418 $\mu$A mM cm$^{-2}$) in the linear lower concentration range of (1–100 $\mu$M) and higher concentration range of (100–200 $\mu$M) with excellent correlations (i.e., $R^2 = 0.999$ and $R^2 = 1.000$, respectively). The sensor exhibits a very low detection limit (LOD = 0.145 $\mu$M) and limit of quantification (LOQ = 0.482 $\mu$M) based on $S/N = 3$ and 10, respectively.

Shukla and Ben-Yoav [63] studied CS-CNT modified microelectrode for the in situ detection of schizophrenia medicine clozapine in a microliter volume of blood. This sensor is based on a microelectrode modified with micrometer thick chitosan encapsulated carbon nanotube (Fig. 7.26).

The sensor detects the clozapine oxidation current. Various species present in the blood generate overlapping electrochemical signals, as shown in Fig. 7.26. The sensor detects clozapine in the blood with a sensitivity of 32 ± 3.0 $\mu$A cm$^{-2}$ $\mu$mol$^{-1}$ L. The limit of detection of clozapine was 0.5 ± 0.03 $\mu$mol L$^{-1}$. Therefore, this sensor can be used for rapid and minimally invasive clozapine detection at the point-of-care.

Tonegawa and his coworkers [64–66] utilized chitosan membranes for pH measurement of solutions. Chitosan composite membrane was prepared by a casting method from an aqueous acetic acid solution of chitosan and cellulose acetate (1:3) and was hydrolyzed in 0.1 N KOH aqueous solution. The hydrolyzed membrane was dyed with a pH indicator, e.g., benzopurpurine 4B, and coated with polyvinyl

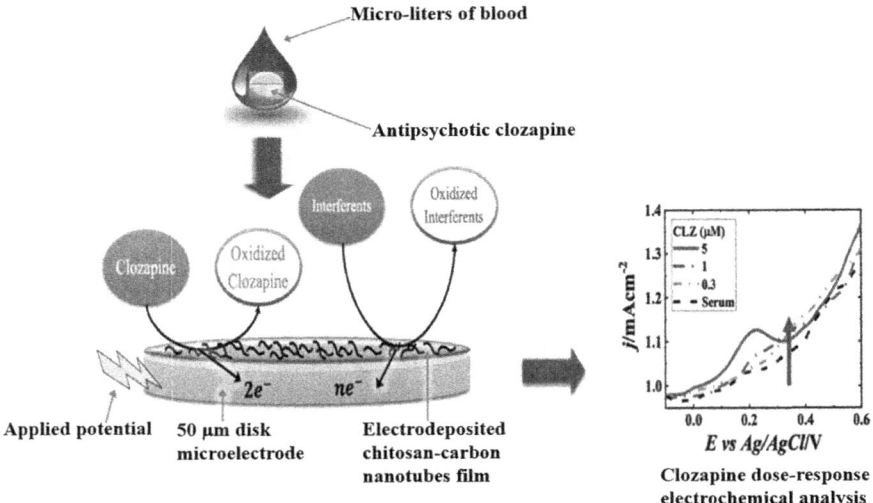

**Fig. 7.26** Scheme describing clozapine in situ detection in microliter volumes of a blood sample [63]. Copyright © 2019 Shukla, R.P., and Ben-Yoav, H. Published by WILEY-VCH Verlag GmbH & Co., Adv. Healthcare Mater. 2019, 8, 1900462, 1–14. Reproduced from Ref. [63] with permission

alcohol-containing $TiO_2$ for reflection. The membrane was further stuck on a transparent polyester support film with cyanoacrylate adhesive. The effect of pH on the reflection spectra of the sensor membrane was measured through the bundle of bifurcated fiber optic [64]. Later a three-layer membrane as a fiber optic pH sensor was prepared from benzopurpurine 4B immobilized chitosan-cellulose blend membrane and with PVA (polyvinyl alcohol)-$TiO_2$ for a reflection layer and PET (polyethylene terephthalate) for the support layer [65]. The reflection spectra of this three-layer membrane were measured by a spectrophotometer equipped with the bifurcated optical fiber bundle. The sensor membrane shows a color change in the pH 2–6 range. The response time was approximately 10 min. Further modification of the pH indicators was made by treating the chitosan-cellulose blend membrane with cyanuric chloride, followed by the reaction with pH indicators such as bromophenol blue, bromocresol green, bromocresol purple, and bromothymol blue. The obtained optodes were laminated with PVA (poly (vinyl alcohol))-$TiO_2$ for a reflection layer and with PET (poly (ethylene terephthalate)) for the support layer. The reflection spectra of the optodes were measured with a spectrophotometer equipped with the bifurcated optical fiber bundle. The fiber optic sensor showed the color change in the pH 2–7 regions [66].

The capability of chitosan for binding with heavy metal ions is well known. This characteristic of chitosan has been exploited by several researchers for the development of sensors for metal ions. Kurauchi et al. [67] fabricated a bifurcated fiber optic sensor for $Zn^{2+}$, $Cd^{2+}$, and $Ga^{3+}$ ions (Fig. 7.27). Chitosan modified with 5-formyl-3-hydroxy-4-hydroxymethyl-2-methylpyridine was immobilized on an agarose gel

## 7.3 Chitosan-Based Sensors

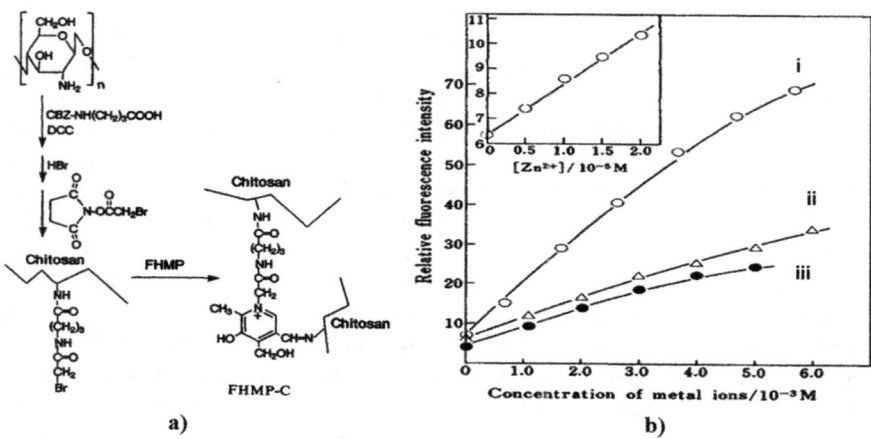

**Fig. 7.27** a Schematic representation of Schiff base reaction of FHMP as a sensing agent with chitosan and **b** calibration curves for (1) $Zn^{2+}$, (2) $Cd^{2+}$, and (3) $Ga^{3+}$ in 80% methanol. The inset is a calibration curve for $Zn^{2+}$ in a lower concentration range. Copyright © 1992 Japan Science and Technology Agency (JST), Analytical Sciences (1992), 8(6), 837–40. Creative Commons Attribution 4.0 International License, Reproduced from Ref. [67]

and used as a fluorogenic probe. The reproducibility of the response to $Zn^{2+}$ was within 5% in eight successive measurements at $5.0 \times 10^{-5}$ M. A linear relationship with a correlation coefficient of 0.994 was obtained in the $Zn^{2+}$ concentration range of $0$–$2.0 \times 10^{-5}$ M, and the detection limit was $1.0 \times 10^{-6}$ M ($S/N = 3$). $Cd^{2+}$ and $Ga^{3+}$ ions were also detected, though with lower sensitivities than $Zn^{2+}$. Chitosan films were employed to detect various ions, including $Cr^{3+}$, $Cr^{6+}$, $Cu^{2+}$, by DuMaurier et al. [68]. Although chitosan absorbed/adsorbed both cations and anions, there was no selectivity present in the mixture. The films have been doped with tetraarylporphyrins in an attempt to improve the selectivity. At times, there is excessive use of the metal salts in the chrome tanning process. It is preferable to add chromium through a recurrent fashion rather than continuous measurement of its concentration. Campanella et al. [69] noted that a sensor based on chitosan was selective toward chrome in the tanning bath constituents, such as metal ions and synthetic and natural tannins. This sensor was used in a simulated tanning process on skin samples. Gallocyanin immobilized in the chitosan membrane was studied as an optical sensor for lead using a flowing system [70]. The optical sensor detected lead in the concentration range from $1.0 \times 10^{-1}$ to $1.0 \times 10^{3}$ ppm with a detection limit of 0.075 ppm. It was reported that the response of the sensor for the lead was reproducible. The sensor can be regenerated by using an acidified $KNO_3$ solution. To determine Co (II), a chitosan-based transducer was prepared by immobilizing 2-(4-pyridylazo) resorcinol (PAR) in the chitosan membrane [71]. This transducer was placed into the distal end of a bifurcated optical fiber and connected to stop flow FIA instrumentation. This optical sensor detects Co(II) in the concentration range of $1.0 \times 10^{-3}$ to

$1.0 \times 10^4$ ppm with a detection limit of $7.9 \times 10^{-6}$ ppm. The sensor can be regenerated using a mixed HCL solution (0.2 M) and saturated KCl. Bao and Nomura [72] prepared a silver-selective sensor using an electrode-separated piezoelectric quartz crystal modified with a chitosan derivative. In this work, a quartz plate was modified using $N$-(2-Pyridylmethyl) chitosan (PMC). Ag(I) adsorbed selectively onto these PMC-modified quartz crystals from the coexisting metal ions in an ammonium chloride buffer solution containing EDTA [72]. This modification provided a selective response for Ag(I). The quartz plate frequency also increased due to the incorporation of $N$-(2-pyridylmethyl) chitosan. The increase in frequency was attributed to the desorption of $H_2O$ from the chitosan derivative. The frequency shifts due to Ag(I) adsorption were proportional to the concentration over the range of 10–80 nM of Ag(I). The detection limit was 6 nM. The relative standard deviation (at 50 nM five times) was 3.4%. GNp capped with chitosan was employed for sensing ions of heavy metals such as $Zn^{2+}$ and $Cu^{2+}$ in $H_2O_5$. Acidic anions such as glutamate ions capped the nanoparticle surfaces. The polycationic chitosan enabled its attachment to the negatively charged Au nanoparticle surfaces through electrostatic interactions. The chelating properties of chitosan and the optical properties of Au nanoparticles to agglomerate were employed to detect a low concentration of $Zn^{2+}$ and $Cu^{2+}$ in $H_2O$ (Fig. 7.28).

Figure 7.28a shows changes in normalized absorption value vs. the concentration of Cu (II) ions between chitosan treated and as-prepared GNp. Chitosan treatment makes the response approximately linear, while untreated nanoparticles tend to agglomerate even under exposure to low concentrations of ions. Hence, estimating the concentration of Cu (II) ions in the solution is much simplified when the nanoparticles are capped with chitosan. Figure 7.28b shows the comparison of optical absorption of 650 nm light for a varying concentration of Zn (II) ions in the colloid.

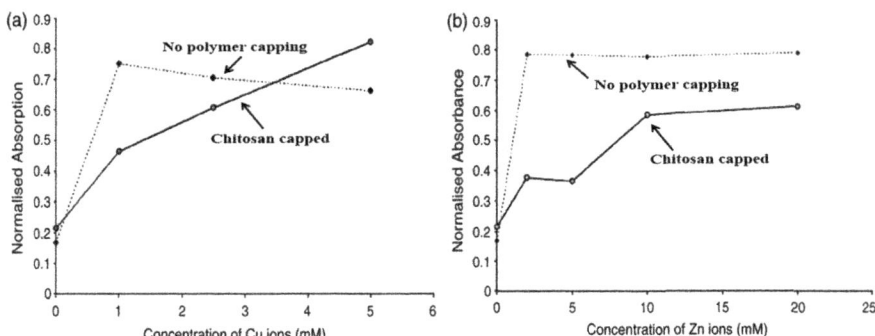

**Fig. 7.28** Optical properties of chitosan gold-based sensor to detect **a** copper and **b** zinc (optical absorption of 650 nm light). Copyright © 2005 Elsevier Science Ltd, Science and Technology of Advanced Materials (2005), 6(3–4), 335–340. Creative Commons Attribution 4.0 International License, Reproduced from Ref. [73]

## 7.3 Chitosan-Based Sensors

Several studies investigated chitosan as a sensing layer for metal ion detection using a surface-sensitive technique such as surface plasmon resonance (SPR) spectroscopy [74–76]. In surface plasmon resonance (SPR)sensing, the affinity between the analytes and the immobilized ligands of an SPR sensor can be estimated by the affinity constant. Kamaruddin et al. [74] investigated the possibility of a highly sensitive Au/Ag/Au/chitosan-graphene oxide sensor as a potential environmental monitoring detector for heavy metals. Based on the binding affinity constant $K$, which is $7 \times 10^5$ $M^{-1}$ for $Pb^{2+}$ and $4 \times 10^5$ $M^{-1}$ for $Hg^{2+}$, it was suggested that this Au/Ag/Au/chitosan-graphene oxide sensor could be used as a heavy metal detector in environmental monitoring [74].

Zhao's group utilized electrochemiluminescence (ECL) properties of chitosan-modified membranes [77, 78] to detect organic molecules, particularly oxalate acid. An ECL sensor having a platinum electrode coated with a tris(2,2′-bipyridine) ruthenium (II) (Ru (II) complex)-modified chitosan found to be very sensitive to oxalic acid. The ECL intensity to oxalic acid was reproducible within 5% (relative standard deviation) on ten repeated runs. A calibration curve for oxalic acid gave a straight line in the concentration range of 0.1 to 10 mM (correlation coefficient, 0.997), and the detection limit ($S/N = 3$) is $3 \times 10^{-5}$ M (Fig. 7.29).

The high selectivity in the presence of trimethylamine can be explained partly by the electrostatic repulsion of the chitosan membrane [77]. Electrostatic repulsion occurs due to the interaction of the positive charge of chitosan with trimethylamine.

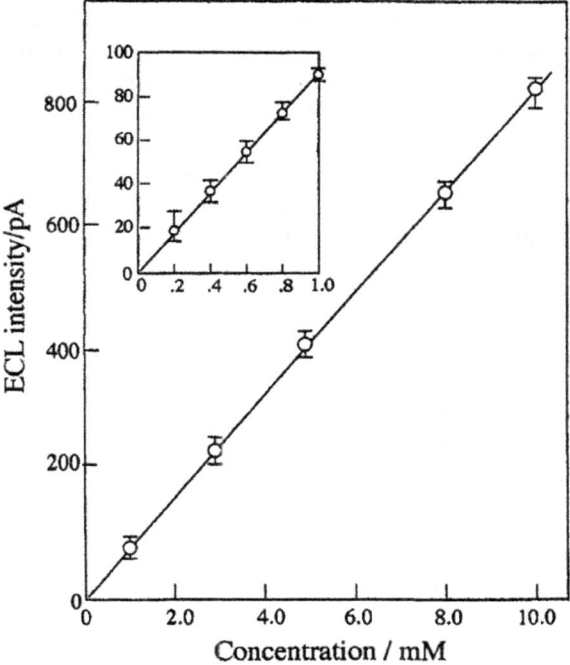

**Fig. 7.29** Calibration curve for oxalic acid in 50 mM phosphate buffer (pH 6.8) obtained at 1.10 V using Ag/AgCl as the reference electrode. Copyright © 1997 Elsevier Science Ltd. Electrochimica Acta (1998), 43(14–15), 2167–2173. Reproduced from Ref. [79] with permission

An ECL sensor with a Pt electrode probe coated doubly with Ru(bpy)$_3^{2+}$-modified chitosan and silica gel membranes was prepared. The electrode response to oxalate ion reached a constant value within approximately 20 s. The response of the electrode in the concentration range of 0.1–10 mM was linear. The detection limit was 0.02 mM (S/N = 3). The response to oxalate was considerably stronger than the response to trimethylamine or proline. The sensor worked stably for over one month. The higher selectivity and stability were attributed to the silica matrix [78].

Zhou et al. [79] studied a reagentless chemiluminescence (CL) flow sensor to determine hydrogen peroxide. The sensor contained two parts combined to construct the reagentless flow biosensor for $H_2O_2$. The first part was the analytical reagent luminol immobilized on anion-exchange resins and packed in a glass tube. The second part was the horseradish peroxidase (HRP) immobilized on the chitosan (CS) membrane and formed on the glass coil. The sensor has two working regions ($1.0 \times 10^{-7}$ to $1.0 \times 10^{-5}$ mol/L and $1.0 \times 10^{-5}$ to $2.0 \times 10^{-4}$ mol/L), and the detection limit is $4.0 \times 10^{-8}$ mol/L (S/N = 3). A relative standard deviation of 1.1% was for the $4.0 \times 10^{-6}$ mol/L $H_2O_2$ solution in 31 repeated measurements using this sensor. Zhang et al. [80] fabricated electrogenerated chemiluminescence (ECL) biosensor. An ECL reagent Ru(bpy)$_3^{2+}$ and alcohol dehydrogenase were immobilized in a gel of chitosan/poly (sodium 4-styrene sulfonate) (PSS) to prepare the film. The presence of chitosan prevented the cracking of the film. The presence of chitosan also provided a biocompatible microenvironment for alcohol dehydrogenase. The detection limit of this ECL biosensor was $9.3 \times 10^{-6}$ M for alcohol (S/N = 3). The response was linear in the range from $2.79 \times 10^{-5}$ to $5.78 \times 10^{-2}$ M. Zhang et al. [81] developed a CBAS for detecting pyrogallol. The core–shell luminol-doped SiO$_2$ nanoparticles were immobilized on the surface of chitosan film, which was coated on a graphite electrode by the self-assembled technique to prepare ECL biosensor. This ECL biosensor was used to detect pyrogallol. The detection limit of the sensor was $1.0 \times 1.0^{-9}$ mol/L for pyrogallol. The linear range extended from $3.0 \times 10^{-9}$ mol/L to $2.0 \times 10^{-5}$ mol/L for pyrogallol. Ding et al. [82] noted that fluorescent chitosan films could be prepared by modifying the chitosan membrane with a fluorophore. They prepared a fluorescent chitosan film containing dansyl as a fluorophore, as shown in Fig. 7.30.

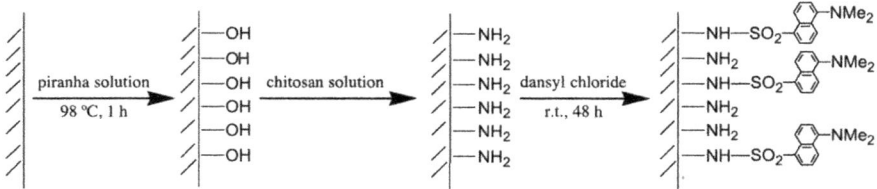

**Fig. 7.30** Schematic representation of chemical binding of dansyl onto chitosan film. Copyright © 2004 Elsevier B.V. Thin Solid Films (2005), 478(1–2), 318–325. Reproduced from Ref. [82] with permission

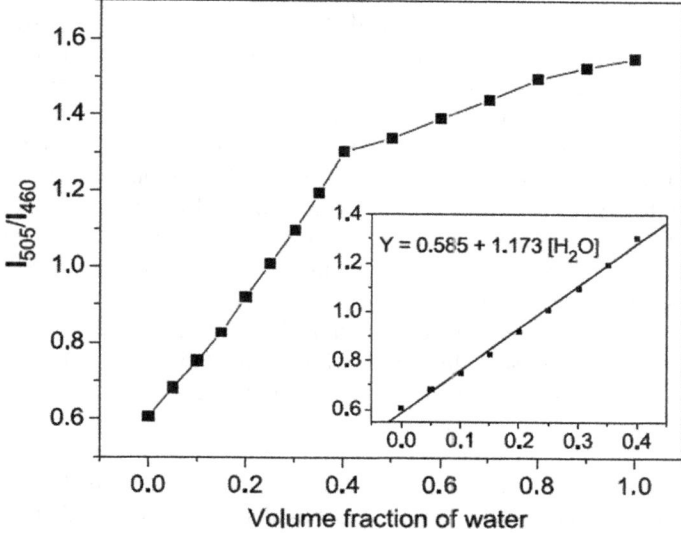

**Fig. 7.31** Plot of the intensity ratio ($I_{505}/I_{460}$) of the film and water concentration in the ethanol–water mixture. Copyright © 2004 Elsevier B.V. Thin Solid Films (2005), 478(1–2), 318–325. Reproduced from Ref. [82] with permission

Due to the twisted intramolecular charge transfer (TICT) in the excited state of the fluorophore, the film shows a dual fluorescence. The maximum emission of the film shifted from 460 nm in ethanol to 505 nm in water [82]. Figure 7.31 shows that the intensity ratio, $I_{505}/I_{460}$, was linear up to a concentration of 40% ethanol in the ethanol–water mixture, and it was reversible.

Gong et al. [83] developed a chitosan-based fluorescence sensor based on supramolecular recognition by glycosylated metalloporphyrin. This sensor was developed for the levamisole (LEV) assay. An optode membrane was prepared using glycosylated metalloporphyrin in chitosan matrices, incorporated an LEV-sensitive active material, (5,10,15,20-tetrakis [2-(2,3,4,6-tetraacetyl-β-D-glucopyranosyl)-1-O-phenyl]porphyrin). The glycosylated metalloporphyrin/chitosan optode membrane showed a linear response over the range of $1.3 \times 10^{-5}$–$3.5 \times 10^{-7}$ mol/L. It had a detection limit of $3.5 \times 10^{-7}$ mol/L for LEV. Excellent selectivity and stability of this sensor toward LEV were reported. The sensor was applied to determine LEV in pharmaceutical preparations. The results agreed with the values obtained by the pharmacopeia method.

Mao et al. [84] modified the microcantilever by coating it with a modified chitosan membrane and noted that it can be used to track changes in the pH of a solution and can provide a sensitive platform for chemical and biological sensors. Chitosan film was crosslinked with tripolyphosphate (TPP) and coated with gels on one side of the microcantilevers. The coated microcantilevers deflected upon exposure to the pH range of 6 to 7.4. A significant 1200 nm bending response was observed, while pH was changed from 7.45 to 6.17. The deflection increased as the pH decreased in the range from 6 to 7.45. The authors envisioned that such a chitosan-coated microcantilever could be used as biological sensors as molecular recognition agents (antibodies/enzymes). Koev et al. [85, 86] noted that sensitivity and selectivity of microcantilevers for mechano-detection could be enhanced by electrodepositing a chitosan film onto the cantilever surface. The objective was to detect nucleic acid hybridization. The chitosan film was biofunctionalized with an oligonucleotide probe. It was used for the detection of DNA hybridization by cantilever bending in solution (static mode) or resonant frequency shifts in the air (dynamic mode). Selective detection of the neurotransmitter dopamine was also investigated in this study. A chitosan-coated cantilever was biased to electrochemical oxidation of dopamine solution. It is reported that the chitosan film reacts with the oxidation products of dopamine and creates tensile stress of approximately 1.7 MPa, causing substantial cantilever bending. The interference from the ascorbic acid solution was negligible.

Powers et al. [87] noted that chitosan could be utilized to fabricate microelectromechanical (MEMS)-based cell sensors. The authors developed a new system for the optical detection of biomolecules based upon chitosan. Chitosan served two purposes: pH-dependent solubility allowed electrochemical deposition, while its functional groups enabled facile coupling of proteins, oligonucleotides, and other biomolecules through covalent bonds. The authors developed multimode waveguides and fluidic channels on a Pyrex wafer using a single layer of SU-8 (epoxy-based negative photoresist.). This new system demonstrated the possibility of the optical detection of biomolecules based upon chitosan.

Attempts were made to detect organic compounds in the solution by using various functional groups present in chitosan. To determine ethanol in alcoholic beverages, a fiber optic sensor was reported by Kurauchi's group [88, 89]. A chitosan/poly (vinyl alcohol) composite membrane was used as the cladding and a Teflon protective coating to fabricate the sensor. It was reported that the crosslinking of the cladding membrane with glutaraldehyde enhanced the sensor sensitivity in the 0–70 vol./vol.% ethanol content range. The Teflon coating prevented interferences from sugars and organic acids present in alcoholic beverages. The response time was within 2 min for 50 vol./vol.% ethanol. The relative standard deviation was 1.7% in 20 repeated determinations during a two-week period. A linear relation between the response and the ethanol content was obtained in the 0–70 vol./vol.% range. The sensor was used to determine ethanol in shochu, sake, wine, whiskey, and beer. This sensor matched results with the certified values and values obtained by the conventional specific gravity method [88].

Hikima et al. [90] reported an amperometric enzyme sensor. This sensor is composed of a mercury film electrode and an enzyme-immobilized chitosan membrane. The mercury film electrode detects the consumption of dissolved dioxygen following enzymatic reaction. The chitosan membrane provides an excellent permselectivity and excludes electro-active interferences. The detection range of this biosensor was $1.0 \times 10^{-5}$–$3.0 \times 10^{-4}$ mol/L. The relative standard deviation at $5.0 \times 10^{-5}$ mol/L was 1.4% ($n = 3$). This biosensor was useful in the direct determination of L-lactate in human serum.

**Fig. 7.32** Responses to aqueous (●) formic acid, (□) acetic acid, and (○) propionic acid. Copyright © 1996 Japan Science and Technology Agency (JST), Analytical Sciences (1996), 12(1), 55–9. Creative Commons Attribution 4.0 International License, Reproduced from Ref. [91]

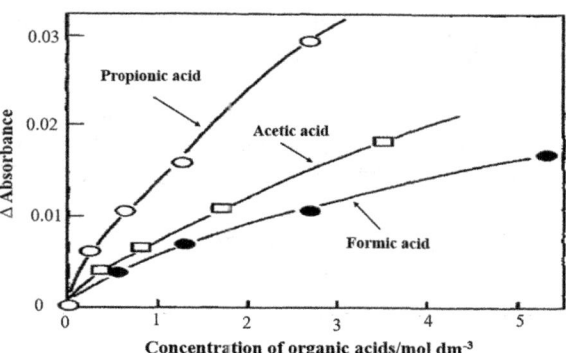

Kurauchi et al. [91] constructed a fiber optic sensor having a chitosan/polyvinyl alcohol) cladding to detect various organic acids. Figure 7.32 shows relationships between the response and the concentration of organic acids, such as formic, acetic, and propionic acids. The response increased as the carbon number increased. The increased response is probably due to the hydrophobic methylene moiety removing water from the membrane and increasing the refractive index. To reduce the interference from ethanol, the crosslinking of the cladding with glutaraldehyde is controlled, and the cladding is modified with 5′,5″-dibromopyrogallolsulfonphthalein. The response time for 5 vol./vol.% acetic acid was within 1 min. The relative standard deviation was approximately 2% for ten successive measurements. The coating of the cladding with an amorphous fluoropolymer increased its durability and removed interferences from inorganic acids and non-volatile compounds. Later Kurauchi et al. [92] developed a total internal reflection-type fiber optic sensor. They used a crosslinked chitosan/polyvinyl alcohol composite membrane as the cladding. This sensor responded reversibly to organic solvents such as alcohols, acetone, acetonitrile, dioxane, and chloroform in water. The response was based on an increase in the refractive index of the cladding membrane by organic solvents. Larger responses were observed for solvents having lower dielectric constants. When the cladding membrane was coated with a thin Teflon protective membrane, response time depended on solvent vapor pressure. Response time was 30 s for acetone with high vapor pressure. The response to chloroform was approximately 10 times larger than that of ethanol. A linear relationship was obtained for a concentration range of 5–50 mM of chloroform.

Carbaryl insecticide is commonly used in agriculture due to its high insecticidal activity. However, it poses a danger to human health due to its inhibitory effect on acetylcholinesterase (AChE). The AChE is essential for the central nervous system [93]. Song et al. [94] developed a biosensor for the determination of carbaryl insecticide. This amperometric acetylcholinesterase (AChE) biosensor was prepared based on PB–chitosan (CS) hybrid film. This biosensor was fabricated by the first electrodepositing PB–chitosan hybrid film on a glassy carbon electrode (GCE). Then AChE was assembled on PB–chitosan hybrid film leading to a stable AChE sensor.

The carbaryl can be detected by measuring the decline of the oxidation current of thiocholine. The sensitivity of the sensor depends strongly on the distribution and immobilization of AChE [93]. The function of PB as a redox mediator promotes electrocatalytic activity toward the oxidation of thiocholine. Thiocholine is a product from the hydrolysis of acetylthiocholine catalyzed by AChE. The oxidation reactions of thiocholine by PB nanoparticles are as follows:

$$(CH_3)_3N^+CH_2CH_2SC(O)CH_3 \xrightarrow{H_2O,\ AChE} (CH_3)_3N^+CH_2CH_2SH + CH_3COOH, \tag{7.9}$$

$$2(CH_3)_3N^+CH_2CH_2SH \rightarrow (CH_3)_3N^+CH_2CH_2SSCH_2CH_2N^+(CH_3)_3 + 2PB_{Red}, \tag{7.10}$$

$$PB_{Red} - 2e^- \rightleftharpoons PB_{OX}. \tag{7.11}$$

Figure 7.33 shows the inhibition by carbaryl on the activity of AChE. This inhibition was proportional to carbaryl concentration in the ranges from 0.01 to 0.4 M (inset A) and from 1.0 to 5.0 M (inset B). The correlation coefficients were 0.9996 and 0.9997, respectively, for the two ranges of carbaryl concentrations [89]. The detection limit was about 3 nM. The biosensor provided a new promising tool for

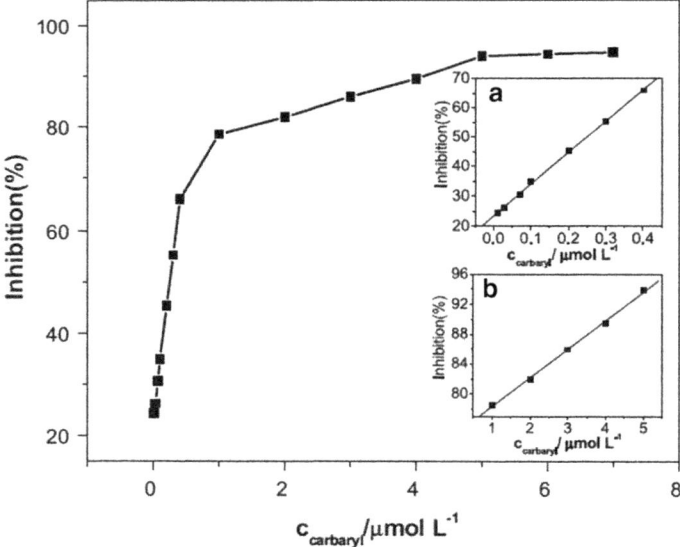

**Fig. 7.33** Inset: calibration curves in the range of 0.01–0.4 μM (**a**) and 1.0–5.0 μM (**b**), respectively. Copyright © 1991 The Chemical Society of Japan, Chemistry Letters (1991), (8), 1411–12. Reproduced from Ref. [89] with permission

## 7.3 Chitosan-Based Sensors

insecticide detection. Bolat and Abaci [95] described a composite film consisting of ionic liquid (IL), chitosan (CS), and electrochemically synthesized GNp (AuNPs) on single-use pencil graphite electrodes (PGEs). The film was prepared to detect organophosphorus compounds such as malathion. They reported that the AuNP-CS-IL/PGE-based sensor showed an affinity for malathion. It can be used as a very simple, fast, ultra-sensitive, and inexpensive sensor device for the detection of malathion.

Several studies reported CBASs for detection of chloroform [96], ammonia [97], acetone [98], methylamine [99], hexanal[100] etc. in different medium. Chandrasekaran et al. [96, 97] investigated the sensing properties of the chitosan sensor. These sensing properties included sensitivity, stability, recovery, and repeatability for the detection of chloroform and ammonia. Nasution et al. [101] studied the sensing mechanism of a CBAS for low concentration acetone. The proposed chitosan-based acetone sensor (CBAS) can operate at room temperature with high performance. CBAS can be used as a gold standard in diagnosing diabetes since it provides quick and early detection. Nainggolan et al. [98] investigated the effect of ferredoxin on sensing properties of chitosan-based acetone sensors. The ratio of ferredoxin to chitosan was varied at 5:95, 10:90, 15:85, and 20:80. The sensors were exposed to acetone vapor with acetone concentrations of 0.1, 1, 10, 20, 50, and 100 ppm. Among the sensors, the chitosan sensor with 15% ferredoxin to detect acetone exhibited the highest response.

Shantini et al. [100] investigated the chitosan film sensor (CFS) to detect volatile gas hexanal. The response of the chitosan film sensor (CFS) was tested via electrical testing by exposing it to different hexanal concentrations. The concentration of hexanal was tested ranged from 20 to 300 ppm using air as a carrier gas. Figure 7.34 shows the hexanal sensing mechanism by the chitosan film sensor (CFS).

Figure 7.34 shows that the hexanal gas molecules interact with the chemisorbed oxygen, which is deposited on the sensor (CFS) surface, and the electrons then tend to be released within the chitosan structure as explained below:

$$C_6H_{12}O + O_2^- \rightarrow C_6H_{12}O_2 + H_2O + e^-. \tag{7.12}$$

The reaction will release electrons and water molecules, as indicated by Eq. 7.12. The released electrons become free electrons in the conduction band. These electrons induce the increment of the electrical response (%) of the sensor. Meanwhile, the water molecules might form hydrogen bonding with the surface of the sensing layer. This bonding might enhance the electron movements that contribute to increasing the response of the sensor. Overall, it was suggested that hexanal sensors based on chitosan could perform well at room temperature as demonstrated by the good response, good recovery, good repeatability, good stability, and good selectivity [100].

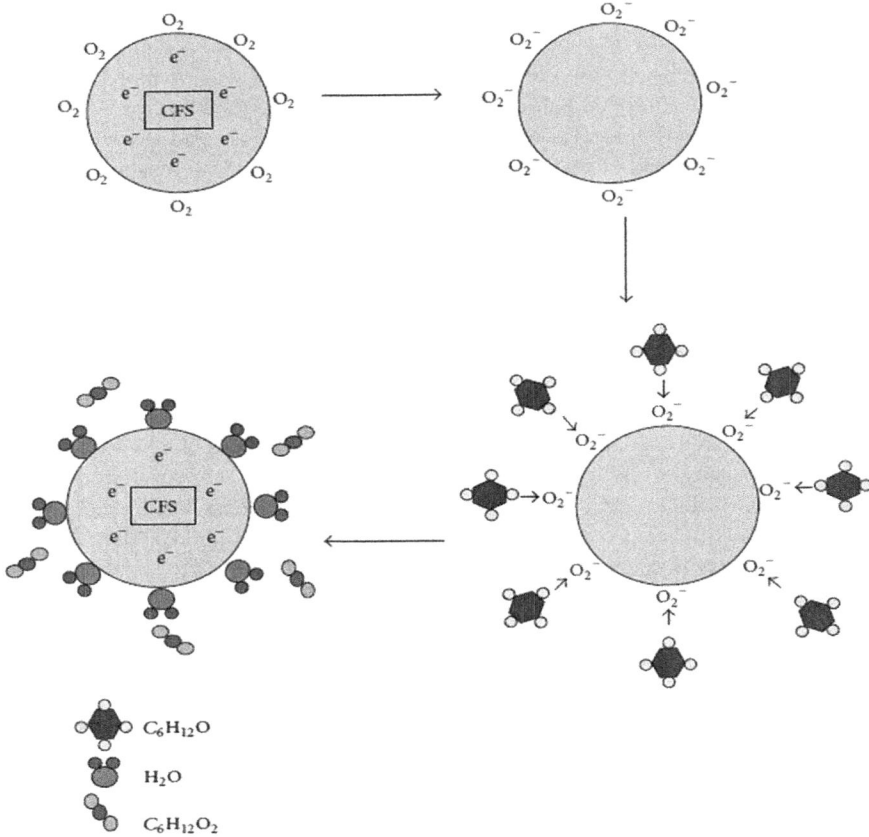

**Fig. 7.34** Schematic of hexanal detection mechanism by the chitosan film sensor (CFS). Copyright © 2016 Devi Shantini et al. Creative Commons CC BY license, Journal of Sensors, 2016, Article ID 8539169, 7 pages, http://dx.doi.org/10.1155/2016/8539169. Reproduced from Ref. [100]

## 7.4 Summary

Chitosan and chitosan derivatives have been used to prepare thin films, membranes, and three-dimensional structures as a direct sensing medium. The electrical and chemical properties of chitosan can be modified by reacting to the functional groups of chitosan and the targeted compound. Chitosan mainly serves as a matrix to entrap components within its film, which is then used to support the development of the bioreceptor part of the sensor chip. Incorporation of nanoparticles into the chitosan matrix is beneficial for the development of sensors to detect various chemical compounds, including complex biomolecules. For example, a chitosan-based biosensor can be used as a practical, feasible sensor to detect glucose in clinical practice. Chitosan-based composite materials can also be considered a potential candidate for constructing a biosensor to detect environmental contaminants.

## References

1. Coté GL, Pishko MV (2003) Emerging biomedical sensing technologies and their applications. Sens J IEEE 3(3):251–266
2. Kim KM, Hanna MA, Xu YX, Nag D (2005) Chitosan-starch composite film: preparation and characterization, Ind Crops Prod 21(2):185–192.
3. Wan Y, Creber KAM, Peppley B, Bui VT (2006) Chitosan-based solid electrolyte composite membranes I. Preparation and characterization. J Membr Sci 280(1–2):666–674
4. Jegal J, Lee K-H (1999) Chitosan membranes crosslinked with sulfosuccinic acid for the pervaporation separation of water/alcohol mixtures. J Appl Polym Sci 71(4):671–675
5. Ang TH, Sultana FSA, Hutmacher DW et al (2002) Fabrication of 3D chitosan-hydroxyapatite scaffolds using a robotic dispensing system. Mater Sci Eng, C 20(1–2):35–42
6. Hatami J, Silva SG, Oliverra MB et al (2017) Multilayered films produced by layer-by-layer assembly of chitosan and alginate as a potential platform for the formation of human adipose-derived stem cell aggregates. Polymers 9:440. https://doi.org/10.3390/polym9090440www.mdpi.com/journal/polymers
7. Constantine CA, Gatta's-Asfura K, Mello SV et al (2003) Layer-by-layer films of chitosan, organophosphorus hydrolase and thioglycolic acid-caped CdSe quantum dots for the detection of Paraoxon. J Phys Chem B 107:13762–13764
8. Barsan MM, David M, Florescu M et al (2014) A new self-assembled layer-by-layer glucose biosensor based on chitosan biopolymer entrapped enzyme with nitrogen-doped graphene. Bioelectrochemistry 99:46–52
9. Sousa MP, Cleymand F, Mano JF (2016) Elastic chitosan/chondroitin sulfate multilayer membranes. Biomed Mater 11:035008. https://doi.org/1088/1748-6041/11/3/035008
10. Cheng W, Huang J, Liu C et al (2017) High sensitivity chlorogenic acid detection based on multiple layer-by-layer self-assembly films of chitosan and multi-walled carbon nanotubes on a glassy carbon electrode. RSC Adv 7:6950
11. Calvo EJ, Etchenique R, Pietrasanto L, Wolosiuk A (2001) Layer-by-layer self-assembly of glucose oxidase and Os $(BPy)_2ClPyCH_2NH$-Poly(allylamine) bioelectrode. Anal Chem 73:1161–1168
12. Zhang W, Li X, Zou R et al (2015) Multifunctional glucose biosensors from $Fe_3O_4$ nanoparticles modified chitosan/graphene nanocomposites. Sci Rep 5:11129. https://doi.org/10.1038/srep11129
13. Postisatityuenyong A, Dubas ST, Sukwattanasinitt M, Layer-by-layer deposition of chitosan/polydiacetylene vesicles for convenient preparation of colorimetric sensing film, NSTI-Nanotech 2006, vol 1, pp 27–30. www.nsti.org, ISBN 0-9767985-6-5
14. David M, Barsan MM, Florescu M, Brett CM (2015) Acidic and basic functionalized carbon nanomaterials as electrical bridges in enzyme loaded chitosan/poly (styrene sulfonate) self-assembled layer-by-layer glucose biosensors. Adv Mater Electroanal 27(9):2139–2149
15. Caseli L, Santos DS Jr, Aroca RF, Oliveira ON Jr (2009) Controlled fabrication of gold nanoparticles biomediated by glucose oxidase immobilized on chitosan layer-by-layer films. Mater Sci Eng, C 29(5):1687–1690
16. Ang LF, Por LY, Yam MF (2013) Study on different molecular weights of chitosan as an immobilization matrix for a glucose biosensor. PLOS ONE 8(8):1–13. www.plosone.org
17. Shrestha BK, Ahmad R, Mousa HM et al (2016) High-performance glucose sensor based on chitosan-glucose oxidase immobilized polypyrrole/nafion/functionalized multi-walled carbon nanotubes bio-nanohybrid film. J Colloid Interface Sci 482:39–47

18. Susanto H, Samsudin AM, Rokhati N, Widiasa IN (2013) Immobilization of glucose oxidase on chitosan-based porous composite membranes and their potential use in biosensors. Enzyme Microb Technol 52:386–392
19. Zare H, Najafpour GD, Jahanshahi M, Rahimnejad M, Rezvani M (2017) Highly stable biosensor based on glucose oxidase immobilized in chitosan film for diagnosis of diabetes. Roman Biotechnol Lett 22(3):12611–12619
20. Gotoh M, Chen CY, Isao K (1993) Glucose biosensor using chitosan membrane. Sens Mater 4(4):187–193
21. Kulys J, Stupak R (2008) Glucose biosensor based on chitosan-gold and Prussian blue-gold nanoparticles. Open Nanosci J 2:34–38
22. Zhu J, Zhu Z, Lai Z et al (2002) Planar amperometric glucose sensor based on glucose oxidase immobilized by chitosan film on Prussian blue layer. Sensors 2(4):127–136. https://doi.org/10.3390/s20400127
23. Wang Y, Zhu J, Zhu R et al (2003) Chitosan/Prussian blue-based biosensors. Measure Sci Technol 14(6):831–836
24. Tan X-C, Tian Y-X, Cai P-X, Zou X-Y (2005) Glucose biosensor based on glucose oxidase immobilized in sol-gel chitosan/silica hybrid composite film on Prussian blue modified glass carbon electrode. Anal Bioanal Chem 381(2):500–507
25. Chen X, Jia J, Shaojun D (2003) Organically modified sol-gel/chitosan composite based glucose biosensor. Electroanalysis 15(7):608–612
26. Chen Q, Han J, Shi H et al (2004) Use of chitosan for developing layer-by-layer multilayer thin films containing glucose oxidase for biosensor applications. Sens Lett 2(2):102–105
27. Chen P-C, Hsieh B-C, Chen RLC et al (2006) Characterization of natural chitosan membranes from the carapace of the soldier crab *Mictyris brevidactylus* and its application to immobilize glucose oxidase in amperometric flow-injection biosensing system. Bioelectrochemistry 68(1):72–80
28. Caseli L, Santos DS Jr, Foschini M et al (2006) The effect of the layer structure on the activity of immobilized enzymes in ultrathin films. J Colloid Interface Sci 303(1):326–331
29. Miscoria SA, Desbrieres J, Barrera GD, Labbe P, Rivas GA (2006) Glucose biosensor based on the layer-by-layer self-assembling of glucose oxidase and chitosan derivatives on a thiolated gold surface. Analytica Chimica Acta 578(2):137–144
30. Kang X, Wang J, Wu H et al (2009) Glucose oxidase-graphene-chitosan modified electrode for direct electrochemistry and glucose sensing. Biosens Bioelectron 25(4):901–905
31. Qiu J-D, Huang J, Liang R-P (2011) nanocomposite film based on graphene oxide for high performance flexible glucose biosensor. Sens Actua B 160:287–294
32. Wu H, Wang J, Kang X et al (2009) Glucose biosensor based on immobilization of glucose oxidase in platinum nanoparticles/graphene/chitosan nanocomposite film. Talanta 80:403–406
33. Rassas I et al (2019) Highly sensitive voltammetric glucose biosensor based on glucose oxidase encapsulated in a chitosan/kappa-carrageenan/gold nanoparticle bionanocomposite. Sensors 19:154. https://doi.org/10.3390/s119010154
34. Jiang L, Wu X, Gong C et al (2002) Electrochemical behaviors of chitosan modified electrode and its application for biosensor. Huaxue Chuanganqi 22(4):18–23
35. Xu J-J, Luo X-L, Du Y, Chen H-Y (2004) Application of $MnO_2$ nanoparticles as an eliminator of ascorbate interference to amperometric glucose biosensors. Electrochem Commun 6(11):1169–1173
36. Zhang F-F, Wan Q, Wang X-L (2004) Amperometric sensor based on ferrocene-doped silica nanoparticles as an electron transfer mediator for the determination of glucose in rat brain coupled to in vivo microdialysis. J Electroanal Chem 571(2):133–138
37. Liu Y, Wang M, Zhao F et al (2005) The direct electron transfer of glucose oxidase and glucose biosensor based on carbon nanotubes/chitosan matrix. Biosens Bioelectron 21(6):984–988
38. Abdul Amir Al-Mokaram AMA et al (2016) One-step electrochemical deposition of polypyrrole–chitosan–iron oxide nanocomposite films for non-enzymatic glucose biosensor. Mater Lett 183:90–93

39. Yang M, Jiang J, Yang Y et al (2006) Carbon nanotube/cobalt hexacyanoferrate nanoparticle-biopolymer system for the fabrication of biosensors. Biosens Bioelectron 21(9):1791–1797
40. Zhai X, Wei W, Zeng J et al (2006) New nanocomposite based on Prussian blue nanoparticles/carbon nanotubes/chitosan and its application for assembling of amperometric glucose biosensor. Anal Lett 39(5):913–926
41. Mani V et al (2017) Core-shell heterostructured multi-walled carbon nanotubes@ reduced graphene oxide nanoribbons/chitosan, a robust nano biocomposite for enzymatic biosensing of hydrogen peroxide and nitrite. Sci Rep 7:11910. https://doi.org/10.1038/541598-017-120 50-x
42. Zhang W, Du Y, Wang ML (2015) On-chip highly sensitive saliva glucose sensing using multilayer films composed of single-walled carbon nanotubes, gold nanoparticles, and glucose oxidase. Sens Bio-Sens Res 4:96–102
43. Amemori S, Matsusaki M, Akashi M (2010) Biocompatible and highly sensitive nitric oxide sensor particles prepared by the layer-by-layer assembly. Chem Lett 39:42–43
44. Chen G, Zhang J, Lu W et al (1998) Application and preparation of amperometric microsensor for determination of nitric oxide. Daxue Xuebao 25(6):448–451
45. Cruz J, Kawasaki M, Gorski W (2000) Electrode coatings based on chitosan scaffolds. Anal Chem 72(4):680–686
46. Fartas FM, Abdullah J, Yusof NA, Sulaiman Y, Saiman MI (2017) Biosensor based on tyrosinase immobilized on graphene-decorated gold nanoparticle/chitosan for phenolic detection in aqueous, Sensors 17:1132. www.mdpi.com/journal/sensors
47. Wang G, Xu J-J, Ye L-H, Zhu J-J, Chen H-Y (2002) Highly sensitive sensors based on the immobilization of tyrosinase in chitosan. Bioelectrochemistry 57(1):33–38
48. Constantine CA, Gatta's-Asfura KM, Mello SV et al (2003) Layer-by-layer biosensor assembly incorporating functionalized quantum dots. Langmuir 19:9863–9867
49. Wang H, Fang Y, Ding L, Gao L, Hu D (2003) Preparation, and nitromethane sensing properties of chitosan thin films containing pyrene and β-cyclodextrin units. Thin Solid Films 440(1, 2):255–260
50. Dubas S, Iamsamai C, Potiyaraj P (2006) Optical alcohol sensor based on dye-chitosan polyelectrolyte multilayers. Sens Actuat B: Chem B113(1):370–375
51. Wang S, Fang Y, Zhang Y et al (2003) Preparation of chitosan/CdS composite films and their sensing properties towards pyridine. Wuli Huaxue Xuebao 19(6):514–518
52. Zhang M, Smith A, Gorski W et al (2004) Carbon nanotube-chitosan system for electrochemical sensing based on dehydrogenase enzymes. Anal Chem 76(17):5045–5050
53. Martinez Y, Retuert J, Yazdani-Pedram M et al (2004) Sensing properties of hybrid polymeric films obtained by sol-gel. J Chil Chem Soc 49(2):127–131
54. Lin M, Shan L, Hoang J et al (2005) A $Fe_3O_4$-based chemical sensor for cathodic determination of hydrogen peroxide. Electroanalysis 17(22):2068–2073
55. Darder M, Aranda P, Hernandez-Velez M et al (2006) Encapsulation of enzymes in alumina membranes of controlled pore size. Thin Solid Films 495(1–2):321–326
56. Darder M, Colilla M, Eduardo R-H (2005) Chitosan-clay nanocomposites: application as electrochemical sensors. Appl Clay Sci 28(1–4):199–208
57. Darder M, Colilla M, Eduardo R-H (2003) Biopolymer-clay nanocomposites based on chitosan intercalated in montmorillonite. Chem Mater 15(20):3774–3780
58. Zhang M, Gorski W (2005) Electrochemical sensing platform based on the carbon nanotubes/redox mediators-biopolymer system. J Am Chem Soc 127(7):2058–2059
59. Schauer CL, Cathell MD, Bui F, William (2006) Thin films of modified chitosan and gold nanoparticles. In: Abstracts, 38th middle Atlantic regional meeting of the American Chemical Society, Hershey, PA, US, 4–7 June 2006, MRM-118
60. Wang Z, Yang Y, Li J et al (2006) Organic-inorganic matrix for electrochemical immunoassay: Detection of human IgG based on ZnO/chitosan composite. Talanta 69(3):686–690
61. Ben Aoun S (2017) Nanostructured carbon electrode modified with N-doped graphene quantum dots-chitosan nanocomposite: a sensitive electrochemical dopamine sensor. R Soc Open Sci 4:171199. https://doi.org/101098/rsos.171199

62. Siqueira JR Jr, Gasparotto LHS, Crespilho FN, Carvalho AJF, Zucolotto V, Oliveira ON Jr (2006) Physicochemical properties and sensing ability of metallophthalocyanines/chitosan nanocomposites. J Phys Chem B 110(45):22690–22694
63. Shukla RP, Ben-Yoav H (2019) A chitosan-carbon nanotube-modified microelectrode for in situ detection of blood levels of the antipsychotic clozapine in a finger-pricked sample volume. Adv Healthcare Mater 8(1900462):1–14
64. Tonegawa M Kikuchi A, Kasuya T et al (1995) Optical fiber pH sensor based on chitosan composite membranes. Kitosan Kenkyu 1(2):104–105
65. Tonegawa M, Masuda T, Nishimura Y et al (1998) Functional membrane for optical fiber pH sensor. 3. Preparation of chitosan-cellulose blend membrane. Tokyo Ika Daigaku Kiyo 24:21–30
66. Tonegawa M, Ikeda T, Ito T et al (2000) Fiber optic sensor with chitosan-cellulose blend membrane immobilized pH indicator. Tokyo Ika Daigaku Kiyo 26:29–37
67. Kurauchi Y, Hayashi R, Egashira N et al (1992) Fluorometric determination of zinc, cadmium and gallium ions with a fiber-optic sensor having a pyridoxal isomer-modified chitosan/agarose gel as a sensing probe. Anal Sci 8(6):837–40
68. Du Maurier SA, Richards RA, Thompson AN et al (1997) Fabrication of metal ion sensors: Properties of polymeric chitosan-porphyrin thin films. In: Book of abstracts, 213th ACS national meeting, San Francisco, 3–17 Apr 1997
69. Campanella L, Favero G, Giovannini P, Aturki Z (1998) Characterization of a chitosan membrane sensor for determination of chromium (III) in tannery waters. Inquinamento 40(2):36–43
70. Yusof NA, Ahmad M (2002) A flow cell optosensor for lead based on immobilized gallocynin in chitosan membrane. Talanta 58(3):459–466
71. Yusof NA, Ahmad M (2002) A flow cell optosensor for determination of Co (II) based on immobilized 2-(4-pyridylazo) resorcinol in chitosan membrane by using stopped-flow, flow injection analysis. Sens Actuat B: Chem B86(2–3):127–133
72. Bao S, Nomura T (2002) Silver-selective sensor using an electrode-separated piezoelectric quartz crystal modified with a chitosan derivative. Anal Sci 18(8):881–885
73. Sugunan A, Thanachayanont C, Dutta J, Hilborn JG (2005) Heavy-metal ion sensors using chitosan-capped gold nanoparticles. Sci Technol Adv Mater 6(3–4):335–340
74. Kamaruddin NH, Bakar AA et al (2017) Binding affinity of a highly sensitive Au/Ag/Au/chitosan-graphene oxide sensor based on direct detection of $Pb^{2+}$ and $Hg^{2+}$ ions. Sensors 17:2277. https://doi.org/10.3390/s17102277
75. Yahya M et al (2019) Optical constant determination of crosslinked chitosan-polyethylene glycol (PEG) using attenuated total reflection method by surface plasmon resonance phenomenon. IOP Conf Ser: Mater Sci Eng 546:042051. https://doi.org/10.1088/1757-899X/546/4/042051
76. Fen YW, Yunus WMM, Yusof NA (2011) Detection of mercury and copper ions using surface plasmon resonance optical sensor. Sens Mater 23(6):325–334
77. Zhao C-Z, Egashira N, Kurauchi Y, Ohga K (1998) Electrochemiluminescence oxalic acid sensor having a platinum electrode coated with chitosan modified with a ruthenium (II) complex. Electrochim Acta 43(14–15):2167–2173
78. Zhao C-Z, Egashira N, Kurauchi Y et al (1998) Electrochemiluminescence sensor having a Pt electrode coated with a $Ru(bpy)_3^{2+}$-modified chitosan/silica-gel membrane. Anal Sci 14(2):439–441
79. Zhou G-J, Wang G, Xu J-J, Chen H-Y (2002) Reagentless chemiluminescence biosensor for determination of hydrogen peroxide based on the immobilization of horseradish peroxidase on biocompatible chitosan membrane. Sens Actuat B: Chem B81(2–3):334–339
80. Zhang L, Xu Z, Dong S (2006) Electrogenerated chemiluminescence biosensor based on $Ru(bpy)_3^{2+}$ and dehydrogenase immobilized in sol-gel/chitosan/poly (sodium 4-styrene sulfonate) composite material. Anal Chim Acta 575(1):52–56
81. Zhang L, Zheng X (2006) A novel electrogenerated chemiluminescence sensor for pyrogallol with core-shell luminol-doped silica nanoparticles modified electrode by the self-assembled technique. Anal Chem Acta 570(2):207–213

82. Ding L, Fang Y, Jiang L et al (2005) Twisted intra-molecular electron transfer phenomenon of dansyl immobilized on chitosan film and its sensing property to the composition of ethanol-water mixtures. Thin Solid Films 478(1–2):318–325
83. Gong F-C, Wu D-X, Cao Z et al (2006) A fluorescence enhancement-based sensor using glycosylated metalloporphyrin as a recognition element for levamisole assay. Biosens Bioelectron 22(3):423–428
84. Mao J, Kondu S, Ji H-F, McShane MJ (2005) Response of chitosan/gelatin-coated microcantilever to small pH change. In: Abstracts of papers, 230th ACS national meeting, Washington, DC, United States, 28 Aug–Sept 1 2005, vol 93 p 595. PMSE-354. PMSE Preprints
85. Koev ST, Powers MA, Yi H et al (2006) Mechano-transduction of DNA hybridization and dopamine oxidation through electrodeposited chitosan network. Lab Chip 7(1):103–111
86. Koev ST, Powers MA, Yi H et al (2007) Mechano-transduction of DNA hybridization and dopamine oxidation through electrodeposited chitosan network. Lab Chip 7(1):103–111
87. Powers MA, Koev ST, Schleunitz A et al (2005) Toward a biophotonic MEMS cell sensor. In: Proceedings of SPIE—Bioengineered and bioinspired systems II, vol 5839. SPIE—The International Society for Optical Engineering, pp 119–126
88. Kurauchi Y, Yanai T, Egashira N, Ohga K (1994) Fiber-optic sensor with a chitosan/poly (vinyl alcohol) cladding for the determination of ethanol in alcoholic beverages. Anal Sci 10(1):213–217
89. Kurauchi Y, Yanai T, Ohga K (1991) Determination of ethanol in aqueous solutions using a fiber-optic sensor with a chitosan/poly (vinyl alcohol) cladding. Chem Lett 8:1411–12
90. Hikima S, Kakizaki T, Taga M, Hasebe K (1993) Enzyme sensor for L-lactate with a chitosan-mercury film electrode. Fresenius J Anal Chem 345(8–9):607–609
91. Kurauchi Y, Ogata T, Egashira N, Ohga K (1996) Fiber-optic sensor with a dye-modified chitosan/poly (vinyl alcohol) cladding for the determination of organic acids. Anal Sci 12(1):55–59
92. Kurauchi Y, Nagase M, Egashira N, Ohga K (1997) Response of a fiber-optic sensor with a chitosan/poly (vinyl alcohol) cladding to organic solvents in water. Anal Sci 13(6):987–990
93. Li P, Song Y, Chen S, Zhang M, Wang L (2013) A novel biosensor based on acetylcholinesterase/chitosan-graphene oxide modified electrode for detection of carbaryl pesticides. Asian J Chem 25(8):4444–4448
94. Song Y, Zhang M, Wang L et al (2011) A novel biosensor based on acetylcholinesterase/Prussian blue-chitosan modified electrode for detection of carbaryl pesticides. Electrochim Acta 56:7267–7271
95. Bolat G, Abaci S (2018) Non-enzymatic electrochemical sensing of malathion pesticide in tomato and apple samples based on gold nanoparticles-chitosan-ionic liquid hybrid nanocomposite. Sensors 18:773. https://doi.org/10.3390/s18030773
96. Chandrasekaran DS et al (2014) Chloroform gas sensor based on chitosan biopolymer. Appl Mech Mater 679:45–49
97. Chandrasekaran DS et al (2015) Ammonia gas sensor based on chitosan biopolymer. Mater Sci Forum 819:429–434
98. Nainggolan I, Nasution TI, Ahmad KR (2018) The effect of ferredoxin in enhancing the sensing properties of chitosan-based acetone sensors IOP Conf Ser: J Phys: Conf Ser 1116:042023. https://doi.org/10.1088/1742-6596/1116/4/042023
99. Mironenko AY et al (2017) Highly sensitive chitosan-based optical fluorescent sensor for gaseous methylamine detection. In: Progress on chemistry and application of chitin and its derivatives, 2017, vol XXII, pp 159–165. https://doi.org/10.15259/PCACD.22.16
100. Shantini D et al (2016) Hexanal gas detection using chitosan biopolymer as sensing material at room temperature. J. Sens Article ID 8539169:7 p. https://doi.org/10.1155/2016/8539169
101. Nasution TI et al (2013) The sensing mechanism and detection of low concentration acetone using chitosan-based sensors. Sens Actuat B 177:522–528

# Chapter 8
# Application of Chitosan in the Medical and Biomedical Field

**Abstract** Chitosan and its derivatives are considered suitable materials for many medical and biomedical applications due to their biological nature. The application of chitosan and its derivatives as physiological materials is widely reported. This chapter presents an overview of potential applications of chitosan and its derivatives for medical applications, including treatment of wounds, surgical sutures, orthopedics, drug delivery, anticancer, and tissue engineering. The application of chitosan-based resin for medical isotope separations is also discussed in this chapter.

## 8.1 Skin Wound

Skin, accounting for approximately 15% of total body mass [1], plays a crucial role in regulating body temperature, controlling water loss, and protecting our internal organs from the external environment [2]. The skin structure consists of the epidermis, dermis, and hypodermis, as shown in Fig. 8.1 [3].

The hypodermis has subcutaneous fat tissues. The primary constituent of the epidermis layer is the keratinocyte cell. The keratinocytes cell is the source of fibrous structural proteins such as keratin. The dermis tissues, located under the epidermis, are composed of extracellular matrix (ECM). Collagen accounts for 70% of the dermis tissue, and it is the main component of ECM, providing firmness to the skin [3]. The subcutaneous tissues are composed of loose connective tissues, and they are mainly fatty layers located under the dermis. The subcutaneous tissue generally connects the skin to the underlying muscle or organs It also acts as a shock absorber. It stores energy and protects the body from outside cold and heat environments. Any break or defect in the skin due to physiological or thermal damage is considered a wound. Wounds related to minor accidental cuts or surgery, generally heal within 4 to 6 weeks, are termed acute wounds. Chronic wounds take 4–6 weeks to heal [6]. The prophylaxis of wound sepsis and fluid loss control is the essential feature of severe burn wounds [7]. Figure 8.2 shows the overview of the wound-healing process. Proper wound-healing treatment for the wound resulting from either a traumatic or surgical

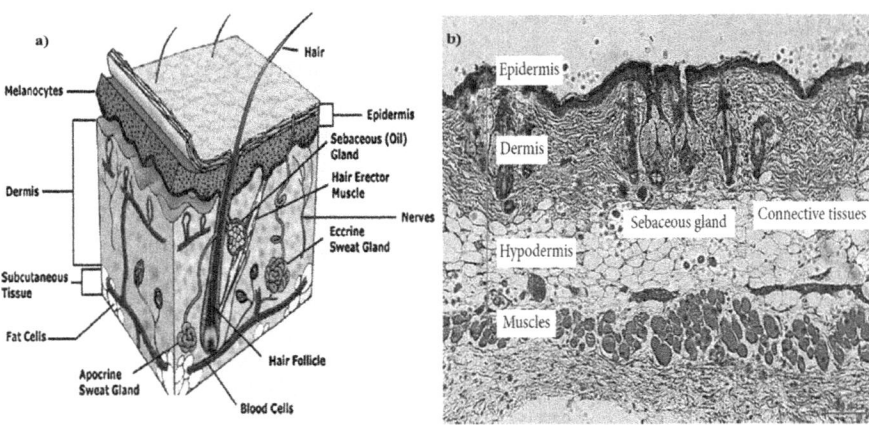

**Fig. 8.1** Schematic representation of **a** normal skin structure and **b** histological image of normal skin tissue stained with hematoxylin and eosin. Copyright © 2017 Production and hosting by Elsevier B.V., Journal of Advanced Research, 8, 2017, 217–233; Reproduced from Ref. [4] with permission. Copyright © 2018 Le Hang Dang et al. Creative Commons open access license (CC BY), Journal of Healthcare Engineering, Volume 2018, Article ID 5754890, https://doi.org/10.1155/2018/5754890. Reproduced from Ref. [5]

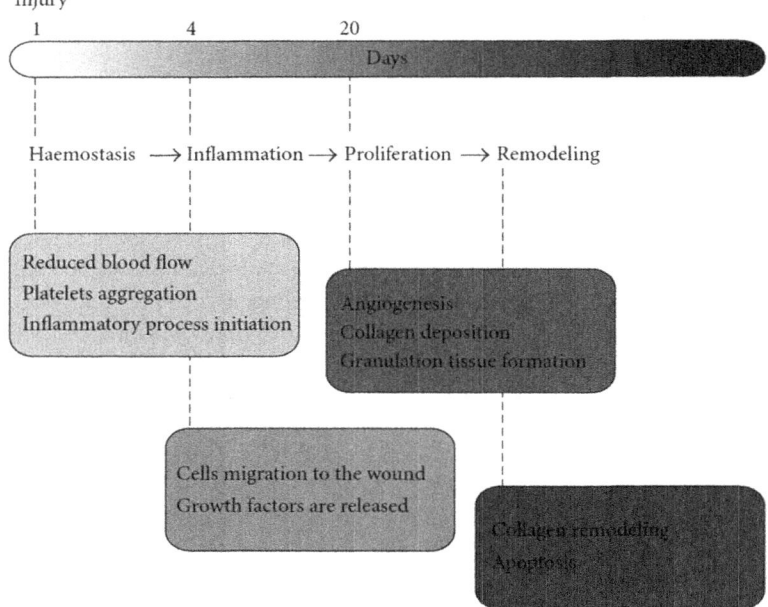

**Fig. 8.2** Overview of the wound-healing process [9]. Copyright © 2015 Francesco Piraino and Šeila Selimović. Creative Commons open access license (CC BY). BioMed Research International, Volume 2015, Article ID 403,801, 10 pages, http://dx.doi.org/10.1155/2015/403801. Reproduced from Ref. [9]

procedure is necessary to repair the wound after injury. The goals of wound healing are preventing infection, maintaining a moist environment, wound protection, and minimum scar formation [8].

## 8.2 Application of Biopolymer in Wound Healing

Wound healing is a complex biological process. The wound-healing process involves molecular and cellular responses to produce skin layers and appendages that are physiologically fit as native skin [10]. It is reported that biomaterials can assist in the proper physiological reconstruction of the skin [11]. Shahana and Rekha [12] reviewed the sources, mechanism of action, and properties of the biopolymers such as cellulose, alginate, hyaluronic acid, collagen, and chitosan, and their commercial applications for the current wound care market. Figure 8.3 depicts the role of a biopolymer in different phases of the wound-healing process.

Wound-healing steps usually pass through hemostasis (blood clotting), inflammation, proliferation, and remodeling (maturation) phases [4, 13]. A biopolymer provides an optimal healing environment for the wound by preventing further infection through bacterial invasion [7]. Wound dressing can protect the injured skin area from fluid and protein loss and maintain a moist environment next to the wound surface. Figure 8.3 shows that blood platelets are bound by fibrin in the hemostasis phase and adhered to the wound, which helps blood to clot [14]. Fibrin is a network of fibers that is formed from the plasma protein fibrinogen [15]. Pogoreilov and Sikora [16] reported that there are three possible ways to arrest or control bleeding: (1)

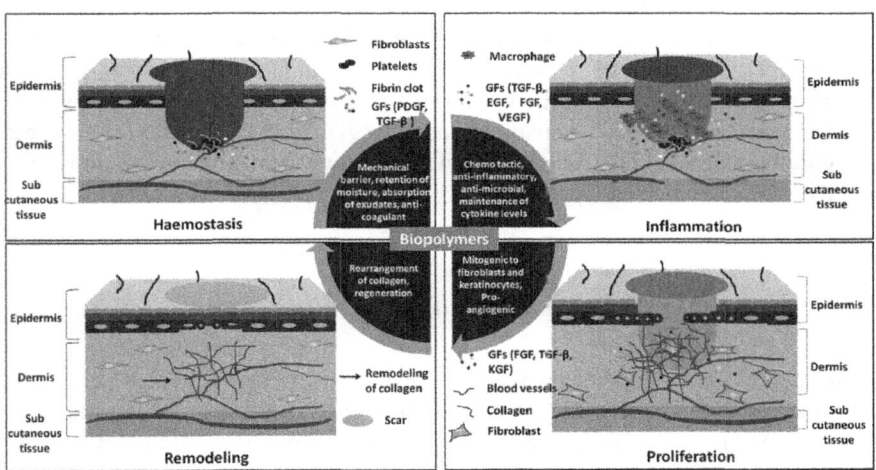

**Fig. 8.3** Different phases of the wound-healing process and biopolymers' role during wound healing Copyright © 2018, Springer Nature B.V. Molecular Biology Reports, 2018, https://doi.org/10.1007/s11033-018-4296-3. Reproduced from Ref. [12] with permission

sorption of plasma, (2) erythrocytes coagulation, and (3) platelet adhesion, aggregation, and activation. It is important to note that erythrocytes are the red blood cells that transport oxygen and carbon dioxide to and from the tissues. Platelets are found in large numbers in the blood that helps blood clotting. Several reports discuss the hemostatic dressing improvements to reduce bleeding from tactical combats [17–19].

In the inflammation phase, a biopolymer can play a role by maintaining an anti-inflammatory, antimicrobial environment. In this phase, bacteria and tissue debris from the damaged site are removed through the phagocytosis process. In the phagocytosis process, white blood cells clear the damaged and dead cells and bacteria, pathogens, and debris in the wound [14]. The proliferative phase is clinically known as the repair phase. In this phase, the wound surface protection is accomplished by forming new blood vessels (angiogenesis), granulation of tissue, matrix (collagen) deposition, and epithelialization. This way, the vascular network to nourish the new tissues is restored [14, 20]. The final stage of wound healing is the remodeling or maturation phase. This phase starts as soon as the wound fully closes. In the remodeling stage, collagen fibers are laid on the fibrin framework, increasing the wound's tensile strength, and maturation [6].

## 8.3 Application of Chitosan in Medical and Biomedical Fields

Medical applications of chitosan are evolving. The biological properties of chitosan make it an excellent candidate for various medical and biomedical research. These biological properties include antimicrobial, antioxidant, analgesic, biodegradability, biocompatibility, and bio-adhesiveness. Chitosan has been investigated extensively in some areas, including the treatment of wounds [21], surgical sutures [22], ophthalmology [23], orthopedics [24], drug delivery [25], and tissue engineering [26]. For instance, chitosan has been widely used in tissue engineering for its ability to be shaped into microspheres, nanoparticles, nanofibers, hydrogel, membranes, sponges, and porous scaffolds [27]. Table 8.1 shows different chitin and chitosan derivatives and their possible medical application.

### *8.3.1 Wound Dressing and Healing*

Chitosan is extensively used in wound healing and burns management. The potential use of chitosan has received very significant attention due to its hemostatic, anti-infection, and non-toxic properties [4, 7, 28]. For example, chitosan-based scaffolds are widely used in the cutaneous wound-healing process [29, 30]. Chitosan-based biomaterials simulate the migration of polymorphonuclear (PMN) and mononuclear

## 8.3 Application of Chitosan in Medical and Biomedical Fields

**Table 8.1** Typical applications of chitin and chitosan in pharmaceutical and biomedical areas

| Form | Method of preparation | Application |
|---|---|---|
| Bead | Precipitation/coacervation, phase inversion | Drug delivery |
| Microsphere | Emulsion crosslinking, coacervation, precipitation, spray drying, ionic gelation, sieving method | Enzyme immobilization, drug delivery vehicle |
| Nanoparticle | Emulsion droplets, coalescence, coacervation, precipitation | Encapsulation of sensitive drug |
| Fibers | Electrospinning | Medical textile, suture |
| Films | Solution casting | Semipermeable film for wound dressing and wound care |
| Powder | Spray drying | Adsorbents for pharmaceutical and medical devices, surgical glove powder, and enzyme immobilization |
| Sponge | Freeze drying | Mucosomal hemostatic dressing, wound dressing, drug delivery, enzyme entrapment, artificial skin |
| Gels | Crosslinking | Drug delivery vehicle, implants, coating, tissue engineering, wound dressing for wet treatment |
| Tablets | Coating mixture | Compressed diluent, disintegrating agent, excipient |
| Capsules | Encapsulate | The delivery vehicle |

Copyright © 2015 Younes, I and Rinaudo, M.; licensee MDPI, Creative Commons open access license CC BY. Mar. Drugs 2015, 13, 1133–1174; https://doi.org/10.3390/md13031133, Reproduced from Ref. [28]

cells. Chitosan-based biomaterials accelerate re-epithelialization and regeneration of normal skin [31].

Skin substitutes' primary role is to stimulate the host to produce various cytokines and promote granulation tissue formation during the wound healing [13]. Ramya et al. [32] summarize the essential information on chitosan's bioactivity properties and their various biomedical field applications. Whang et al. [20] reviewed the literature on new hemostatic agents. They described the use of chitosan in medical and surgical procedures. They reported that chitosan's hemostasis possibility and binding ability depend upon their chemical properties such as molecular mass, ionization, counter ion, degree of deacetylation (DD), and crystallinity. In addition, the protonated amino groups of chitosan can react with the negative part of the red blood cell membrane. This reaction promotes the hemostasis activity of chitosan [33]. Khan and Mujahid [34] reviewed recent advances in the chitosan-based composite for hemostatic dressings. They reported that the composite dressing demonstrates both antibacterial and hemostasis properties.

The first step in the early wound-healing process is hemostasis due to blood coagulation. The direct interaction with erythrocytes and platelets starts the hemostasis mechanism of chitosan. It does not depend on host coagulation pathways. In the case of composite dressing, this hemostasis mechanism makes it ideal for coagulopathic patients. Okamoto et al. [19] reported that reducing blood coagulation time (BCT) by chitosan is dose dependent and related to platelets and erythrocyte aggregation. They further reported that chitin and chitosan enhanced the release of the platelet-derived growth factor-AB (PDGF-AB). Chitin and chitosan also transformed growth factor-b1(TGF-b1) from the platelets. The specific mechanism for the action of chitosan as a hemostatic agent remains unresolved. Rao and Sharma [35] suggested that chitosan's hemostatic mechanism is the interaction between erythrocytes and chitosan cell membrane. It is independent of the classical coagulation cascade. Pogoreilov and Sikora [16] suggested that the hemostatic mechanism starts with plasma sorption by chitosan, causing the blood cell concentration in the injured place (wound). They further added that the interaction of chitosan with erythrocytes could be part of its hemostatic function. Mercy et al. [36] reported that the agglutination of erythrocytes occurs in the presence of chitosan. Erythrocytes can be repolymerized to form a lattice that captures cells creating an artificial clot. The agglutination potential of chitosan can be attributed to both its polymeric structure and molecular mass. Due to its polycationic properties, chitosan interacts with negatively charged residues of molecules at the cell surface, causing agglutination of cells [35, 37]. Fibroblast formation is the hallmark of the healing process. Chitosan increases the tensile strength of tissue by forming fibroblasts in the wound area [11]. Cheung et al. [18] reported that the polymorphonuclear leukocytes, macrophages, and fibroblasts enhance granulation and the organization of tissue repair. Chitosan promotes these activities and accelerates the wound-healing process [18]. Fibroblasts and keratinocytes are the primary cell components of the dermal and epidermal layers. A biopolymer can increase mitogenesis in fibroblast and keratinocyte. Chitosan contributes to the recovery of dermal and epidermal tissues [38]. Chitosan adheres to fibroblasts and favors keratinocytes' proliferation and, thereby, epidermal regeneration [17]. Once placed on the wound, chitosan degradation occurs under the influence of enzymes present in body fluids, such as lysozyme and *N*-acetyl-glucosoaminidase. For instance, lysozyme is an antimicrobial enzyme in the human body that is part of the innate immune system. Chitosan can be hydrolyzed slowly by lysozyme, which increases the epithelialization rate and collagen deposition [39]. *Macrophages* are known as immune system cells formed in response to an infection. Chitosan has a stimulatory effect on fibroblasts and activates macrophages [36]. This mechanism rebuilds normal physiological tissues by chitosan and thus accelerates the wound-healing process [31, 40]. In the absence of a graft, re-epithelialization is imperfect leading to increased scar formation. Porporatto et al. [41] suggest that chitosan's healing activity relies on the enhanced activity of enzymes, such as arginase, in an inflammatory milieu [41]. Chitin and chitosan form granulation tissue with angiogenesis [17]. Hilmi et al. [10] reported that chitosan accelerates full thickness wound healing in irradiated rats. The use of chitosan significantly decreases the size of the scar. Table 8.2 shows a list of hemostatic dressing products and their application in the wound-healing process.

8.3 Application of Chitosan in Medical and Biomedical Fields

**Table 8.2** Hemostatic dressing based on chitin and chitosan derivatives

| Product | Description | Mode of action | Regulatory approval |
|---|---|---|---|
| HemCon® Bandage | HemCon product is freeze-dried chitosan acetate salt. It is made of positively charged chitosan by the lyophilization process | Positively charged chitosan salt has a strong affinity to bind with red blood cells. This salt activates the platelets and forms a clot that stops massive bleeding. Therefore, it is mainly used to stop blood loss and enhance platelets | FDA approved this product in 2002 |
| ChitoFleX® Hemostatic Dressing | An antibacterial and biocompatible wound dressing | It is designed to reduce moderate to severe bleeding by adhering firmly to tissue surfaces, forming a flexible barrier that seals off and stabilizes the wound surface | ChitoFleX® Hemostatic dressing approved by the FDA in 2007 |
| Clo-Sur® Pad | Clo-Sur® is made as a non-woven pad. A soluble form of chitosan seals it | Used topically to stimulate wound healing at sites of vascular injury | Approved by the FDA and CE-certificated |
| ChitoSeal® | ChitoSeal® is made from soluble chitosan salt | Supported with a cellulose coating for hemorrhage wounds, reduces compressible timing. It is intended for temporary external use to control moderate to severe bleeding | Approved by the FDA and CE-certificated |
| Traumastat® | Traumastat® is freeze-dried chitosan containing highly porous silica. It is made on the non-woven substrate of porous polyethylene fibers filled with precipitated silica | It is proposed for temporary external use to control moderate to severe bleeding | Approved by FDA and CE-certificated |

(continued)

**Table 8.2** (continued)

| Product | Description | Mode of action | Regulatory approval |
|---|---|---|---|
| Syvek® Patch | It is poly-*N*-acetylglucosamine (pGlcNAc) isolated in a unique fiber crystalline structural form | Syvek® Patch controls bleeding at vascular access sites in interventional cardiology and radiology procedures. Achieves faster hemostasis by activating the platelets, agglutinating red blood cells, and controls bleeding following catheter removal in diagnostic operations | Approved by FDA and CE-certificated |
| ChitiPack® S | This wound dressing is made with freeze-dried chitin. The chitin used is prepared from a squid pen | The dressings are widely used for traumatic wounds and surgical tissue defects. No scar formation upon usage was observed, supported on polyethylene terephthalate, treats large skin defects, suitable for defects that are difficult to suture | |
| Chitipack P | This wound dressing was designed by drifting the chitin suspension onto poly-[ethylene terephthalate] non-woven fabric | Chitosan acetate salt was spun into a coagulating ethylene glycol, ice, and sodium hydroxide. This pack can be used to reconstruct the body tissues by rebuilding normal subcutaneous tissues. It can also be used for regular regeneration of the skin | |
| Chitipack C | The wound dressing was made by spinning chitosan acetate solution in a bath containing a mixture of ethylene glycol, cold water, sodium, or potassium hydroxide | It is used to regenerate and reconstruct body tissue, subcutaneous tissue, and skin | |

(continued)

**Table 8.2** (continued)

| Product | Description | Mode of action | Regulatory approval |
|---|---|---|---|
| Chitodine | Powdered chitosan containing elemental iodine | Chitodine is used as a primary wound dressing for the disinfection and cleaning of wounded skin and surgical dressings | Wound dressing has got CE-mark |
| QuikClot® | Adsorbent hemostatic agent | It speeds up the coagulation profile, stops blood loss, and is suitable for more extensive wounds | |
| ExcelArrest® | A hemostat is manufactured in foam by the lyophilization process from the chitin and polysaccharides suspension | ExcelArrest® is comprised of modified chitin particles and polysaccharide binders | Approved by FDA in 2007 |

Copyright © 2018, Springer Nature B.V. Advanced Drug Delivery Reviews, 56 (2004), 1467–1480. Reproduced from Ref. [17] with permission.; Copyright ©TESMA, Regeneration Research, 1(1), 2012, 38–46. Reproduced from Ref. [36] a with permission

Hydrogel has a very flexible porous structure of crosslinked polymers with hydrophilic nature. It can promote excellent healing environment for cellular metabolism for skin wounds by lowering oxygen tension [5]. Hydrogels can also absorb and retain the wound exudates, which promote fibroblast proliferation and keratinocyte migration necessary for complete epithelialization and healing of the wound [4]. Wu et al. [42] summarize recent advances in hydrogels and their applications in drug delivery and tissue engineering. Collagen is an excellent natural protein that consists of essential molecules such as lysine. Arginine can be used for guided tissue regeneration (GTR) applications [43]. The fibroblast cells are used as wound-healing indicators because they are the dominant cells in the wound-healing process. Agus et al. [21] reported that the chitosan-collagen membrane increased the number of fibroblasts and new blood vessels in the wound-healing process.

Various methods to prepare chitosan nanocomposites and their application in tissue engineering, drug delivery, gene delivery, wound healing, and bioimaging have been reviewed by Mudasir et al. [44] have reviewed bio-imaging. Khorasani et al. [45] studied hydrogels of heparinized polyvinyl alcohol (PVA)/chitosan (CS)/nanozinc oxide (nZnO)-based bionanocomposite as a wound dressing. In this bionanocomposite material, chitosan and ZnO are used as antibacterial agent, and heparin is used as an anticoagulant and anti-inflammatory agent. Heparin is also used to improve the bioactivity and biocompatibility of the wound dressing. They suggested PVA/CS/nZnO/heparin hydrogel-based bionanocomposites can be used as a robust wound dressing. It is reported that chitosan-based functional hydrogels allow absorption of water and, or bioactive compounds without dissolution, thereby allowing drug release by diffusion [46]. Parsa et al. [47] used a freezing–thawing cycle to prepare nanocomposite hydrogel from polyvinyl alcohol (PVA), 5% chitosan nanoparticle (NC), and tetracycline. This nanocomposite hydrogel showed good antibacterial activity against gram-negative and gram-positive bacteria. They further suggested that this nanocomposite hydrogel is useful for wound dressing and drug delivery systems, such as a colon-specific drug.

Omidi et al. [48] reported the application of pH-sensitive carbon dots/chitosan smart hydrogel nanocomposite material for wound dressing. They have discussed the effect of pH on both acute and chronic wound-healing processes in detail. Figure 8.4 shows chronic wounds have less oscillation (in the pH range 7.0–8.0), suggesting a lengthy healing process than the acidic pH range ~5.0–6.0 for an acute wound. Figure 8.4 shows the pH oscillation in the wound-healing process. It is reported that the fluctuation of pH depends upon the colonization of bacteria, type of bacteria, toxicity, type of tissue, angiogenesis, protease activity, and oxygen release [48]. This nanocomposite smart material is reported to be a good candidate for monitoring pH during the wound-healing process.

Hyaluronic acid (HA) is found in the ECM of many tissues, such as skin [49]. The $N$-acetyl-glucose-amine of chitosan, which is similar to that of hyaluronic acid, increases its potential as a wound-healing agent [17] and is reported to contribute to regenerative tissue and fewer scars [10]. Xu et al. [50] investigated chitosan/hyaluronic acid film wound dressing. It was reported that chitosan/hyaluronic acid film showed a faster healing rate of the wound at the initial

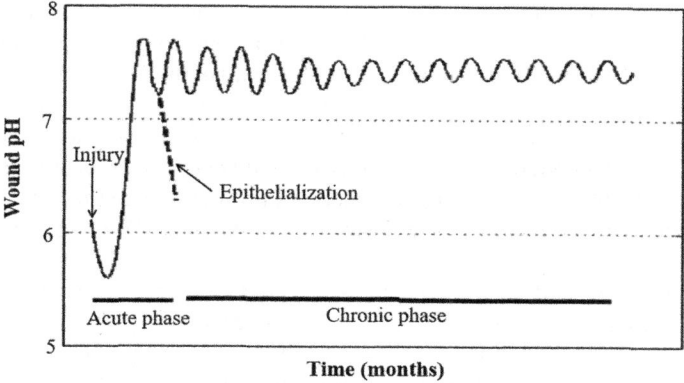

**Fig. 8.4** Typical fluctuation of pH for both acute and chronic wound-healing processes. Copyright © The Royal Society of Chemistry 2017, Creative Commons open access license CC BY, *RSC Adv.*, 2017, 7, 10638–10649. https://doi.org/10.1039/c6ra25340g, Reproduced from Ref. [48]

stage, and the film could be easily peeled off with less re-injury to the wound bed. In another attempt, Iacob et al. [51] studied chitosan and chitosan/hyaluronic acid membranes containing novel arginine derivatives with thiazolidine-4-one scaffold for potential wound dressing materials. They observed that the chitosan-arginine derivatives (CS-6h, CS-6i) and chitosan/hyaluronic acid-arginine derivative (CS-HA-6h) membranes had excellent healing effects on the burn wound in rats. A complete re-epithelialization was observed for these membranes after 15 days of the experiment. It is reported that $N$-carboxymethyl chitosan is superior to hyaluronic acid in terms of its hydrating effects and can reduce fluid and heat loss from the wound sites [13, 36]. Chen et al. [52] incorporated $N,O$-(carboxymethyl) chitosan into the backbone of a collagen-chondroitin sulfate or a collagen-acellular dermal matrix. It was reported that both combinations markedly enhanced wound healing [52] by stimulating the recruitment of fibroblasts. The chitosan graft-acrylic acid-graft-2-hydroxyethyl methacrylate kept adequate swelling capacity without compromising its physical stability, suggesting it to be the best matrix for drug delivery systems [53]. Hydro-active membrane such as polyvinyl alcohol/chitosan lactate blended with hydrogels containing nitrofurazone (a local anti-infective drug) is reported as a potential wound dressing material [54]. Silver sulfadiazine or thiazolidinone drug impregnated chitosan-alginate sponges effectively suppressed bacterial proliferation [55].

### 8.3.2 Scaffolds for Bone Tissue Engineering

Bone is considered a complex organic–inorganic-based strong and flexible dynamic tissue in the human body. It consists of approximately 60% minerals such as calcium phosphate with other materials, including collagen, chondroitin sulfate, keratin sulfate, lipids, and water [56, 57]. In the bone matrix, the inorganic materials provide rigidity to the bones, whereas collagen, the organic part of the bone, plays a significant role in improving fracture resistance and aid in cell growth, proliferation, and differentiation [58]. In response to any mechanical and metabolic process, the primary cells that dictate remodeling and bone formation are osteoblasts, osteoclasts, osteocytes, and cells [59]. For example, the function of osteoblasts in the bone formation is to regulate mineralization and synthesize the unmineralized organic portion (bone matrix protein) [60, 60a].

Chitosan-based scaffolds have been investigated to repair or regenerate an organ in bone tissue engineering [27, 61]. Chitosan is an excellent candidate for scaffolds because of its inherent properties [27, 61]. These inherent properties of chitosan include antibacterial nature, biocompatibility, biodegradability, cellular interaction, and similarities with the ECM. It is reported that chitosan supports the attachment and proliferation of osteoblast cells and the formation of a mineralized bone matrix [26]. Figure 8.5 shows the SEM image of chitosan-based scaffolds. Bone is a vascularized connective tissue [60a]. The interactions between the bone cells and blood vessels impact bone regeneration. Costa-Pinto et al. [27] provided a survey report on in vitro studies with chitosan-based scaffolds for bone tissue engineering. They stated that the main property of the scaffold directly related to vascularization is its porosity. It is reported that an ideal scaffolding should have a porosity of 80 to 90% with a pore size range of 50–250 μm [61]. The interconnected porous scaffolds provide physical support to cells and guide their proliferation and differentiation, facilitating neovascularization. The most common process of producing a porous chitosan scaffold is the lyophilization of a chitosan solution using the freeze-drying method [62]. However, the chitosan-based scaffolds' strength and structural stability are needed to be improved for bone tissue engineering applications [26].

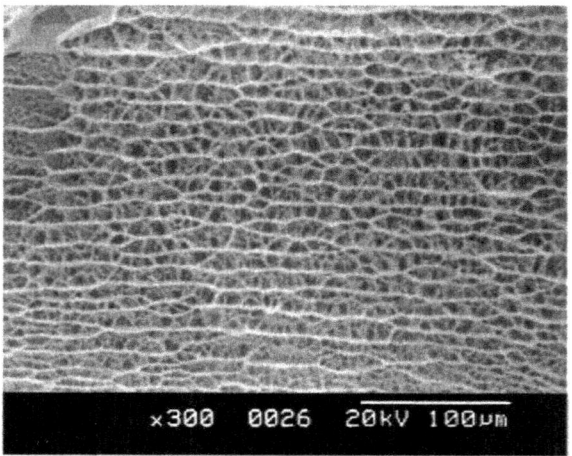

**Fig. 8.5** Scanning electron microscopic (SEM) image of a typical chitosan scaffold. Copyright © 2001 Wiley Periodicals, Inc. J. Biomed Mater. Res. 2002, 59:438–449. Reproduced from Ref. [31] with permission

Oryan and Sahvieh [29] reviewed the effectiveness of chitosan and chitosan-based scaffold in skin, bone, and cartilage healing. They reported that chitosan activates platelets in the inflammatory phase. Chitosan then enhances granulated tissue formation and angiogenesis by proliferating mesenchymal (stem cell) and endothelial cells. Several studies investigated composite chitosan-based scaffolds. Nanoscale ceramic or biocompatible polymers were incorporated to improve the mechanical and biological properties of bone tissue [56, 58, 63].

Dhivya et al. [64] reported that hydroxyapatite (HAp) is a mineral component of the natural bone. Hydroxyapatite (HAp) has osteoconductive and osteoinductive properties and bone-bonding abilities. Hydroxyapatite (HAp) exhibits slow degradation in situ. McCullough et al. [65] prepared a chitosan-based composite scaffold using chitosan, carboxymethyl chitosan, and hydroxyapatite. It is reported that the composite scaffolds show great promise in providing mechanical support for regenerating bone. In another study, Dhivya et al. [64] studied nanohydroxyapatite-reinforced chitosan composite hydrogel for bone tissue repair in vitro and in vivo. They reported zinc-doped chitosan/nanohydroxyapatite/β-glycerophosphate (Zn-CS/nHAp/β-GP) thermosensitive hydrogel accelerates potential clinical application toward bone regeneration. Sharma et al. [58] prepared chitosan-gelatin-alginate-hydroxyapatite nanobiocomposite scaffolds for bone tissue engineering. They reported that this composite scaffold is a suitable scaffold for osteoblast attachment and proliferation, and hence, it can be potentially applied for bone regeneration. Verisqa et al. [66] reported that hydroxyapatite (HAp) formation is essential for bone regeneration. Chitosan–hydroxyapatite collagen composite scaffold shows potential as a maxillofacial reconstruction material since its composition favors HAp formation. Forohberg et al. [67] reported a one-step platform to electrospun nanofibrous scaffolds from chitosan containing hydroxyapatite nanoparticles crosslinked with genipin. They also discussed the scaffold application for non-weight-bearing bone tissue engineering. This scaffold can be used for cranial and maxillofacial reconstruction. Table 8.3 shows chitosan-related applications in tissue engineering.

Croisier and Jérôme [37] reported the preparation and properties of chitosan-based biomaterials and their tissue engineering applications focused on wound healing. They also reported the advantages and disadvantages of chitosan-based biomaterials and their biomedical applications, as shown in Table 8.4.

### 8.3.3 Antimicrobial Application of Chitosan

The equilibrium of traumatized tissue and bacteria is greatly in favor of bacterial growth [68]. Therefore, the suppression of bacterial proliferation in an infected wound is considered a crucial step. Chitosan has been shown to inhibit bacterial proliferation in the treatment of infected wounds [17]. Goy et al. [69] studied the antimicrobial activity of chitosan and its quaternized derivative on the growth of *Escherichia coli* (gram-negative) and *Staphylococcus aureus* (gram-positive). Figure 8.6a shows that the growth of both bacteria increased exponentially during the

**Table 8.3** Chitosan-based scaffold for tissue engineering

| Chitosan combination | Scaffold | Experimental model | Tissue application |
|---|---|---|---|
| Chitosan | Membrane | Embryonal submandibular gland cells | Salivary gland |
| Chitosan | Viscous solution and a monolayer rigid physical hydrogel | Female minipigs third-degree burns | Skin |
| Chitosan + hyaluronan | Hybrid polymer fiber | | Ligament |
| Collagen-chitosan + fibrin glue | Asymmetric porous scaffold | Human dermal fibroblasts and keratinocytes | Skin |
| Chitosan + alginate | Polyelectrolyte multilayer film | C2C12 myoblasts | Muscle |
| Chitosan + aloe vera | Blended membrane | Bovine articular chondrocytes and mesenchymal stem cells | Skin |
| Chitosan + layer of chitosan/gelatin | Sandwich tubular Scaffold | Vascular smooth muscle cells from rabbit aorta | Blood vessel |
| Genipin-crosslinked chitosan, chitosan-nanohydroxyapatite | Framework | Human periodontal ligament tissue, periodontal ligament stem cells | Bone |
| Chitosan + collagen | Hydrogel | Epididymal fat pads cells and subcutaneous pocket of male rat | Adipose tissue |
| Chitosan + polyester | Compressed porous disk | Bovine articular chondrocytes | Cartilage |
| Chitosan + collagen + genipin | Crosslinked porous membrane | Rabbit articular chondrocytes | Cartilage |
| Chitosan + chondroitin sulfate | Bidimensional glass surfaces or 3D packet of paraffin | Bovine articular chondrocytes and human mesenchymal stem cells culture | Cartilage |
| Chitosan + adipose-derived stem cells | Tube nerve conduit | Male, Sprague Dawley rats sciatic nerve transection | Nerve |
| Chitosan + silk fibroin | Thin blended film | Female guinea pig's ventral hernia | Muscle |
| Chitosan + β-sodium glycerophosphate + hydroxyethyl cellulose | Hydrogel | Male and female sheep articular defect | Cartilage |

(continued)

Table 8.3 (continued)

| Chitosan combination | Scaffold | Experimental model | Tissue application |
|---|---|---|---|
| Chitosan + calcium phosphate cement | Chitosan microspheres inside cement paste | Male rabbit femoral defect | Bone |

Copyright © 2015 Martin Rodríguez-Vázquez et al. Creative Commons open access license CC BY, BioMed Research International, Volume 2015, Article ID 821,279, 15 pages. http://dx.doi.org /. Reproduced from Ref. [61]

Table 8.4 Main advantages, disadvantages, and applications of the chitosan biomaterials

| Type | Specification | Advantages | Disadvantages | Applications |
|---|---|---|---|---|
| Hydrogels (3D) | Physically associated (reversible) | Soft flexible non-toxic | Not stable (uncontrolled dissolution may occur) | Tissue replacements/engineering drug/growth factor delivery |
| | Chemically crosslinked (irreversible) | Soft flexible stable controlled pore size | Low mechanical resistance Pore size challenging to control May be toxic | |
| Sponges (3D) | | High porosity | May shrivel | Tissue engineering (filling material) |
| Films (2D) | Freestanding | Soft | Low porosity | Wound dressings skin substitutes |
| | Thin (LB) | Material coating | Laborious for the construction of multilayers | Coatings for a variety of scaffolds wound dressings skin substitutes |
| | Thin (LBL) | Material coating Multilayer construction | Many steps | |
| Porous membranes (2D) | Nanofibers | High porosity mimic skin ECM | ESP of pure chitosan difficult | Coatings for a variety of scaffolds wound dressings skin substitutes |

Copyright © 2013 Elsevier Ltd. European Polymer Journal, 49, 2013, 780–792. Reproduced from Ref. [37] with permission

first 6 to 8 h. It is also reported that the antibacterial effectiveness of chitosan on the reduction of microorganisms is strongly dependent on the concentration of chitosan in the medium (Fig. 8.6b) [68, 69]. A similar observation was reported by Ardila et al. [70] that the antibacterial properties of chitosan depend on the sensitivity of bacterial species. The sensitivity of the *Escherichia coli* was highest, followed by *Listeria innocua* and then *Staphylococcus aureus*. Several mechanisms for the antimicrobial activity of chitosan have been reported [37]. However, the exact mechanism is not fully established [68]. It is reported that the interaction of positively charged

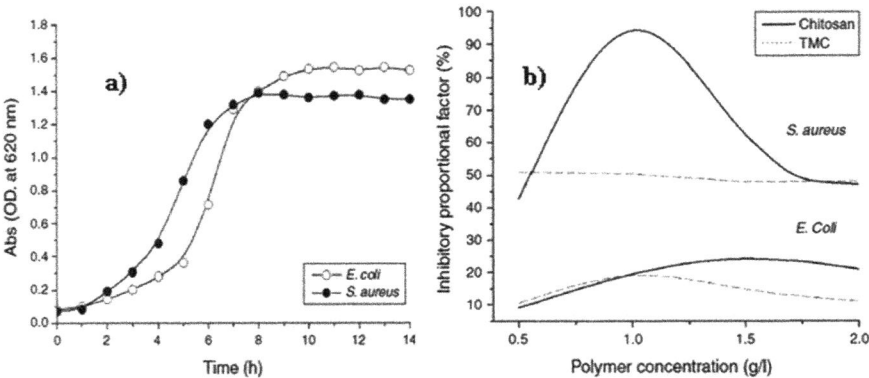

**Fig. 8.6** **a** Growth kinetic of microorganism *E. coli* (gram-negative) and *S. aureus* (gram-positive) and **b** inhibitory growth factor in the presence of chitosan and trimethyl chitosan. Copyright © 2015 Sociedade Brasileira de Farmacognosia. Published by Elsevier, Revista Brasileira de Farmacognosia, 26, 2016, 122–127. Reproduced from Ref. [69] with permission

chitosan ($NH_3^+$) and the negatively charged microbial cell membranes is responsible for cellular lysis (Fig. 8.7). The effectiveness of the interaction of chitosan with microorganisms depends on the species of the target organism [71, 72].

Kim [74] reported that the bioactive properties of chitosan are strongly dependent on the DD and molecular weight (MW) of chitosan. The bioactive properties of chitosan, such as antimicrobial, antioxidant, anticancer, and anti-inflammatory activities, are also discussed. The antimicrobial activity of chitosan is dose dependent and influenced by its MW [75, 76]. For example, the antibacterial activity of low MW is

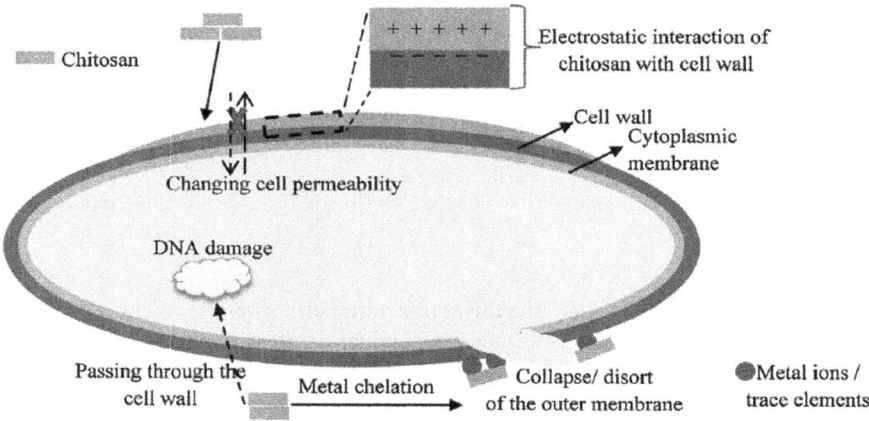

**Fig. 8.7** Schematic representation of the antimicrobial mechanism of chitosan and its derivatives [73]. Copyright © 2016 Elsevier B.V. International Journal of Biological Macromolecules, 2016, 85, 467–475., Reproduced from Ref. [73] with permission

higher than the chitosan sample with high MW [75]. It is reported that the antimicrobial activities of chitosan are much dependent on its physical characteristics, most notably MW and DD as well as particular bacterium [77, 78]. Chang et al. [79] reported how the antimicrobial activities of chitosan are affected by MW, temperature, and pH. They observed that the chitosan activity increased with increasing MW in acidic pH conditions, irrespective of the temperature and bacteria tested. However, at neutral pH, chitosan activity increased as the MW decreased. Tavaria et al. [80] assessed the antimicrobial activity of chitosan upon skin microorganisms (*S. aureus*, *S. epidermidis*, and *E. coli*) in vitro when subjected to a combination of different abiotic factors. These abiotic factors include pH, ionic strength, organic acids, and free fatty acids. It was suggested that free fatty acids, ionic strength, and pH significantly affected chitosan's capability of reducing the viable numbers of *S. aureus*. A higher ionic strength (0.4% NaCl) favored chitosan's action upon reducing viable numbers of *S. epidermidis* and *E. coli*. Benhabiles et al. [77] reported that the minimal inhibitory concentrations (MIC) of chitin and chitosan depended on the bacterium studied. The MIC values of chitin generally ranged from 0.006 to 0.01%, except *P. aeruginosa*, *S. typhimurium*, and *P. melaninogenica*, which required more than 0.1%. In contrast, chitosan MIC values ranged from 0.006% to 0.03%, except for *P. aeruginosa* (0.05%) and the foodborne pathogen *S. typhimurium* (more than 0.1%). Table 8.5 shows the minimum inhibitory concentration (MIC) value of chitosan for microorganisms.

### 8.3.4 Anticancer and Antitumor Activity of Chitosan and Its Derivatives

Chitosan is also considered a biologically safe polymer for the controlled delivery of therapeutic agents [82], gene delivery [83], and influenza vaccine delivery [84]. Adhikari and Yadav [85] reviewed the anticancer activity of chitosan and its derivatives. They discussed the effect of permeation, antiangiogenic, immunoenhancement, cellular apoptotic, and sustained release mechanisms on anticancer activities. They suggested that chitosan-based hybrid materials can be used for anticancer therapeutic applications. The mucoadhesive properties of chitosan make it capable of increasing the residence time of dosage forms at mucosal sites [86]. In addition, it has a permeation enhancing effect for the paracellular route of absorption, which is essential for transporting hydrophilic compounds such as (poly)peptides across the mucosal membrane.

A genetic mutation has been recognized as one of the critical steps in cancer development [87]. This mutation results from the reactive oxygen species (ROS) related to oxidative stress and other free radicals. Several studies focused on the antitumor activity of chitosan and its cytotoxicity and antiproliferative properties [88–91]. Li et al. [91] investigated immune-modulatory or anticancer properties of chitosan and concluded that chitosan enhanced the antitumor activity of natural killer cells (first

**Table 8.5** Typical example of a MIC value of chitosan against several microorganisms (concentration normalized to ppm)

| Types of microorganism | Microorganism | MIC (ppm) |
|---|---|---|
| Gram-negative | Escherichia coli | 1000 |
| | Xanthomonas campestris | 500 |
| | Salmonella enterica | 2000 |
| | Salmonella tiphymurium | 1000, 1500, 2000 |
| | Pseudomonas aeruginosa | >200, 1700 |
| | Aeromonas hydrophila | 1000 |
| | Shigella dysenteriae | >200 |
| | Vibrio cholera | 200 |
| | Vibrio parahaemolyticus | 1000 |
| | Pseudomonas fluorescens | ~1000 |
| | Enterobacter aerogenes | 250 |
| Gram-positive | Bacillus cereus | 1000 |
| | Bacillus megaterium | 800 |
| | Staphylococcus aureus | 20, 100, >800, 700, >1250 |
| | Listeria monocytogenes | 150, 250, 800 |
| | Candida lambica | 250 |
| | Lactobacillus plantarum | <1000, 2000 |
| | Lactobacillus brevis | 1000 |
| | Lactobacillus bulgaricus | >1000 |
| Fungi | Aspergillus fumigatus | >2000 |
| | Aspergillus parasiticus | >2000 |
| | Fusarium oxysporum | 100 |
| | Botrytis cinerea | 10 |
| | Byssochlamys spp. | 1000–5000 |
| | Candida albicans | 500, 600, >1250 |

(continued)

## 8.3 Application of Chitosan in Medical and Biomedical Fields

**Table 8.5** (continued)

| Types of microorganism | Microorganism | MIC (ppm) |
|---|---|---|
| | *Drechstera sorokiana* | 10 |
| | *Microsporum canis* | 1100 |
| | *Trichophyton mentagrophytes* | 2200 |

Goy, R. C. et al. A Review of the antimicrobial activity of chitosan, Polimeros, 19(3), 2009, 241–247. https://doi.org/10.1590/S0104-14282009000300013. Creative Commons open access CC BY license. Reproduced from Ref. [72]

line of defense against viral infections and malignant cells) by activating dendritic cells (innate immune cells). Chang et al. [87] studied the effect of chitosan's MW on its antioxidant and antimutagenicity properties. They reported that the effect of MW of chitosan on antimutagenicity was similar to that of antioxidant activity. It was reported that the antioxidant activities of chitosan were inversely proportional to its MW. The highest antioxidant activities were obtained with chitosan with 2.2-kDa. Chitosan samples also exhibited strong antimutagenic effects against direct (4-nitroquinoline 1-oxide) and indirect (benzo[α]pyrene) mutagens. The antimutagenic activity of chitosans against both direct and indirect mutagens increased as the MW of the chitosans decreased. Chitosan showed high antimutagens activity (>50.0%) at low MW (<30.0 kDa). It was suggested that chitosan might interact with the free radicals produced by the mutagens or inactivate the mutagens directly before the mutagens react with DNA [87]. Another study [92] reported that a low molecular mass chitosan-coated liposome for oral delivery of calcitonin showed better performance over high molecular mass chitosan. The immobilization of porcine pancreatic lipase, cardosin A on chitin, and chitosan demonstrated chitosan as a right vehicle for applying the enzyme [93, 94]. Tan et al. [89] studied the antitumor activity of chitosan from mayflies. They observed that the chitosan, with low molecular mass, had a cytotoxic effect at a 500 μg/mL concentration on cancer cells. They also suggested that mayflies' dead bodies are useful in producing low molecular mass chitosan with antiproliferative activity. Water-soluble chitosan with high molecular mass is useful in preventing and treating sterility [95]. The use of chitin/Se nanocomposites in improving immunity, delaying senescence, and preventing and treating neoplasm and cardiovascular disease is the subject of a paper by Zhang et al. [96]. Gao et al. [97] discussed chitin's bioactivity and pharmacological effects. They also discussed the utilization of chitosan in drug film [97]. Aranaz et al. [33] reported that chitosan could activate macrophages to mediate the antitumor effects in vivo. They further added that chitosan could inhibit growth of tumor cells, mainly due to an immune stimulation effect.

In the pharmaceutical area, particle size is a critical factor in determining appropriate drug administration routes. The powdered chitosan has been reported as an efficient vehicle for pulmonary gene delivery. The oral LD50 for mice, 16 g/Kg, indicates a low-toxicity potential for chitosan [98]. The powdered form of chitosan is reported to be the most effective formulation for nasal delivery of insulin in the sheep model [99]. The chitosan-coated submicron-sized liposomes appeared to be promising in the oral administration of peptide drugs for their retentive property in the intestinal tract [100].

The success of gene therapy for the permanent cure of cancer depends on the efficient delivery of nucleic acid into the target cells [101]. Li et al. [102] reviewed different nanomaterials generated from chitosan and its derivatives for controlled drug delivery. Chitosan nanoparticles serve as a promising gene delivery carrier for targeted gene therapy for cancer [103–105]. For example, Tahamtan et al. [106] studied nanochitosan (NCS) as a carrier system for intramuscular administration using a recombinant DNA vaccine expressing HPV-16 E7 (NCS-DNA E7 vaccine). They suggested that chitosan nanoparticles can be considered an efficient carrier to improve the immunogenicity of DNA vaccination. Zhang et al. [103] reported hyaluronic acid-modified chitosan nanoparticles loaded with cyanine 3 (Cy3)-labeled siRNA (nucleotides for gene therapy) as a promising gene delivery carrier for targeted gene therapy for cancer. Copper-loaded chitosan nanoparticles can be used effectively to treat bone cancer osteosarcoma [107]. Zhong et al. [90] describe the preparation of a polymeric micelle using doxorubicin (DOX) conjugated trimethyl chitosan (TMC) with Beclin-1 siRNA (Si-Beclin-1/DOX-TMC). In preclinical studies, they indicated that the nanocarrier could effectively suppress drug-resistant bladder cancer. However, the specific mechanism of chitosan nanoparticles and their effects on tumor cells is still unclear. Jiang et al. [105] reported that the ROS-mediated cell apoptosis regulates cell death. They demonstrated that the induction of intracellular ROS through the internalization of chitosan nanoparticles in cells directly resulted in cell apoptosis by ROS-mediated mitochondrial damage and endoplasmic reticulum (ER) stress. A summary of chitosan and its derivatives' antitumor activity is given in Table 8.6.

## 8.4 Chitosan-Based Medical Isotope Separation

In the radiopharmaceutical area, $^{99m}$Tc ($t_{1/2} = 6$ h) is considered the most widely used radioisotope in diagnostic medicine. $^{99m}$Tc is the decay product of parent $^{99}$Mo ($t_{1/2} = 66$ h). $^{99m}$Tc is a pure gamma emitter (0.143 meV). $^{99m}$Tc is ideal for medical applications due to its short half-life (6 h). It is used in 80–85% of about 25 million diagnostic nuclear medicine procedures performed each year. Currently, most of the world's $^{99}$Mo supply comes from the thermal fission of highly enriched uranium (HEU). However, this process generates large quantities of radioactive waste and does not permit reprocessing unused uranium targets due to proliferation concerns. The use of low enriched uranium (LEU, 20% $^{235}$U or less) as a substitute would yield large volumes of waste due to the presence of un-useable $^{238}$U. Production of $^{99}$Mo via the neutron capture method is an alternative to fission-derived $^{99}$Mo. The main concern with neutron capture-produced $^{99}$Mo (n, $\gamma$) over the common

## 8.4 Chitosan-Based Medical Isotope Separation

**Table 8.6** Summary of the antitumor activity of chitosan and its derivatives

| Compound | In vivo model or target cell lines | Results |
|---|---|---|
| Chitosan | Meth-A solid tumor transplanted into BALB/c mice | Increased production of interleukins 1 and 2, leading to the antitumor effect. Cytolytic T-lymphocytes proliferated with the optimum inhibition ratio at the dose of 10 mg/kg |
| Chitosan | Aberrant tumor lesions in the colon of mice | Increased lymphokine production. Cytolytic T-lymphocytes proliferated at the dose of 5 mg/kg |
| Chitosan | Cultures with A375, SKMEL28, and RPMI7951 cell lines | Chitosan was coated in culture wells with cultures of A375, SKMEL28, and RPMI7951 Decreased adhesion of A375 cells Decreased proliferation of SKMEL28 cells, inhibited specific caspases, upregulated Bax, and downregulated Bcl-2 and Bcl-XL in RPMI7951 cells Induced CD95 receptor expression in the RPMI7951 cell surface renders them more susceptible to FasL-induced apoptosis |
| Carboxymethyl chitosan | Hydrogen peroxide-induced apoptosis models of Schwann cells | The cell viability was improved in a dose-dependent manner with a maximum effect of 2.02 ± 0.16-fold at the dose of 200 μg/mL carboxymethyl chitosan Decreased caspase-3, -9, and Bax activities and increased Bcl-2 activity |
| Carboxymethyl chitosan | BEL-7402 cell line Hepatoma-22 cells in Kunming mice | Reduced the expression of MMP-9 in a dose-dependent manner Inhibited the lung metastasis in a mouse model with the highest inhibition of 66.56% at a dose of 300 mg/kg |
| Chitosan | PC3 A549 and HepG2 cell line | Suppressed cancer cell growth of PC3 A549 and HepG2 cells for 50% cell death at 25 μg/mL, 25 μg/mL, and 50 μg/mL, respectively |

(continued)

**Table 8.6** (continued)

| Compound | In vivo model or target cell lines | Results |
|---|---|---|
| Chitosan | HepG2 and LCC cell line HepG2 and LCC xenografts in a mouse model | Inhibited MMP-9 expression, reduced cells in S-phase and decreased the rate of DNA synthesis, upregulated p21 and downregulated PCNA, cyclin A and CDK-2 with the highest inhibition at the dose of 1 mg/kg Inhibited tumor growth and decreased the number of metastatic colonies at a dose of 500 mg/kg |

Copyright © 2015 Cheung, R.C.F., Ng, T.B., Wong, J.H., et al. licensee MDPI, Basel, Switzerland. Creative Commons open access CC BY license. Mar. Drugs, 13, 2015, 5156–5186, https://doi.org/10.3390/md13085156. Reproduced from Ref. [18]

fission-produced involves lower Curie yield and lower specific activity. However, the limitation can be overcome by using adsorbent with a higher capacity for molybdenum. The success of adsorption processes in the $^{99}$Mo/$^{99m}$Tc generators depends mainly on the cost and capacity of the adsorbents and the ease of $^{99m}$Tc release from the generator. The main problem with the neutron capture method from a radiation safety standpoint involves the breakthrough or partial elution of the $^{99}$Mo parent and the $^{99m}$Tc from the generator, which must be kept within Nuclear Regulatory Commissions (NRC) standards.

### *8.4.1 Separation of $^{99m}$Tc from $^{99}$Mo Using Chitosan-Based Adsorbent*

Typical adsorption of molybdenum onto sorbent depends on several factors [108]:

- The pH of the solution defines the dominant species present in the solution.
- The concentration of Mo in the solution.
- The surface charge of sorbent.
- The pore structure of the adsorbent materials.

Currently, alumina is widely used as an adsorbent in fission molybdenum-based $^{99}$Mo/$^{99m}$Tc generator. Molybdate in acid forms relatively stable heteropoly complexes with many cations, including Al$^{3+}$ [109]. The adsorption capacity of alumina for molybdenum is reported to be in the range of 2 to 26 mg/g of alumina [110]. A US patent is reported by Hasan [111] described chitosan-based alternative adsorbent of alumina in neutron activation (n, γ) $^{99}$Mo-based $^{99m}$Tc/$^{99}$Mo generator. This work describes preparing chitosan-based microporous composite material

## 8.4 Chitosan-Based Medical Isotope Separation

(MPCM) resin. It is reported that the MPCM resin, porous in nature, is found to be resistant to radiation exposure (50 MRad), extreme pH conditions, potent oxidizing agents, and temperatures exceeding 100 °C without substantial physical degradation of the resin [111].

The potential for this MPCM resin as an adsorbent for the preparation of a $^{99}$Mo/$^{99m}$Tc generator has been evaluated by exposing it to a 1% neutron-captured produced molybdenum solution with an activity of 10 mCi/mL. It was reported that the chitosan-based composite (MPCM) resin absorbs >60 wt% molybdenum at solution pH 3.0 [111]. Molybdenum-99, in the form of molybdate, $MoO_4^{2-}$, is absorbed onto the resin so that when it decays, the resulting pertechnetate, $TcO_4^-$, is less tightly bound to the resin surface and can be removed via saline flush [112]. Figure 8.8 shows the possible reaction mechanism of molybdenum uptake onto MPCM resin.

The adsorption mechanism of $^{99}$Mo onto resin and simultaneous release of $^{99m}$Tc from the column prepared from resin have been described based on surface charge analysis of the resin. Figure 8.9 shows the surface charge pattern for Mo (VI) loaded MPCM sample with or without oxidization. In the case of a non-oxidized MPCM sample loaded with Mo (VI), the protonation of the surface appeared to be increased

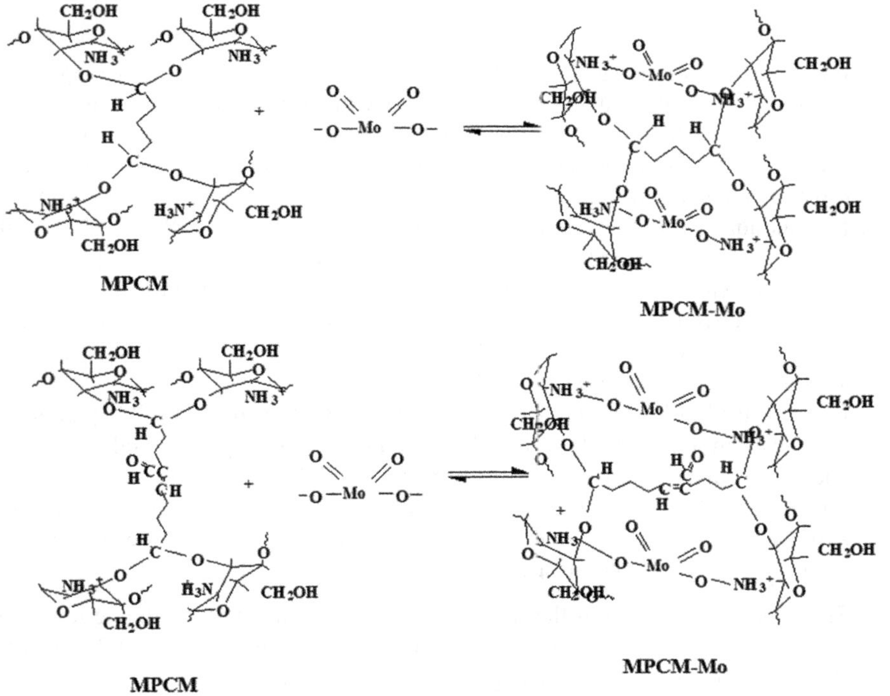

**Fig. 8.8** Reaction mechanisms for the adsorption of Mo (VI) on composite resin from aqueous solution. Copyright © 2020, Springer Nature Switzerland AG, SN Appl. Sci. **2**, 1782 (2020). https://doi.org/10.1007/s42452-020-03524-1. Reproduced from Ref. [108] with permission

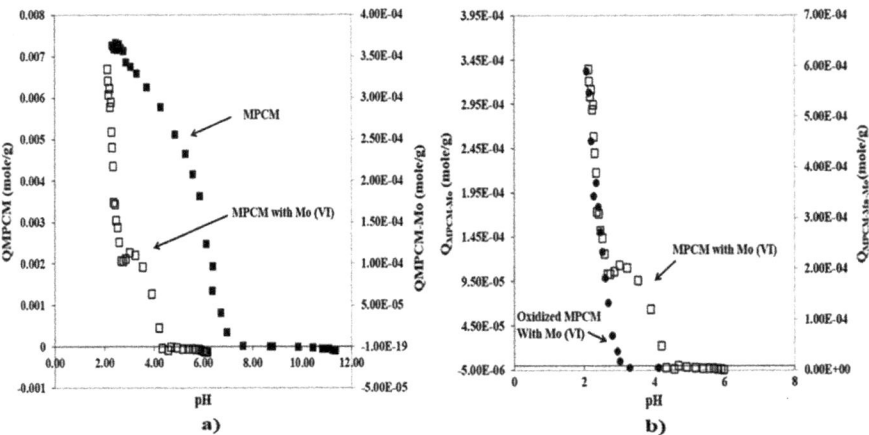

Fig. 8.9 Surface charge of a MPCM and MPCM exposed to 1% of Mo (VI) in solution in the presence of 1M NaNO$_3$, respectively, and b oxidized and non-oxidized MPCM exposed 1% of Mo (VI) in solution in the presence of 1N NaNO$_3$, respectively. Copyright © 2020, Springer Nature Switzerland AG, SN Appl. Sci. **2**, 1782 (2020). https://doi.org/10.1007/s42452-020-035 24-1. Reproduced from Ref. [108] with permission

gradually at the pH range of 4.5–3 (Fig. 8.9a). Therefore, at this pH range, the formation of covalent bonding by pertechnetate with the positive surface sites of Mo-loaded MPCM surface is possible. However the surface charge of Mo (VI) loaded, oxidized MPCM shows almost zero charges in the pH range of 3–4.5, compared to Mo (VI) loaded onto non-oxidized MPCM sample (Fig. 8.9b). It was reported that technetium did not adsorb on to Mo (VI) loaded oxidized MPCM, whereas it shows a strong affinity for the Mo (VI) loaded non-oxidized MPCM sample [111]. The MPCM resin demonstrates the capacity to adsorb $^{99}$Mo and release the daughter product $^{99m}$Tc simultaneously under batch and equilibrium conditions. It was also reported that $^{99m}$Tc, the decay product of $^{99}$Mo, was eluted with standard (0.9%) saline solution to yield more than 80% elution [111].

In the case of molybdenum (Mo) uptake onto MPCM resin, Hasan [111] reported that the adsorption follows Type-I isotherm and the Mo adsorption mainly occurs at the monolayer of active surface sites of the resin. In another attempt, Hasan [108, 113] pointed out that the critical structures of the resin that absorb irradiated molybdenum need to be protected from the negative impact of higher radiation dose. To address this issue, the following points are considered [108]:

- Since the active region of the resin structure is assumed to be thin due to range consideration, a new type of shielding concept can be used to protect the critical structure of the resin from the negative impact of absorbed dose.
- The possibility of microshielding on the resin's critical surface needs to be investigated, reducing the impact of high radiation flux and minimizing the radiolytic effect on the resin surface.

**Table 8.7** Physicochemical quality control data of pertechnetate Na[$^{99m}$Tc]

| pH | 5–6 |
|---|---|
| Clarity | Clear |
| Radiochemical purity | >99% |
| $^{99}$Mo breakthrough (with guard column) | |
| Radionuclidic purity | 99.99% |
| Al | <10 ppm |
| Mo | <1 ppm |

Copyright © 2017, Akadémiai Kiadó, Budapest, Hungary, J. Radioanal. Nucl. Chem., 313, 2017, 647–653. Reproduced from Ref. [114] with permission

- High Z elements, for instance, Hf can be used as a microshielding candidate who may cut down the dose on the active surface sites of the resin.
- Other high Z elements with higher stopping power are also good microshielding materials.
- Since the main constituents of MPCM resin are low Z elements (with less stopping power), the negative impact of high-energy particles can also be minimized by maintaining a proper aspect ratio of the column.

It was reported that the radiation tolerance limit and selectivity of the MPCM resin for certain isotopes were further enhanced by the high Z element crosslinked MPCM resin as it was not being limited by the radiolytic-driven reaction [113]. The microshielded resin was termed MPCM-Z resin. Chattopadhyay et al. [114] reproduced Hasan's work [111] and concluded that chitosan resin-based $^{99}$Mo/$^{99m}$Tc generator using low-specific activity (n, γ)$^{99}$Mo may find an application in nuclear medicine [114]. Table 8.7 shows the data of pertechnetate Na[$^{99m}$Tc]TcO$_4^-$ obtained from a typical chitosan-based adsorbent in the $^{99m}$Tc/$^{99}$Mo generator using an alumina guard column.

## 8.5 Summary

Chitosan has received significant attention in medical and pharmaceutical applications due to its high biocompatibility, biodegradability, antioxidant, antimicrobial, hemostatic, anti-infection, and non-toxic properties. Chitosan and its derivatives are extensively investigated to improve understanding of their mechanism of action and potential applications in wound care, burn management, antimicrobial, anticancer, and bone tissue engineering. Scaffolds are used to prevent infection, maintain a moist environment, protect the wound, and achieve minimum scar formation. Chitosan-based scaffolds contribute to the healing processes hemostasis, inflammation, proliferation, and remodeling phases. The inherent safe biological properties of chitosan, such as antibacterial nature, biocompatibility, biodegradability, mucoadhesive, cellular interaction, and ECM, are also investigated to repair or regenerate bone tissue. The potential use of chitosan nanoparticles and chitosan-based derivatives has also been investigated. These studies led to some achievements in the controlled delivery of therapeutic agents, anticancer formulation, gene delivery, etc. The chitosan-based composite materials can also be used to separate medical isotope $^{99m}$Tc from $^{99}$Mo from a low activity $^{99m}$Tc/$^{99}$Mo (n, γ) generator. The chitosan-based composite provides high sorption capacity for molybdenum ($^{99}$Mo) while simultaneously providing the selective elution of pertechnetate ($^{99m}$Tc) by saline (0.9% NaCl) solution. The applications of chitosan and chitosan-based materials in biomedical and pharmaceutical areas are evolving. However, the specific mechanisms for the action of chitosan in wound dressing, bone tissue engineering, and gene delivery technologies remain unresolved.

## References

1. Kolarsick PAJ, Kolarsick MA, Goodwin C (2011) Anatomy and physiology of the skin, Chapter-1. J Dermatol Nurses' Assoc 3(4):203–213. https://doi.org/10.1097/JDN.0b013e318 2274a98
2. Murray RZ, West ZE, Cowin AJ, Farrugia BL (2019) Development and use of biomaterials as wound healing therapies. Burns Trauma 7:2
3. Yagi M, Yonei Y (2018) Review article, Glycative stress and anti-aging: 7 Glycative stress and skin aging. Glycative Stress Res 5(1):050–054
4. Kamoun EA, Kenawy E-RS, Chen X (2017) A review on polymeric hydrogel membranes for wound dressing applications: PVA-based hydrogel dressings. J Adv Res 8:217–233
5. Dang LH et al (2018) Injectable nanocurcumin-formulated chitosan-g-pluronic hydrogel exhibiting a great potential for burn treatment. J Healthcare Eng 2018(Article ID 5754890). https://doi.org/10.1155/2018/5754890.
6. Wallace HA, Zito PM (2019) Wound healing phases [updated 13 May 2019]. In: StatPearls [Internet]. StatPearls Publishing, Treasure Island (FL), 2019 Jan. Available from https://www.ncbi.nlm.nih.gov/books/NBK470443/
7. Yu H, Xu X, Chen X, Hao J, Jing X (2006) Medicated wound dressings based on poly (N-vinyl pyrrolidone)/chitosan hydrogels. J Appl Poly Sci 101:2453–2463
8. Mezzana P (2008) Clinical efficacy of a new chitin nanofibrils-based gel in wound healing. Acta Chir Plast 50(3):81–84
9. Piranio F, Selimovic S (2015) A current view of functional biomaterials for wound care, molecular and cellular therapies. BioMed Res Int 2015(Article ID 403801):10 p. https://doi.org/10.1155/2015/403801
10. Hilmi ABM, Halim AS, Jaafar H et al (2013) Chitosan dermal substitute and chitosan skin substitute contribute to accelerated full-thickness wound healing in irradiated rats. BioMed Res Int 2013(Article ID 795458):13 p. https://doi.org/10.1155/2013/795458
11. Azad AK, Sermsintham N, Chandrkrachang S, Stevens WF (2004) Chitosan membrane as a wound-healing dressing: characterization and clinical application. J Biomed Mater Res Part B: Appl Biomater 69B:216–222
12. Shahana TG, Rekha PD (2018) Biopolymers: applications in wound healing and skin tissue engineering. Mol Biol Rep. https://doi.org/10.1007/s11033-018-4296-3
13. Lim CK, Halim AS (2010) Biomedical grade chitosan in wound management and its biocompatibility in vitro. In: Elnashar M (ed) Biopolymers. InTech. ISBN 978-953-307-109-1.

Available from http://www.intechopen.com/books/biopolymers/biomedical-grade-chitosan-in-wound-management-andits-biocompatibility-in-vitro
14. Akula S (2016) Fluorinated methacrylamide chitosan hydrogel improves cellular wound healing processes. M.S. thesis, December 2016. The University of Akron, Ohio
15. Falvo MR, Gorkun OV, Lord ST (2010) The molecular origins of the mechanical properties of fibrin. Biophys Chem 152(1–3):15–20. https://doi.org/10.1016/j.bpc.2010.08.009
16. Pogoreilov MV, Sikora VZ (2015) Chitosan as a hemostatic agent: current State. Eur J Med Ser B 2(1):24–33
17. Senel SS, McClure SJ (2004) Potential applications of chitosan in veterinary medicine. Adv Drug Deliv Rev 56:1467–1480
18. Cheung RCF, Ng TB, Wong JH, Chan WY (2015) Chitosan: an update on potential biomedical and pharmaceutical applications. Mar Drugs 13:5156–5186. https://doi.org/3390/md13085156
19. Okamoto Y, Yano R, Miyatake K, Tomohiro I, Shigemasa Y, Minami S (2003) Effects of chitin and chitosan on blood coagulation. Carbohyd Polym 53:337–342
20. Wang HS, Kirsch W, Zhu YH, Yang CZ, Hudson SM (2005) Hemostatic agents derived from chitin and chitosan. J Macromol Sci Part C Polymer Rev 45(4):309–323
21. Agus S et al (2019) The effect of the chitosan-collagen membrane on wound healing process in rat mandibular defect. J Ind Soc Periodontol 23(2):113–118
22. Altinel Y, Chung SS, Okay G, Uğraş N, Işık AF, Öztürk E et al (2018) Effect of chitosan coating on surgical sutures to strengthen the colonic anastomosis. Ulus Travma Acil Cerrahi Derg 24:405–411
23. Irimia T, Ghica MV, Popa L et al (2018) Strategies for improving ocular drug bioavailability and corneal wound healing with chitosan-based delivery systems. Polymers 10:1221. https://doi.org/10.3390/polym10111221
24. Raftery R, O' Brien FJ, Cryan S-A (2013) Chitosan for gene delivery and orthopedic tissue engineering applications. Molecules 18:5611–5647
25. Al-Jabour ND, Mohammad D, Gimbun J, Moshiul Alam AKM (2019) An overview of chitosan nanofibers and their applications in the drug delivery process. Current Drug Deliv 16(4):272–294
26. Levengood SL, Zhang M (2014) Chitosan-based scaffolds for bone tissue engineering. J Mater Chem B Mater Biol Med 2(21):3161–3184
27. Costa-Pinto AR, Reis RL, Neves NM (2011) Scaffolds based bone tissue engineering: The role of chitosan. Tissue Eng Part B 17(5). https://doi.org/10.1089/ten.teb.2010.0704
28. Younes I, Rinaudo M (2015) Chitin and chitosan preparation from marine sources, structure, properties, and applications. Mar Drugs 13:1133–1174
29. Oryan A, Sahvieh S (2017) Effectiveness of chitosan scaffold in skin, bone and cartilage healing. Int J Biol Macromol 104:1003–1011
30. Niekraszewicz A (2005) Chitosan medical dressings. Fibres Textiles Eastern Eur 6(54):16
31. Mi F-L, Wu Y-B, Shyu S-S et al (2002) Control of wound infections using a bilayer chitosan wound dressing with sustainable antibiotic delivery. J Biomed Mater Res 59:438–449
32. Ramya R, Jayachandran V, Kim SK, Sudha PN (2012) Biomedical applications of chitosan: an overview. J Biomater Tissue Eng 2(2):100–111
33. Aranaz I, Mengibar M, Harris R et al (2009) Functional characterization of chitin and chitosan. Curr Chem Biol 3:203–230
34. Khan M, Mujahid M (2019) A review on recent advances in chitosan-based composite for hemostatic dressings. Int J Biol Macromol 124:138–147
35. Rao SB, Sharma CP (1997) Use of chitosan as a biomaterial: studies on its safety and hemostatic potential. J Biomed Mater Res 34:21–28
36. Mercy HP, Halim AS, Hussein AR (2012) Chitosan-derivatives as hemostatic agents: Their role in tissue regeneration. Regener Res 1(1):38–46
37. Croisier F, Jérôme C (2013) Chitosan-based biomaterials for tissue engineering. Eur Polymer J 49:780–792

38. Mustafa A, Cadar E, Sirbu R (2015) Pharmaceutical uses of chitosan in the medical field. Eur J Interdisc Stud 3(1):35–40
39. Mi F-L, Shyu S-S, Wu Y-B, Lee S-T, Shyong J-Y, Huang R-N (2001) Fabrication and characterization of a sponge-like asymmetric chitosan membrane as a wound dressing. Biomaterials 22(2):165–173
40. Adekogbe I, Ghanem A (2005) Fabrication and characterization of DTBP-crosslinked chitosan scaffold for skin tissue engineering. Biomaterials 26:7241–7250
41. Porporatto C, Bianco ID, Riera CM, Correa SG (2003) Chitosan induces different L-arginine metabolic pathways in resting and inflammatory macrophages. Biochem Biophys Res Commun 304(2):266–272
42. Wu T, Li Y, Lee DS (2017) Chitosan-based composite hydrogels for biomedical applications. Macromol Res 25:480
43. Kasaj A, Reichert C, Gotz H et al (2008) In vitro evaluation of various bioabsorbable and nonresorbable barrier membranes for guided tissue regeneration. Head Face Med 4:22. https://doi.org/10.1186/1746-160X-
44. Mudasir A, Kaiser M, Singh S et al (2017) Chitosan centered bio nanocomposites for medical specialty and curative applications, a review. Int J Pharmaceut 529(1–2):200–217
45. Khorasani MT, Joorabloo A, Moghaddam A et al (2018) Incorporation of ZnO nanoparticles into heparinized polyvinyl alcohol/chitosan hydrogels for wound dressing application. Int J Biol Macromol 114:1203–1215
46. Liu H, Wang C, Li C et al (2018) A functional chitosan-based hydrogel as a wound dressing and drug delivery system in the treatment of wound healing. RSC Adv 8:7533–7549
47. Parsa P, Paydayesh A, Davachi SM (2019) Investigating the effect of tetracycline addition on nanocomposite hydrogels based on polyvinyl alcohol and chitosan nanoparticles for specific medical applications. Int J Biol Macromol 121:1061–1069
48. Omidi M, Yadegari A, Tayebi L (2017) Wound dressing application of pH-sensitive carbon dots/chitosan hydrogel. RSC Adv 7:10638–10649
49. Khunmanee S, Jeong Y, Park H (2017) Crosslinking method of hyaluronic-based hydrogel for biomedical applications. J Tissue Eng 8:1–16
50. Xu H, Ma H, Gao C, Han C (2007) Chitosan-hyaluronic acid hybrid film as a novel wound dressing: in vitro and in vivo studies. Polym Adv Technol 18:869–875
51. Iacob A-T, Dragan M, Ghet N et al (2018) Preparation, characterization and wound healing effects of new membranes based on chitosan, hyaluronic acid and arginine derivatives. Polymers 10:607. https://doi.org/10.3390/polym10060607
52. Chen RN, Wang GM, Chen C-H, Ho H-O, Sheu M-T (2006) Development of N,O-(Carboxymethyl)chitosan/collagen matrix as a wound dressing. Biomacromolecules 7(4):1058–1064
53. Ferreira P, Coelho JFJ, dos Santos KSCR, Ferreira EI, Gil MH (2006) Thermal characterization of chitosan-grafted membranes to be used as wound dressing. J Carbohydr Chem 25(2):233–251
54. Alencar De Queiroz, Alvaro A, Ferraz HG, Abraham GA, Del Mar FM, Bravo AL, San RJ (2003) Development of new hydro active dressings based on chitosan membranes: characterization and in vivo behavior. J Biomed Mater Res Part A 64A(1):147–154
55. Yu S-H, Mi F-L, Wu Y-B, Peng C-K, Shyu S-S, Huang R-N (2005) Antibacterial activity of chitosan-alginate sponges incorporating silver sulfadiazine: effect of ladder-loop transition of inter polyelectrolyte complex and ionic crosslinking on the antibiotic release. J Appl Polym Sci 98(2):538–549
56. Venkatesan J, Kim S-K (2010) Chitosan composites for bone tissue engineering—An overview. Mar Drugs 8:2252–2266. https://doi.org/10.3390/md8082252
57. Rao SH, Harini B, Kumar RP et al (2017) Natural and synthetic polymers/bioceramics/bioactive compounds-mediated cell signaling in bone tissue engineering. Int J Biol Macromol. https://doi.org/10.1016/j.ijbiomac.2017.09.029
58. Sharma C, Dinda AK, Potdar PD, Chou CF, Mishra NC (2016) Fabrication and characterization of novel nano-bio composite scaffold of chitosan-gelatin-alginate-hydroxyapatite for bone tissue engineering. Mater Sci Eng C Mater Biol Appl 64:416–427

59. Florencio-Silva R, da Silva Sasso GR, Sasso-Cerri E et al (2015) Biology of bone tissue: structure, function, and factors that influence bone cells. BioMed Res Int 2015(Article ID 421746):17 p. https://doi.org/10.1155/2015/421746
60. Raggatt LJ, Partridge NC (2010) Cellular and molecular mechanisms of bone remodeling. J Biol Chem 285(33):25103–25108. (a) Filipowska J, Tomaszewsk KA, Wiedzki ŁN et al (2017) The role of vasculature in bone development, regeneration and proper systemic functioning. Angiogenesis 20:291–302. https://doi.org/10.1007/s10456-017-9541-1
61. Rodríguez-Vázquez M, Vega-Ruiz B, Ramos-Zúñiga R et al (2015) Chitosan and its potential use as a scaffold for tissue engineering in regenerative medicine. BioMed Res Int 2015(Article ID 821279):15 p. https://doi.org/10.1155/2015/821279
62. Costa-Pinto AR, Reis RL, Neves NM (2011) Scaffolds based bone tissue engineering: the role of chitosan. Tissue Eng Part B 17(5):1–17
63. Geetha B, Premkumar J, Pradeep JP, Krishnakumar S (2019) Synthesis and characterization of bioscaffolds using freeze-drying technique for bone regeneration. Biocatal Agric Biotechnol 20:101184
64. Dhivya S, Saravanan S, Sastry TP, Selvamurugan N (2015) nanohydroxyapatite-reinforced chitosan composite hydrogel for bone tissue repair in vitro and in vivo. J Nanobiotechnol 13:40. https://doi.org/10.1186/s12951-015-0099-z
65. McCullogh MBA, Gomes M, Sankar J, Bhattarai N (2017) Development of chitosan-based scaffolds for bone regeneration: a preliminary report. EC Orthopaed 8(1):15–25
66. Varisqa F, Triaminingsih S, Corputty JEM (2017) Composition of chitosan-hydroxyapatite-collagen composite scaffold evaluation after simulated body fluid immersion as reconstruction material. J Phys: Conf Ser 884:012035
67. Frohbergh ME, Katsman A, Botta GP et al (2012) Electrospun hydroxyapatite-containing chitosan nanofibers crosslinked with genipin for bone tissue engineering. Biomaterial 33(36):9167–9178. https://doi.org/10.1016/j.biomaterials.2012.09.009
68. Dai T, Tanaka M, Huang Y-Y et al (2011) Chitosan preparations for wounds and burns: antimicrobial and wound healing effects. Expert Rev Anti Infect Ther 9(7):857–879. https://doi.org/10.1586/eri.11.59
69. Goy RC, Morris STB, Assis OBG (2016) Evaluation of the antimicrobial activity of chitosan and its quarternized derivative on *E. Coli* and *S. aureus* growth. Revista Brasileira de Farmacognosia 26:122–127
70. Ardila N, Daigle F, Heuzey M-C, Ajji A (2017) Antibacterial activity of neat chitosan powder and flakes. Molecules 22:100. https://doi.org/10.3390/molecules22010100
71. Rabea EI, Badawy MET, Stevens CV et al (2003) chitosan as antimicrobial agent: applications and mode of action. Biomacromol 4(6):1457–1465
72. Goy RC, De Britto D, Assis OBG (2009) A review of the antimicrobial activity of chitosan. Polimeros 19(3):241–247
73. Hosseinnejad M, Jafari SM (2016) Evaluation of different factors affecting antimicrobial properties of chitosan. Int J Biol Macromol 85:467–475
74. Kim S (2018) Competitive biological activities of chitosan and its derivatives: antimicrobial, antioxidant, anticancer, and anti-inflammatory. Int J Polymer Sci 2018(Article ID 1708172). https://doi.org/10.1155/2018/1708172
75. Liu N, Chen X-G, Park H-J et al (2006) Effect of MW and concentration of chitosan on antibacterial activity of *Escherichia coli*. Carbohyd Polym 64:60–65
76. Raafat D, Bargen KV, Haas A, Sahl H-G (2008) Insights into the mode of action of chitosan as an antibacterial compound. Appl Environ Microbiol 74(12):3764–3773
77. Simõesa D, Miguel SP, Ribeiro MP et al (2018) Recent advances on antimicrobial wound dressing: a review. Eur J Pharm Biopharm 127:130–141
78. Mohammadi A, Hashemi M, Hosseini SM (2016) Effect of chitosan molecular weight as micro and nanoparticles on antimicrobial activity against some soft rot pathogenic bacteria. LWT Food Sci Technol 71:347–355
79. Chang S-H, Lina H-T, Wu G-J et al (2015) pH effects on solubility, zeta potential, and correlation between antibacterial activity and molecular weight of chitosan. Carbohyd Polym 134:74–81

80. Tavaria FK, Costa EM, Gens EJ et al (2013) Influence of abiotic on the antimicrobial activity of chitosan. J Dermatol 40:1014–1019
81. Benhabiles MS, Salah R, Loinici H et al (2012) Antibacterial activity of chitin, chitosan and its oligomers prepared from shrimp shell waste. Food Hydrocoll 29:48–56
82. Chaudhury, Das S (2011) Recent advancement of chitosan-based nanoparticles for oral controlled delivery of insulin and other therapeutic agents. AAPS Pharm Sci Tech 12(1):10–19
83. Borchard G (2001) Chitosan for gene delivery. Adv Drug Deliv Rev 52:145–150
84. Read RC, Naylor SC, Potter CW, Bond J, Jabbal-Gill I, Fisher A, Illum L, Jennings R (2005) Effective nasal influenza vaccine delivery using chitosan. Vaccine 23:4367–4374
85. Adhikari HS, Yadhav PN (2018) Anticancer activity of chitosan, chitosan derivatives, and their mechanism of action. Int J Biomater 2018(Article ID 2952085):29 p. https://doi.org/10.1155/2018/2952085
86. Hejazi R, Amiji M (2002) Chitosan-based delivery systems: physicochemical properties and pharmaceutical applications. In: Dumitriu S (ed) Polymeric biomaterials, 2nd edn. Marcel Dekker, Inc., New York, NY, pp 213–237
87. Chang S-H, Wub C-H, Tsai G-J (2018) Effects of chitosan molecular weight on its antioxidant and antimutagenic properties. Carbohyd Polym 181:1026–1032
88. Kaya M, Akyuza B, Buluta E et al (2016) DNA interaction, antitumor and antimicrobial activities of three-dimensional chitosan ring produced from the body segments of a dipod. Carbohyd Polym 146:80–89
89. Tan G, Kaya M, Tevlek A et al (2018) Antitumor activity of chitosan from mayfly with comparison to commercially available low, medium and high molecular weight chitosans. In Vitro Cellular Dev Biol Animal 54:366–374. https://doi.org/10.1007/s11626-018-0244-8
90. Zhong Z, Cheng Z, Su D et al (2018) Synthesis, antitumor activity and molecular mechanism of doxorubicin conjugated trimethyl chitosan polymeric micelle loading Beclin1 siRNA for drug-resisted bladder cancer therapy. RSC Adv 8:35395–35402
91. Li X, Dong W, Nalin AP et al (2018) The natural product chitosan enhances the antitumor activity of natural killer cells by activating dendritic cells. Oncoimmunology 7(6):e1431085 (13 p). https://doi.org/10.1080/2162402X.2018.1431085
92. Thongborisute J, Tsuruta A, Kawabata Y, Takeuchi H (2006) The effect of particle structure of chitosan-coated liposomes and type of chitosan on oral delivery of calcitonin. J Drug Target 14(3):147–154
93. Kilinc A, Teke M, Oenal S, Telefoncu A (2006) Immobilization of pancreatic lipase on chitin and chitosan. Prep Biochem Biotechnol 36(2):153–163
94. Pereira AO, Cartucho DJ, Duarte AS, Gil MH, Cabrita AMS, Patricio JA, Barros MMT (2005) Immobilization of cardosin A in chitosan sponges as a novel implant for drug delivery. Curr Drug Discov Technol 24:231–238
95. Choi HG, Choo YG, Jang TS, Jung GY (2001) Method for producing high molecule water-soluble chitosan and its composition useful for prevention and treatment of sterility by reducing body fat. Republic of Korean Kongkae Taeho Kongbo, KR 2001085046 A 20010907 Patent written in Korean. Application: KR 2001-44981 20010725
96. Zhang S, Tian Y, Gao H (2003) Faming Zhuanli Shenqing Gongkai Shuomingshu
97. Gao N, Dou H, Zeng H (2004) Huaxi Yaoxue Zazhi 19(3):209–210
98. Huang YC, Vieira A, Huang KL, Yeh MK, Chiang CH (2005) Pulmonary inflammation caused by chitosan microparticles. J Biomed Mater Res 75A:283–287
99. Dyer AM, Hinchcliffe M, Watts P, Castile J, Jabbal-Gill I, Nankervis R, Smith A, Illum L (2002) Nasal delivery of insulin using novel chitosan-based formulations: a comparative study in two animal models between simple chitosan formulations and chitosan nanoparticles. Pharm Res 19(7):998–1008
100. Takeuchi H, Matsui Y, Sugihara H, Yamamoto H, Kawashima Y (2005) Effectiveness of submicron-sized, chitosan-coated liposomes in oral administration of peptide drugs. Int J Pharm 303(1–2):160–170
101. Tahamtan A, Tabarraei A, Moradi A et al (2015) Chitosan nanoparticles as a potential nonviral gene delivery for HPV-16 E7 into mammalian cells. Artif Cells Nanomed Biotechnol 43:366–372

102. Li J, Cai C, Li J et al (2018) Chitosan-based nanomaterials for drug delivery. Molecules 23:2661. https://doi.org/10.3390/molecules23102661
103. Zhang W, Xu W, Lan Y et al (2019) Antitumor effect of hyaluronic-acid-modified chitosan nanoparticles loaded with siRNA for targeted therapy for non-small cell lung cancer. Int J Nanomed 14:5287–5301
104. Zhang H, Wu F, Li Y et al (2016) Chitosan-based nanoparticles for improved anticancer efficacy and bioavailability of mifepristone. Beilstein J Nanotechnol 7:1861–1870
105. Jiang Y, Yu X, Su C et al (2019) Chitosan nanoparticles induced the antitumor effect in hepatocellular carcinoma cells by regulating ROS mediated mitochondrial damage and endoplasmic reticulum stress. Artif Cells Nanomed Biotechnol 47(1):74–756. https://doi.org/10.1080/21691401.2019.1577876
106. Tahamtan A, Ghaemi A, Gorji A et al (2014) Antitumor effect of therapeutic HPV DNA vaccines with chitosan-based nano delivery systems. J Biomed Sci 21:69
107. Ai J-W, Liao W, Ren Z-L (2017) Enhanced anticancer effect of copper-loaded chitosan nanoparticles against Osteosarcoma. RSC Adv 7:15971–15977
108. Hasan S, Prelas MA (2020) Molybdenum-99 production pathways and the sorbents for $^{99}Mo/^{99m}Tc$ generator systems using (n, $\gamma$) $^{99}Mo$: a review. SN Appl Sci 2:1782. https://doi.org/10.1007/s42452-020-03524-1
109. Steigman J (1982) Chemistry of the alumina column. Int J Appl Radiat Isot 33:829–834
110. Chakravarty R, Ram R, Dash A, Pillai MRA (2012) Preparation of clinical-scale $^{99}Mo/^{99m}Tc$ column generator using neutron activated low specific activity $^{99}Mo$ and nanocrystalline $\gamma$-$Al_2O_3$ as column matrix. Nucl Med Biol 39:916–922
111. Hasan S (2014) Preparation of chitosan-based microporous composite material and its applications. US Patent 8,911,695B2, 16 Dec 2014
112. Cardenas G, Cabrera G, Taboada E et al (2006) Synthesis and characterization of ionically cross-linked chitosan hydrogels for biomedical applications. Eur J Pharm Biopharm 57(1):19–34
113. Hasan S (2019) Preparation of chitosan-based microporous composite material. US Patent No. US 10,500,564 B2, 10 Dec 2019
114. Chattopadhyay S, Das SS, Alam MN, Madhusmita (2017) Preparation of $^{99}Mo/^{99m}Tc$ generator based on cross-linked chitosan polymer using low-specific activity (n, $\gamma$) $^{99}Mo$. J Radioanal Nucl Chem 313:647–653

# Chapter 9
# Application of Chitosan in Textiles

**Abstract** The utility of chitosan-based materials in various phases of textile manufacturing is promising. The biopolymer chitosan has been investigated for use in textile industries due to several inherent properties. These properties include biocompatibility, biodegradability, antimicrobial, and non-toxicity. The reaction mechanisms and applications of chitosan as a substitute for synthetic polymers in various end-uses in textile industries are discussed.

## 9.1 Chitosan Use in Textiles

Cotton, a vital natural fiber, represents approximately forty percent of the total fibers used for textile products [1]. As the global demand for textile products increases, cotton is expected to play a vital role. The increasing textile demand is mainly met by synthetic fibers such as polyester. However, the comfort properties of cotton are not matched by polyester [2]. Moreover, the management and disposal of the pollutants generated from the production of synthetic fibers pose significant environmental concerns [3]. The critical environmental impact arises from the cotton used in dyeing and spinning yarn production phases [1].

Chitosan is a promising biopolymer in textile industries due to its eco-friendly nature. In recent years, it has been utilized in recent years as a substitute for synthetic polymer in textile industries because of its biocompatibility, biodegradability, antimicrobial, and non-toxic properties. Chitosan is reported to be used in various applications in textile industries such as sizing and desizing agent, textile dyeing processes, and as a finishing agent for durable press [4]. Moattari et al. [5] reported that chitosan is also used to manufacture fibers, antimicrobial finishing, antistatic finishing, deodorizing finishing, binders for textile printing, textile inkjet printing, and medical textiles.

In textile processing, sizing agents provide the required strength and hydrophilicity for a successful weaving. Sizing agents are used to form a coating on the surface of the yarn to increase its tensile strength and abrasion resistance. The most common sizing agents include starch and its derivatives. The most common synthetic sizing agent used is polyvinyl alcohol (PVA) [6]. It is important to note that

© The Author(s), under exclusive license to Springer Nature Switzerland AG 2022
S. Hasan et al., *Chitin and Chitosan*, Engineering Materials and Processes,
https://doi.org/10.1007/978-3-031-01229-7_9

the sizing agents used in the yarn must be removable from the fabrics entirely to avoid complications in the subsequent treatment of fibers or fabric [7]. In the desizing step, most of the sizing aids removed from the fabrics are usually released as effluents. This may increase biological oxygen demand (BOD) and chemical oxygen demand (COD) in the receiving stream. Increased BOD and COD cause environmental problems, especially when non-biodegradable sizing agents such as PVA are used [6]. Chitosan is considered a potential sizing agent for textile yarn due to its favorable properties. These properties include biodegradability, biocompatibility, non-toxicity, antibacterial activity, high moisture absorbency, and environmental friendliness. It is reported that chitosan modifies the surface energy of yarn. This modification facilitates textiles product's hydrophilicity, wettability, and dyeability properties [8]. Hebeish et al. [9] reported that hydrolyzed chitosan and carboxymethyl chitosan could be used as a textile sizing agent. It was reported that the sized fabrics exhibited an average increment in tensile strength of about 55% and an average decrease in elongation at break of about 3%. A hundred percent desizing of fabrics was achieved with the hydrolyzed chitosans irrespective of the sizing agent solution concentration [9].

Dyeing of textile fabrics is a critical process carried out in acidic baths in the pH range of 4–6 [10]. Cotton textiles are dyed using direct (substantive) dyes, sulfur, vat, reactive, and azoic combinations [11]. The interaction between the dye anions and the cellulose macromolecules mostly occurs through van der Waals forces. Different salts are used to reduce negative charges on the fiber surface in the textile dyeing process. These salts also enhance the exhaustion of the dyes. Talukdar et al. [12] reported that the salt and the alkali (soda ash) push the color toward polysaccharide molecules [12]. Therefore, the salts and alkali function as hydrolyzing/fixing agents for the reactive dye and make the dyeing process more economical. The effluent containing large amount of salts from the textile dyeing process needs to be treated before discharging to rivers and streams. Otherwise, there is a possibility that salts being detrimental to aquatic life.

Sharma and Sayed [13] reported that the affinity of anionic dyes for cotton could be improved by adding cationic sites to the fiber through chemical modification. This chemical modification would enhance dye uptake onto the fibers from the dye bath. However, the majority of the chemicals used for the cationization of cotton cause serious environmental concerns [14]. Chitosan, a cationic polymer, is considered an ideal eco-friendly fixing agent for anionic dyes [15]. Chitosan improves the surface properties of the fiber. Chitosan reduces the Coulomb repulsion between the fiber and the anionic dyes. Therefore, dye uptake on the fiber treated with chitosan may occur due to the ionic reaction. This reaction occurs between the cationic group of chitosan and the anionic dye group in an acidic medium. Several studies indicated that chitosan treatment increased the dye uptake of cotton fabric in a concentration-dependent manner [13, 14]. The effect of chitosan concentration on cotton dying was studied based on the K/S value (the color strength) [13]. This study revealed that at a concentration of 1 wt% of chitosan or above, the adsorption of chitosan on the fabric was not uniform, as indicated by the unlevel [uneven is better] dyeing [13]. Chatha et al. [14] reported a decline in dye exhaustion beyond the 2% concentration of chitosan treatment for reactive dyes. Houshyar and Amirshahi [16] treated

## 9.1 Chitosan Use in Textiles

cotton fabric with chitosan using exhaustion, pad-dry, pad-batch, pad-steam, and pad-dry steam methods. They investigated the effect of chitosan on cotton dying based on different properties, e.g., K/S value (the color strength), washing fastness, light fastness, and rubbing fastness of the dyed cotton. It was reported that chitosan pretreatment increases the exhaustion of reactive dyes and the highest dye uptake. Therefore, chitosan application in cotton dyeing may reduce the amount of dye in the dying process due to increased dye exhaustion. This may ultimately reduce the amount of unused dye in the wastewater that is generated from the dyeing process [14].

Sadeghi-Kiakhani and Safapour [11] investigated salt-free dyeing of cotton fabric modified with prepared chitosan-poly(propylene) imine (CS-PPI) dendrimer using direct dyes. The direct dye is the salts of sulfonic acids and strong electrolytes. It almost absolutely dissociated in dye baths into colored anions and sodium cations.

Figure 9.1 shows that dye anions are attached to cellulose macromolecules primarily through van der Waals forces. The amine groups of CS-PPI were mainly responsible for dye uptake (Fig. 9.1). It was reported that cotton treatment with CS-PPI significantly improved dye uptake required to obtain a given color depth.

Different surface modification methods are used for wool fabrics to improve their hydrophilicity, dyeability, antimicrobial, and shrink-proofing properties [17]. The surface of the wool fiber plays a vital role in the textile finishing processes. Wool fiber surface is highly hydrophobic due to covalently bound fatty acids and a high amount of disulfide [18]. The enzymatic treatment of wool is considered environmentally friendly [19], and it enhances whiteness and shrink resistance to wool. However, the physicomechanical properties of wool deteriorate with an increase in enzyme concentration in the treatment process [20]. Pascal and Julia [21] investigated the role of chitosan in wool processing [21]. Using chitosan and its derivatives in woolen fabric treatment improves wool's shrinkage, antimicrobial, and antifelting properties

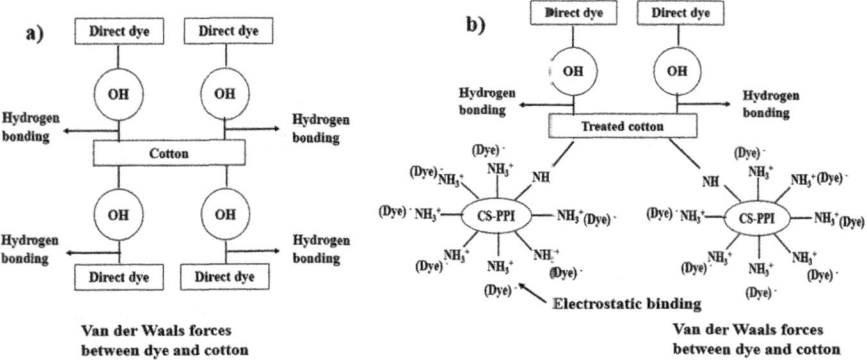

**Fig. 9.1** Possible reaction mechanism of direct dyes with **a** untreated cotton and **b** CS-PPI treated cotton. Copyright © Institute for Color Science and Technology. Creative Commons open access license CC BY, Prog. Color Colorants Coat. 11 (2018). https://doi.org/10.30509/PCCC.2018.75734, Reproduced from Ref. [11]

[22, 23]. Chitosan viscosity has a remarkable influence on wool shrink-resistance and the dyeing processes [24]. The application of chitosan/non-ionic surfactant mixture in reactive dyes for dyeing wool fabrics is reported by Yen [25]. Yen reported that the color strength was maximum when 0.5% chitosan and 1.0% surfactant were used. Periolatto et al. [26] studied the surface modification of wool fibers by chitosan-UV grafting. This grafting conferred a multifunctional finishing to the fabrics without damaging the comfort properties of wool fabrics. This study applied chitosan in acetic acid solution on wool by padding and then grafted by ultraviolet (UV) radiation. It was reported that 2% chitosan-grafted wool fabrics showed a strong dyeability toward acidic dye. The grafting was enough to confer good antimicrobial activity (67% reduction of *Escherichia coli*) after an oxidative wool pretreatment and impregnation at 50 °C for an hour.

Safapour et al. [27] used chitosan-cyanuric chloride hybrid (Ch-Cy) to treat wool fiber. First, they investigated the effect of Ch-Cy on the wool surface for dyeing, colorfastness, and antibacterial properties. Figure 9.2 shows the reaction mechanism of Ch-Cy with wool.

It was reported that the application of Ch-Cy on wool improved dye uptake. Also, it decreased optimum dyeing time and temperature. Antibacterial finish against gram-negative *E. coli* and gram-positive *Staphylococcus aureus* bacteria was durable up to 20 washing cycles. Sadeghi-Kiakhani et al. [28] reported chitosan-acrylamide (Ch-Ac) grafting to enhance reactive dye uptake and functional finishing of the wool yarn. Ch-Ac-grafted wool fabrics showed excellent antibacterial activity against gram-negative (*E. coli*) and gram-positive (*S. aureus*) bacteria [28]. Scanning electron microscopy (SEM) images show that the chitosan-acrylamide hybrid compound was efficiently grafted on wool yarn (Fig. 9.3).

Chitosan improves the dye coverage of immature fibers in cotton dyeing [29]. It can be used as a binder and thickener in the pigment printing of polyester and polyester-cotton blend fabrics with limitations [30]. Wax-impregnated chitosan microcapsules are useful in manufacturing heat-storage finishes for textiles [31]. The effects of detergents on chitosan-coated fabrics showed that chitosan is resistant to

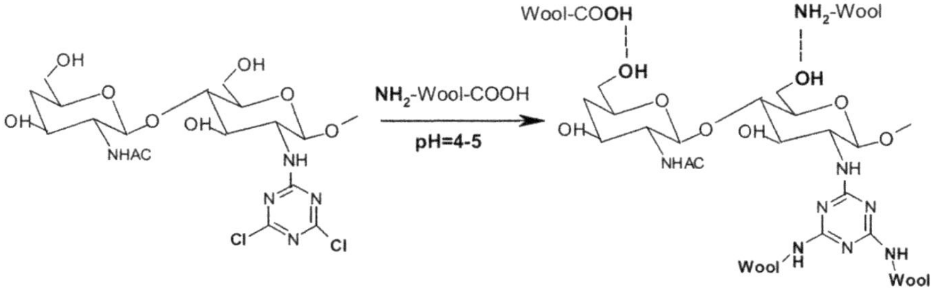

**Fig. 9.2** Grafting between wool yarn and modified chitosan. Copyright © 2019 Rights managed by Taylor & Francis, The Journal of the Textile Institute, (2019), 110, 81–88. Reproduced from Ref. [27] with permission

**Fig. 9.3** SEM images of **a** raw wool and **b** chitosan-acrylamide (Ch-Ac)-treated wool yarn Copyright © 2019 Published by Elsevier B.V., International Journal of Biological Macromolecules, (2019), 134, 1170–1178. Reproduced from Ref. [28] with permission

washing agents that do not contain sodium perborate or, presumably, other oxidative bleaches [32].

Cotton fabrics have a tendency to wrinkle [33]. Chakraborty et al. [34] reported that the wrinkling behavior of cellulosic fabrics is linked to the mobility of free hydroxyl groups. Crease properties of the fabrics depend on the hydrogen bond of the cellulose molecule [12] that can be reduced by decreasing or masking the hydrogen bond formation capacity of hydroxyl groups [34]. Dimethylol dihydroxy ethylene urea (DMDHEU) and 1, 2, 3, 4-Butane tetracarboxylic acid (BTCA) are the most widely used crosslinking agent in cotton finishing for prominent wrinkle-resistance property [35]. DMDHEU is a crosslinker that improves wrinkle recovery angle (WRA) and durable press (DP) finishing, but it may release formaldehyde, a known carcinogen, during processing, storage, and consumer use [13]. It is reported that BTCA with sodium hypophosphite (SHP) monohydrate provides good finishing durability. However, the high cost of BTCA is an obstacle [36]. Chitosan has been crosslinked to cotton using dimethylol dihydroxy ethylene urea (DMDHEU), citric acid, 1, 2, 3, 4-Butane tetracarboxylic acid (BTCA), or glutaric dialdehyde to improve antimicrobial durability. These chemicals can be crosslinked with chitosan to cotton through hydroxyl groups [37]. Ye et al. [38] reported a chitosan-based durable antibacterial finish on cotton fabric without any chemical binder [38].

The functional properties of chitosan can be modified by crosslinking with different fiber reactive groups to improve the wrinkle resistivity of the fabrics [39]. The affinity of chitosan for oppositely charged molecules or surfaces has been utilized to alter the characteristics of the fibers. These characteristics include friction, moisture adsorption [40, 41]. Verma et al. [42] reported chitosan citrate as a non-formaldehyde durable press finish that produces wrinkle resistance in textiles. Each cotton sample was first impregnated in a solution containing citric acid as a crosslinking agent (10%), chitosan (4%), silicon softener, and disodium hypophosphite 6% as standardized proportion. The maximum retention of the finish was found to be satisfactory after 20 washing cycles. The crease recovery angle was also improved. Wang et al. [43] treated cotton fabric with chitosan and epoxy-silicone finishing agents to enhance its wrinkle-resistant property. The optimum processing condition for cotton fabric wrinkle-resistance finishing by chitosan and organosilicon microemulsion was: chitosan 0.5% (w/w), epoxy-organosilicon 30% (v/v), baking at 150 °C for 120 s. The study showed that a mixed finish by two agents could provide higher wrinkle resistance than a single-finish cotton fabric due to their synergistic effect.

The molecular structure of chitosan is almost similar to that of cellulose used in manufactured fiber. Textiles made from cellulose and protein fibers have inherent properties such as hydrophilic porous structure and moisture transport characteristics [44] to grow microorganisms. Some common microorganisms and their influence on textile materials and human health are given in Table 9.1.

Ramachandran et al. [46] reported the benefits of antimicrobial treatment of textile materials, and they are as follows:

Table 9.1 Common microbes and their influence

| Microbes | Kind of microbes | Odor | Influence on the human body | Garments mostly stained | | |
|---|---|---|---|---|---|---|
| | | | | Hosiery | Underwear | Trousers |
| Bacteria | *Staphylococcus* | + | – Acute<br>– Ability to affect internal organ | + | + | – |
| | *Bacillus subtilis* | – | – It is considered a benign organism | + | + | + |
| | *Escherichia coli* | + | Ulcer | + | + | + |
| | *Pseudomonas a* | – | Thorax, otitis | – | + | + |
| | *Proteus vulgaris* | + | | + | – | – |
| | *Klebsiella pneumoniae* | | | – | + | + |
| Yeast | *Candida* | | | | | |
| Fungi | *C. albicans* | | – Thrust<br>– Causing of infections, from superficial mucosal to hematogenously disseminated candidiasis | + | – | – |
| Mold | *Trichophyton* | | Cause of itchy and crack skin in athletes foots | | | |
| | *Aspergillus niger* | | Discoloration of clothing | | | – |

Copyright © IJES, The International Journal of Engineering and Science (IJES), Volume 2 (8), 2013, pp. 9–13. Reproduced from Ref. [45] with permission
− Negative influence
+ Active influence

## 9.1 Chitosan Use in Textiles

- To avoid crossinfection by pathogenic microorganisms.
- To control the infestation by microbes.
- To arrest metabolism in microbes to reduce the formation of odor.
- To safeguard the textile products from staining, discoloration, and quality deterioration.

It is reported that the microorganisms may cause deterioration in the fabrics; therefore, antishrinkage and antimicrobial treatment are essential for the wearability of fabrics [47]. Many antimicrobial chemicals are used to protect fabrics from microbial attack [18]. The antimicrobial agent can be functional on microbes by binding to the surface or through a controlled release mechanism. The antimicrobial mode of action (diffusion or contact) depends on the concentration of the active substance in the textile. Therefore, active substance concentration should be above the minimum inhibitory concentration (MIC). The inhibitory activity depends on different factors. These factors include solid surface characteristics and the morphology of the solid [48]. It is reported that most antimicrobial agents utilize a controlled release mechanism through gradual and persistent release from the textile into their surroundings in the presence of moisture [49]. Figure 9.4 shows a schematic diagram of

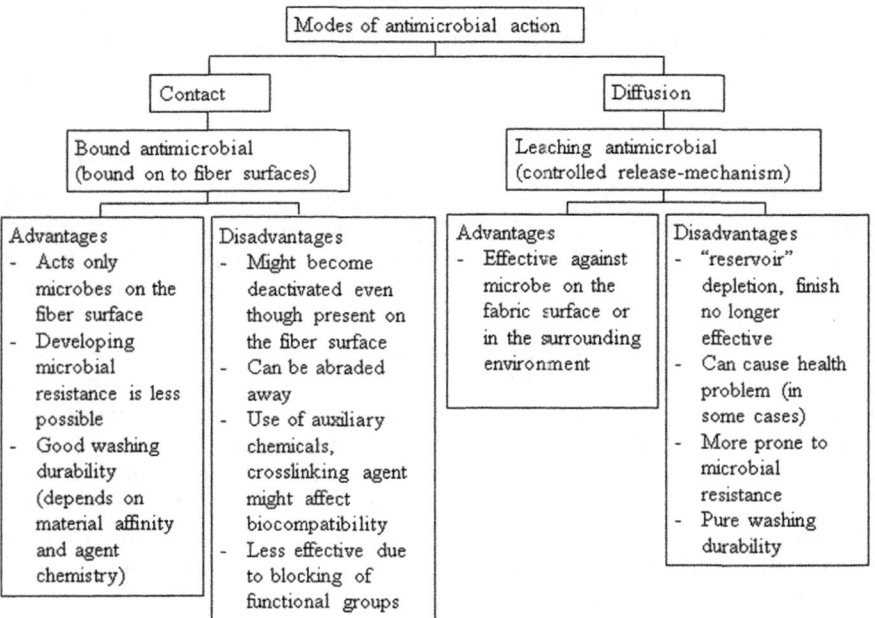

**Fig. 9.4** Advantages and disadvantages of the mode of action of antimicrobial agents. Copyright © FORMATEX 2011, science against microbial pathogens: communicating current research and technological advances, Méndez-Vilas A. (Ed), 2011, Badajoz, Spain, Formatex (pp. 36–51) Reproduced from Ref. [50] with permission

the advantages and disadvantages of the mode of action of antimicrobial agents on microbes.

Chitosan and its derivatives have been reported to be used in textile industries for (1) the production of chitin and chitosan fibers for medical use such as wound dressing, textiles for sensitive skin, and ion-binding capacities and (2) the treatment of fibrous materials in classical or alternative finishing procedures [50, 51].

Several studies reviewed the versatile uses of chitosan in textile applications [18, 51–53]. Marzendorfer and Cohen summarized the treatment of textile fabrics using chitosan derivatives and described their conceivable potential for the textile industry [54]. Chitosan with high molecular weights is more effective in inhibiting bacterial growth on treated cotton fabric than chitosan with low molecular weights [55]. The mechanisms by which antimicrobial substances control microbial growth are varied and depend on the type of agent used. Generally, antimicrobial agents prevent cell reproduction, damage cell walls or cell permeability, denature proteins, block enzymes, and make cell survival impossible [50]. As an antimicrobial agent, chitosan adsorbs on the cell surface of the microorganism and disturbs its physiological activities through ionic interaction [56], which eventually kills the cell. Chitosan is soluble in acidic aqueous solutions. Chitosan loses its antimicrobial properties in an alkaline environment because of its cationic properties [57].

The use of chitosan has been widespread as an antimicrobial agent, either alone or blended with other natural polymers [48]. Rastic et al. [50] reported that cotton fabric treated with chitosan derivative, O-acrylamidomethyl-N-[(2-hydroxy-3-trimethylammonium) propyl] chitosan chloride, water-soluble carboxymethyl showed good antibacterial activity against *E. coli* and *S. aureus*. Several studies reported the application of nanotechnology in the antimicrobial finishing of textiles [37, 58, 59]. Zille et al. [60] reviewed the antimicrobial activities of metal, metal oxide, chitosan-based nanoparticles, and their application in biomedical textiles. Table 9.2 shows antimicrobial nanomaterials and their application on textiles. It was observed that the application of nanomaterials is very useful to reduce microorganisms in textile fabrics.

Sunder et al. [61] reported that chitosan with polycarboxylic acid (PCA) exhibits good protection against *E. coli* and *S. aureus* bacterial strains. It also improves other textiles' functional properties, such as crease recovery, flame retardant, and soil release. Arik et al. [62] studied on antibacterial and wrinkle-resistance improvement of nettle biofiber using chitosan and BTCA. The combination treatment (1% chitosan +6% BTCA +4% sodium hypophosphite) resulted in 15% wrinkle-resistance improvement [62]. They also reported that this combination exhibited good antibacterial activity against *S. aureus* and *E. coli, and it was* higher against gram-positive than gram-negative bacteria.

Habeeba et al. [63] studied chitosan immobilized cotton fibers for antibacterial textile materials. The results showed that the chitosan particle was extensively dispersed in the cellulose matrix with strong interaction. Chitosan-coated cotton is shown to have excellent antibacterial activity against *E. coli* and *S. aureus.* Therefore, it was concluded that chitosan-coated cotton fabric could be used for surgical fabrics and high durable antibacterial finishing textile products. Li and

**Table 9.2** Antimicrobial nanomaterials applied to textiles

| Textile fabric and fiber (nanomaterials) | Nanoparticles average size (nm) (concentration) | Method | % bacterial reduction (strain) | % of bacterial reduction after washing (strain) (washing cycles) |
|---|---|---|---|---|
| Acrylic (Ag) | 1–7 | Photocured Carboxymethyl Starch NPs | 20* (S. aureus) 15* (E. coli) | 1* (S. aureus) (15) 0.5* (E. coli) (15) |
| Cotton (alginate/TSA) | 99 (70 μg g$^{-1}$) | Colloid NPs impregnated on fabric | 99.9 (E. coli) 99.9 (S. aureus) | 99 (E. coli) (30) |
| (Ag/chitosan) | 257 (34.5 wt%) 40 (1 wt%) 50–175 | NPs impregnated on fabric Pad-dry-cure method Colloid NPs impregnated on fabric | 99.9 (E. coli) 31* (E. coli) 26* (S. aureus) 3* (E. coli) | 199.9 (E. coli) (10) 15* (E. coli) (20) 17* (S. aureus) (20) n.a |
| (Ag/chitosan/TiO$_2$) | 5000 (7 wt%) | Pad-dry-cure method | 98 (E. coli) 100 (S. aureus) | n.a |
| (Chitosan) | 5–180 (0.5 wt%) | Colloid NPs impregnated on fabric | 99.9 (E. coli) 99.9 (S. aureus) | 65 (E. coli) (20) 78 (S. aureus) (20) |
| (Chitosan/alginate) | 35 | Pad-dry-cure method of NPs loaded with leaf extract | 100 (B. cereus) 98 (E. coli) 100 (P. aeruginosa) 100 (S. aureus) | 95 (B. cereus) (30) 87 (E. coli) (30) 98 [P. aeruginosa] (30) 98 (S. aureus) (30) |
| (ZnO/chitosan) | 15–60 (0.3 wt%) 28–100 (6 wt%) | Colloid NPs impregnated on fabric by the US Pad-dry-cure method | 99.9 (E. coli) 98.5 (S. aureus) 22* (E. coli) 25* (S. aureus) | 85 (E. coli) (10) 70 (S. aureus) (10) n.a |
| Polyacrylonitrile (chitosan) | 1000 (15 wt%) | Electrospun nanofibers containing chitosan | 100 (E. coli) 100 (S. aureus) 99.8 (P. aeruginosa) 100 (M. luteus) | n.a |

(continued)

**Table 9.2** (continued)

| Textile fabric and fiber (nanomaterials) | Nanoparticles average size (nm) (concentration) | Method | % bacterial reduction (strain) | % of bacterial reduction after washing (strain) (washing cycles) |
|---|---|---|---|---|
| (Ag/chitosan) | 166 (0.2 wt%) | Colloid NPs impregnated on PVP-treated fabric | 100 (*S. aureus*) | n.a |
| (Chitosan) | 5–180 (0.5 wt%) | Colloid NPs impregnated on fabric | 90 (*E. coli*) 99.9 (*S. aureus*) | 50 (*E. coli*) (20) 75 (*S. aureus*) (20) |
| (Chitosan) | 115 (0.2 wt %) | Colloid NPs impregnated on PVP-treated fabric | 90 (*S. aureus*) | n.a |
| Polyethylene/chitosan (Ag) | 12–18 (1.3 wt%) | Electrospinning nanofibers containing NPs | 15* (*E. coli*) 20* (*S. aureus*) 18* (*P. aeruginosa*) 12* (*C. albicans*) | n.a |
| (Ag/chitosan) | 2–10 (1 wt%) | Electrospun nanofibers containing NPs | 99.9 (*E. coli*) | n.a |
| (Ag/chitosan) | 20 (0.6 wt%) | Electrospun PVA nanofibers containing NPs | 100 (*E. coli*) | n.a |
| (TiO$_2$/Ag/chitosan) | 100 (0.04 wt%) | Electrospun nanofibers containing NPs | 99 (*E. coli*) 98 (*S. aureus*) | n.a |
| (Chitosan) | 20.8 (1 wt%) | Colloid NPs impregnated on fabric | 95 (*S. aureus*) | 90 (*S. aureus*) (20) |
| (Chitosan) | 5–180 (0.5 wt%) | Colloid NPs impregnated on fabric | 99.9 (*E. coli*) 99.9 (*S. aureus*) | 55 (*E. coli*) (20) 76 (*S. aureus*) (20) |

Copyright © 2014 IOP Publishing Ltd., *Mater. Res. Express*, **1**, 032,003 2014. Reproduced from Ref. [60] with permission
* Inhibition zone in mm, *n.a* Not available, *PVP* Polyvinylpyrrolidone

Tang [64] prepared chitosan fiber with high antioxidant and acid-resistant capability for healthy and hygienic textiles. Chitosan fiber was first crosslinked by a water-soluble aziridine crosslinker and then dyed with natural lac dye consisting of polyphenolic anthraquinone compounds. It was reported that the crosslinked chitosan fibers were suitable for bio-active, healthy, and hygienic medical textiles products as the crosslinking and dyeing had no impact on the good inherent antibacterial activity of chitosan fiber.

Petkova et al. [65] studied the sonochemical coating of textiles with hybrid ZnO/chitosan antimicrobial nanoparticles. This work generated hybrid antimicrobial coatings on cotton fabrics using a one-step simultaneous sonochemical deposition of ZnO nanoparticles (NPs) and chitosan. It was reported that the combination of ZnO nanoparticles and chitosan resulted in enhanced antibacterial efficiency against *S. aureus* and *E. coli* even at low ZnO concentration, compared to the individual ZnO and chitosan coatings. It was further noted that the treated fabrics' antimicrobial effect was resistant to multiple washing cycles at 75 °C.

Şahan and Demir [17] reported the surface modification of wool fabrics using chitosan nanoparticles loaded with silver. The effects of surface modification of wool fabrics were investigated in terms of hydrophilicity, antibacterial activity, dyeing, air permeability, surface morphology, and tensile strength. The effect of wool fabrics' enzyme and atmospheric plasma treatments was studied, and their data are given in Table 9.3.

Table 9.3 shows that both chitosan and nanochitosan particles have a more significant impact on the antibacterial activity. It was reported that enzyme and plasma treatments, along with chitosan, showed significant contributions to wool properties. These properties included antibacterial, hydrophilicity, dyeability, tensile strength, fastness, and air permeability of wool fabrics. In terms of the dyeing process, the

**Table 9.3** Antibacterial activity (%)

| Treatment | Bacterial reduction (%) | | |
|---|---|---|---|
| | Types of bacteria | | |
| | *S. aureus* | *E. coli* | *K. pneumoniae* |
| Untreated | – | – | – |
| Chitosan | 73.15 | 81.83 | 33.73 |
| Plasma + chitosan | 77.02 | 83.16 | 40.42 |
| Enzyme + plasma + chitosan | 80.64 | 89.21 | 48.54 |
| Nanochitosan | 84.86 | 89.56 | 51.16 |
| Plasma + nanochitosan | 86.99 | 91.55 | 60.66 |
| Enzyme + plasma + nanochitosan | 90.35 | 93.98 | 68.62 |
| Chitosan-silver nanoparticles (Ag-NCHT) | 96.38 | 97.66 | 66.62 |
| Enzyme + plasma + Ag-NCHT | 99.27 | 99.99 | 84.24 |

Copyright © 2016 TEKSTIL ve KONFEKSIYON, TEKSTIL ve KONFEKSIYON, (2016), 26(4), pp. 414–420. Reproduced from Ref. [17] with permission

synergetic effect of enzyme, plasma, and nanochitosan treatments led to 2.5 times higher K/S values than the untreated fabric.

The development of chitosan-based biocompatible and biodegradable microfiber yarn for medical applications has always remained a critical research objective [66]. A review on the area of chitosan-coated textile substrates for wound care applications is reported by Chellamani et al. [67]. Shabunin et al. [68] reported a bilayer wound dressing consisting of aliphatic copolyamide nanofibers and a layer of chitosan-based composite. The chitosan-based composite was prepared using chitosan nanofibers and chitin nanofibrils filler. Experimental studies demonstrated that up to 97.8% epithelialization of the third-degree burn wound surface was achieved within 28 days [68]. The obtained data suggested that the composite chitosan-copolyamide wound dressings can be used efficiently to treat burn wounds.

## 9.2 Summary

Chitosan and its derivatives are useful in improving the surface properties of textile fibers. Chitosan and its derivatives improve mechanical properties, hydrophilicity, wettability, dyeability, and electric conductivity of textiles. In addition, chitosan has been used to improve the wrinkle resistance of textile fabrics to some extent. Chitosan-based textile products are gaining attention in biomedical applications due to their bacteriostatic and antimicrobial properties. Chitosan-based wound dressings are also considered to treat burn wounds due to their biocompatibility, biodegradability, and non-toxic nature.

## References

1. Bevilacqua M, Ciarapica FE, Mazzuto G, Paciarotti C (2014) Environmental analysis of a cotton yarn supply chain. J Clean Prod 82:154–165
2. Bjorquist S, Aronsson J, Henriksson G, Persson A (2018) Textile qualities of regenerated cellulose fibers from the cotton waste pulp. Text Res J 88(21):2485–2492
3. Sandin G, Peters GM (2018) Environmental impact of textile reuse and recycling—a review. J Clean Prod 184:353–365
4. Vikele L, Laka M, Sable I et al (2017) Effect of chitosan on properties of paper for packaging. Cellul Chem Technol 51(1–2):67–73
5. Moattari M et al (2018) Application of chitosan in textiles. Ann Mater Sci Eng 3(1):1032
6. Zhang Y (2013) Environmentally friendly sizing agent from corn distillers dried grains. M.S. thesis, The University of Nebraska-Lincoln. http://digitalcommons.unl.edu/cehsdiss/167
7. Roy J, Salaun F, Giraud S et al (2017) Chitosan-based sustainable textile technology: process, mechanism, innovation, and safety, Chapter 12. In: Biological activities and application of marine polysaccharides. InTechOpen, pp 251–278. https://doi.org/10.5772/65259
8. Ferrero F, Periolatto M (2015) Modification of surface energy and wetting of textile fibers, Chapter 6. Intech Open, pp 139–168. https://doi.org/10.5772/60812

9. Hebeish A, Higazy A, Shafei AE (2005) New sizing agents and flocculants derived from chitosan. In: 8th Arab international conference on polymer science & technology, 27–30 Nov 2005, Cairo-Sharm El-Sheikh, Egypt
10. El-Sayed H, El-Gabry LK, Kantouch A (2005) Replacement of acetic acid with citric acid in dyeing of textile fiber. In: 2nd international conference of textile research division, 11–13 Apr 2005, NRC, Cairo, Egypt. Text Process State Art Future Dev 2(1):61–68
11. Sadeghi-Kiakhani M, Safapour S (2018) Salt-free dyeing of cotton fabric modified with prepared chitosan-poly(propylene) imine dendrimer using direct dyes. Prog Color Colorants Coat 11:21–32
12. Teli MD, Sheikh J, Shastrakar P (2013) Exploratory investigation of chitosan as mordant for eco-friendly antibacterial printing of cotton with natural dyes. J Text 2013, Article ID 320510, p 6. https://doi.org/10.1155/2013/320510
13. Sharma R, Sayed U (2016) Surface modification of cellulosic fabric. Int J Adv Chem Eng Biol Sci (IJACEBS) 3(1)
14. Chatha SAS et al (2016) Significance of chitosan to improve the substantive of reactive dyes. J Chil Chem Soc 61(2):2895–2897
15. Huang L, Xiao L, Yang G (2018) Chitosan application in textile processing. Curr Trends Fashion Technol Textile Eng 4(2):555635. https://doi.org/10.19080/CTFTTE.2018.04.555635
16. Houshyar S, Hossein S (2002) Treatment of cotton with chitosan and its effect on dyeability with reactive dyes. Iran Polym J 11(5):295–301
17. Şahan G, Demir A (2016) A green application of nanosized chitosan in textile finishing. TEKSTIL ve KONFEKSIYON 26(4):414–420
18. Enescu D (2008) Use of chitosan in surface modification of textile materials. Rom Biotechnol Lett 13(6):4037–4048
19. Onar N, Sariisik M (2004) Application of enzymes and chitosan biopolymer to the antifelting finishing process. J Appl Polym Sci 93:2903–2908
20. Velchez S, Jovancic P, Manich AM, Julia MR, Erra P (2005) Chitosan application on wool before enzymatic treatment. J Appl Polym Sci 98:1938–1946
21. Pascual E, Julia MR (2001) The role of chitosan in wool finishing. J Biotechnol 89(2,3):289–296
22. Shih C-Y, Huang K-S (2003) Synthesis of a polyurethane-chitosan blended polymer and a compound process for shrink-proof and antimicrobial woolen fabrics. J Appl Polym Sci 88(9):2356–2363
23. Roberts GAF, Wood FA (2001) A study of the influence of structure on the effectiveness of chitosan as an anti-felting treatment for wool. J Biotechnol 89(2,3):297–304
24. Julia MR, Pascual E (2000) Chitosan viscosity influences both wool shrink-resistance and dyeing. Agro-Food-Ind Hi-Tech 11(6):34–37
25. Yen M-S (2001) Application of chitosan/nonionic surfactant mixture in reactive dyes for dyeing wool fabrics. J Appl Polym Sci 80(14):2859–2864
26. Periolatto M, Ferrero F, Vineis C, Rombaldon F (2013) Multifunctional finishing of wool fabrics by chitosan UV-grafting: an approach. Carbohyd Polym 98:624–629
27. Safapour S, Sadeghi-Kiakhani M, Dustmohammdi S (2019) Chitosan- cyanuric chloride hybrid as an efficient novel bio-mordant for improvement of cochineal natural dye absorption on wool yarns. J Text Inst 110:81–88
28. Sadeghi-Kiakhani M, Safapour S, Ghanbari-Adivi F (2019) Grafting of chitosan-acrylamide hybrid on the wool: characterization, reactive dyeing, antioxidant and antibacterial studies. Int J Biol Macromol 134:1170–1178
29. Jocic D, Vilchez S, Yopalovic T, Navarro A, Jovancic P, Julia MR, Erra P (2005) Chitosan/acid dye interaction in wool dyeing system. Carbohyd Polym 60:51–59
30. Arab-Bahmani S, East GC, Holmes I (2000) Application of chitosan in textile printing. Adv Chitin Sci 4 (EUCHIS'99):136–142
31. Petit J-LV, Garces JG, Tacies A (2002) Use of chitosan microcapsules for finishing textiles. Eur Pat Appl 11, pp EP 1223243 A1 20020717 Application: EP 2001-100629 20010111
32. Roberts GAF, Wood FA (2000) The effects of detergents on chitosan. Adv Chitin Sci 4(EUCHIS'99):466–472

33. Schramm S, Amann A (2019) Formaldehyde-free, crease-resistant functionalization of cellulosic material modified by a hydrolyzed dicarboxylic acid based alkoxysilane/melamine finishing system. Cellulose 26:4641–4654
34. Chakraborty JN et al (2018) Performance of durable press finish on cotton with modified DMDHEU. Citric acid BTCA Maleic Acid Tekstilec 61(4):289–297
35. Qi H, Huang Y, Ji B et al (2016) Anti-crease finishing of cotton fabrics based on crosslinking of cellulose with acryloyl malic acid. Carbohyd Polym 135:86–93
36. Kittinaovarat S, Kantuptim P, Singhaboonpong T (2006) Wrinkle resistant properties and antibacterial efficacy of cotton fabrics treated with glyoxal system and with the combination of glyoxal and chitosan system. J Appl Polym Sci 100:1372–1377
37. Vellingiri T, Ramachandran T, Senthilkumar M (2013) Eco-friendly application of nano chitosan in antimicrobial coatings in the textile industry. Nanosci Nanotechnol 3(4):75–89
38. Ye W, Xin JH, Li P, Lee K-LD, Kwong T-L (2006) Durable antibacterial finish on cotton fabric by using chitosan-based polymeric core-shell particles. J Appl Polym Sci 102:1787–1793
39. Chen C-W, Hsieh S-H, Cheng C-H, Chang T-T (2004) Wrinkle-resistance of cotton fabrics resulting from crosslinking chitosan through citric acid. Cellul Chem Technol 38(5–6):431–444
40. Hakeim OH, Abou-Okeil A, Abdou LAW, Waly A (2005) The influence of chitosan and some of its depolymerized grades on natural color printing. J Appl Polym Sci 97:559–563
41. Suzuki K, Oda D, Shinobu T, Saimoto H, Shigemasa Y (2000) New selectively N-substituted quaternary ammonium chitosan derivatives. Polym J (Tokyo) 32(4):334–338
42. Verma M, Khambra K, Yadav N, Singh R (2013) Effect of crease resistance finish on crease recovery properties of chitosan fabrics. Int J Text Fashion Technol 3(4):9–14
43. Wang H, Huang C, Liu N (2011) Wrinkle-resistant property of cotton fabric treated by chitosan and epoxy-silicon micro-emulsion. Adv Mater Res 331:382–385
44. Ye W, Leung MF, Xin J, Kwong TL, Lee DKL, Li P (2005) Novel core-shell particles with poly (n-butyl acrylate) cores and chitosan shells as an antibacterial coating for textiles. Polymers 46:10538–10543
45. Boryo DEA (2013) The effect of microbes on textile material: a review on the way-out so far. Int J Eng Sci (IJES) 2(8):9–13
46. Ramachandran T, Rajendra Kumar K, Rajendran R (2004) Antimicrobial textile an overview. IE(I) J-TX 84:42–47
47. Hsieh S-H, Zhang F-R, Li H-S (2006) Anti-ultraviolet and physical properties of woolen fabrics cured with citric acid and $TiO_2$/chitosan. J Appl Polym Sci 100:4311–4319
48. Kong M, Chen XG, Xing K, Park HJ (2010) Antimicrobial properties of chitosan and mode of action: a state of the art review. Int J Food Microbiol 144:51–63
49. Simoncic B, Tomsic B (2010) Structures of novel antimicrobial agents for textiles—a review. Text Res J 80(16):1721–1737
50. Ristić T, Zemljič LF, Novak M et al (2011) Antimicrobial efficiency of functionalized cellulose fibers as potential medical textiles. In: Méndez-Vilas A (ed) Science against microbial pathogens: communicating current research and technological advances, Badajoz, Spain, Formatex, pp 36–51
51. Knittel D, Schollmeyer E (2006) Chitosan for permanent antimicrobial finish on textiles. Lenzinger Ber 85:124–130
52. Lim SH, Hudson SM (2003) Review of chitosan and its derivatives as antimicrobial agents and their uses as textile chemicals. J Macromol Sci-Polym Rev C43(2):223–269
53. Inamdar MS, Chattopadhyay DP (2006) Chitosan and its versatile applications in textile processing. Man-Made Text India 49(6):211–216
54. Marzendorfer H, Cohen E (2019) Chitin/chitosan: versatile ecological, industrial, and biomedical applications, Chapter 14. In: Cohen E, Merzendorfer H (eds) Extracellular sugar-based biopolymers matrices, biologically-inspired systems, p 559. https://doi.org/10.1007/978-3-030-12919-4_14
55. Shin Y, Yoo DI, Jang J (2001) Molecular weight effect on antimicrobial activity of chitosan treated cotton fabrics. J Appl Polym Sci 80(13):2495–2501

56. El- KF, El- MA, Elhendawy AG, Hudson SM (2005) The antimicrobial activity of cotton fabrics treated with different crosslinking agents and chitosan. Carbohyd Polym 60:421–430
57. Lim SH, Hudson SM (2004) Application of a fiber-reactive chitosan derivative to cotton fabric as an antimicrobial textile finish. Carbohyd Polym 56:227–234
58. Arif D, Niazi MBK, Haq N et al (2015) Preparation of antibacterial cotton fabric using chitosan-silver nanoparticles. Fibers Polym 16(7):1519–1526
59. Rivero PJ, Urrutia A, Giocoechea J et al (2015) Nanomaterials for functional textiles and fibers. Nanoscale Res Lett 10(501):1–22
60. Zille A, Almeida L, Amorim T et al (2014) Application of nanotechnology in antimicrobial finishing of biomedical textiles. Mater Res Express 1:032003
61. Sunder AE, Nalankilli G, Swamy NKP (2014) Multifunctional finishes on cotton textiles using a combination of chitosan and polycarboxylic acid. Indian J Fiber Text Res 39:418–424
62. Arik B, Yavas A, Avinc O (2017) Antibacterial and wrinkle resistance improvement of nettle biofiber using chitosan and BTCA. Fibers Text East Europe 25, 3(123):106–111
63. Habeeba AAU, Reshmi CR, Sujith A (2017) Chitosan immobilized cotton fibers for antibacterial textile materials. Polym Renew Resour 8(2):61–70
64. Li X-Q, Tang R-C (2016) Crosslinked and dyed chitosan fiber presenting enhanced acid resistance and bioactivities. Polymers 8:119. https://doi.org/10.3390/polym8040119
65. Petkova P et al (2014) Sonochemical coating of textiles with hybrid ZnO/chitosan antimicrobial nanoparticles. ACS Appl Mater Interfaces 6:1164–1172
66. Toskas G et al (2013) Pure chitosan microfibers for biomedical applications. AUTEX Res J 13(4). https://doi.org/10.2478/v10304-012-0041-5
67. Chellamani KP (2013) Chitosan treated textile substrates for wound care applications. J Acad Ind Res (JAIR) 2(2):97–102
68. Shabunin AS (2019) Composite wound dressing based on chitin/chitosan nanofibers: processing and biomedical applications. Cosmetics 6:16. https://doi.org/10.3390/cosmetics6010016

# Chapter 10
# Chitosan for the Agricultural Sector and Food Industry

**Abstract** Chitosan and its derivatives are drawing interest in the food and agricultural sectors due to their eco-friendly properties. The potential application of chitosan and its derivatives in various agricultural sectors and food industries is discussed in this chapter. The antimicrobial activities of chitosan-based materials, their mode of action, and possible reaction mechanisms are also discussed.

## 10.1 Introduction

Sustainable food production is necessary to meet the growing nutritional demand of the world population. Food production, however, can be significantly hampered by the adverse effect of both natural causes and anthropogenic activities. In crop production, the germination of seeds plays a significant role. Optimum seed germination can be threatened by adverse environmental conditions and pathogen attacks, decreasing crop productivity [1]. Several strategies such as seed coating and plant protection have been developed to improve crop resistance against natural causes and diseases [2–5]. Most of the strategies designed for increased food production use toxic chemicals for crop pest and disease control. Seed-coating technology uses pesticides on the seed during coating that acts as a phytosanitary agent against pests and diseases, enhancing the germination rate and crop yield [2, 3]. Although the current strategies enhance food production multifold, toxic chemicals in the agricultural field lead to various environmental issues. For example, toxic fungicides such as carbofuran have been used in the past for seed coating. A dose of 7 mg carbofuran is lethal to rats, and the period of residual toxicity in the soil is up to 50 years [3]. Moreover, the use of such toxic compounds increases public concern regarding the continuous contamination of agricultural products. It is reported that the contamination of agricultural products could occur from pesticide residues and the increased resistance in the pest population [5]. Several studies reported sustainable, eco-friendly agricultural farming as an alternative to agrochemical compounds [6, 7]. The application of chitosan in the agricultural research sector is receiving increased attention. This is due to chitosan's antibacterial, antioxidant, antiviral, and immunostimulatory properties. This chapter discusses the applications of chitosan in the agricultural sector. Besides, chitosan and its derivatives for the food industry are also discussed in this chapter.

## 10.2 Application of Chitosan in Agriculture and Agro-based Industry

a. Seed coating/germination

The effect of chitosan on seed germination [2], plant disease resistance mechanisms against fungal growth [8], plant growth rate [9, 10], and crop yield [11] are encouraging, which suggest chitosan as a potential eco-friendly farming material. The application of chitosan as antimicrobial, plant pathogens, and plant protectant agents for pre- and post-harvest disease control in the agricultural sector has been investigated [12, 13]. Chitosan has been broadly used in agriculture to reduce harmful effects on plants during unfavorable conditions and enhance plant growth [14]. Chitosan possesses a symbiotic relation with rhizobacteria. Rhizobacteria fix nitrogen in the plant roots and convert into an available nutrient for the host plant, supporting germination rate and plant growth by improving nutrient uptake [15].

Several studies investigated chitosan to decrease the seed's disease susceptibility during its germination in soil [3, 4, 16]. Chitosan has excellent film-forming properties, making it easy to form a semi-permeable film on the seed surface, maintaining the seed moisture and absorbing the soil moisture, thus promoting seed germination [4]. Pandey et al. [16] reported that chitosan could alter the seed plasma membrane's permeability, change the concentration of sugars and proline (amino acid), and enhance defense enzyme activities. These chitosan properties make it an ideal candidate as a seed-coating material for cereals, nuts, fruits, and vegetables. Siddaiah et al. [17] investigated the efficacy of chitosan nanoparticles (CNP) against downy mildew disease Perl millet by *Sclerospora graminicola*. It was reported that seed treatment with CNP changed gene expression profiles and induced systemic and durable resistance. This seed treatment also showed significant protection against downy mildew under greenhouse conditions compared to the untreated control conditions. In another attempt, Zeng et al. [4] studied the effect of chitosan coating on soybean seeds. They observed that the seed germination, plant growth, and soybean yield increased substantially with chitosan-coated soybean seed. Zeng et al. [3] studied chitosan-based novel corn seed-coating agent (NCSCA) and compared it with the conventional seed-coating agent AMULET (20% Thiram/carbofuran seed-coating agent for corn). Table 10.1 summarizes the antifungal activities of various seed-coating agents as measured by the inhibitory index.

As shown in Table 10.1, the chitosan-based seed-coating agent has better antifungal activity than a conventional seed-coating agent. The inhibitory index by NCSCA is 73.45, 78.35, 82.73, and 91.37% at the concentration of 500, 1000, 1500, and 2000 μg/mL, respectively. González et al. [18] studied the effect of chitosan on seed treatment in a saline medium. They reported that salinity adversely affects

**Table 10.1** Inhibition effect of seed-coating agent against *Sphacelotheca reiliana*

| Agent | Concentration (μg/mL) | Colony diameter (mm) | Inhibitory index % |
|---|---|---|---|
| AMULET | 0 | 77.0 | |
| | 500 | 42.5 | 44.8 |
| | 1000 | 32.5 | 57.79 |
| | 1500 | 19.7 | 74.42 |
| | 2000 | 10.2 | 86.75 |
| NCSCA | 0 | 77.6 | – |
| | 500 | 20.6 | 73.45 |
| | 1000 | 16.8 | 78.35 |
| | 1500 | 13.4 | 82.73 |
| | 2000 | 6.7 | 91.37 |

Copyright © 2010 Academic Journals, Journal of Agricultural Biotechnology and Sustainable Development 2010, Vol. 2(6), pp. 108–112. Copyright © 2022 Zeng, D., Mei, X., and Wu, J. This article is published under the terms of the *Creative Commons Attribution License 4.0*. Reproduced from Ref. [3]

seedling growth. The application of chitosan can enhance catalase and peroxidase enzyme activities, which can reverse the salinity effects in rice seedling growth. It was further reported that the seed treated with chitosan (100 mg/L) in the saline medium lowered malondialdehyde (MDA) and increased proline levels. Pandey et al. [16] reported that the decline of malondialdehyde (MDA) in the seed indicates antioxidant activity and lipid peroxidation.

b. Plant protection

Approximately 20–40% of the crops are lost each year due to pathogen and pest-related plant diseases [19]. Plants respond naturally against biological and environmental stress conditions, but sometimes induced defense is needed against more challenging threats [16]. Table 10.2 shows the elaborate defense mechanisms of plants to hinder pathogens.

Chemical pesticides are currently used to control plant pathogens [5]. However, most pesticides used are not biodegradable and are highly toxic to humans and animals [21]. Several studies reported that chitosan and its derivatives have plant defense response-related properties and stimulate beneficial microbes' growth and activity [22–24]. When exposed to chitosan, plants usually activate their defense system by releasing various defense enzymes and pathogenesis-related (PR) proteins [8, 25]. Hassan and Chang [26] reported a possible reaction mechanism of chitosan as a biopesticide (Fig. 10.1).

It is reported that chitin and chitosan have elicited activities (inducer of plant resistance), leading to various defense responses against microbial infection and pathogen attack in host plants [27, 28]. This eliciting property of chitosan makes it a potential antimicrobial agent to control plant disease caused by the pathogen [29]. Singh et al.

**Table 10.2** Plant defense mechanisms

|  | Structural | Chemical |
|---|---|---|
| Constitutive (passive performed) | Anatomical barriers (trichomes cuticle cell wall) | Preformed inhibitors (phytoanticipins: glycosides saponins alkaloids), antifungal proteins (lectins), and ribosome-inactivating proteins (RIP) |
| Inducible (active) | Cell wall strengthening (callose lignin and suberin appositions; oxidative extension crosslinking) | Oxidative burst hypersensitive response (HR) phytoalexins (phenylpropanoids) pathogenesis-related (PR) proteins |

Copyright © 2014, Springer-Verlag Berlin Heidelberg, Environ Sci. Pollut. Res., 2015, 22, 2935–2944. Reproduced from Ref. [20] with permission

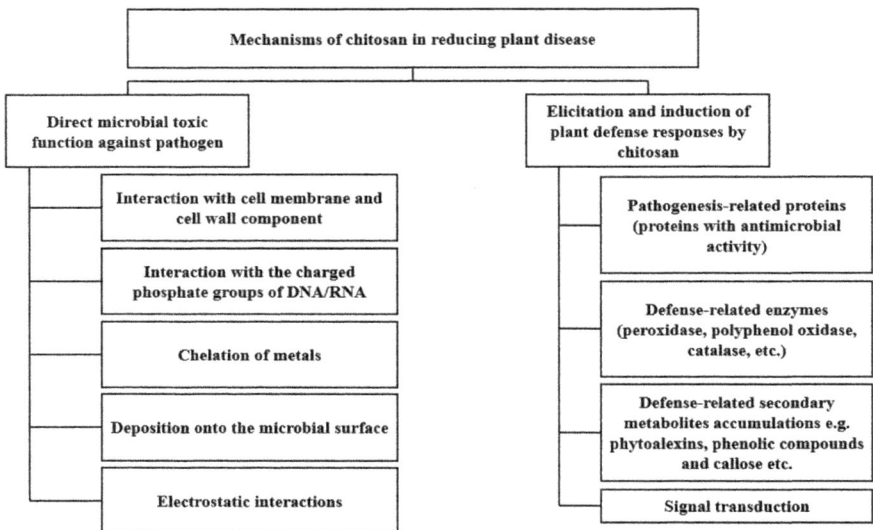

**Fig. 10.1** Mode of action of chitosan and its derivatives as antimicrobial compounds and their ability to elicit natural plant defense response. Copyright © 2017 Oliul Hassan and Taehyun Chang, Creative Commons open access license, Asian J. Plant Pathol., 2017, 11(2), 53–70, https://doi.org/10.3923/ajppaj.2017.53.70. Reproduced from Ref. [26]

[30] reviewed modes of action and the plant defense responses from chitosan use. Elicitors, classified as biotic and abiotic, have a huge role in plant defense mechanisms [5]. Chitosan has gained consideration as an elicitor and antifungal agent for plants due to its ability to induce phenolic compounds [2]. The biogenic elicitors of an oligosaccharide nature are the critical signal molecules in plants [31]. These elicitors generate resistance to phytopathogen [9]. It is reported that the elicitation of

**Table 10.3** Oligochitins/chit oligosaccharides as stimulators and elicitors of plant defenses

| Plants | Effects |
|---|---|
| Rice | Induction of phytoalexin |
| Wheat | Increase phenolic compounds |
| Pea | phytoalexin production |
| Tomato | Proteinase inhibitor synthesis |
| Soybean | Synthesis of callose |
| Parsley | Synthesis of callose |
| Potato | Enhance tube size |
| Strawberry | Increase fruits yields |
| Barley | Increase phenolic compounds |
| Maize | Increase seed weight |
| Rape | Increase chlorophyll |
| Basil | Increase phenolic compounds |

Copyrights © 2019 Brasselet, C., Guillaume, P., Dubessay, P. et al. Licensee MDPI, Basel, Switzerland. Creative Commons Attribution (CC BY) license, *Appl. Sci.* 2019, *9*, 1321. https://doi.org/10.3390/app9071321. Reproduced from Ref. [28]

plant defense responses by chitosan may be related to various pathogenesis-related (PR) proteins, defense-related enzymes, and secondary metabolites accumulation, as well as the complex plant defense signaling pathways [14, 15]. Table 10.3 shows the chitin/chitosan-based elicitors for plant defense [28].

It is reported that the disease resistance in the crop is correlated with the production of defense enzymes. These defense enzymes include phenylalanine ammonia-lyase (PAL), polyphenol oxidase (PPO), and peroxidase (POD) [16]. The enzyme PAL is used in the biosynthesis of phenolics, phytoalexins, and lignin, which increase plant resistance [23]. Enzymes POD and PPO contribute to forming defense barriers for reinforcing the cell structure [23]. These phenolic compounds are associated with the expression of resistance in plants against biotic (polysaccharides, microorganisms, glycoproteins) or abiotic (temperature, fungicides, antibiotics, heavy metals, pH) stressors [32]. Xing et al. [33] reviewed various growth inhibitory properties against bacteria, fungi, and viruses. They also discussed the modes of action for antimicrobial mechanisms. These mechanisms included electrostatic interactions, plasma membrane damage mechanism, chitosan-DNA/RNA interaction, deposition onto the microbial surface, and chitosan's metal chelation capacity. Zeng et al. [3] reported that the binding of chitosan with DNA and inhibition of mRNA synthesis occur through chitosan penetration toward the nuclei of the microorganisms and interference with the synthesis of mRNA and proteins. Chitosan interacts with cellular DNA generating multiple biochemical reactions in the plant, showing a rapid response against pathogens. Hence, it is considered an elicitor (a defense mechanism activator in plants) [16]. The mode of action of chitosan may also be related to its ability to chelate some essential nutrients, metal ions, and trace elements necessary to grow bacteria and fungi [26]. Chitosan can control pathogenic microorganisms and induce various defense responses in host plants by inhibiting different biochemical activities during the plant-pathogen interaction [22, 30]. Typical examples of chitosan used in plant-pathogen control are given in Table 10.4.

Kulikov et al. [31] reported that the resistance to phytopathogens in the plant might occur in vivo upon the destruction of cell walls leading to the inhibition of systemic propagation of viruses or viroids over a plant and enhancing the hypersensitivity response of plants to viral infection.

In another review, Kong et al. [24] reported that chitosan-mediated inhibition is affected by several factors such as (1) microbial factors related to microorganism species and cell age; (2) intrinsic factors of chitosan, including positive charge density, molecular weight, concentration, hydrophilic/hydrophobic characteristic, and chelating capacity; (3) physical state, namely water soluble and solid state of

Table 10.4 Chitosan effects on plant-pathogen

| Plant species | Chitosan formulation and administration | Pathogen/pest |
|---|---|---|
| Tomato | 0.4% (seeds soaking, fruits spraying) | *Phytophthora infestans, Alternaria solani* |
| Wheat | 0.1–0.5% (spikelets spraying) | *Fusarium graminearum* |
| Green bean | 0.025–0.2% (seeds soaking, foliar spraying) | *Fusarium solani, Rhizoctonia solani* |
| Cucumber | 0.05–0.1% (foliar spraying) | *Colletotrichum* spp. |
| Tea | 0.01% (foliar spraying) | *Exobasidium vexans* |
| Soybean | 0.5% (soil treatment) | *Heterodera glycines* |
| Tomato | 0.01% (plant irrigation) | *Meloidogyne* spp. |
| Banana | 1% (w/v) (in vivo) | *Anthracnose* |
| Carrots | 2 or 4% (w/v) (in vitro) | *Sclerotinia sclerotiorum* |
| Eggplant | 20 mL (cotton leaf disk elicitation method) | *Ralstonia solanacearum* |
| Mango | 1% (w/v) (post-harvest coating) | *Colletotrichum gloeosporioides* |
| Orange | 2% (w/v) (post-harvest coating) | *Penicillium italicum and Penicillium digitatum* |
| Pear | 25 g/L (post-harvest treatment) | *A. kikuchiana and P. piricola* |
| Papaya | 1.5% (w/v) (in situ) | *C. gloeosporioides* |
| Peach | 0.5 g/L (dipping in solution) | *Monilinia fructicola* |

Copyrights © 2018 Sharif, R., Mujtaba, M., Rahman, M., et al. Licensee MDPI, Basel, Switzerland. Creative Commons Attribution (CC BY) license, Molecules, 2018, 23, 872, https://doi.org/10.3390/molecules23040872. Reproduced from Ref. [15]; Copyright © 2018 Malerba, M., and Cerana, R. Licensee MDPI, Basel, Switzerland. Creative Commons Attribution (CC BY) license, Polymers (Basel). 2018 Feb; 10(2): 118. https://doi.org/10.3390/polym10020118. Reproduced from Ref. [34]

chitosan; (4) environmental factors, involving ionic strength in medium, pH, temperature, and reactive time. The growth of various fungal and bacterial plant pathogens inhibited by chitosan and its derivatives is given in Table 10.5.

Malerba and Cerana [35] reported that chitosan's mode of action for the antimicrobial mechanism is not yet known completely. Figure 10.2 shows the parameters that modulate chitosan's antimicrobial activity reported by Brasselet et al. [28].

Badawy and El-Aswad [36] suggested that chitosan metal ion complexes can perform as an alternative to pesticides. For example, chitosan complexes with metals of Ag(I), Cu (II), Ni (II), and Hg (II) showed good insecticidal activity against larvae of cotton leafworm *S. littoralis* and oleander aphid *A. nerii*. It was further reported that chitosan-Ni and chitosan-Hg were effective on cotton leaf larvae *S. littoralis* after

**Table 10.5** Typical example of the minimum growth inhibitory concentrations (MIC) of native chitosan or its derivatives against plant pathogens

| Microorganisms | Chitosan samples | MIC (ppm) |
|---|---|---|
| **Fungi** | | |
| *Botrytis cinerea* | Chitosan | 10 |
| *Drechstera sorokiana* | Chitosan | 10 |
| *Fusarium oxysporum* | Chitosan | 100 |
| *Micronectriella nivalis* | Chitosan | 10 |
| *Piricularia oryzae* | Chitosan | 5000 |
| *Rhizoctonia solani* | Chitosan | 1000 |
| *Trichophyton equinum* | Chitosan | 2500 |
| **Bacteria** | | |
| *Agrobacterium tumefaciens* | N- (o, o-dichlorobenzyl) chitosan | 500 |
| *Agrobacterium tumefaciens* | Quaternary N-(benzyl) chitosan | 500 |
| *Agrobacterium tumefaciens* | N-(benzyl) chitosan | 800 |
| *Clavibacter michganensis* subsp. *michganensis* | Chitosan | 1000 |
| *Erwinia carotovora* | Chitosan | 200 |
| *Erwinia carotovora* | N-(o, o-dichlorobenzyl) chitosan | 480 |
| *Erwinia carotovora* | Quaternary N-(benzyl) chitosan | 600 |
| *Erwinia carotovora* | N-(benzyl) chitosan | 700 |
| *Erwinia carotovora* | N-($\alpha$-methylcinnamyl) chitosan | 1025 |
| *Erwinia carotovora* subsp. *carotovora* | Chitosan | 5000 |
| *Xanthomonas campestris* | Chitosan | 500 |

Copyrights © 2014 Hemantaranjan et al., Creative Commons Attribution license, Adv Plants Agric Res 1(1): 00,006, https://dx.doi.org/10.15406/apar.2014.01.00006. Reproduced from Ref. [32] with permission; Xing, K. et al. Qin. Chitosan antimicrobial and eliciting properties for pest control in agriculture: a review. Agronomy for Sustainable Development. 2015, 35 (2), pp. 569–588. Reference [33]

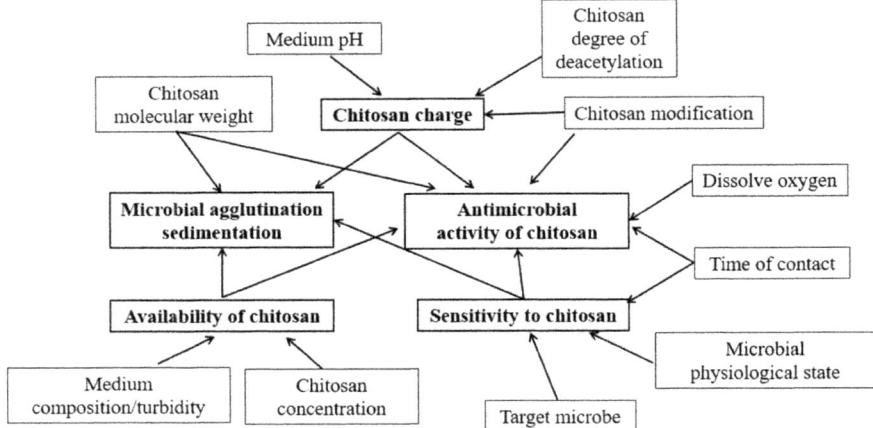

**Fig. 10.2** Overall main parameters modulating the antimicrobial activity of chitosan. Copyrights © 2019 Brasselet, C., Guillaume, P., Dubessay, P. et al. Licensee MDPI, Basel, Switzerland. Creative Commons Attribution (CC BY) license, *Appl. Sci.* 2019, *9*, 1321. https://doi.org/10.3390/app907 1321. Reproduced from Ref. [28]

three days of application. The possible different mechanisms for its antimicrobial effect are summarized in Table 10.6.

c. Chitosan and its derivatives for fertilizer

In general, fertilizers are used in agricultural sectors to provide essential nutrients such as nitrogen, phosphorus, and potassium to promote plant and fruit growth [37]. The growing demand for food production has increased the use of fertilizers for crop production. Fertilization of crop fields can be accomplished mainly through the soil (uptake by plant roots) or foliar feeding (uptake through leaves) [38]. The efficiency of conventional fertilizer uptake by the plant in the agricultural sectors is relatively low [39]. It is reported that traditional agrochemicals never achieved the objectives. About 40–70% of nitrogen, 80–90% of phosphorus, and 50–70% of potassium of the applied conventional fertilizers cannot be absorbed by the plant [37]. Therefore, excessive fertilizer application is necessary to increase crop productivity. Unutilized fertilizers have adverse environmental consequences [29]. The gradual demand for fertilizers in agricultural sectors has increased the interest for controlled release formulations to reduce wider environmental contamination by decreasing the toxicity to non-targeted organisms [40]. Controlled release formulations for agrochemicals may provide an optimum concentration of the ingredient in agriculture over an extended period of time [41, 42]. The benefits for the controlled release of fertilizer formulations are as follows:

1. The slow release of ingredients to the target minimizes environmental impact by reducing agrochemical leaching, volatilization, and degradation.
2. Controlled release formulation can increase the water-holding capacity of the soil.

## 10.2 Application of Chitosan in Agriculture and Agro-based Industry

**Table 10.6** Antimicrobial activities of chitosan

| Sample | Microorganism | Mechanism/action | Effect |
|---|---|---|---|
| Chitosan | Escherichia coli, Staphylococcus aureus | The phospholipids of the microbial cell plasma membrane have interacted | Leakage of intracellular substances disrupts the bacterial cell membrane |
| Chitosan | Escherichia coli, Staphylococcus aureus | Proteins of the microbial cell plasma membrane have interacted | Bacterial cell membrane integrity is disrupted |
| Chitosan | Escherichia coli | The charged phosphate groups of DNA/RNA have interacted | The synthesis of mRNA and proteins is inhibited |
| Chitosan | Escherichia coli, Staphylococcus aureus | Kill bacteria | Kill bacteria through cell membrane damage |
| Chitosan | Escherichia coli, Staphylococcus aureus | Electrostatic interaction | Destroy cell structure, induce the leakage of enzymes and nucleotides |
| Chitosan nanoparticle | Escherichia coli, Staphylococcus aureus | Interaction with cell wall | Damage cell membrane structure and putative bind to extracellular or intracellular targets |
| Chitosan | Alternaria alternata | Chelation of metals | Inhibition of toxin production and microbial growth |
| Chitosan | Escherichia coli, Bacillus cereus | Deposition of bacterial surface (high molecular weight of chitosan) | Blockage of nutrient flow |
| Chitosan | Pseudomonas syringae | Electrostatic interaction | Morphological changes and damage in bacterial surfaces |
| Chitosan | Rhizopus stolonifer | Negatively charged components of the cell surface have interacted | $H^+$-ATPase activity and chemiosmotic-driven transport are inhibited |
| Chitosan | Gram-negative and gram-positive bacteria | The phospholipids of the microbial cell plasma membrane (chitosan concentration <0.2 mg/mL) have interacted | Agglutination |

Copyrights © INRA and Springer-Verlag France 2014, Agron. Sustain. Dev. (2015) 35:569–588. Reproduced from Ref. [33] with permission; Copyrights © 2016 Malerba, M., and Cerana, R. licensee MDPI, Basel, Switzerland. Creative Commons Attribution (CC-BY) license, Int. J. Mol. Sci. 17, 996, 2016. https://doi.org/10.3390/ijms17070996. Reproduced from Ref. [35]

3. Controlling weeds in the long run.
4. Controlled release of polymer–clay formulations can store ionic plant nutrients.
5. Polymer hydrogel formulations can reduce compaction, erosion, and water run-off.

Several studies investigated the chitosan application to encapsulate bio-active materials for crop production [39, 43, 44]. Chitosan-coated fertilizer [43] improves soil properties by slowly releasing fertilizers and nutrients into the soil [35]. It also enhances the productivity and quality of plants without additional fertilizers and agricultural chemicals [43, 45]. Chitosan is used to microencapsulate water-soluble chemicals, which are potential carriers for the controlled release of agrochemicals [46, 47]. Maharani et al. [48] investigated the fertilizer release rate from zeolite and chitosan composite. They reported that zeolite-chitosan fertilizer has the slowest nitrogen release than zeolite-fertilizer composite. Followings are some advantages of encapsulating agrochemicals and genetic materials in a chitosan matrix [47]:

- Ability to function as a protective reservoir for the active ingredients.
- Protect the ingredients from the surrounding environment.
- Controlled release of the ingredients and allow them to serve as:
    - An efficient gene delivery system.
    - An efficient release of fertilizers system.
    - An efficient release of the pesticide system.

Table 10.7 shows a typical example of fertilizer and micronutrient encapsulated in a chitosan-based controlled release matrix.

Several studies investigated chitosan nanoparticles based on nano- and subnanocomposites. They also analyzed their application as carriers in agriculture to improve nutrient delivery [14, 45, 47]. In the case of chitosan nanoparticles coated with the NPK nutrients, depending on the reaction conditions, the –COO and –$NH_2$ functional groups of chitosan nanoparticles can interact electrostatically with the nutrients N-, P-, and K-based fertilizer. It is reported that chitosan nanoparticles-coated NPK fertilizers improve fertilizer degradation rate and slow-release pattern of the nutrients [16]. Abdel-Aziz et al. [46] investigated wheat plants' growth and productivity rate in sandy soil by the foliar uptake of chitosan nanoparticles-coated NPK nutrients. The plant growth and productivity increased significantly compared to the control wheat plant treated with or without normal fertilized NPK. Al-Tawaha et al. [14] stated that foliar application of chitosan nanoparticles can improve plant growth for non-saline treated plants. This foliar application of chitosan also alleviated the adverse effects of salinity on the shoot and root growth. Kashyap et al. [47] described different processes of chitosan nanoparticle synthesis for possible application in agriculture. Figure 10.3 shows a typical example of chitosan-based nanomaterials and their possible applications in agriculture.

Grilli et al. [49] prepared nanoparticles using chitosan (CS) and sodium tripolyphosphate (TPP) as a safer herbicide formulation to control weeds. They concluded that the CS/TPP nanoparticles showed good encapsulation efficiency (62%) and were stable for at least 60 days. In another attempt, Choudhary et al.

**Table 10.7** Example of fertilizers and micronutrients encapsulated in chitosan-based controlled release matrix

| Matrices | Active ingredient | Releasing rate |
| --- | --- | --- |
| Chitosan nanoparticles | NPK fertilizer | 15% by the 3rd day and 75% by the 30th day |
| Chitosan-methacrylic acid particles (diameter ~78 nm) | NPK fertilizer | n.d |
| Chitosan microsphere (diameter ~200 mm) | Urea | n.d |
| Chitosan-montmorillonite microspheres (diameter ~200 mm) | $KNO_3$ | Fast for the first three days. Then, continuous K release for at least 60 days |
| Chitosan-EDTA | Urea | n.d |
| Chitosan-suberoyl chloride particles, crosslinking densities ranges from 0 to 7.4% | $Zn^{2+}$; $Cu^{2+}$ | 40 mg release after 6 h (0% crosslink density) 15 mg release after 6 h (7.4% crosslink density) |
| Chitosan-phthalic anhydride | 1-naphthyl acetic acid | Slow continuous release for several weeks. For example, at 20 °C, 10% and 25% by the 10th and 60th day, respectively |

Copyrights © 2016 Malerba, M., and Cerana, R. Licensee MDPI, Basel, Switzerland. Creative Commons Attribution (CC-BY) license, Int. J. Mol. Sci. 17, 996, 2016. https://doi.org/10.3390/ijms17070996. Reproduced from Ref. [35]

*n.d.* Not detectable

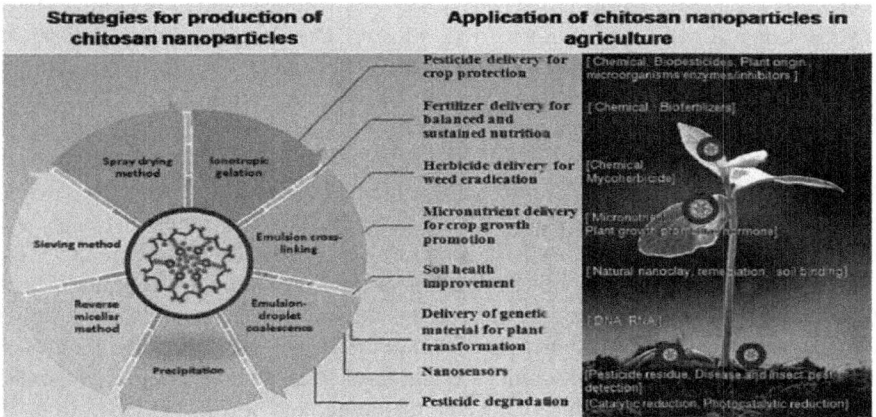

**Fig. 10.3** Strategies to produce chitosan nanoparticles and their applications as a delivery system in agriculture. Copyright © 2015 Elsevier B.V. International Journal of Biological Macromolecules, 2015, 77, 36–51. Reproduced from Ref. [47] with permission.

[50] studied the potential use of Cu-chitosan nanoparticles (NPs) ability to boost defense against Curvularia leaf spot (CLS) disease of maize and plant growth. It was reported that plants treated with Cu-chitosan NPs showed significant defense response through higher antioxidant activities (superoxide dismutase and peroxidase) and defense enzymes (polyphenol oxidase and phenylalanine ammonia-lyase). In addition, significant control of maize's CLS disease was recorded at 0.04–0.16% of Cu-chitosan nanoparticles treatments in a pot and 0.12–0.16% of Cu-chitosan nanoparticles treatments in field conditions.

Chitin or chitosan and their derivatives show an enhancement of fungicidal and insecticidal activities to control blight and harmful insects [5, 8]. The versatile uses of chitin and chitosan in animal farming and agricultural product processing make it a promising biomaterial. Świątkiewicz et al. [51] reviewed chitosan and its oligosaccharide derivatives (chitooligosaccharides) as feed supplements in poultry and swine nutrition. Chitosan possesses immunomodulatory, antioxidative, antimicrobial, and hypocholesterolemic properties. Therefore, chitosan has some benefits as a feed additive for poultry and pig [51]. Chitosan is used as semi-permeable antifungal coatings on fresh fruits to extend shelf life and control mold growth during cold storage by minimizing the rate of respiration and reducing water loss [52, 53].

d. Chitosan and its derivatives for Food Industries

Chitin and chitosan are not soluble in inorganic acids, restricting their use in physiological functional foods [54]. However, chitosan modified with a chemical or enzyme functional group improves its solubility in water and biocompatibility with animal and human cells and tissues [55]. Hence, the modified chitosan is acceptable in food technology and for functional food production [56]. Chitosan and its derivatives have a wide range of applications in the food industry. The potential use of chitosan as a dietary weight loss supplement, lipid-lowering [57], cholesterol-lowering [58], oil or fat entrapping [59], and edible film [60] quality has been investigated for food industries.

Fiber is an essential compound in the diet [61]. Fiber binds bile acids and lipids in the small intestine [62]. Fiber influences protein metabolism and limits the cholesterol level in the blood [58]. Bile acids act as biological surfactants that help food digestion and absorption of lipids in the gastrointestinal (GI) tract. Bile acids are the cholesterol-based amphiphilic molecules produced in the liver [63]. It has been reported that the bile acid-binding capabilities of chitosan or insoluble dietary fiber are moderate or low [64]. Increasing fecal excretion of bile acids has long been recognized as a mechanism to reduce cholesterol [62]. Chitosan can bind fat and cholesterol through electrostatic effects, embedding, and adsorption [58]. Su et al. [65] reported that chitosan could form positively charged gel in an acidic stomach environment. Negatively charged molecules such as fats, fatty acids, lipids, bile acids can be attached firmly to the positively charged chitosan, avoiding their reabsorption and causing fecal excretion. Chitosan-associated *Brassica olearaceae L.* was found to be most efficient for lowering the total cholesterol, LDL-cholesterol, VLDL-cholesterol, and triglycerides levels in blood serum [66].

Chitosan possesses properties similar to dietary fiber. Once ingested, chitosan can pass into the large intestine intact without being digested in the small intestine. Jin et al. [58] reported an increase in the pH of the duodenum and intestine solutions as the digestion progresses. Chitosan molecules tend to lose their charge and precipitate with trapped molecules when the pH rises above 6.5. Maezaki et al. [67] reported that the bile acids could combine with chitosan in the digestion tract, and the combined products finally excrete into the feces. This excretion reduces molecules, including cholesterol, sterols, and triglycerides pool in the body [58]. Chitosan has been reported to improve lipid metabolism by regulating total cholesterol (TC) and low-density lipoprotein cholesterol (LDL-C) by upregulating (stimulating) hepatic LDL receptor mRNA expression and increasing the excretion of fecal bile acids [68, 69]. Vakili et al. [70] reported molecular imprinting cholic acid to chitosan reduced blood cholesterol levels significantly. Kim et al. [71] suggested that chitosan oligosaccharides could improve the altered blood glucose metabolism in diabetic rats. They suggested accelerated proliferation, neogenesis of $\beta$-cells, and increased insulin secretory capacity are the reasons for improving the altered blood glucose metabolism [71].

A diet with high dietary fiber (DF) is considered healthy. The dietary intake with sufficient dietary fiber (DF) is recommended worldwide [61]. The growing demand for dietary supplements for weight loss has drawn interest in chitosan as a valuable alternative source of fiber due to its capability to adsorb lipophilic and acidic substances. Gades and Stern [72] showed that chitosan supplementation does not affect fat absorption in healthy males fed a high fat diet. However, chitosan is reported to have the capacity to modify fat absorption. Therefore, chitosan is promoted primarily for use in obesity and hyperlipidemia [73, 74].

The antimicroorganism properties of chitosan are useful to control fungal growth in post-harvest products and as natural food preservatives [75]. The chitosan-based edible film has been widely used as coating and packaging materials to enhance the shelf life of ready-to-eat foods [60, 76]. Chitosan kills bacteria through cell membrane damage [77]. The inhibitory activity of chitosan-based coatings has been investigated for food pathogen [78] and spoilage microorganisms [79] in foods and laboratory media. It is noted that chitosan can be used as a preservative in certain types of food. The antimicrobial activity following partial hydrolysis is too small to justify the extra processing involved [80].

## 10.3 Summary

In the agricultural sector, chitosan and its derivatives receive attention for an eco-friendly use as a plant growth promoter, pathogen control, and extending the post-harvest life of fruits and vegetables. These chitosan-based processes can be considered sustainable alternatives to harmful agrochemicals due to their biodegradability, biocompatibility, antibacterial, antiviral, antioxidant, and immunostimulatory properties. Chitosan has the potential to stimulate seed germination and plant immune systems as it can alter the permeability of the seed plasma membrane, change the concentration of sugars and proline (amino acid), and enhance defense enzyme activities. The eliciting activities of chitosan are useful in defense responses against microbial infection and pathogen attack in host plants. Furthermore, chitosan-coated fertilizer appears to improve soil properties by slow release of fertilizers and nutrients into the soil, thereby reducing the side effects of agrochemicals on the environment. The potential use of chitosan in the food industry is also encouraging.

Chitosan and its derivatives are used for dietary weight loss supplements, lipid-lowering, cholesterol-lowering, oil, or fat entrapping. The inhibitory activity of chitosan for food pathogen and spoilage microorganisms has increased its potential as a preservative in certain types of food processing applications. Chitosan-based technologies are promising in fulfilling the current demand for sustainable, eco-friendly farming and food processing.

## References

1. Lizarrage-Paulin E-G et al (2013) Maize seed coatings and seedling sprayings with chitosan and hydrogen peroxide: their influence on some phenological and biochemical behaviors. J Zhejiang Univ-Sci B (Biomed Biotechnol) 14(2):87–96
2. Silva-Castro L, Diez JJ, Martin-Ramos P et al (2018) Application of bioactive coatings based on chitosan and propolis for pinus ssp. Prot Against *Fusarium circinatum* Forests 9:685. https://doi.org/10.3390/f9110685
3. Zeng D, Mei X, Wu J (2010) Effects of an environmentally friendly seed coating agent on combating head smut of corn caused by *Sphacelotheca reiliana* and corn growth. J Agric Biotechnol Sustain Dev 2(6):108–112
4. Zeng D, Luo X, Tu R (2012) Application of bioactive coatings based on chitosan for soybean seed protection. Int J Carbohydr Chem 2012, Article ID 104565, p 5
5. Teixeira de Castro GA (2017) Chitosan as seed soaking agent: germination and growth of Coriandrum sativum and *Solanum lycopersicum*. Masters Degree in Marine Resources Biotechnology, Escola Superior de Turismo e Tecnologia do Mar-Mar -IPL, Portugal
6. Carvalho FP (2017) Pesticides, environment, and food safety. Food Energy Secur 6(20):48–60
7. Brozozowski L, Mazourek M (2023) A sustainable agricultural future relies on the transition on organic agroecological pest management. Sustainability 2018:10. https://doi.org/10.3390/su10062023
8. Kumar GP, Desai S, Moerschbacher BM et al (2019) Seed treatment with chitosan synergizes plant growth promoting ability of *Pseudomonas aeruginosa*-P17 in sorghum (*Sorghum bicolor* L.), bioRxiv preprint first posted online 7 Apr 2019. https://doi.org/10.1101/601328.
9. Winkler AJ, Dominguez-Nunez JA, Aranaz I et al (2017) Short-chain chitin oligomers: promoters of plant growth. Mar Drugs 15(40). https://doi.org/10.3390/md15020040, www.mdpi.com/journal/marinedrugs
10. Choudhary RC, Kumaraswamy RV, Kumari S et al (2017) Cu-chitosan nanoparticle boost defense responses and plant growth in maize (*Zea mays* L). Sci Rep 7:9754. https://doi.org/10.1038/s41598-017-08571-0
11. Sharp RG (2013) A review of the applications of chitin and its derivatives in agriculture to modify plant-microbial interactions and improve crop yields. Agronomy 3:757–793. https://doi.org/10.3390/agronomy3040757
12. Badawy MEI, Rabea EI (2011) A biopolymer chitosan and its derivatives as promising antimicrobial agents against plant pathogens and their applications in crop protection. Int J Carbohydr Chem, Article ID 460381, p 29. https://doi.org/10.1155/2011/460381

13. El Hadrami A, Adam LR, El Hadrami I, Daayf F (2010) Chitosan in plant protection. Mar Drugs 8:968–987
14. Al-Tawaha AR, Turk MA, Al-Tawaha et al (2018) Using chitosan to improve growth of maize cultivars under salinity conditions. Bul J Agric Sci 24(3):437–442
15. Sharif R, Mujtaba M, Rahman M et al (2018) The multifunctional role of chitosan in horticultural crops, a review. Molecules 23:872. https://doi.org/10.3390/molecules23040872
16. Pandey P, Verma MK, De N (2018) Chitosan in agricultural context- a review. Bull Env Pharmacol Life Sci 7(4):87–96
17. Siddaiah CN et al (2018) Chitosan nanoparticles having a higher degree of acetylation induce resistance against Perl millet downy mildew through nitric oxide generation. Sci Rep 8:2485. https://doi.org/10.1038/s41598-017-19016-z
18. González LM et al (2015) Effect of seed treatment with chitosan on the growth of rice (*Oryza sativa* L.) seedlings cv. Inca LP-5 in saline medium. Cultivos Troicales 36(1):136–142
19. Worrall EA, Hamid A, Mody KT et al (2018) nanotechnology for plant disease management. Agronomy 8:285. https://doi.org/10.3390/agronomy8120285
20. Iriti M, Varoni EM (2015) Chitosan-induced antiviral activity and innate immunity in plants. Environ Sci Pollut Res 22:2935–2944
21. Zhang M et al (2003) Insecticidal and fungicidal activities of chitosan and oligo-chitosan. J Bioact Compat Polym 18:391–400
22. Katiyar D, Hemantaranjan A, Singh B (2015) Chitosan as a promising natural compound to enhance potential physiological responses in a plant: a review. Ind J Plant Physiol 20(1):1–9. https://doi.org/10.1007/s40502-015-0139-6
23. Duan C, Yu J, Bai J et al (2014) Induced defense responses in rice plants against small brown planthopper infection. Crop J 2:55–62
24. Kong M, Chen XG, Xing K, Park HJ (2010) Antimicrobial properties of chitosan and mode of action: a state of the art review. Int J Food Microbiol 144:51–63
25. Orzali L, Corsi B, Forni C, Riccioni L (2017) Chitosan in agriculture: a new challenge for managing plant disease, Chapter-2. In: Chitosan in agriculture: a new challenge for managing plant disease. INTECH, pp 17–36. https://doi.org/10.5772/66840
26. Hasan O, Chang T (2017) Chitosan for eco-friendly control of plant disease. Asian J Plant Pathol 11(2):53–70
27. Pusztahelyi T (2018) Chitin and chitin related compounds in plant-fungal interactions. Mycology 9(3):189–201
28. Brasselet C, Guillaume P, Dubessay P et al (2019) Modification of chitosan: how generating new functional derivatives? Appl Sci MDPI 9:1321. https://doi.org/10.3390/app9071321www.mdpi.com/journal/applsci
29. Xing K, Zhu X, Peng X, Qin S (2015) Chitosan antimicrobial and eliciting properties for pest control in agricultural: a review. Agron Sustain Dev 35:569–588
30. Singh A, Gairola K, Upadhyay V, Kumar J (2018) Chitosan: an elicitor and antimicrobial bio-resource in plant protection. Agric Rev 39(2):163–168
31. Kulikov SN, Chirkov SN, Il'ina AV et al (2006) The effect of the molecular weight of chitosan on its antiviral activity in plants. Appl Biochem Microbiol 42(2):200–203
32. Katiyar D, Hemantaranjan A, Bharti S, Nishant Bhanu A (2014) A future perspective in crop protection: chitosan and its oligosaccharides. Adv Plants Agric Res 1(1):00006. https://doi.org/10.15406/apar.2014.01.00006
33. Xing K, Zhu X, Peng X, Qin S (2015) Chitosan antimicrobial and eliciting properties for pest control in agricultural: a review. Agron Sustain Dev 35(2):569–588. https://doi.org/10.1007/s13593-014-0252-3 (Springer Verlag/EDP Sciences/INRA)
34. Malerba M, Cerana R (2018) Recent advances of chitosan applications in plants. Polymers 10:118. https://doi.org/10.3390/polym10020118
35. Malerba M, Cerana R (2016) Chitosan effects on plant systems. Int J Mol Sci 17:996. https://doi.org/10.3390/ijms17070996 www.mdpi.com/journal/ijms
36. Badawy MEI, El-Aswad AF (2012) Insecticidal activity of chitosans of different molecular weight and chitosan-metal complexes against cotton leafworm *Spodoptera littoralis* and oleander aphid *Aphis nerii*. Plant Protect. Sci. 48(3):131–141

37. Corradini E, de Moura MR, Mattoso LHC (2010) A preliminary study of the incorporation of NPK fertilizer into chitosan nanoparticles. eXPRESS Polym Lett 4(8):509–515
38. Abdel-Aziz HMM (2019) Effect of priming with chitosan nanoparticles on germination, seeding growth and antioxidant enzymes of broad beans. Catrina 18(1):81–86
39. Kusumastuti Y, Rochmadi AI, Purnomo CW (2019) Chitosan-based polyion multilayer coating on NPK fertilizer as controlled released fertilizer. Adv Mater Sci Eng 2019, Article ID 2958021. https://doi.org/10.1155/2019/2958021
40. Maruyama CR, Guilzer M, Pascoli M et al (2016) Nanoparticles based on chitosan as carriers for the combined herbicides Imazapic and Imazapyr. Sci Rep 6:19768. https://doi.org/10.1038/srep19768
41. Campos EVR, de Oliveira JL, Fraceto LF et al (2015) Polysaccharides as safer release systems for agrochemicals. Sustain Dev 35:47–66
42. Liu Y, Sun Y, He S et al (2013) Synthesis and characterization of gibberellin-chitosan conjugate for controlled-release applications. Int J Biol Macromol 57:213–173
43. Roshanravan B, Soltani SM, Rashid SA (2015) Enhancement of nitrogen release properties of urea-kaolinite fertilizer with chitosan binder. Chem Speciat Bioavailab 27(1):44–51
44. Abdul Hamid NN, Mohamad N, Hing LY et al (2013) The effect of chitosan content on physical and degradation properties of biodegradable urea fertilizer. J Sci Innov Res 2(5):893–902
45. Naderi MR, Danesh-Shahraki A (2013) Nano fertilizers, and their roles in sustainable agriculture. Int J Agri Crop Sci 5(19):2229–2232
46. Abdel-Aziz HMM, Hasaneen MNA, Omer AM (2016) Nano chitosan-NPK fertilizer enhances the growth and productivity of wheat plants grown in sandy soil. Spanish J Agric Res 14(1):e0902. https://doi.org/10.5424/sjar/2016141-8205
47. Kashyap PL, Xiang X, Heiden P (2015) Chitosan nanoparticle-based delivery systems for sustainable agriculture. Int J Biol Macromol 77:36–51
48. Maharani D, Dwiningsih K, Savana R, Andika P (2018) Usage of zeolite and chitosan composites as slow-release fertilizer. https://doi.org/10.2991/icst-18.2018.38
49. Grille R, Pereira AES, Nishisaka CS et al (2014) Chitosan/tripolyphosphate nanoparticles loaded with paraquat herbicide: an environmentally safer alternative for weed control. J Hazard Mater 278:163–171
50. Choudhary RC, Kumaraswamy RV, Kumari S et al (2017) Cu-chitosan nanoparticle boost defense responses and plant growth in maize (*Zea mays* L.). Sci Rep 7:9754. https://doi.org/10.1038/s41598-017-08571-0
51. Swiatkiewicz S, Swiatkiewicz M, Arczewska-Wlosek A, Jozefiak D (2015) Chitosan and its oligosaccharides derivatives (Chito-oligosaccharides) as feed supplements in poultry and swine nutrition. J Animal Physiol Anim Nutr 99:1–12
52. No HK, Kim Soon D, Prinyawiwatkul W, Meyers SP (2006) Growth of soybean sprouts affected by chitosans prepared under various deproteinization and demineralization times. J Sci Food Agric 86(9):1365–1370
53. Park S-I, Stan SD, Daeschel MA, Zhao Y (2005) Antifungal coatings on fresh strawberries (Fragaria × ananassa) to control mold growth during cold storage. J Food Sci 70(4):M202–M207
54. Jeon Y-J, Shahidi F, Kim S-K (2000) Preparation of chitin and chitosan oligomers and their applications in physiological functional foods. Food Rev Int 16(2):159–176
55. Synowiecki J, Al-Khateeb NA (2003) Production, properties, and some new applications of chitin and its derivatives. Crit Rev Food Sci Nutr 43(2):145–171
56. Seiichi T, Hiroshi T (2004) Functions of chitin and chitosan acceptable for food science. Foods Food Ingredients J Jpn 209:4
57. Wu S, Pan H, Tan S et al (2017) In vitro inhibition of lipid accumulation induced by oleic acid and in vivo pharmacokinetics of chitosan microspheres (CTMS) and chitosan-capsaicin microspheres (CCMS). Food Nutr Res 6:1331658
58. Jin et al (2017) The effect of the molecular weight of water-soluble chitosan on its fat-/cholesterol-binding capacities and inhibitory activities to pancreatic lipase. Peer J 5: e3279. https://doi.org/10.7717/peerj.3279

59. Badwan AA, Amro BI, Nazzal HG (2002) Use of a chitosan solution for entrapping oil and/or fat. Eur Pat Appl 11 pp. EP 1226810 A1. Application: EP 2001-101601 20010125
60. Otoni CG, Avena-Bustillos RJ, Azeredo HMC et al (2017) Recent advances on edible films based on fruits and vegetables—a review. Compr Rev Food Sci Food Saf 16:1151–1169
61. Capuano E (2017) The behavior of dietary fiber in the gastrointestinal tract determines its physiological effect. Crit Rev Food Sci Nutr 57(16):3543–3564
62. Jesch ED, Carr TP (2017) Food ingredients that inhibit cholesterol absorption. Prev Nutr Food Sci 22(2):67–80
63. Mendonca PV, Serra AC, Silva CL et al (2013) Polymeric bile acid sequestrants- synthesis using conventional methods and new approaches based on controlled/living radical polymerization. Prog Polym Sci 38:445–461
64. van Bennekum AM, Nguyen DV, Schulthess G, Hauser H, Phillips MC (2005) Mechanisms of cholesterol-lowering effects of dietary insoluble fibers: relationships with intestinal and hepatic cholesterol parameters. Br J Nutr 94(3):331–337
65. Su C-Y, Ho H-O, Chen Y-C et al (2018) Complex hydrogels composed of chitosan with ring-opened polyvinyl pyrrolidone as a gastroretentive drug dosage form to enhance the bioavailability of bisphosphonates. Sci Rep 8:8092. https://doi.org/10.1038/s41598-018-26432-2
66. Geremias R, Pedrosa RC, Locatelli C, de Favere VT, Coury-Pedrosa R, Laranjeira MCM (2006) Lipid-lowering activity of hydrosoluble chitosan and association with *Aloe vera* L. and *Brassica olearaceae* L. Phytotherapy Res 20(4):288–293
67. Maezaki Y, Tsuji K, Nakagawa Y et al (1993) Hypocholesterolemic effect of chitosan in adult males. Biosci Biotech Biochem 57(9):1439–1444
68. Xu G, Huang X, Qiu L et al (2007) Mechanism study of chitosan on lipid metabolism in hyperlipidemic rats. Asia Pac J Clin Nutr 16(1):313–317
69. Jiang Y, Fu C, Liu G et al (2018) Cholesterol-lowering effects and potential mechanisms of chitooligosaccharide capsules in hyperlipidemic rats. Food Nutr Res 62:1446. https://doi.org/10.29219/fnr.v62.1446
70. Vakili M-H, Khodaee A, Kordi S (2003) Molecular imprinting of cholic acid to chitosan to decrease blood cholesterol level. In: 8th Iranian national chemical engineering congress, Mashhad, Islamic Republic of Iran, 19–21 Oct 2003, 138/1–138/7
71. Kim JN, Chang LY, Kim HI, Yoon SP (2009) Long-term effects of chitosan oligosaccharide in streptozotocin-induced diabetic rats. Islets 1:2:111–116
72. Gades MD, Stern JS (2002) Chitosan supplementation does not affect fat absorption in healthy males fed a high-fat diet, a pilot study. Int J Obes 26(1):119–122
73. Zhang H-L, Zhong X-B, Tao Y et al (2012) Effects of chitosan and water-soluble chitosan micro-and nanoparticles in obese rats fed a high-fat diet. Int J Nanomed 7:4069–4076
74. Haitao P, Qingyun Y, Guidong H et al (2016) Hypolipidemic effects of chitosan and its derivatives in hyperlipidemic rats induced by a high-fat diet. Food Nutr Res 60:1:31137. https://doi.org/10.3402/fnr.v60.31137
75. Jiang HJ, Sun Z, Jia R et al (2016) the effect of chitosan as an antifungal and preservative agent on post-harvest blueberry. J Food Qual 39:516–523
76. Santonicola S, Ibarra VG, Sendon R et al (2017) Antimicrobial films based on chitosan and methylcellulose containing natamycin for active packaging applications. Coatings 7:177. https://doi.org/10.3390/coatings7100177
77. Xing Y, Xu Q, Li X et al (2016) Chitosan-based coating with antimicrobial agents: Preparation, property, mechanism, and application effectiveness fruits and vegetables. Int J Polym Sci 2016, Article IDD 4851730. https://doi.org/10.1155/2016/4851730
78. Karsli B, Caglak E, Li D et al (2019) Inhibition of selected pathogens inoculated on the surface of catfish fillets by high molecular weight chitosan coating. Int J Food Sci Technol 54:25–33
79. Brown SRB, Kozak SM, D'Amico DJ (2018) Applications of edible coatings formulated with antimicrobials inhibit *Listeria monocytogenes* growth on Queso Fresco. Front Sustain Food Syst 2:1. https://doi.org/10.3389/fsufs.2018.00001

80. Rhoades J, Roller S (2000) Antimicrobial actions of degraded and native chitosan against spoilage organisms in laboratory media and foods. Appl Environ Microbiol 66(1):80–86

# Chapter 11
# Applications of Chitosan in Fuel Cells

**Abstract** Polymer electrolyte membranes play an important role in a fuel cell. Chitosan-based polyelectrolyte membrane can be used in fuel cells as a substitute for conventional Nafion membrane. This chapter discusses the types of different fuel cells and the performance of various chitosan-based polyelectrolyte membranes in fuel cells.

## 11.1 Proton Exchange Membrane Fuel Cells

Fuel cells convert chemical energy directly into electrical energy without combustion. This chemical energy is generated from a redox reaction in the fuel cells. The term "fuel cell" was used in 1889 by Ludwig Mond and Charles Langer, who built a practical device using air and industrial coal gas that became the basis for modern-day fuel cells. Fuel cells are considered a promising alternative energy source. Fuel cells can offer many advantages, such as high efficiency, high energy density, quiet operation, and environmental friendliness [1–3]. The proton exchange membrane (polymer electrolyte membrane) is composed of non-conductive electron material in fuel cell operation. This membrane is responsible for the internal transport of protons from the anode to the cathode electrode [4]. The movement of the protons occurs from the anode to the cathode of the fuel cells through the ionic channels of the proton exchange membrane (PEM). These ionic channels are formed by micro- or nanophase separation between the hydrophilic proton exchange sites and the hydrophobic domains [5]. This proton movement mechanism is essential in fuel cell operation. A single fuel cell is the typical basic structure of most fuel cells and is shown in Fig. 11.1.

Vaghari et al. [1] described different types of fuel cells based on the electrolyte material. These cells consist of an electrolyte material packed between two thin electrodes (porous anode and cathode). The six generic fuel cells in various stages of development are as follows [7]: (i) fuel cells with PEM; (ii) direct $CH_3OH$ fuel cells (DMFCs); (iii) alkaline fuel cells; (iv) phosphoric acid fuel cells; (v) molten carbonate fuel cells; and (vi) solid oxide fuel cells. Hydrogen fuel cell, fuel cells with polymer electrolyte membrane (PEMFCs), and direct methanol fuel cells (DMFCs)

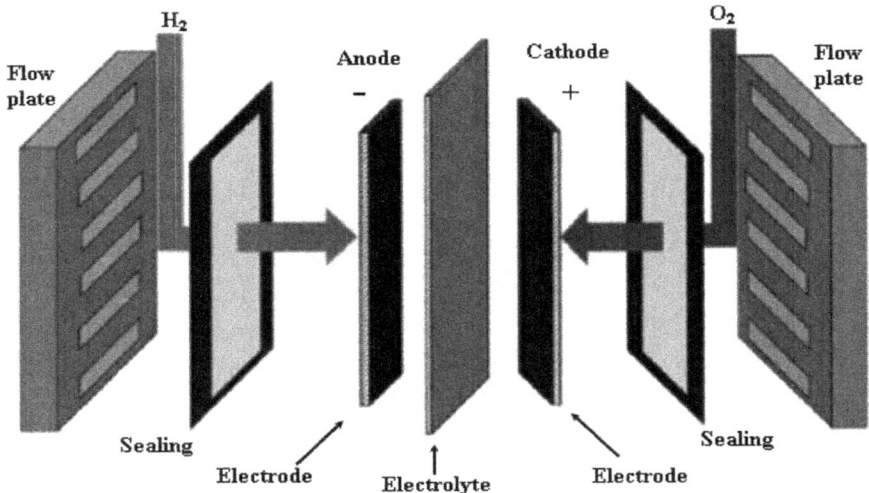

**Fig. 11.1** Basic components of a single cell fuel cell. Copyright © The Royal Society of Chemistry 2005, Green Chem 7: 2005, 132–150. Reproduced from Ref. [6] with permission

can be used for energy supply in portable electronic devices due to their capability to provide high energy density, simplified system design, and convenient fuel [8].

## 11.2 Proton Exchange Membrane (PEM) Fuel Cell

Proton exchange membrane (PEM) fuel cells, also known as fuel cells with polymer electrolyte membrane, are developed mainly for automobile applications. The growth of the fuel cell industry depends to a great extent on the PEM fuel cells. PEM fuel cells can be used in automobiles and as an important component for renewable energy sources. The essential elements of a PEM fuel cell may be categorized as follows:

- Membrane/electrode assembly.
- Catalyst.
- Backing layer.
- Bipolar plates.
- Stacking.
- Water and heat management systems.

A solid thin polymer membrane in PEM fuel cells serves as the electrolyte. Hydrogen is fed in the anode containing platinum as a catalyst. Platinum breaks down hydrogen atoms into protons and electron. The polymer membrane is permeable to protons but must be moist for proton transportation. The membrane does not conduct electrons. Instead, the electrons move through an external circuit to the

## 11.2 Proton Exchange Membrane (PEM) Fuel Cell

cathode and produce electric power. Oxygen, usually from the air, is supplied to the cathode. Oxygen combines with electrons and protons to produce water. The overall PEM fuel cells reactions are as follows:

Anode Reaction: $2H_2 \rightarrow 4H^+ + 4e^-$.
Cathode Reaction: $O_2 + 4H^+ + 4e^- \rightarrow 2H_2O$.
Overall Cell Reaction: $2H_2 + O_2 \rightarrow 2H_2O$.

PEM fuel cells generate more power for a given volume or weight than any other fuel cells. PEM fuel cells are compact, lightweight, and most suitable for automobile applications. The operating temperature is less than 100 °C, which allows rapid start-up. The PEM fuel cells use solid electrolyte, which makes the sealing of the components simpler and manufacturing costs lower. However, the low temperature is not high enough for useful cogeneration. Both electrodes must be moist for proton transport; therefore, careful water management is necessary.

### 11.2.1 Membrane

The polymer electrolyte membrane is the most critical component of the fuel cell. The most widely used membrane is called Nafion$^{TM}$, which is manufactured by DuPont, USA. The chemical name of Nafion is polyperfluorosulfonic acid, which is a solid organic polymer. Its chemical structure is shown in Fig. 11.2.

The sulfonic acid ions, $SO_3^-$ $H^+$, play a crucial role in proton transport through the membrane. The membrane must be hydrated for the proton to be transported through it. The proton transported on the catalyst surface attaches to the water molecule by forming the hydrogen bonds which creates hydronium ion ($H_3O^+$). These hydronium ions attach themselves to $SO_3^-$ sites releasing $H^+$ ions, which then again attach to

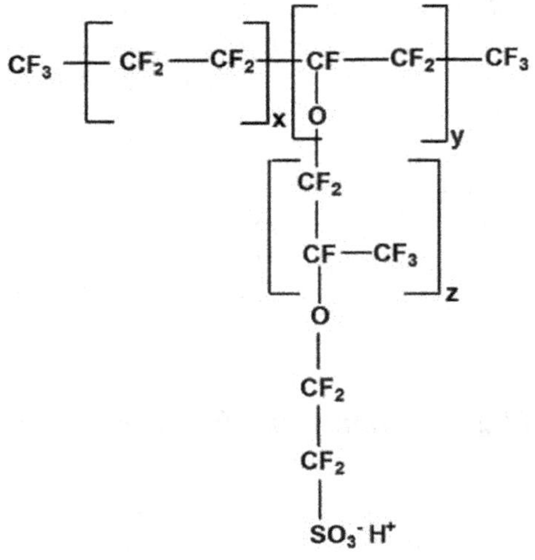

**Fig. 11.2** Chemical structure of Nafion membrane ($x = 5$–$10$, $y = 1000$, $z = 1$–$2$)

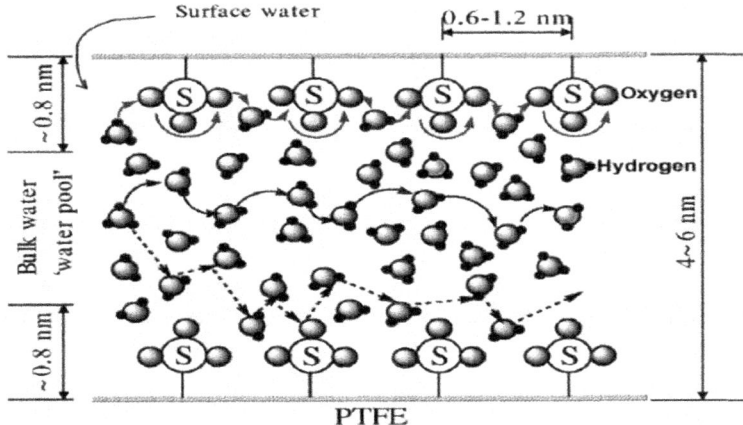

**Fig. 11.3** Simplified representation of proton transfer in the Nafion membrane. Copyright © 2005 ECS—The Electrochemical Society, Pyoungho Choi et al. 2005 *J. Electrochem. Soc.* **152** E123. Reproduced from Ref. [9] with permission

water molecules to form hydronium ions. The newly formed hydronium ions hop to a nearby $SO_3^-$ site releasing $H^+$ ions. The process continues until $H^+$ ions pass through the membrane and react with oxygen ions at the cathode. The process is shown in Fig. 11.3.

Liquid water is essential for proton transport. The operating temperature of the Nafion membrane must be kept below 100 °C, around 80 °C. If it is operated above 100 °C, water will evaporate, reducing the proton transport capability. A conversely, too low a temperature can flood the membrane with water reducing the proton transport through the membrane. Therefore, careful water management is critical in the PEM fuel cell operation. The membrane serves two other purposes. First, it acts as a gas separator and prevents mixing the hydrogen fuel with air fed at the cathode. Second, the membrane only allows proton conduction and does not conduct electrons. The electrons, therefore, must travel through an external circuit to the cathode.

The Nafion membrane is considered a gold standard for PEM fuel cells. Any new membranes must perform better than the Nafion membrane to be acceptable by the industry. Shortcomings of the Nafion membrane are its lower operating temperature and very careful heat and water management. Keeping the membrane hydrated and yet not flood it with liquid water is challenging.

## 11.2.2 Membrane/Electrode Assembly

The membrane/electrode assembly is shown in Fig. 11.4. The thickness of the membrane varies from 50 to 175 μm (0.05 to 0.175 mm). Both sides of the membrane

## 11.2 Proton Exchange Membrane (PEM) Fuel Cell

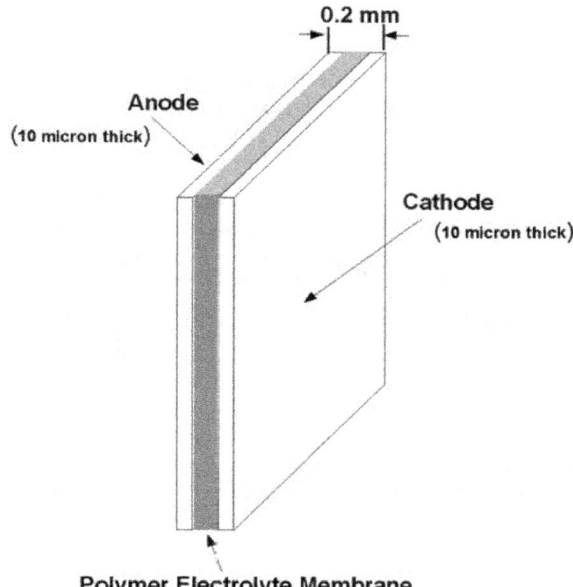

**Fig. 11.4** Membrane electrode assembly for a polymer electrolyte membrane fuel cell [10]. Copyright © 2006 Los Alamos National Security, LLC under Contract No. DE-AC52-06NA25396, Zalbowitz, T. S. (2009) Fuel cells Green Power. Los Alamos National Laboratory LA-UR-99-3231, https://www.lanl.gov/orgs/mpa/mpa11/Green%20Power.pdf. Reproduced from Ref. [10] with permission

contain a catalyst layer that includes the catalyst dispersed on carbon-based supports. Platinum is used as a catalyst in this membrane/electrode assembly, one side of the membrane functions as an anode, and the other side functions as a cathode.

The use of platinum as a catalyst accelerates the hydrogen oxidation reaction (HOR) at the anode and oxygen reduction reaction (ORR) at the cathode [11]. Platinum catalysts increase the cost of fuel cells. The platinum catalyst cost can be 57–60% of the total cost of the fuel cell. Therefore, significant effort is underway to reduce the usage of platinum catalysts in a fuel cell. Several non-platinum materials have been suggested as catalysts [12–15]. However, platinum is still the best catalyst for the fuel cell with a polymer electrolyte membrane.

### 11.2.2.1 Backing Layer/Gas Diffusion Layer

Another layer is placed on the top of the catalyst layer on both sides. This layer is called the backing layer or gas distribution layer. The backing layers are usually made of porous carbon paper or carbon cloth, typically 100–300 μm thick. Carbon allows the conduction of electrons exiting the anode and entering the cathode. The porous nature of the layers ensures the effective distribution of each reactant gas to the catalyst. An enlarged cross section of a membrane/electrode assembly is shown in Fig. 11.5. These layers are designed for water management during the fuel cell operation. These layers are waterproofed with Teflon coating. Also, the gas diffusion

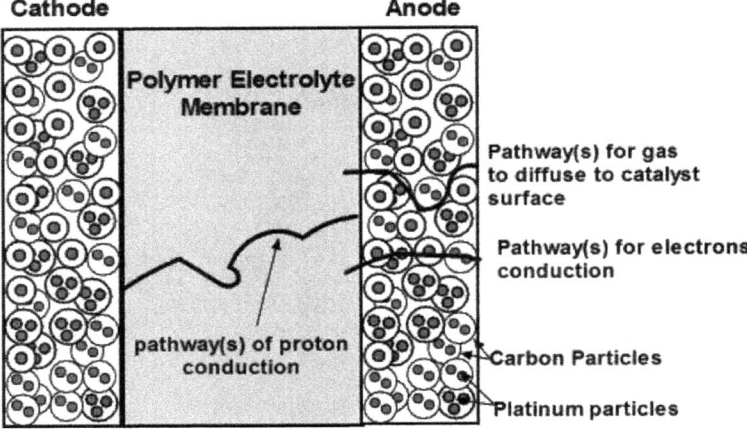

**Fig. 11.5** Polymer electrolyte membrane with porous electrodes containing platinum catalysts supported on carbon. Copyright © 2006 Los Alamos National Security, LLC under Contract No. DE-AC52-06NA25396, Zalbowitz, T. S. (2009) Fuel cells Green Power. Los Alamos National Laboratory LA-UR-99-3231, http://www.lanl.gov/orgs/mpa/mpa11/Green%20Power.pdf. Reproduced from Ref. [10] with permission

layer should be conductive so that electrons can flow through them to the current collector.

Finally, a gasket is used on each side to provide the necessary gas seal to prevent leaking of the reactant gases to the atmosphere. The gasket seals are generally made of Teflon film. A graphite plate is used for a single cell construction as the end plate on both sides, as shown in Fig. 11.6. These end plates serve both as a flow field and a current collector. The graphite plate is pressed on the outer surface of each backing layer. The graphite plates are the last components making up the cell. The

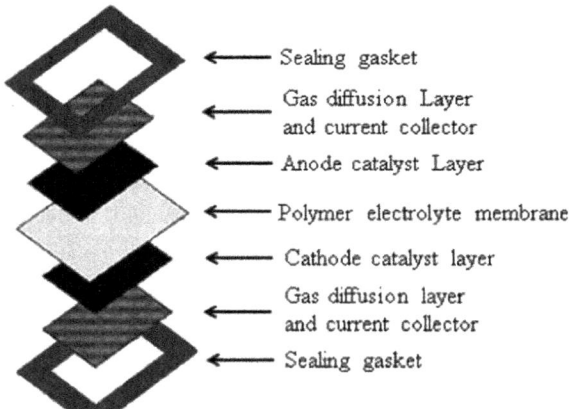

**Fig. 11.6** Various components of a single cell for the proton exchange membrane. Copyright © 2014 John Wiley & Sons, Ltd. International Journal of Energy Research 38(11):1367–1390. https://doi.org/10.1002/er.3163. Reproduced from Ref. [7] with permission

## 11.2 Proton Exchange Membrane (PEM) Fuel Cell

plate next to the backing layer contains channels machined into the plate for gas flow. The channels carry the reactant gas into the backing layers of the cell for further gas distribution.

### 11.2.3 Efficiency, Power, and Energy of Polymer Electrolyte Membrane Fuel Cell

The maximum voltage or the theoretical electrical energy available from the fuel cell with polymer electrolyte membrane fuel cell can be calculated from Eq. 11.1. The maximum cell voltage ($E$) at a constant pressure of 1 atm and the temperature of 298 K (25 °C) is calculated as:

$$E = -\frac{G}{nF} = -\frac{-237,200 \text{ J}}{2 \times 90,487 \text{ J/V}} = 1.23 \text{ V} \quad (11.1)$$

$G$ = Gibbs free energy $\Delta S$,
$n$ = the number of electrons transferred per mole of fuel,
$F$ = the Faraday constant.

At the normal operating temperature of the PEM fuel cells, 80 °C (353 K), the Gibbs free energy ($G$) is approximated as follows:

$$G = -285,800 \text{ J/mol} - (353 \text{ K})(-163.2 \text{ J/mol K}) = -228,200 \text{ J/mol}.$$

It is assumed that enthalpy and entropy changes are negligible for Gibbs free energy $\Delta S$ ($G$) calculation.

Therefore, the maximum cell voltage at the standard operating conditions of a fuel cell (80 °C and 1 atm) is given by:

$$E = -\frac{-228,200 \text{ J}}{2 \times 96,487 \text{ J/V}} = 1.18 \text{ V}. \quad (11.2)$$

Air is used instead of pure oxygen at the cathode for economic reasons. Hence, a drop in cell voltage occurs. Also, both the air and hydrogen stream must be humidified. This humidification further reduces the maximum voltage output from the cell. Considering all these losses, the ideal open-circuit voltage reduces to about 1.16 V at 80 °C and 1 atm. The voltage drop corresponding to a load is shown in Fig. 11.7. Generally, a fuel cell operates at 0.7 V under a load. Therefore, the chemical to electrical energy conversion efficiency may be assumed to be about 60% (0.7/1.16 = 60%). The overall fuel cell reaction is exothermic, and the remaining 40% of the energy appears as heat.

A single fuel cell cannot deliver the required power for transportation applications. High specific power and power density are necessary for transportation. In a typical fuel cell, the term specific power is the ratio of the power produced by a cell to the

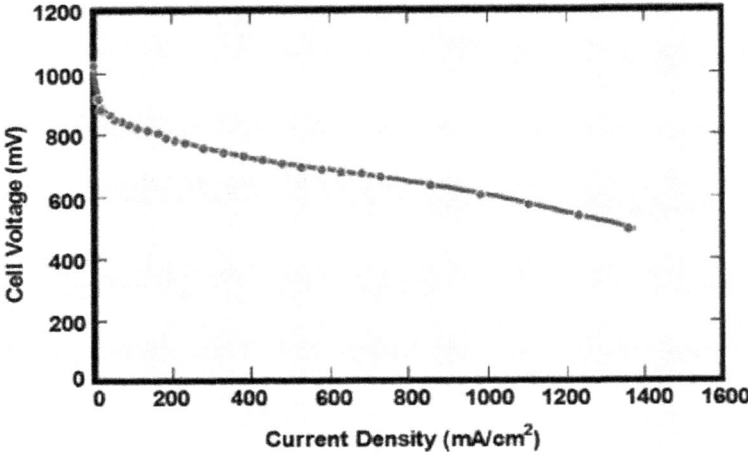

**Fig. 11.7** Polarization curve for a fuel cell with a polymer electrolyte membrane. Copyright © 2006 Los Alamos National Security, LLC under Contract No. DE-AC52-06NA25396, Zalbowitz, T. S. (2009) Fuel cells Green Power. Los Alamos National Laboratory LA-UR-99-3231, http://www.lanl.gov/orgs/mpa/mpa11/Green%20Power.pdf. Reproduced from Ref. [10] with permission

mass of the cell. The power density is the ratio of the power produced by a cell to the volume of the cell. All the power generated by the cell is not available as electrical power. A typical example of the amount of power converted to heat energy may be calculated [10].

***Example 11.1*** To generate approximately 2.5 kJ electric energy/minute, a 100 cm² fuel cell generates 0.6 A/cm² of current for a total current of 60 A. The fuel cell operates under typical conditions of one-atmosphere pressure and 80 °C at 0.7 V. The following equation can estimate the excess heat generated by this cell:

$$\text{Power due to heat} = \text{Total power generated} - \text{Electrical power}, \tag{11.3}$$

$$\begin{aligned}
P_{\text{heat}} &= P_{\text{total}} - P_{\text{electrical}} \\
&= (V_{\text{ideal}} \times I_{\text{cell}}) - (V_{\text{cell}} \times I_{\text{cell}}) \\
&= (V_{\text{ideal}} - V_{\text{cell}}) \times I_{\text{cell}} \\
&= (1.16 - 0.7)\, V \times 60\, A \\
&= 0.46\, V \times 60\, C/s \times 60\, s/\text{min} \\
&= 1650\, J/\text{min} = 1.65\, kJ/\text{min}.
\end{aligned} \tag{11.4}$$

Therefore, the cell is generating about 1.65 kJ of excess heat per minute. This heat energy must be removed continuously to maintain a constant temperature in the cell.

### 11.2.4 The Polymer Electrolyte Membrane Fuel Cell Stack

A single PEM fuel cell generates about 0.7 V, which is insufficient for most practical applications. The required voltage is obtained by stacking individual fuel cells over each other, and the anode and cathode current collectors are placed side by side. When stacking fuel cells, instead of using two current collectors (the anode and the cathode pressed together), only one plate is used with a flow field cut into each side of the plate. The use of only one plate decreases the overall volume and weight of the stack. This type of plate is called a bipolar plate, which has some unique characteristics and serves several purposes [16]:

1. It separates one cell from the next.
2. Flow channels grooved on both sides of the plate allow feeding of the reactant gases into one single plate. Hydrogen gas is fed on one side and air on the other side of the bipolar plate.
3. The bipolar plate should be non-porous. This plate should also be impermeable to hydrogen, air, and other impurities present in the gases.
4. The bipolar plate must be electronically and thermally conductive.
5. The bipolar plate should not corrode.

A stack may consist of a few cells to a hundred or more cells. The function of bipolar plates is to connect the cells in series and facilitate the transport of generated electrons from anode to cathode. Figure 11.8 shows the typical arrangement of a fuel

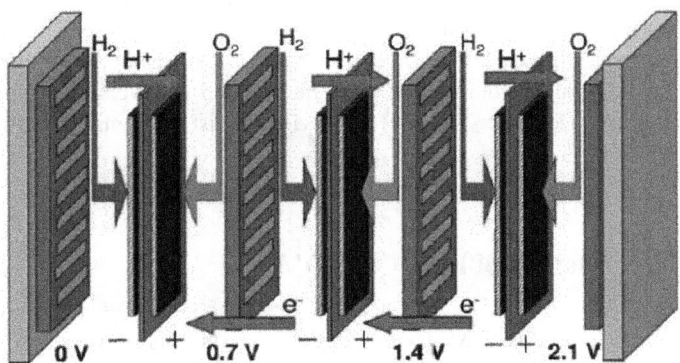

**Fig. 11.8** Arrangements of a fuel cell stack showing voltage addition. Copyright © The Royal Society of Chemistry 2005, Green Chem 7: 2005, 132–150. Reproduced from Ref. [6] with permission

cell stack. Many stacks can be used in series or parallel combinations to generate large amounts of power.

## *11.2.5 Water Management in a Fuel Cell*

Water management is essential for the effective operation of a fuel cell with a polymer electrolyte membrane. Both the fuel and air entering the cell must be humidified before coming in contact with the membrane. However, the degree of humidification of the streams is very important. If the membrane is not adequately moist, it will hinder the conduction of $H^+$ ions, and therefore, the current density will be reduced. At higher humidification, water will flood the electrodes. This flooding will block the pores of the porous gas distribution layer. Water produced from the reaction must be drained continuously to prevent water build-up. The gas flow rate and operating temperature play an essential role in water management. If the cathode reaction is too slow, the air cannot carry all the water produced at the cathode out of the fuel cell. Cell performance is decreased because not enough oxygen can pass through the excess water to reach the cathode. The temperature in the fuel cell should be below 80 °C. Otherwise, the water loss will be too much and will prevent adequate membrane humidification. If the temperature is too low, not only will it reduce the catalyst activity, but water will condense out from the entering gas streams flooding the electrodes.

## 11.3 Fuel

Both hydrogen and air streams for a fuel cell with polymer electrolyte membrane (PEMFC) should be clean for better performance and long life of cell components, particularly that of the membrane, cathode, and anode. The presence of impurities could result in a dramatic performance drop. Carbon monoxide and sulfur are especially problematic contaminants and must be reduced to levels of around 10 and 1 ppm or less, respectively. Table 11.1 provides a list of contaminants and their sources [16].

## 11.4 Direct Methanol Fuel Cell (DMFC)

Direct methanol fuel cells (DMFCs) are getting lots of attention because methanol can be stored as a liquid and used as a fuel directly without any prior reforming. Figure 11.9 shows the working principle of a DMFC [17]. The structure of a DMFC and PEM fuel cell is very similar. In DMFC, a solid polymer membrane is used as the electrolyte. The fuel is liquid methanol, which is fed into the anode along with

## 11.4 Direct Methanol Fuel Cell (DMFC)

**Table 11.1** Major contaminants that are identified in the operation of PEM fuel cells

| Impurity source | Type |
|---|---|
| Air | $N_2$, $NO_x$ (NO, $NO_2$), $SO_x$ ($SO_2$, $SO_3$) $NH_3$, $O_3$ |
| Reformate hydrogen | CO, $CO_2$, $H_2S$, $NH_3$, $CH_4$ |
| Bipolar metal plates (end plates) | $Fe^{3+}$, $Ni^{2+}$, $Cu^{2+}$, $Cr^{3+}$ |
| Membranes (Nafion®) | $Na^+$, $Ca^{2+}$ |
| Sealing gasket | Si |
| Coolants, DI water | Si, Al, S, K, Fe, Cu, Cl, V, Cr |
| Battlefield pollutants | $SO_2$, $NO_2$, CO, propane, benzene |
| Compressors | Oils |

Copyright © 2006 Elsevier B. V. All rights reserved. Journal of Power Sources 165(2), 2007, 739–56. Reproduced from Ref. [16] with permission

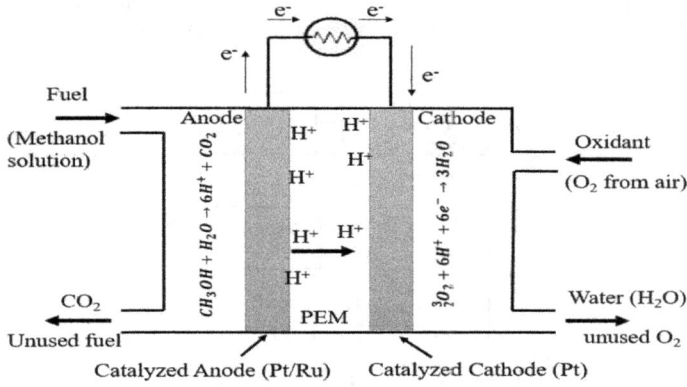

**Fig. 11.9** Working principle of direct methanol fuel cell (DMFC). Copyright© 2020 Junoh, H., Jaafar, J., Nordin, N. A. H. M., et al. Licensee MDPI, Basel, Switzerland. Creative Commons Attribution (CC BY) license. Membranes 2020, 10, 34; https://doi.org/10.3390/membranes10030034. Reproduced from Ref. [17] with permission

water. Methanol is dissociated into $CO_2$, protons, and electrons by the catalyst on the anode. Protons transport through the solid electrolyte and react with oxygen from the air and the electrons from an external circuit to form water at the anode. This process completes the circuit to produce electric power.

The reactions at the anode and the cathode are given below.
Anode reaction: $CH_3OH + H_2O \rightarrow CO_2 + 6H^+ + 6e^-$.
Cathode reaction: $\frac{3}{2}O_2 + 6H^+ + 6e^- \rightarrow 3H_2O$.
Overall reaction: $CH_3OH + \frac{3}{2}O_2 \rightarrow CO_2 + 2H_2O$.

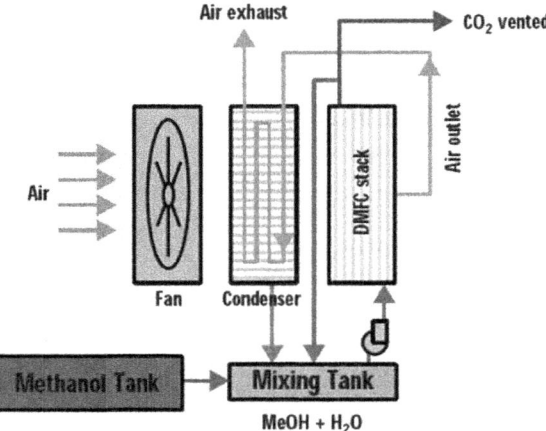

**Fig. 11.10** Flow diagram of a direct methanol fuel cell stack. Copyright © 2006 Los Alamos National Security, LLC under Contract No. DE-AC52-06NA25396, Zalbowitz, T. S. (2009) Fuel cells Green Power. Los Alamos National Laboratory LA-UR-99-3231, http://www.lanl.gov/orgs/mpa/mpa11/Green%20Power.pdf. Reproduced from Ref. [10] with permission

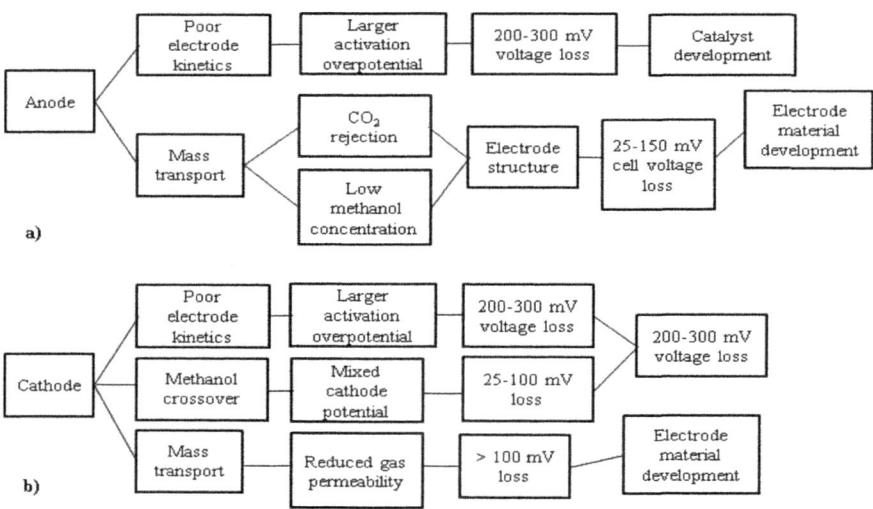

**Fig. 11.11** Technological limitation with the **a** anode and **b** cathode of DMFC. https://doi.org/10.1002/er.3163. Reproduced from Ref. [7] with permission Copyright © 2014 John Wiley & Sons, Ltd. International Journal of Energy Research 38(11):1367–1390

The schematic diagram of a DMFC system is shown in Fig. 11.10. The working temperature of a DMFC ranges from 50 to 120 °C. The efficiency of DMFCs is around 40%. Figure 11.11 shows the issues that need to be addressed before their commercial applications.

Figure 11.11 summarizes the key limiting issues related to DMFC electrodes [7]. Their use in cell phones, computer batteries, and other electronic devices requires low operating temperature. A catalyst capable of methanol oxidation to hydrogen ions and carbon dioxide at low temperatures requires high activity. Platinum is the best catalyst for a DMFC, but it increases the cost significantly. In addition, methanol crossover is another major issue for a DMFC. Shaari and Kamaruddin [18] identified the following limitations for a functional polymer electrolyte membrane fuel cell technology:

- Fuel crossovers (e.g., methanol or hydrogen leaking across fuel cell membranes).
- CO poisoning.
- Low durability.
- High cost.

The membrane must have very low methanol permeability while maintaining high proton mobility. Although Nafion is a suitable candidate membrane for a DMFC, a thicker membrane is necessary to prevent *methanol crossover*. The increase in the thickness of the membrane can cause voltage and power losses through ohmic polarization (IR drop). Methanol crossover causes cathode depolarization and the poisoning of the catalyst from methanol oxidation products [19]. Another concern with methanol is its toxicity requiring better seals to prevent leakage from the cells.

## 11.5 Application of Chitosan in Fuel Cells

Chitosan is considered a basic polyelectrolyte due to its hydrophilic nature [20]. The chitosan structure's hydrophilic group presence plays a vital role in preferential water sorption and diffusion through the chitosan membrane [21]. Furthermore, the chemical and thermal resistance properties of chitosan facilitate energy-generating electrochemistry. Hence, chitosan can be a potential PEM for fuel cells [2, 22]. Chitosan is a semi-crystalline polymer in its natural state and has very low electrical conductivity [23]. Ma and Sahai [24] reviewed the structure and property of chitosan, considering recent achievements and prospects of their applications in fuel cells. It is reported that the biopolymer chitosan can be used as membrane electrolyte and electrode in various fuel cells such as alkaline polymer electrolyte fuel cells, direct methanol fuel cells, and biofuel cells [1, 24]. Table 11.2 shows the properties of chitosan-based membranes and their application in fuel cells.

Luptani et al. [25] reported that the chitosan membrane's proton conductivity depends on the number of available $-NH_2$ groups present in chitosan. Available $-NH_2$ groups are dependent on the degree of deacetylation and molecular weight of chitosan. The conductivity of chitosan occurs via OH– transport across the amorphous areas of chitosan [26]. Chitosan can perform as a polyelectrolyte through the protonation of the $-NH_2$ groups in an acidic medium.

**Table 11.2** Properties of chitosan-based membranes for fuel cell applications

| Membrane | The ionic conductivity (S cm$^{-1}$) | Methanol permeability (cm$^2$ s$^{-1}$) |
|---|---|---|
| CS–H$_2$SO$_4$ | $2 \times 10^{-2}$, hydrated, 60 °C | $8.0 \times 10^{-7}$, 12 mol L$^{-1}$ MeOH, 20 °C |
| CS–glutaraldehyde sulfosuccinic acid | $4.52 \times 10^{-2}$, hydrated, 25 °C | $9.6 \times 10^{-7}$, 25 °C |
| Sulfonated CS | $3.1 \times 10^{-2}$ at 80 °C | $4.7 \times 10^{-7}$ 2 M MeOH |
| PVA/CS (glutaraldehyde) | $9.9 \times 10^{-3}$ at 100% R.H | $9.45 \times 10^{-8}$, 50% MeOH |
| Phosphonic CS/PVA (formaldehyde, Na$_2$SO$_4$, H$_2$SO$_4$) | $2.48–4.29 \times 10^{-2}$ | $0.49–1.03 \times 10^{-7}$, 30% MeOH |
| CS/PAA | $3.8 \times 10^{-2}$, hydrated, RT | $3.9 \times 10^{-8}$, 30 °C, 50% MeOH |
| CS/sodium alginate | $4.2 \times 10^{-2}$, 100% R.H., 30–32 °C | $4.6 \times 10^{-8}$, 30–32 °C, 50% MeOH |
| CS/P(AA-AMPS) (H$_2$SO$_4$) | $3.59 \times 10^{-3}$, hydrated, 30 °C | $2.41 \times 10^{-7}$, 5 M MeOH |
| CS/PVP (H$_2$SO$_4$, glutaraldehyde) | $2.4 \times 10^{-2}$, hydrated, RT | $7.3 \times 10^{-8}$, 30 °C, 50% MeOH |
| CS/silica (H$_2$SO$_4$) | $1.6–2.9 \times 10^{-2}$, hydrated, 20 °C | $6.31–11.4 \times 10^{-7}$, 2 M MeOH |
| CS/phosphorylated titanate nanotube (H$_2$SO$_4$) | $1.58–1.75 \times 10^{-2}$, hydrated, 20 °C | $6.42–8.64 \times 10^{-7}$, 2 M MeOH |
| CS/beta zeolite–SO$_3$H (H$_2$SO$_4$) | $1.17 \times 10^{-2}$ to $1.49 \times 10^{-2}$ hydrated, 20 °C | $5.8–9.55 \times 10^{-7}$, 2 M MeOH |
| CS/STiO$_2$ (H$_2$SO$_4$) | $1.14–1.86 \times 10^{-2}$, hydrated, 20 °C | $5.69–7.62 \times 10^{-7}$, 2 M MeOH |
| CS/phosphomolybdic acid | $1.5 \times 10^{-2}$, 25 °C | $2.7 \times 10^{-7}$, ambient temperature |
| CS functionalized montmorillonite/Nafion | $4.5–8.3 \times 10^{-2}$, 95% R.H., 25 °C | $0.57–1 \times 10^{-7}$, 25 °C |
| CS/Nafion triple layer (glutaraldehyde, sulfosuccinic acid) | $8.8 \times 10^{-2}$, 95% R.H., 25 °C | $2.52 \times 10^{-7}$, 25 °C |
| Quaternized PVA/CS (glutaraldehyde) | $10^{-3}–10^{-2}$, hydrated, 30 °C | $5.68–4.42 \times 10^{-7}$, 30 °C, 1 M MeOH |

Copyright © 2012 Elsevier Ltd. Carbohydrate Polymers, 92, 955–975, 2013. Reproduced from Ref. [24] with permission

Chitosan forms a complex with inorganic salts due to its high content of the amino functional groups. The ionic conductivity of chitosan can be enhanced by crosslinking with suitable plasticizers and fillers. The chitosan membrane's crystallinity can also be significantly decreased by slightly crosslinking the membrane [3]. Navaratnam et al. [27] investigated the transport mechanism of chitosan-based

## 11.5 Application of Chitosan in Fuel Cells

electrolyte systems. They used the chitosan-ethylene carbonate/propylene carbonate (chitosan-EC/PC) system with lithium acetate ($LiCH_3COO$) and lithium triflate ($LiCF_3SO_3$) as salts to study the conduction mechanism of ions. It was reported that lithium ions form complexes with chitosan and provide a platform for ion hopping [27]. At room temperature, the electrolyte system using $LiCF_3SO_3$ salt showed better ionic conductivity and greater dielectric constant and dielectric loss value than $LiCH_3COO$.

Chitosan biopolymer has been used in direct methanol fuel cells (DMFCs) as a potential replacement to conventional Nafion membranes for its considerably reduced methanol crossover. Soontarapa and Intra [28] reported that chitosan-based complex membranes showed better performance over commercially available Nafion membranes in ion exchange capacity and hydrogen gas permeability. However, the proton conductivity of the Nafion membrane was reported to be better than chitosan membranes. Based on the structural investigation, Bahlakeh et al. [4] reported that increased methanol loading weakened the water interactions with chitosan functionalities. In contrast, it improved the methanol affinities toward chitosan, reflecting higher methanol sorption capability of chitosan at enhanced concentrations. They also suggested that supplying concentrated methanol or acidic feed solutions into DMFCs with chitosan PEM could lower membrane performance due to the significant methanol transport dynamics.

For application as PEMs for fuel cells, different methods such as copolymerization, grafting, polymer blending, or crosslinking are being explored to improve the proton conductivity without sacrificing mechanical strength or vice versa. The polymer selection for PEMs for DMFCs is a critical consideration because proton conductivity and methanol permeability are properties that mainly depend on the polymers [2]. Mukoma et al. [29] evaluated the chitosan membranes in sulfuric acid for methanol permeability at high to medium methanol concentrations. It was observed that the chitosan membrane has almost three times lower methanol permeability than methanol permeability in Nafion 117 membranes at 20 °C. Ma et al. [30] studied chitosan-based cost-effective alternative material to Nafion® for application in direct borohydride fuel cells (DBFC). They reported that a combination of chitosan (CS) membrane and chitosan hydrogel (CH) as anode binder in a direct borohydride fuel cell (DBFC) exhibited better performance than Nafion®. The chitosan membrane was prepared using sulfuric acid as a crosslinking agent, and the chitosan hydrogel was prepared from nickel or palladium. The maximum peak power density was 450 mW $cm^{-2}$ at 60 °C for DBFC employing CS membrane and CH binder-based anode.

Hasani-Sadrabadi et al. [8] tested a PEM comprising two thin layers of structurally modified chitosan as methanol barrier layers with both sides coated with Nafion® 105 for high-performance direct methanol fuel cell applications. They observed that proton conductivity and methanol permeability measurements improved transport properties for the multilayer membrane compared to Nafion® 117 with approximately the same thickness. Binsu et al. [31] describe the modification of chitosan by introducing the phosphonic acid group and preparing its composite membranes with polyvinyl alcohol (PVA) of different compositions. The membrane exhibited a comparable proton transport number and conductivity as the Nafion 117 membrane but relatively lower methanol permeability than Nafion 117. Meenakshi et al. [32] prepared the CS-PVA-SPES membrane using chitosan (CS) and polyvinyl alcohol (PVA) with sulfosuccinic acid (SSA), which was further modified with sulfonated polyethersulfone (SPES), for the application in direct methanol fuel cells (DMFCs). The methanol crossover in these membranes is found to be about 33% lower than the methanol crossover in the Nafion 117 membrane. The DMFC employing a CS-PVA-SPES mixed-matrix membrane with an optimum content of 25 wt% SPES delivers a peak power density of 5.5 mW cm$^{-2}$ at a load current density of 25 mA cm$^{-2}$ while operating at 70 °C. Oliveira and Mendes [33] prepared chitosan and polyvinyl alcohol-based membranes with or without nylon. They use sulfosuccinic acid as a crosslinking and sulfonating agent. The proton conductivity of the membrane was around $10^{-2}$ S cm$^{-1}$. The membrane showed lower hydrogen and methanol permeability than the standard Nafion® 115 membrane [33]. Kraytsberg and Ein-Eli [26] reported that hydrated chitosan's intrinsic conductivity is approximately $10^{-4}$ S/cm. This level of intrinsic conductivity is inadequate for fuel cell applications. Jayakumar et al. [34] reported that phosphorylated-chitosan membranes were almost non-conductive in their dry states. However, the hydrated membranes showed ionic conductive properties that were one order of magnitude higher than unmodified chitosan membranes. Yamada and Honma [35] observed that the chitosan membrane crosslinked with methane phosphonic acid showed higher proton conductivity of $5 \times 10^{-3}$ S cm$^{-1}$ at 150 °C under anhydrous (water-free) conditions. The thermal stability of this composite material was found to increase with the increase of the methane diphosphonic molecules. Chitosan membrane crosslinked with polyvinyl sulfonic acid [36] and polyacrylic acid [37] has been investigated. These ionically crosslinked membranes exhibited high ion exchange capacity, high proton conductivity, low methanol permeability, and adequate thermal and mechanical stability. Yang and Chiu [38] reported that the glutaraldehyde crosslinked polyvinyl alcohol/chitosan membrane has lower methanol permeability than the Nafion membrane. Zhou et al. [39] studied chitosan membranes modified with polymeric reactive dyes containing quaternary ammonium groups (PRDQA) to use in alkaline fuel cells. They reported that the CTS/PRDQA membrane (1:0.5 in mass) exhibited a high OH conductivity of $8.17 \times 10^{-3}$ S cm$^{-1}$ at room temperature. At a current density of 57.4 mA cm$^{-2}$, this membrane achieved a power density of 29.1 mW cm$^{-2}$ with an open-circuit voltage (OCV) of 991.6 mV in an $H_2/O_2$ system. This CTS/PRDQA membrane was more stable after one week of testing in the 8.0 M of KOH solution at 80 °C in terms of both integrity and OH conductivity than the chitosan membrane by itself. Mohanapriya et al. [40] prepared a membrane (CS-HEC) by incorporating phosphotungstic acid (PTA) into chitosan (CS)-hydroxyethyl cellulose (HEC) for the application in direct methanol fuel cells (DMFCs). They reported that the DMFC with 3 wt% stabilized PTA-CS-HEC mixed-matrix membrane has a lower methanol crossover. It delivers a peak power density of 58 mW/cm$^2$ at a load current density of 210 mA/cm$^2$ compared to DMFC operating with a Nafion membrane electrolyte.

Smitha et al. [41] prepared membranes using chitosan (with 84% DD) and sodium alginate biopolymers for a direct methanol fuel cell. They observed that these membranes were suitable for direct methanol fuel cell (DMFC) applications because of their low methanol permeability, excellent physicomechanical properties, and relatively high proton conductivity. Wairoa et al. [42] prepared a sulfonated chitosan-calcium oxide (CaO) composite membrane for the proton exchange. The optimum characteristics obtained in the presence of 25% CaO refer to ionic exchange capacity of 1.517 meq/g, swelling of 28.207%, methanol permeability of $1.211 \times 10^{-4}$ kg/m$^2$ s, and proton conductivity of $6.415 \times 10^{-5}$ S/cm. Garcia-Cruz et al. [43] studied graphene oxide (GO)-doped chitosan and polyvinyl alcohol mixed matrix (MMM) as a polymer electrolyte membrane. The estimated diffusion coefficient value was $3.38 \times 10^{-7}$ and $2.43 \times 10^{-7}$ cm$^2$ S$^{-1}$ after 60 and 120 min, respectively, avoiding additional alcohol crossover. They reported that the GO-based-(MMM) membrane exhibited a low conductivity of 0.19 mS cm$^{-1}$. The GO-based-(MMM) membrane appears to be an excellent physical barrier for alcohol permeability.

## 11.6 Summary

The polymer electrolyte membrane is the most critical component of a fuel cell. It plays a vital role in transporting protons from anode to cathode electrodes. Nafion™, a poly perfluorosulfonic acid-based organic polymer, is the most widely used polymer electrolyte membrane for PEM fuel cells. The shortcomings of the Nafion membrane are its low operating temperature and intensive heat and water management. Keeping the membrane hydrated and yet not flood it with water is a difficult task.

Chitosan is considered a basic polyelectrolyte due to its hydrophilic nature. The hydrophilic group present in the chitosan structure plays an essential role in preferential water sorption and diffusion through the chitosan membrane. The chemical and thermal resistance properties of chitosan and its derivatives facilitate energy-generating electrochemistry and make it a potential PEM for fuel cells. Chitosan can be used as a membrane electrolyte and electrode in various fuel cells. These fuel cells include polymer electrolyte fuel cells (alkaline), direct methanol fuel cells, and biofuel cells. Chitosan biopolymer has been extensively investigated in direct methanol fuel cells (DMFCs) as a potential replacement to conventional Nafion membrane for its considerably reduced methanol crossover. Chitosan membrane has almost three times lower methanol permeability than methanol permeability in Nafion 117 membranes at 20 °C. The performances of chitosan-based composite materials as the polyelectrolyte membrane appear to be promising in fuel cell applications compared to the Nafion membrane. However, mechanical strength, simple fabrication techniques, operating reliability, and cost-effectiveness are essential to successful PEM fuel cells.

# References

1. Vaughan H, Jafarizadeh-Malmiri H, Berenjian A, Anarjan N (2013) Recent advances in application of chitosan in fuel cells. Sustain Chem Process 1:16. http://www.sustainablechemicalprocesses.com/content/1/1/16
2. Ye Y-S, Rick J, Hwang B-J (2012) Water-soluble polymers as proton exchange membranes for fuel cells. Polymers 4. https://doi.org/10.3390/polym4020913
3. Wan Y, Creber KAM, Peppley B, Tambui V, Halliop E (2005) New solid polymer electrolyte membranes for alkaline fuel cells. Polym Int 54:5–10
4. Bahlakeh G, Hasani-Sadrabadi MM, Jacob KI (2017) Morphological and transport characteristics of swollen chitosan-based proton exchange membranes studied by molecular modeling. Biopolymers 107:5–19
5. Smitha B, Sridhar S, Khan AA (2006) Chitosan-poly (vinyl pyrrolidone) blends as membranes for direct methanol fuel cell applications. J Power Sourc 159:846–854
6. de Bruijn F (2005) The current status of fuel cell technology for mobile and stationary applications. Green Chem 7:132–150
7. Kumar P, Dutta K, Das S, Kundu PP (2014) An overview of unsolved deficiencies of direct fuel cell technology: factors and parameters affecting its widespread use. Int J Energy Res © 2014. Wiley. https://doi.org/10.1002/er.3163
8. Hasani-Sadrabadi MM, Dastimoghadam E, Mokarram N et al (2012) Triple-layer exchange membranes based on chitosan biopolymer with reduce methanol crossover for high-performance direct methanol fuel cells application. Polymer 53:2643–2651
9. Choi P, Jalani NH, Datta R (2005) Thermodynamics and proton transport in Nafion II. Proton diffusion mechanisms and conductivity. J Electrochem Soc 152(3):E123–E130
10. Zalbowitz TS (2009) Fuel cells green power. Los Alamos National Laboratory LA-UR-99-3231. http://www.lanl.gov/orgs/mpa/mpa11/Green%20Power.pdf
11. Holton OT, Stevenson JW (2013) The role of platinum in proton exchange membrane fuel cells. PlatinMetals Rev 57(4):259–271
12. Strickland K, Miner E, Jia Q et al (2015) Highly active oxygen reduction non-platinum group metal electrocatalyst without direct metal-nitrogen coordination. Nat Commun 6:7343. https://doi.org/1038/ncomms8343
13. Pavlicek R, Barton SC, Leonard N et al (2018) Resolving challenges of mass transport in non-Pt-group metal catalysts for oxygen reduction in proton exchange membrane fuel cells. J Electrochem Soc 165(9):F589–F596
14. Gavidia LMR, Sebastian D, Pastor E et al (2017) Carbon-supported Pd and PdFe alloy catalysts for direct methanol fuel cell cathodes. Materials 10:580. https://doi.org/10.3390/ma10060580
15. Iranzo A, Arredondo CH, Kannan AM, Rosa F (2020) Biomimetic flow fields for proton exchange membrane fuel cells: a review of design trends. Energy 190:116435
16. Cheng X, Shi Z, Glass N et al (2007) A review of PEM hydrogen fuel cell contamination: impacts, mechanisms, and mitigation. J Power Sourc 165(2):739–756
17. Junoh H, Jaafar J, Nordin NAHM et al (2020) Performance of polymer electrolyte membrane for direct methanol fuel cell application: perspective on morphological structure. Membranes 10:34. https://doi.org/10.3390/membranes10030034
18. Shaari N, Kamaruddin SK (2015) Chitosan and alginate types of bio-membrane in fuel cell application. J Power Source 289:71–80
19. Jiang R, Chu D (2004) Comparative studies at methanol crossover and cell performance for DMFC. J Electrochem Soc 151(1):A69–A76
20. Wan Y, Katherine AM, Creber BP, Tambui V (2006) Chitosan-based electrolyte composite membranes II. Mechanical properties and ionic conductivity. J Membrane Sci 284:331–338

21. Clasen C, Wilhelms, Kulicke WM (2006) Formation and characterization of chitosan membranes. Biomacromolecules 7:3210–3222
22. Mukoma P, Jooste BR, Vosloo HCM (2004) Synthesis and characterization of crosslinked chitosan membranes for application as alternative proton exchange membrane materials in fuel cells. J Power Sourc 136:16–23
23. Marroquin JB, Rheea KY, Park SJ (2013) Chitosan nanocomposite films: enhanced electrical conductivity, thermal stability, and mechanical properties. Carbohyd Polym 92:1783–1791
24. Ma J, Sahai Y (2013) Chitosan biopolymer for fuel cell applications. Carbohyd Polym 92:955–975
25. Lupatini KN, Schafer JV, Machado B et al (2016) Development of chitosan membranes for use in PEM fuel cells. In: II S3IE 2016 2nd international seminar on industrial innovation in electrochemistry, pp 103–116
26. Kraytsberg A, Ein-Eli Y (2014) Review of advanced materials for proton exchange membrane fuel cells. Energy Fuels 28:7303–7330
27. Navaratnam S, Ramesh K, Sanusi A et al (2015) Transport mechanism studies of chitosan electrolytes systems. Electrochim Acta 175:68–73
28. Soontarapa K, Intra U (2006) Chitosan-based fuel cell membranes. Chem Eng Commun 193(7):855–868
29. Mukoma P, Jooste BR, Vosloo HCM (2004) A comparison of methanol permeability in chitosan and Nafion-117 membranes at high to medium methanol concentrations. J Membr Sci 243:293–299
30. Ma J, Choudhury NA, Sahai Y, Buchheit RG (2011) A high performance direct borohydride fuel cell employing cross-linked chitosan. J Power Sourc 196:8257–8264
31. Bindu VV, Nagarale RK, Shahi VK, Ghosh PK (2006) Studies on $N$-methylene phosphonic chitosan/poly (vinyl alcohol) composite proton-exchange membrane. React Funct Polym 66:1619–1629
32. Meenakshi S, Bhat SD, Sahu AK et al (2012) Chitosan-polyvinyl alcohol-sulfonated polyether sulfone mixed-matrix membranes as methanol-barrier electrolytes for DMFCs. J Appl Polym Sci 124:E73–E82
33. De Oliveira PN, Mendes AMM, Preparation and characterization of an eco-friendly polymer electrolyte membrane (PEM) based in a blend of sulfonated poly (vinyl alcohol)/chitosan mechanically stabilized by Nylon 6, 6. Mater Res. https://doi.org/10.1590/1980-5373-MR-2016-0387
34. Jayakumar R, Reis RL, Mano JF (2006) Chemistry and applications of phosphorylated chitin and chitosan. e-Polymers 35:1–16. http://www.e-polymers.org
35. Yamada M, Honma I (2005) Anhydrous proton conductive membrane consisting of chitosan. Electrochimia Acta 50:2837–2841
36. Congo LC, Battisti MV, Pereira-da-Silva MA, Oliver ON Jr, Nart FC, Huguenin F (2006) Layer-by-layer films of chitosan, poly (vinyl sulfonic acid), and platinum for methanol electrooxidation and oxygen electroreduction. J Power Sourc 158:160–163
37. Smitha B, Sridhar S, Khan AA (2004) Polyelectrolyte complexes of chitosan and poly (acrylic acid) as proton exchange membranes for fuel cells. Macromolecules 37:2233–2239
38. Yang JM, Chiu HC (2012) Preparation and characterization of polyvinyl alcohol/chitosan blended membrane for alkaline direct methanol fuel cells. J Membr Sci 419–420:65–71
39. Zhou T, He X, Song F, Xie K (2016) Chitosan modified by polymeric reactive dyes containing quaternary ammonium groups as a novel anion exchange membrane for alkaline fuel cells. Int J Electrochem Sci 11:590–608
40. Mohonapriya S, Bhat SD, Sahu AK et al (2009) A new mixed-matrix membrane for DMFCs. Energy Environ Sci 2:1210–1216
41. Smitha B, Sridhar S, Khan AA (2005) Chitosan-sodium alginate polyion complexes as fuel cell membranes. Eur Polymer J 41:1859–1866

42. Warfare S, Abdollah, Warden WK (2017) Production and characterization of sulfonated chitosan-calcium oxide composite membrane as a proton exchange fuel cell membrane. J Chem Technol Metall 52(6):1092–1096
43. Garcia-Cruz L, Casado-Coterillo C, Irabien A et al (2016) High performance of alkaline anion-exchange membranes based on chitosan/poly (vinyl) alcohol doped with graphene oxide for the electrooxidation of primary alcohols. J Carbon Res C2016 2:10. https://doi.org/10.3390/c2020010

# Chapter 12
# Chitosan Uses in Cosmetics

**Abstract** The use of chemical-based cosmetic products for various target organs is becoming an essential part of our daily life. Chitosan and its derivatives can be labeled as natural polymers with antimicrobial and anti-inflammatory properties. The application of chitosan and its soluble derivatives in various cosmetic products is discussed in this chapter.

## 12.1 Introduction

Cosmetic products are manufactured using various chemical compounds. Cosmetic products are currently consumed by both men and women worldwide for hygiene, beauty, improving appearance, and enhancing attractive features [1, 2]. Many cosmetics products such as soap, shampoo, toothpaste have become part of our daily hygiene [3]. The applications of cosmetic products include skin moisture maintenance, acne treatment, and skin protection, becoming part of our daily lives. Skin treatment applications including whitening, tanning, antiwrinkle, antiaging, and artificial skin are also gaining popularity to improve the impression of the skin. Other widely used cosmetic products are nail and hair care, sprays for hair softening, dandruff removal, and hair suppleness [4]. Oral diseases, like periodontal dental diseases, impact considerably on self-esteem and quality of life [5]. In this chapter, the applications of chitosan in cosmetic-related skincare and hair care have been discussed. The chitosan applications and its derivatives for oral care products and oral disease treatment are also discussed in this chapter.

## 12.2 Target Organs for Cosmetics Products

Skin health is an essential aspect of the human body as it is a primary barrier for the entry of microbes into the body [6]. Skin health can be affected by several internal and external damaging factors that affect the skin [7]. Therefore, there is a growing need for active ingredients to preserve the health of the skin. Healthy skin is also the precondition for a youthful appearance. One of the objectives of using cosmetics is to moisturize the skin and wipe out unwanted dirt to facilitate healthy skin [6]. Cosmetic creams are helpful in moisturizing hard, dry, and chapped skin. Cosmetics applied to the face to enhance facial appearance are often called makeup. Common makeup items for everyday use include lipstick, mascara, eye shadow, and foundation. Other everyday cosmetics include skin cleansers, body lotions, shampoos and conditioners, hair styling products (i.e., gel, hair spray), perfumes, and colognes. Products like hair gels, oils, and lotions have been on the market to protect hair fall and remove dandruff [8]. Oral care products toothpaste and mouthwash contribute to esthetic expectations.

## 12.2.1 Skin

The skin has a network of significant enzymatic and non-enzymatic protective antioxidants. These antioxidants prevent oxidative damage of the cell components by the reactive oxygen species (ROS), such as oxygen, peroxide, or free radicals [9–11]. The ability of the antioxidants to regulate ROS decreases as the body ages. Therefore, the oxidative stress of tissue related to wrinkle, skin laxity, abnormal pigmentation, and skin dryness may increase with age [11]. It is reported that the skin damage related to photoaging is primarily dependent on the degree of ultraviolet radiation (UVR) and the amount of melanin in the skin (skin phototype) [9]. Collagen is one of the most abundant proteins in the human body and one of the skin's essential components [12]. In the skin photoaging process, the vital components of the skin are depolymerized by the free radicals. This polymerization causes fine wrinkles or hyperpigmentation by decreasing the natural elasticity of the skin [9, 13]. As shown in Fig. 12.1, the UVA (long wave) and UVB (short wave) photons from sunlight can be directly absorbed by the DNA. This process can induce the formation of reactive oxygen species (ROS), which may indirectly cause damage to DNA [14, 15]. Although the UVB (short wave ultraviolet B) irradiation of the skin is the primary source of vitamin D, excessive UVB exposure causes skin cancer, which can be prevented by prior antioxidant treatment [9, 15].

It is reported that acne vulgaris is one of the most typical skin diseases affecting approximately 80% of adults aged from 11 to 30 years [16]. Commercial antibiotics including benzoyl peroxide, tetracycline, erythromycin, and lincomycin are commonly used to treat acne vulgaris-related cutaneous pathogens *Propionibacterium acnes, Staphylococcus epidermidis, Staphylococcus aureus,* and *Pseudomonas aeruginosa* [16–18]. The treatment of acne vulgaris by antibiotics is reported to cause several side effects [16]. In addition, the pathogens related to acne vulgaris have become increasingly resistant to antibiotics [17].

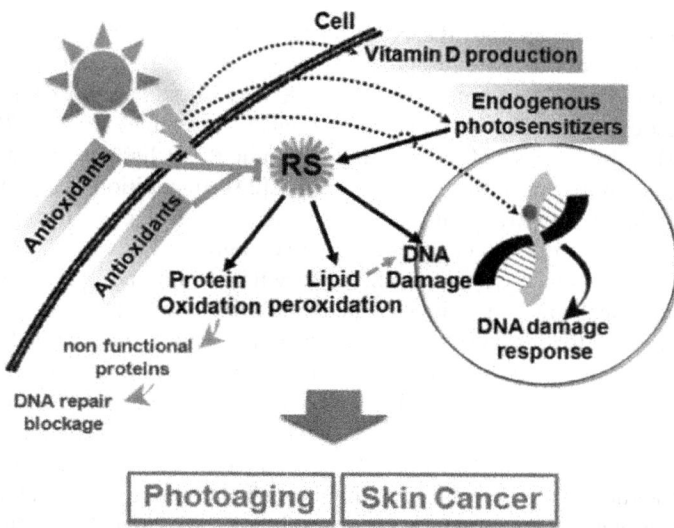

**Fig. 12.1** DNA damage and redox process by solar UV photons. Copyright © 2017 Schuch, A. P., Moreno, N. C., Schuch, N. J. et al. Published by Elsevier Inc., Free Radical Biology and Medicine, 107 (2017), [110–124]. Reproduced from Ref. [15] with permission

### 12.2.2 Hair

Hair can protect the scalp from sunburn, mechanical abrasion and provides thermoregulation [19]. The human hair shaft, a complex structure of morphological components, consists of a cuticle, cortex, and medulla, which in general act as a hair unit [20, 21]. The hair structure is divided into two distinct parts: (a) the hair follicle, which is deeply buried in the skin, and (b) the visible hair fiber [19]. Figure 12.2 shows the schematic cross section of the hair fiber.

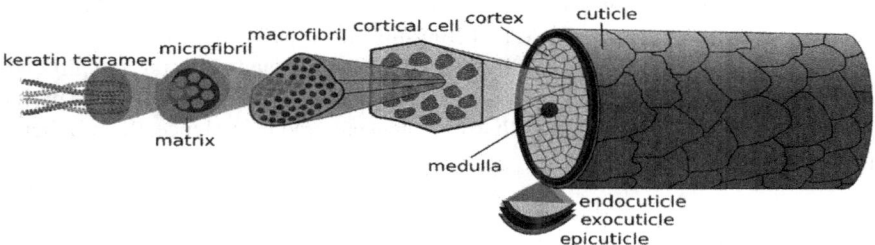

**Fig. 12.2** Schematic cross section of a hair fiber. Copyright © 2016 Cruz, C. F., Costa, C., Gomez, A. C., and Matama, T. Licensee MDPI, Basel, Switzerland. Creative Commons Attribution (CC-BY) license, *Cosmetics* 2016, *3*(3), 26; https://doi.org/10.3390/cosmetics3030026, Reproduced from Ref. [19]

Approximately 65–95% of the constituents of hair fibers are keratin, and the remaining constituents are represented by other proteins, water, lipids, pigments, and trace elements [22]. A multilayered cuticle surrounds the cortex with/without a central medulla [23]. The outer part of the hair (i.e., cuticle) (Fig. 12.2) consists of flap overlapping scales [21]. The individual cuticular cells (scale) comprised of proteins, lipids, polysaccharides, and ceramides [23]. In addition, there are protein and cysteine-based thin membrane coating on the outer surface of the cuticle [22]. These coatings are endocuticle, exocuticle, and epicuticle. The cortex, the principal component of the hair, consists of cells filled with keratin and other proteins that provide mechanical strength, hair color, and it controls water uptake [4, 23]. Franca et al. [24] reported that the cortical cells are rich in cystine, amino acids, lysine, and histidine. Figure 12.2 shows that the keratin microfibrils are aligned in the direction of the hair strand. Hair gloss depends on the cuticles' smoothness and the integrity of the cortex. It is reported that the cuticle plays a significant role in regulating chemicals/water to and from the cortex [23]. It is reported that the surface of untreated hair has an acidic pH of 4.5 to 5.5 [23]. In this acidic pH range, the cuticle cells shut themselves, protecting the cortex from weathering. In alkaline pH, hair cuticle cells may open, which causes damage to the cortex. Cuticle primarily ensures chemical resistance and keeps the hair surface smooth and glossy [24]. Therefore, the acidic pH (4.5 to 5.5) preserves the hair health. The appearance of the hair depends on the health of the cuticle.

The medulla is the hair's innermost region, usually present in coarser hair, as shown in Fig. 12.2. The absence of medulla does not interfere with the hair structure, for example, the fine hair of children [21]. The medulla serves as a pigment source and contribute to the hair's brightness [24].

The hair cosmetics include shampoos, conditioners, hair sprays, waxes, gels, and mousses. They can act on the part of the hair shaft that projects beyond the scalp surface [21]. The hair shaft can be damaged due to physical and chemical causes. For example, hair cuticles can be damaged in some procedures that require disruption of the cortex, such as bleaching, coloring, perming, and repeated activity such as rough combing and brushing. The morphological changes of the hair surface structure may occur due to the frequent use of cosmetic chemicals [20]. This may change the hair structure. The loss of cuticle or lifting-up of the cuticular plates exposes the cortex. The exposed cortex is readily susceptible to damage resulting in split ends (trichoptilosis) and frayed end. Hair weathering can damage the cuticle over a period of time, causing moisture loss rapidly from the weathered hair. Hence, the weathered hair looks dull and dry [23].

### 12.2.3 Teeth

Teeth are composed of phosphate-based mineral hydroxyapatite (HA) in the enamel, collagen in the dentine, and living tissues [25]. Enamel is the wear-resistant hardest substance that covers the entire clinical crown [26]. Dental caries and tooth wear

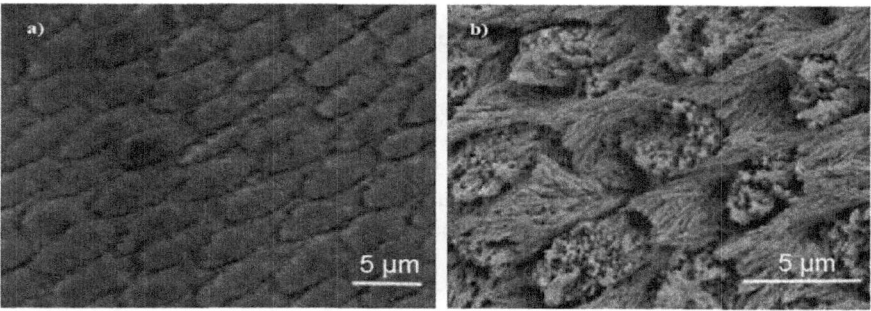

**Fig. 12.3** Scanning electron microscopic (SEM) image of **a** normal enamel and the **b** enamel etched with phosphoric acid with concentration same as found in drinks. Copyright © 2016 Abou Neel et al. This work is published and licensed by Dove Medical Press Limited, International Journal of Nanomedicine, 2016, 11, 4743–4763. Reproduced from Ref. [25] with permission

are the main factors related to mineral loss and enamel demineralization [4]. Dental caries occurs due to the demineralization of the enamel and dentine. It is a dietary carbohydrate-modified bacterial infectious disease [27]. The demineralization of teeth is caused by the acidic attack from dietary acid consumed through food or drink and the microbial attack from bacteria present in the mouth [25]. The pH at which the demineralization of the tooth occurs is often referred to as the critical pH. The critical pH for tooth demineralization is approximately 5.5 [5]. A typical example of a scanning electron microscopic image of normal and demineralized teeth enamel is shown in Fig. 12.3.

The role of bacteria, such as *Streptococcus mutans* and *Streptococcus sobrinus*, is well known in dental caries [28]. Dental caries can be caused by bacteria and sugar interaction on the tooth enamel surface [5, 25]. In general, saliva becomes supersaturated with calcium and phosphate at pH 7. Saliva can promote remineralization of the tooth by deposition of calcium [5]. Therefore, dental caries occurs when demineralization of the tooth exceeds remineralization.

## 12.3 Chitin and Chitosan and Their Derivatives in the Cosmetics

Approximately 41.4% of chitosan and its derivatives currently used in the cosmetic industry are mainly skincare products [29]. The chemical structure of chitin is close to mucopolysaccharides (heparin and hyaluronic acid). Hyaluronic acid (HA) is unique for its moisture retention capability and influences several cellular functions such as migration, adhesion, and proliferation [30]. Chitosan is non-toxic and non-immunogenic and is used extensively in cosmetic formulations. It possesses gelling properties, enhancing emulsion stability [31], and antimicrobial activities [32]. Besides, chitosan forms a protective film on the surface of the skin. The films are permeable to the air. Stable and pH-sensitive chitosan gel has applications in the cosmetic and pharmaceutical fields [33]. Chitosan with 3% in solution from pH 3 to 5 forms a clear gel. The viscosity of gel decreases with the decrease of pH [34]. In an acidic medium, chitosan forms a cationic network that can interact with negatively charged residues of the skin surface [13]. These desirable properties of chitin and chitosan have made them an essential ingredient in skin creams, shampoos, lacquers, and varnishes. Chitosan and chitin are being researched for making antimicrobial agents used in cosmetic, food, and pharmaceutical applications.

## 12.3.1 Synthesis of Chitosan Derivatives

Chitosan is a naturally occurring polysaccharide composed of $\beta$-(1-4)-2-amino-2-deoxy-D-glucose and 2-acetamido-2-deoxy-D-glucose units [35]. It is soluble in dilute aqueous acids such as hydrochloric acid, nitric acid, phosphoric acid, formic acid, acetic acid, and 100% citric acid [36]. It is insoluble in neutral water and common organic solvents [36]. The rigid crystalline domains, formed by intra- and/or intermolecular hydrogen bonding on chitosan structure, are reported to be responsible for chitosan's low solubility in near-neutral aqueous solution [37]. Chitosan has amino functional groups and also primary and secondary hydroxyl groups on its structure. The amino groups are considered the most reactive functional groups.

The amino groups of chitosan get protonated at acidic pH and make the chitosan soluble, but chitosan precipitates when the amino groups are neutralized. The growing interest in the modification of chitosan with improved solubility has increased its potential for new applications in the pharmaceutical and cosmetic fields [38, 38a]. The synthesis of several water-soluble chitosan derivatives expanded their use further into cosmetics and toiletries. Lang and Clausen [39] noted that for using chitin or chitosan for cosmetics, water-soluble compounds should be synthesized and compatible with anionic detergents. They described several processes for making water-soluble chitin and chitosan-based compounds. Figure 12.4 shows the possibility of chemical modification of chitosan using functional groups to improve water solubility. These modifications include attaching another functional group to $NH_2$ or replacing OH of the $CH_2OH$ group of chitosan with another functional group.

## 12.3.2 Skincare Cosmetics

Chitosan can be chemically modified to improve its solubility in aqueous solutions. When chitosan is chemically modified, many useful derivatives of chitosan are obtained [40]. Water-soluble chitosan is prepared by substituting hydrophilic groups onto chitosan and is readily soluble with various compounds in an aqueous solution [41]. The chitosan derivatives, based on their water solubility, hold a great promise in the field of cosmetics [14]. Several techniques are used to modify chitosan. These

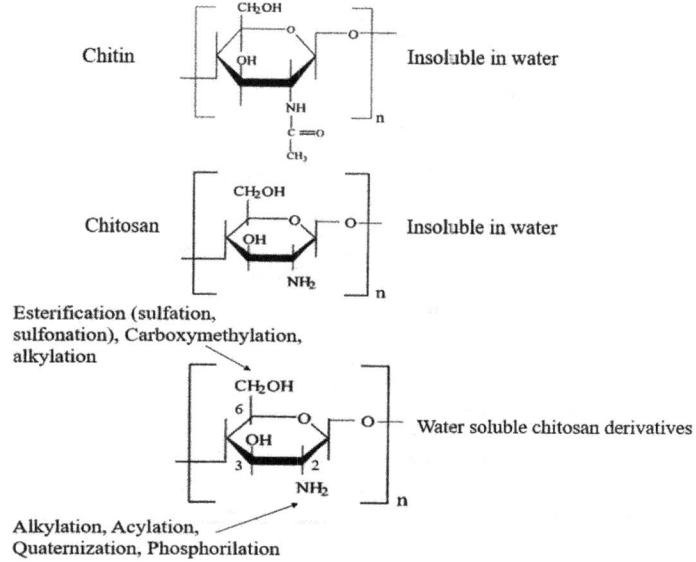

**Fig. 12.4** Chitin and chitosan and their soluble derivatives for use in pharmaceutical and cosmetic fields. Copyright © 2020 Wang, W., Meng, Q., Li, Q., et al. Licensee MDPI, Basel, Switzerland. Creative Commons Attribution (CC BY) license. Int. J. Mol. Sci. 2020, 21, 487; https://doi.org/10.3390/ijms21020487. Adapted from Ref. [38a]

techniques include acylation, alkylation, sulfation, hydroxylation, quaternization, esterification, graft copolymerization, and etherification [42]. Chitosan derivatives prepared from these modification processes are biocompatible, biodegradable, and non-toxic. Succinyl chitosan, carboxymethyl chitosan, N-sulfofurfuryl chitosan, N-trimethyl acetate chitosan, and mercapto-chitins are the typical example of a few chemically modified water-soluble chitosan derivatives [43].

It is worth noting that chitin and chitosan and their derivatives, such as N-Carboxybutyl chitosan and glycol chitosan, have properties similar to hyaluronic acid [44, 45] as far as their chemical and biochemical structures are concerned. As an excellent hydrating agent, chitin supplies water and avoids dehydration. The moisture retention ability of pure chitosan depends on its molecular weight [46]. An alkyl chain's introduction onto water-soluble chitosan opens a new perspective in the cosmetic field [31, 47]. The moisturizing effect of n-succinyl-chitosan derivative in cosmetic cream performed better than hyaluronic acid [48]. It is also compatible with anionic and non-ionic surfactants present in cosmetics. The N-Carboxymethyl chitosan, the carboxymethylated soluble form of chitosan, was first developed by Muzzarelli [49]. The chemical structure of carboxymethyl chitosan derivatives prepared by the direct alkylation of different reactive functional groups of chitosan at different conditions is shown in Fig. 12.5 [14, 36].

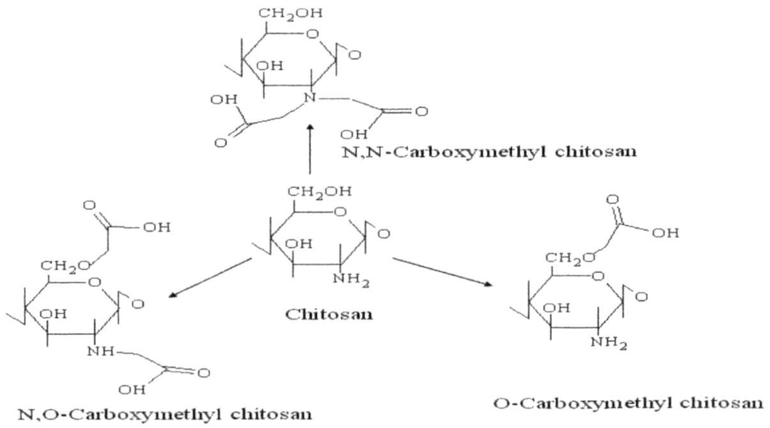

Fig. 12.5 Chemical structure of carboxymethyl derivatives of chitosan. Copyright © 2017 Elsevier B. V. All rights reserved. Materials Science and Engineering C 77 (2017) 1349–1362. Reproduced from Ref. [14] with permission

Chitosan possesses an amino group ($NH_2$) at the $C_2$ position and the primary and secondary hydroxyl (OH) groups at its structure's $C_6$ and $C_3$ positions. Fonseca-Santos and Chorilli [14] reported that introducing a carboxymethyl group at the C6 position creates an active site for moisture absorption and retention. However, carboxymethylation at the $C_3$–OH or $C_2$–$NH_2$ positions is not essential. Carboxymethyl-based chitosan derivatives are used in cosmetics due to their film-forming, thickening, and emulsion stabilization properties [43, 49–51]. Chaiwong et al. [47] have investigated the effect of molecular weight of chitosan on the antioxidant and moisturizing properties of carboxymethyl chitosan (CMC). The CMC prepared from high molecular weight chitosan (310–375 kDa) is termed as CMC-H showed good water solubility (89%) and high viscosity. Tzaneva et al. [38] reported that the viscosity of 0.25% carboxymethyl chitosan is almost equal to that of hyaluronic acid (HA). As far as the long-lasting hydrating effect on human skin is concerned, the application of carboxymethyl-chitin(chitosan) is reported to be superior to hyaluronic acid (HA) as an ingredient in cosmetic and clinical medicine [38, 46, 50].

Uses of chitosan derivatives in cosmetics include glycol chitin ether, carboxymethyl-chitin ether, and chitin sulfate in skin lotions, glycosaminoglycans polymer (obtained through oxidizing chitosan) in skin treatment, and any water-soluble chitosan derivative in soap. A thermosensitive hydrogel containing less than 5% polysaccharide is prepared from chitosan with glycerophosphate salt. The gel has significant use in biomedical and cosmetic fields [52]. Carboxymethyl-hexanoyl chitosan hydrogel showed excellent water absorption and retention ability under neutral conditions, and it can be used as a carrier for delivering amphipathic agents [53] in cosmetics. Vanlerberghe and Sebag prepared chitosan derivatives by reacting

## 12.3 Chitin and Chitosan and Their Derivatives in the Cosmetics

chitosan with maleic acid anhydride [54]. The product was used as a skin moisturizing agent in a cosmetic composition.

Chitosan beads loaded with essential oils are used for cosmetic formulations [55]. Cosmetic composition comprising mesoporous silica with chitosan-lipase conjugates inhibits an excessive secretion of sebum and shows excellent sebum decomposition capacity and mattifying effect [56]. Liping et al. [57] reported chitosan/vitamin C complex (CSVC) as a multifunctional raw material for cosmetics applications. This chitosan derivative has good antioxidant, moisturizing, antibacterial, and film-forming properties. Chitosan containing ascorbic acid and niacinamide increases the cellular growth of fibroblasts. Chitosan with linoleic acid and retinyl palmitate increases skin penetration. These chitosan substances are considered as potential antiaging skincare products [58, 58a]. Casadidio et al. [59] reviewed chitin and chitosan's physicochemical and biological properties. They also reviewed their applications in the cosmetic field. Du and Vuong [10] reported that the antioxidant activity of water-soluble chitosan (WSC) was higher than that of insoluble chitosan. WSC with medium molecular weight (18–90 kDa) has good solubility in water. Chaiwong et al. [47] reported that carboxymethyl chitosan (CMC-L) with low molecular weight (50–190 kDa) has 96% solubility in water. For antioxidant properties, $IC_{50}$ values of DPPH and ABTS radical scavenging activity for CMC-L were 1.70 and 1.37 mg/mL, respectively. Please note that DPPH is the ability of antioxidants to scavenge the 2,2-diphenyl-1-picrylhydrazyl (DPPH) free radical and ABTS is the 2,2-Azino-bis-(3-ethylbenzothiazoline-6-sulfonic acid) (ABTS) radical scavenging activity. Therefore, water-soluble chitosan has the potential to be used as a stabilizer or an antioxidant substance. It is reported that UVR-related oxidation of cellular biomolecules of human skin can be prevented by a prior antioxidant treatment, which depletes endogenous antioxidants simultaneously [9]. Lee et al. [60] reported Chitosonic® acid (i.e., carboxymethyl-hexanoyl chitosan) is a novel chitosan material as a cosmetic ingredient with antibacterial and moderate to good antioxidant activity. Table 12.1 shows the antioxidant activity of Chitosonic® acid compared to arbutin and hyaluronic acid. The antioxidant activity of Chitosonic® acid was tested using the scavenging effect on 2,2-diphenyl-1-picrylhydrazyl (DPPH) radicals.

**Table 12.1** Antioxidant activity of Chitosonic® acid (carboxymethyl-hexanoyl chitosan)

| Antioxidant activity | | |
| --- | --- | --- |
| Ingredient | Free radical[a] | Scavenging rate (%) |
| 1% Arbutin | DPPH | 84 |
| 1% Hyaluronic acid | DPPH | 0 |
| 1% Chitosonic® acid | DPPH | 66 |
| 4% Chitosonic® acid | DPPH | 89 |

Copyright © 2013 Lee, S-M., Liu, K-H., and Liu, Y-Y. Licensee MDPI, Basel, Switzerland. Creative Commons Attribution (CC-BY) license, Materials, 2013, 6, 1391–1402; https://doi.org/10.3390/ma6041391. Reproduced from Ref. [60]

[a] DPPH, 2,2-diphenyl-1-picrylhydrazyl

Microencapsulation techniques are generally used in cosmetics and pharmaceutical applications. Microencapsulation of cosmetic ingredients enhances oxidative ability, thermostability, shelf life, and the ingredients' biological activity [61]. Several research studies have focused on the formulation of chitosan-encapsulated cosmetic ingredients for antiaging [62, 63], skin lightening [64], and skin remodeling and rejuvenating look. Morganti et al. [65] formulated antiaging cosmetic products using chitin nanofibril-hyaluronan (CN-HA) nanoparticles as a carrier for active ingredients. The in vivo studies for the effectiveness and safety of such antiaging cosmetic products were also investigated. Gomaa et al. [66] studied chitosan microparticles incorporated phenyl benzimidazole sulfonic acid (PBSA), a hydrophilic sunscreen agent. Phenyl benzimidazole sulfonic acid (PBSA) is commonly used in cosmetics as a UV filter. It was reported that incorporation of PBSA in chitosan microparticles greatly enhances its performance as a sunscreen product. Chen and Hea [43] reported sunscreen lotions containing water-soluble chitosans exhibit good sun protection, moisture-holding capacitance, and no irritative effects. The sun protection factor (SPF) value of sunscreen lotions containing 0.2% water-soluble chitosans ranged from 5.6 and 7.5 for UVA and 14.5 and 16.2 for UVB [43]. This information will be useful in a wide range of applications in cosmetics. Ito et al. [67] formulated chitin nanofibril-encapsulated UV-absorbing chromophore such as urocanic acid. They evaluated the performance of this chromophore against ultraviolet radiation. It was suggested that the chitin urocanate nanofibers enhanced protective effects against UVB radiation. In another study, Morganti et al. [68] evaluated the property of a pyrrolidonecarboxylic acid (PCA)-chitosan and or a glycochitosan gel (glycolic acid and chitosan)-based cream to increase hydration and to restore the skin lipid content and skin hydration. Use of cream improves skin hydration (+72%, $p < 0.05$), increases skin surface lipids (+38%, $p < 0.05$), and a contemporary decrease in transepidermal water loss (TEWL) ($-37\%, p < 0.05$), and a decrease in the skin redness compared to the placebo. It was also suggested that the chitosan cream decreases inflammation and skin hydroperoxides, alleviating the symptoms of dry skin caused by environmental or pathological reasons. Sakulwech et al. [69] prepared nanoparticles from quaternized cyclodextrin-grafted chitosan associated with hyaluronic acid for cosmetics. The in vitro hydrating ability of these nanoparticles was found to be significantly higher ($P < 0.001$) than that of bulk HA (3.29 ± 0.41 and 1.71 ± 0.05 g water/g sample, respectively). The human skin fibroblasts test shows that these nanoparticles are safe at concentrations range of 0.01 to 0.1 mg/ml. Therefore, the quaternized cyclodextrin-grafted chitosan hyaluronic acid nanoparticles are promising ingredients for cosmetics applications. Many studies confirm using chitosan-based natural antiaging cosmetic products [70, 71]. Rajashree and Rose [70] formulated an antiaging gel by blending three biopolymers, namely collagen (COL) 3%w/v, chitosan (CS) 1.5% w/v, and *aloe vera* (AV) gel 0.21% w/v for natural antiaging cosmetic. It was reported that the prepared AV-blended COL-CS gel helps in the regeneration and rejuvenation of the skin.

Chitosan derivatives used as a face cleanser [72] improve skin hydration and elasticity of subjects affected by skin xerosis due to environmental or pathological reasons [73]. For example, chitosan argininamide is the only chitosan derivative identified, at present, as a skin cleaner [74]. Retinol-encapsulated low molecular water-soluble chitosan nanoparticles are used in the cosmetic formulations [75]. Formulation of various chitosan-based cosmetics has been patented by Lang et al. [76]. They have used desalted water (92.94%), sodium cetyl stearyl sulfate (1%), 10% lactic acid (0.76%), Vaseline (1%) wool fatty alcohol (adeps lanae) (1%), stearyl alcohol (3%), and quaternary chitosan derivatives (0.3%) as composition of a typical chitosan-based skin cream [76].

The cutaneous pathogen *acne vulgaris* is often associated with bacteria such as *Propionibacterium acnes, S. epidermidis, Staphylococcus aureus*, and *P. aeruginosa* [18, 77]. Benzoyl peroxide, tetracycline, erythromycin, and lincomycin are commercial antibiotics commonly used to treat acne vulgaris. Most commercial antibiotics are very useful in treating acne vulgaris. However, they have significant side effects [17, 18]. Chitosan is drawing attention to treatment of acne vulgaris due to its anti-inflammatory and antibacterial properties. The application of chitosan or chitosan derivatives to treat these cutaneous pathogens appears to be less toxic to human cells than commercial antibiotics [16–18]. It is reported that the chitosan's antimicrobial activity against cutaneous pathogens depends on the concentration of chitosan [17]. Table 12.2 shows the minimum inhibitory concentration (MIC) of commercial antibiotics and unmodified chitosan for typical cutaneous pathogens. Table 12.2 shows that *P. acnes* have a higher antibiotic resistance than other strains [18].

It is reported that chitosan combined with antibiotics has a superior antibacterial effect on cutaneous pathogens than chitosan and antibiotics alone [16, 18]. Table 12.2 shows that the chitosan MIC was in the range of 16–512 $\mu$g/mL against acne-related bacteria. Compared to chitosan, the MIC value for the conjugate chitosan-caffeic acid (CCA) was found to range from 8 to 256 $\mu$g/mL against acne-related bacteria [18]. The combination of CCA with antibiotics such as erythromycin, lincomycin, and tetracycline further reduces the MIC value against the acne vulgaris-related bacteria. In another study, Kim et al. [16] noted that oligochitosan (10 kDa) has MIC values on *P. acnes* ranging from 32 to 63 $\mu$g/mL. The combination of tetracycline

**Table 12.2** Minimum inhibitory concentration (MIC) of commercial antibiotics and unmodified chitosan against acne-related bacteria

| Strain | MIC($\mu$g/mL) | | | |
| --- | --- | --- | --- | --- |
| | Erythromycin | Lincomycin | Tetracycline | Chitosan |
| *S. aureus* KCTC 1927 | 2 | 4 | 0.5 | 16 |
| *S. epidermidis* KCTC 1370 | 0.125 | 0.25 | 2 | 64 |
| *P. aeruginosa* KCTC 1637 | 16 | 64 | 0.125 | 32 |
| *P. acnes* KCTC 3314 | 1024 | 1024 | 32 | 512 |
| *P. acnes* isolate 2874 | 1024 | 1024 | 16 | 512 |
| *P. acnes* isolate 2875 | 0.125 | 1 | 0.125 | 256 |

Copyright © 2017 Kim, J-H., Yu, D., Eom, S-H., et al. Licensee MDPI, Basel, Switzerland. Creative Commons Attribution (CC BY) license, Mar. Drugs, 2017, 15, 167; https://doi.org/10.3390/md15060167. Reproduced from Ref. [18]

or erythromycin with oligochitosan (10 kDa) resulted in an antibacterial synergy against *P. acnes*. Friedman et al. [78] reported the benzoyl peroxide encapsulated in the chitosan-alginate nanoparticle demonstrated superior antimicrobial activity against *P. acnes* compared to benzoyl peroxide alone.

## 12.3.3 Hair Care Cosmetics

Hair damage caused by external treatment such as perming, bleaching, and hair dyeing can be minimized by the correct use of hair care products. Hair cosmetics can reduce oxidative damage and increase the hair's tensile strength and hydrophobicity [23]. In addition, hair cosmetics minimize electrical charges and friction of forces [21]. The applications of chitosan have been investigated for hair cosmetics formulation. The film-forming, emulsifier activity, and cationic surface properties of chitosan can be used in hair cosmetics [4]. It is reported that hair care products containing chitosan show superior film-formation ability, compatibility, stiffness, and curl retention compared to synthetic polymers [79].

Synthesis of several water-soluble chitosan derivatives expanded their use into hair care cosmetic products. Several chitosan-based patents have been issued for cosmetics and hair care products [80–82]. These patents describe the preparation of chitosan-based shampoo, conditioner, and hair-setting lotion based on gelling ability, film-forming, and water-soluble properties of chitosan [81, 82]. For example, A US patent by Gross et al. [81] described a process for hair-setting agents using water-soluble chitosan to improve the permanence and shape of the hairstyle. Morganti et al. [83] reported that hair keratin treated by chitin derived compounds retains the right quantity of sebum. Thus, making the hair appear shinier and more flexible compared to the non-treated ones. [83]. The product was used as a skin moisturizing agent in a cosmetic composition. Grollier et al. [84] patented various cosmetic compositions based on amphoteric and cationic polymers. They used these products to develop a hair treatment lotion to prevent the hair from looking oily. A US patent [76] described a process to prepare chitosan-based hair care product. In this process, water (73.8%), isopropanol (25%), 10% formic acid (0.4%), perfume oil (0.2%), and quaternized chitosan derivatives (0.6%) were used as main ingredients to prepare chitosan-based hair straightener.

It is reported that chitosan and chitosan derivatives such as glycerol chitosan, alkyl-hydroxypropyl-substituted chitosan can be used as conditioning agents in cosmetics for hair sprays, hair fixing, and hair conditioning [4, 79]. For example, chitosan glycerol is used as a component in liquid hair strengtheners and hair sprays for its improved solubility and film-forming capacity [45]. Sionkowska et al. [12] outlined a polymer blending process and its use for hair cosmetics. They reported that a blend of chitosan, collagen, and hyaluronic acid exhibits film-forming properties that increase hair thickness. This blend also improves the mechanical properties of hair. The film-forming properties of chitosan/silk fibroin or chitosan/partially hydrolyzed polyacrylamide (HPAM) are suitable for use in the cosmetics field [85].

These chitosan-biopolymer blends improved the cosmetic appearance of damaged hair. Grabska–Zielińska and Sionkowska [86] investigated laser-modified chitosan, hyaluronic acid, and collagen blend for biomedical applications. In another study, Grabska and Sionkowska [87] reported a formulation for hair care conditioner. In a typical hair care conditioner, they have used Steric acid (20%), *Cera alba* (9%), Ceteareth (20.4%), and Argan oil (1%) as an oil phase and glycerol (2%), collagen (0.35%), chitosan (0.3%), and aqua (63%) as water-phase ingredients.

Dandruff affects 50% of the human population. Narshana and Ravikumar [88] reviewed the scalp condition of dandruff. This article discusses the microbial and non-microbial causes of dandruff. Morganti et al. [89] reported shampoo and conditioner formulation enriched with zinc (Zn) and chitin nanofibrils (CN). Zinc influences cysteine-cystine bridge formation. Zinc enhances the reparative activities of CN at the level of the hair cortex. The presence of chitin nanofibril (CN) and Zn in shampoo and conditioner formulation reduces hair flakes. It was further reported that formulation with CN and Zn activates sebum that improves hair shining condition and manageability [89]. Kim et al. [90] describe a process for poly($\gamma$-glutamic acid)/chitosan hydrogel nanoparticles formulation. It was suggested that the hydrogel formulation can deliver hair growth-promoting compounds more efficiently into hair follicles. Therefore, it can be used as a hair growth-promoting product in hair care cosmetics.

Lang et al. [76, 91, 92] synthesized several chitosan derivatives for cosmetic products. Chitosan was reacted with ethylene oxide, propylene oxide, butylene oxide, and glycidol to obtain water-soluble, anionic detergent-compatible products with good film-forming properties. The product may be used in setting lotions, shampoos, and soap. The chitosan derivative obtained by reacting chitosan with glycidyl-trimethyl-ammonium-chloride and ethylene oxide, propylene oxide, and glycidol could be used in nearly all cosmetic products. The chitosan derivative's detergent property is due to the substitution of short-chain non-ionic epoxides by longer chain epoxides. This product was suitable for mousse products for hair setting as foaming and film-forming compounds. Lang et al. [76, 91, 92] modified chitosan by N-hydroxy-propylation of chitosan to prepare N-hydroxy-propyl chitosan. N-hydroxy-propyl chitosan was then reacted with ethyl bromide or ethyl chloride for use in hair sprays. These chitosan-based hair sprays have properties such as solubility in organic solvents, resistance to climatic conditions, and low turbidity points. Acylchitins were found to have excellent application in nail polish. Similarly, sulfopropyl-substituted anionic or ampholytic chitosan prepared by sulfoalkylation of chitosan was suitable for various skin and hair cosmetics. Lang et al. [76, 91, 92] provide the formulation of various cosmetics using chitosan derivatives and their applications in cosmetics, as shown in Fig. 12.6.

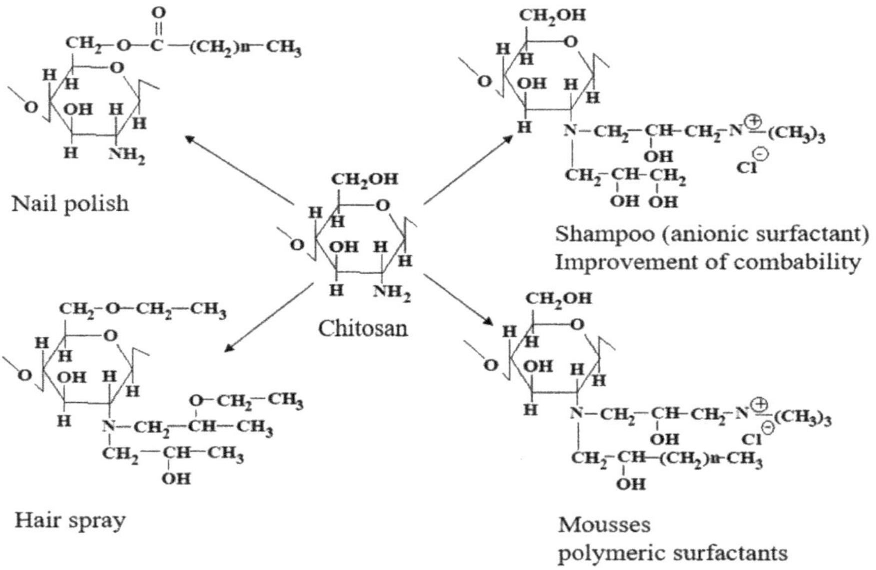

**Fig. 12.6** Chemical formula of chitosan derivatives and their possible uses in cosmeticsz. Copyright © 2020, Springer-Verlag GmbH Germany, part of Springer Nature, Carbohydrates from the sea-chitin, chitosan and algae, Chemistry of Renewables. pp 177–189. Springer, Berlin, Heidelberg. https://doi.org/10.1007/978-3-662-61430-3_9. Reproduced from Ref. [92] with permission

### 12.3.4 Teeth

Chitosan possesses numerous favorable biological properties such as biocompatibility, biodegradability, hydrophilicity, antimicrobial, and anti-inflammatory. The application of chitosan has been explored in various fields of dentistry due to its biological properties. These fields include preventive dentistry, conservative dentistry, endodontic dentistry, periodontology, prosthetic dentistry, orthodontics, and oral surgery [94, 95].

Dental caries, the most prevalent oral infection that can demineralize dental tissue, is caused by organic acids. Dental caries is a pathological process by which organic acids are produced from carbohydrate fermentation by the bacteria such as *S. mutans* in dental plaque [96]. Dental plaque is a dynamic biofilm that can be described as the communities of microorganisms that can survive and colonize onto the dental surface [97]. *Streptococcus mutans* microorganism plays a crucial role in forming biofilm associated with dental caries [28]. The most important biological property of chitosan and its derivatives is the antimicrobial activity that could function through a combination of bacteria cell binding and DNA binding mechanism [94]. It is reported that chitosan has a strong effect against *S. mutans* adherence and biofilm formation and can also disrupt the mature biofilm. Therefore, the role of *S. mutans* in the formation of biofilm on the dental surface can be restricted by the use of chitosan

**Table 12.3** Effect of molecular weight of chitosan and degree of deacetylation on minimum inhibitory concentrations (MIC) against dental pathogens

| Pathogens | Chitosan properties | | MIC (mg/mL) |
|---|---|---|---|
| | MW (kDa) | Deacetylation | |
| S. mutans | 1400 | 0.2 | 0.08 |
| | 1080 | 0.14 | 2.5 |
| | 624 | < 0.25 | 3 |
| | 107.5 | 0.15–0.25 | 5 |
| P. intermedia | 1080 | 0.14 | 2.5 |
| | 624 | < 0.25 | 1 |
| | 107 | 0.15–0.25 | 3 |
| P. buccae | 624 | < 0.25 | 3 |
| | 107 | 0.15–0.25 | 1 |
| T. forsythensis | 624 | < 0.25 | 1 |
| | 107 | 0.15–0.25 | 3 |
| A. actinomycetemcomitans | 1080 | 0.14 | 2.5 |
| | 107 | < 0.25 | 5 |
| | 624 | 0.15–0.25 | 3 |
| P. gingivalis | 1080 | 0.14 | 0.5 |
| | 624 | < 0.25 | 1 |
| | 272 | 0.05 | 3.8 |

Copyright © 2018 Aranaz, I., Acosta, N., Civera, C. et al. Licensee MDPI, Basel, Switzerland. Creative Commons Attribution (CC BY) license, Polymers, 2018, 10, 213, 1–25, https://doi.org/10.3390/polym10020213. Reproduced from Ref. [4]

[76, 98]. Therefore, it can be used to reduce dental plaque formation by inhibiting the growth of pathogens. Chitosan with high molecular weight inhibits bacterial growth more effectively than chitosan with a lower molecular weight. Table 12.3 shows the effect of chitosan's molecular weight on the MIC value of different oral pathogens.

Costa et al. [98] studied chitosan's effect against five periodontal pathogens in biofilm formation in oral cavities. These pathogens were *Porphyromonas gingivalis, Prevotella intermedia, Prevotella buccae, Tanerella forsythensis,* and *Aggregatibacter actinomycetemcomitans.* The prevention of undesirable microbiological colonization following dental treatments was also investigated. The minimum inhibitory concentration (MIC) effect of chitosan on biofilm formation by different oral pathogens is presented in Table 12.4.

Table 12.4 shows that both high molecular weight (HMW) and low molecular weight (LMW) of chitosan with even 1/2 and 1/4 amount of (MIC) value were capable of inhibiting biofilm formation [99]. However, it was suggested that further studies are required to better understand chitosan's activity upon oral pathogen biofilm formation [99].

**Table 12.4** Chitosan subMIC tested for each bacterium (mg/mL) and its inhibitory effect

| Pathogen | Control | HMW (mg/mL) | | | LMW (mg/mL) | | |
|---|---|---|---|---|---|---|---|
| | | MIC | 1/2(MIC) | 1/4(MIC) | MIC | 1/2(MIC) | 1/4(MIC) |
| P. gingivalis | +++ | 1 | 0 | 0 | 1 | 0 | 0 |
| T. forsythensis | +++ | 1 | ++ | ++ | 3 | 0 | ++ |
| P. buccae | +++ | 3 | 0 | + | 1 | 0 | ++ |
| A. aectinomycetemcomitans | +++ | 5 | 0 | 0 | 3 | ++ | ++ |
| P. intermedia | +++ | 1 | ++ | ++ | 3 | ++ | ++ |

Copyright © 2014 Pintado et al. Creative Commons Attribution (CC BY) license, SOJ Microbiol Infect Dis 2(1): 1–6. https://doi.org/10.15226/sojmid.2014.00114. Reproduced from Ref. 98; Copyright © 2012 Elsevier Ltd., Anaerobe, 2012, 18, 305–309. Reproduced from Ref. [99] with permission
0 No biofilm formation
+ Weak biofilm formation,
++ Moderate biofilm formation,
++++ Strong biofilm formation,
*Control* Positive control

Water-soluble chitosan can be effectively used to protect the teeth and prevent demineralization caused by the caries process. Water-soluble chitosan and its derivatives are also useful as dietary supplements and in dental procedures [100, 101]. Chen and Chung [102] reported that the acidogenic *Lactobacilli* spp. and *S. mutans* are the pathogens of dental caries. These pathogens occur in both superficial and deep caries. They observed that the maximal antibacterial activity of water-soluble chitosan occurred at 37 °C. Only five seconds of contact between water-soluble chitosan and oral bacteria attained at least 99.60% antibacterial activity at 500 µg/mL concentration. Abedian et al. [28] stated that chitosan could exert higher antibacterial activities against S. mutans (MIC 0.62 mg/mL) than dental caries pathogen *S. sorbinus* (MIC 1.25). The application of chitosan as a promising antimicrobial agent for mouthwash and toothpaste is evolving [101–104]. For instance, a solution with even 0.2% chitosan was reported to disinfect the root canal [104]. Table 12.5 shows the application of chitosan derivatives in oral health care.

Several studies reported chitosan hydrogel-based compounds for anticaries treatment [27, 105–107]. Ren et al. [27] studied chitosan hydrogel containing amelogenin-derived peptide (CS-QP5 hydrogel) for anticaries treatment. It was reported that the CS-QP5 hydrogel effectively inhibited the growth of the *S. mutans* biofilm, reduced lactic acid production, and decreased metabolic activity over a prolonged time. This hydrogel can promote remineralization of the initial enamel caries even in a biofilm model over a prolonged time. Ruan et al. [105] stated that Amelogenin-chitosan (CS-AMEL) hydrogel could prevent, restore, and treat defective dental enamel. In another attempt, Mukherji et al. [107] investigated the repairing of tooth enamel using leucine-rich amelogenin peptide (LARP)-chitosan hydrogel. Figure 12.7 shows a scanning electron microscopic (SEM) image for a demineralized tooth slice treated

12.3 Chitin and Chitosan and Their Derivatives in the Cosmetics

with amelogenin-chitosan (CS-AMEL) hydrogel. It was suggested that the chitosan-peptide hydrogel system promoted faster mineral induction onto the demineralized tooth surface. In addition, his hydrogel system resulted in organized enamel-like apatite crystals.

**Table 12.5** Use of chitin and chitosan derivatives in oral health care

| Polymer derivative | Bacterial strain | Effect |
|---|---|---|
| Ethylene glycol chitin | S. mutans<br>S. sanguis<br>S. mitis | Reduce bacterial adsorption on S-HA in vitro<br>Dose-dependent effect<br>Better activity on S. mutans |
| Carboxymethyl-chitin | S. mutans<br>S. sanguis<br>S. mitis | Reduce bacterial adsorption on S-HA in vitro<br>Dose-dependent effect<br>Better activity on S. mutans |
| N-Carboxymethyl chitosan | S. mutans | Prevent bacterial adsorption to HA in vitro |
|  | S. sanguis, S. gordonii, S. constellatus, S. anginosus, S. intermedius, S. oralis, S. salivarius, S. vestibularis | Adsorption reduction on HA and S-HA (60%–98%) in vitro |
| Imidazolyl chitosan | S. mutans | Prevent bacterial adsorption to HA in vitro |
|  | S. sanguis, S. gordonii, S. constellatus, S. anginosus, S. intermedius, S. oralis, S. salivarius, S. vestibularis | No effect on bacterial adhesion to HA or S-HA in vitro |
| Sulfated chitosan | S. mutans<br>S. sanguis<br>S. mitis | Reduce bacterial adsorption on S-HA in vitro<br>Dose-dependent effect<br>Better activity on S. mutans |
| Phosphorylated chitosan | S. mutans<br>S. sanguis<br>S. mitis | Reduce bacterial adsorption on S-HA in vitro<br>Dose-dependent effect. Better activity on S. mutans<br>Plaque reduction<br>Slight plaque buffering capacity |
| N-1hydroxy 3 trimethyl ammonium chitosan HCl | P. gingivalis<br>P. intermedia<br>A. actinomycetemcomitans<br>S. mutans | Antibacterial activity in vitro<br>MIC: 0.5–1 mg/mL |
| Glucosamine Maillard chitosan derivative | S. mutans<br>L. brevis | CBM S. mutans 0.4 mg/mL<br>CBM L. brevis 0.5 mg/mL<br>No cytotoxicity in vivo |

(continued)

**Table 12.5** (continued)

| Polymer derivative | Bacterial strain | Effect |
|---|---|---|
| Water-soluble reduced chitosan | S. mutans<br>S. sanguinis | MIC S mutans 1.25 mg/mL<br>MIC S sanguinis 10 mg/mL<br>Reduction plaque index<br>Reduction vital fluorescence |

Copyright © 2018 Aranaz, I., Acosta, N., Civera, C. et al. Licensee MDPI, Basel, Switzerland. Creative Commons Attribution (CC BY) license, Polymers, 2018, 10, 213, 1–25, https://doi.org/10.3390/polym10020213. Reproduced from Ref. [4]
*MIC* Minimum inhibitory concentration
*MBC* Minimum bactericidal concentration
*HA* Hydroxyapatite
*S-HA* Saliva-coated hydroxyapatite

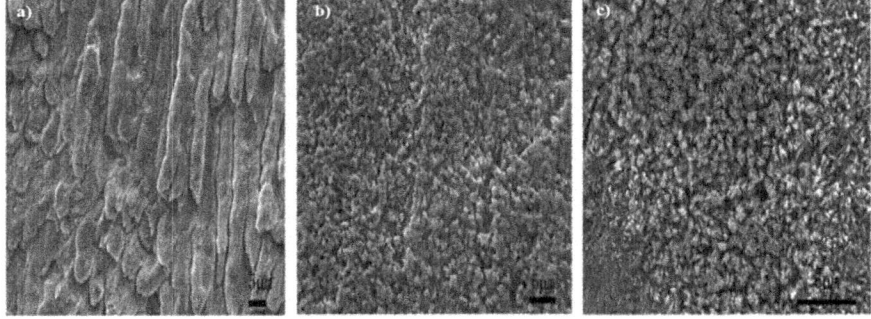

**Fig. 12.7** SEM image of **a** demineralized enamel from the tooth surface. The newly grown layer was treated with two control groups in artificial saliva for three days **b** chitosan only and **c** CS-AMEL. Copyright © Materials Research Society 2016, J. Mater. Res., 2016, 31(5), 556–563, Reproduced from Ref. [107] with permission

The effect of chitosan nanoparticles as an antimicrobial agent has also been investigated against microorganisms associated with dental caries. Ikono et al. [108] reported that nanochitosan exhibited antimicrobial activity by decreasing the survival rate of microbial cells against *S. mutans* and *Candida albicans*. In endodontics, persistent pulpal inflammation is usually associated with *Enterococcus faecalis* microorganisms. The application of sodium hypochlorite is widely used to irrigate the root canal infected by *E. faecalis,* but this procedure has many disadvantages. For example, a few notable disadvantages associated with sodium hypochlorite are irritation of periapical tissues, unpleasant taste, high toxicity, and burning of surrounding tissues [104]. Ibrahim et al. [109] reported that chitosan nanoparticles could eliminate *E. faecalis,* pathogens present in a planktonic state. Chitosan nanoparticles cause a significant reduction of bacteria in the biofilm state. Table 12.6 shows the typical antimicrobial effect of chitosan nanoparticles on endogenic pathogens.

12.3 Chitin and Chitosan and Their Derivatives in the Cosmetics

Table 12.6 Typical example of antimicrobial effect of chitosan nanoparticles against some endogenic pathogens [109–115]

| Microorganism | Test method | Findings | Reference |
|---|---|---|---|
| *Streptococcus mutans*, *Enterococcus faecalis*, and *Candida albicans* in their planktonic state | The antimicrobial activity of chitosan nanoparticles was evaluated against planktonic cells of *S. mutans*, *E. faecalis*, and *C. albicans*. Test was performed by using broth microdilution technique. The cytotoxicity of chitosan nanoparticles was evaluated using Balb/c 3T3 fibroblast cells with the standard MTT assay | The chitosan nanoparticles with 200 nm average size and 51.7 mV surface charge were evaluated for their antimicrobial activity. Chitosan nanoparticles completely eliminate *S. mutans* and *E. faecalis* in the planktonic state and showed fungistatic activity against *C. albicans*. Significant reduction against *S. mutans* ($p = 0.006$), *E. faecalis* ($p < 0.0001$), and *C. albicans* ($p = 0.004$) in biofilm state is reported. Chitosan nanoparticles shows no cytotoxicity when tested for cytotoxicity using 3T3 cells | 109 |
| *Enterococcus faecalis* and *Streptococcus mutans* and fungicidal activity against *Candida albicans* | The antimicrobial effect of chitosan nanoparticles, silver nanoparticles, and ozonated olive oil ($O_3$-oil) on *S. mutans*, *E. faecalis*, and *C. albicans* was evaluated by routine microbiological techniques such as MIC, MBC, and MFC tests | Chitosan nanoparticles show etter antimicrobial activity against the tested endodontic pathogens when used at its MIC and MBC/MFC compared to both silver nanoparticles and ozonated olive oil. Chitosan nanoparticles inhibit immature single and mixed-species biofilm formations by 97 and 94%, respectively | 110 |

(continued)

**Table 12.6** (continued)

| Microorganism | Test method | Findings | Reference |
|---|---|---|---|
| *Enterococcus faecalis* (*E. faecalis*) associated with endodontic infection [in planktonic and biofilm state] | A colony-forming unit (CFU) assay was used to determine the concentration of residual bacteria after treatment | Chitosan and *N*-(2-hydroxyl) propyl-3-trimethyl ammonium chitosan chloride (HTCC) show the minimum bactericidal concentrations (MBC) 70 and 140 µg/ml, respectively. Chitosan and HTCC have significant antibacterial properties on *E. faecalis* in the planktonic state | 111 |
| Endodontic treatments in the control of *E. faecalis* bacteria | The antibacterial activity increment (AAI) of endodontic sealers containing chitosan nanoparticles and chemical components (chlorhexidine (Chx)) was evaluated against *E. faecalis* by direct contact assays | The combination of chitosan nanoparticles and chlorhexidine (Chx) shows elimination of *E. faecalis* bacteria using [CsNPs-Chx ($p < 0.05$)] | 112 |
| *Enterococcus faecalis* biofilm in root canal | The effectiveness of chitosan-propolis nanoparticle (CPN) was evaluated for *E. faecalis* (*ATCC 29,212*) in human tooth model | Significant reduction of *E. faecalis* with chitosan nanoparticles and chitosan nanoparticles with 2% chlorhexidine (CHX) | 113 |

(continued)

## 12.3 Chitin and Chitosan and Their Derivatives in the Cosmetics

**Table 12.6** (continued)

| Microorganism | Test method | Findings | Reference |
|---|---|---|---|
| Mandibular necrotic premolars | As an irrigate solution, the final flush with 3% chitosan nanoparticles (CNP) was carried out. Samples were collected from root canals before and after canal preparation, then cultured to assess the number of colony-forming units/ml. The data were compared with the irrigate solution with 2% chlorhexidine (CHX), 5.25% sodium hypochlorite (NaOCl), and chitosan nanoparticles incorporated with CHX, respectively | CNPs and CHX/CNPs were significantly more effective than either CHX or NaOCl. There was no significant difference between the irrigates with CNP, CHX, and NaOCl against anaerobic bacteria. All tested irrigates were similarly effective against aerobic bacteria. The irrigate with CNPs and CHX/CNPs was associated with significantly lower post-operative pain levels in the first 24 h after treatment | 114 |
| *Enterococcus faecalis* in planktonic and biofilm state | In the presence of tissue inhibitors, the antibacterial activity of chitosan nanoparticles with anionic photosensitizer (rose bengal) was evaluated under photodynamic therapy | Chitosan-rose bengal nanoparticles (CsRBnp) eliminate bacteria (*Enterococcus faecalis*) in the absence of tissue inhibitors. The presence of tissue inhibitors delayed the antibacterial action of CSRBnp (without photoactivation) significantly ($P < 0.05$) up to 24 h | 115 |
| *Enterococcus faecalis* (ATCC 29,212) in biofilm state at 37 °C | Chitosan nanoparticles with zinc oxide eugenol (ZOE) sealer | Chitosan nanoparticles with zinc oxide eugenol (ZOE) sealer inhibited *Enterococcus faecalis* biofilm formation within the sealer-dentin interface | 115 |

*MIC* Minimum inhibitory concentration
*MBC* Minimum bacterial concentration
*MFC* Minimum fungicidal concentration

Chitosan is a naturally derived polymer with antimicrobial and anti-inflammatory properties. Savitha et al. [106] evaluated the in vivo antibacterial efficacy of 2% chlorhexidine gel (CHX), 2% chitosan (CS) gel, and their combination. This evaluation used the gel combination as an intracanal medicament against *E. faecalis* during the endodontic retreatment procedure. It was reported that 2% CHX with chitosan group showed the highest microbial reduction against *E. faecalis* during retreatment of failed endodontic cases. Arancibia et al. [116] reported that chitosan particles inhibited the growth of periodontal pathogens like *P. gingivalis* and *A. actinomycetemcomitans*. Triclosan is an antibacterial and anti-inflammatory agent currently used in periodontal therapy. Pavez et al. [117] stated that chitosan-triclosan particles could modulate the inflammatory response in gingival fibroblasts. This might be useful in the prevention and treatment of inflammation in periodontal diseases.

## 12.4 Summary

Cosmetic products, ranging from hygienic to improving appearance, have become part of our daily life. Unfortunately, manufacture of cosmetic products uses many chemicals that harm human health in the long run. Hyaluronic acid (HA) is widely used in cosmetics due to its unique moisture retention capability. It influences several cellular functions, such as migration, adhesion, and proliferation. Chitosan is used in cosmetic formulation due to its non-immunogenic and non-toxic properties. It possesses properties of gelling, enhancing emulsion stability, and antimicrobial activities. Chitosan can be chemically modified to improve its solubility in an aqueous solution. Increased solubility further expands its potential use in cosmetics and toiletries. Chitosan possesses numerous desired biological properties such as biocompatibility, biodegradability, hydrophilicity, antioxidant, anti-inflammatory, and antibacterial activities. These biological properties of chitosan pave the way for its use in cosmetics formulation related to antiaging, sunscreen cream, and the treatment of acne vulgaris.

Chitosan is useful for hair cosmetics formulation due to its film-forming, emulsifier activity, and cationic surface properties. Chitosan hydrogel-based hair care formulations can be potential candidates for hair growth-promoting products. Chitosan is useful as an antimicrobial agent in mouthwash and toothpaste to control dental plaque. Chitosan is also very useful in preventing and treating of inflammation in periodontal disease. The applications of chitosan in cosmetics are growing. However, further studies are required to understand better the activity of chitosan in cosmetic formulations and end-uses.

# References

1. Okereke JN, Udebuani AC, Ezeji EU, Obasi KO, Nnoli MC (2015) Possible health implications associated with cosmetics: a review. Sci J Public Health 3(5–1):58–63
2. Pereira JX, Pereira TC (2018) Cosmetics and its health risks. Glob J Med Res B Pharma Drug Discov Toxicol Med 18(2):1–9 (Version 1)
3. Amberg N, Fggarassy C (2019) Green consumer behavior in the cosmetic market. Resources 8(137):1–19. https://doi.org/10.3390/resources8030137
4. Aranaz I, Acosta N, Civera C et al (2018) Cosmetics and cosmeceutical applications of chitin, chitosan and their derivatives. Polymers 10(213):1–25. https://doi.org/10.3390/polym10020213
5. Moynihan P, Petersen PE (2014) Diet, nutrition and the prevention of dental diseases. Public Health Nutr 7(1A):201–226. https://doi.org/10.1079/PHN2003589
6. Nilforoushzadeh MA et al (2018) Skin care and rejuvenation by cosmeceutical facial mask. J Cosmet Dermatol 1–10. https://doi.org/10.1111/jocd12730
7. Miastkowska M, Sikora E (2018) Antiaging properties of plant stem cell extracts. Cosmetics 5:55. https://doi.org/10.3390/cosmetics5040055
8. Mohiuddin AK (2019) An extensive review of cosmetics in use. AJODRR 2(7):1–40
9. Pandel R, Poljsak B, Godic A, Dahmane R (2013) Skin photoaging and the role of antioxidants in its prevention. ISRN Dermatol 2013(Article ID 930164):11. https://doi.org/10.1155/2013/930164
10. Du DX, Vuong BX (2019) Study on preparation of water-soluble chitosan with varying molecular weights and its antioxidant activity. Adv Mater Scie Eng 2019(Article ID 8781013): 8. https://doi.org/10.1155/2019/8781013
11. Brunt EG, Burgess JG (2018) The promise of marine molecules as cosmetic active ingredients. Int J Cosmet Sci 40:1–15
12. Sionkowska A, Kaczmarek B, Michalska M et al (2017) Preparation and characterization of collagen/chitosan/hyaluronic acid thin films for application in hair care cosmetics. Pure Appl Chem 89(12):1829–1839
13. Libio IC, Demori R, Ferrao MF et al (2016) Films based on neutralized chitosan citrate as innovative composition for cosmetic application. Mater Sci Eng, C 67:115–124
14. Fonseca-Santos B, Chorilli M (2017) An overview of carboxymethyl derivatives of chitosan: Their use as biomaterials and drug delivery systems. Mater Sci Eng C 77:1349–1362
15. Schuch AP, Moreno NC, Schuch NJ et al (2017) Sunlight damage to cellular DNA: focus on oxidatively generated lesions. Free Radical Biol Med 107:110–124
16. Kim S-H, Eom S-H, Yu D et al (2017) Oligochitosan as a potential anti-acne vulgaris agent: combined antibacterial effects against *Propionibacteriu acnes*. Food Sci. Biotechnol 26(4):1029–1036
17. Champer J et al (2013) Chitosan against cutaneous pathogens. AMB Express 3:37
18. Kim J-H, Yu D, Eom S-H et al (2017) Synergistic antibacterial effects of chitosan-caffeic acid conjugate against antibiotic-resistant acne-related bacteria. Mar Drugs 15:167. https://doi.org/10.3390/md15060167
19. Cruz CF, Costa C, Gomez AC, Matama T (2016) Human hair and the impact of cosmetic procedures: a review on cleansing and shape-modulating cosmetics. Cosmetics 3:26. https://doi.org/10.3390/cosmetics3030026
20. Lee Y-H, Kim Y-D, Hyun H-J et al (2011) Hair shaft damage from heat and drying time of hair dryer. Ann Dermatol 23(4):455–462
21. Reis MF, Dias G (2015) Hair cosmetics: an overview. Int J Trichol 7(1):2–15
22. Miranda-Vilela AL, Botelho AJ, Muehlemann A (2014) An overview of chemical straightening of human hair: technical aspects, potential risks to hair fiber and health and legal issues. J Cosmet Sci 36:2–11
23. Madani N, Khan K (2013) Hair cosmetics. Indian J Dermatol Venereol Leprol 79:654–667
24. Franca SA, Dario MF, Esteves VB et al (2015) Types of hair dye and their mechanisms of action. Cosmetics 2:110–126. https://doi.org/10.3390/cosmetics2020110

25. Neel EAA, Alijabo A, Strange A et al (2016) Demineralization-remineralization dynamics in teeth and bone. Int J Nanomed 11:4743–4763
26. Klimuszko E, Orywal K, Sierpinska T et al (2018) Evaluation of calcium and magnesium contents in tooth enamel without any pathological changes: in vitro preliminary study. Odontology 106:369–376
27. Ren Q, Li Z, Ding L et al (2018) Anti-biofilm and remineralization effects of chitosan hydrogel containing amelogenin derived peptide on initial caries lesions. Regenerative Biomater 69–76. https://doi.org/10.1093/rb/rby005
28. Abedian Z, Jenabian N, Moghadamnia AA et al (2019) Antibacterial activity of high-molecular-weight and low-molecular-weight chitosan upon oral pathogens. J Conserv Dent 22(2):169–174
29. Chalongsuk R, Sribundit N (2013) Usage of chitosan in Thai pharmaceutical and cosmetics industries. Silpakorn U Sci Tech J 7(1):49–53
30. Choi YS, Lee SB, Hong SR, Lee YM, Song KW, Park MH (2001) Studies on gelatin-based sponges. Part III: a comparative study of cross-linked gelatin/alginate, gelatin/hyaluronate and chitosan/hyaluronate sponges and their application as a wound dressing in full-thickness skin defect of rat. J Mater Sci Mater Med 12:67–73
31. Ramos VM, Rodriguez NM, Rodriguez MS, Heras A, Agullo E (2003) Modified chitosan carrying phosphonic and alkyl groups. Carbohydr Polym 51(4):425–429
32. Huei CR (2002) Functionalities of chitin and chitosan and their applications in cosmetic. Adv Chitin Sci 12:67–73
33. Yu S, Hu J, Pan X, Yao P, Jiang M (2006) Stable and pH-sensitive nanogels prepared by self-assembly of chitosan and ovalbumin. Langmuir 22(6):2754–2759
34. Leelapornpisid P, Leesawat P (2002) Factors affecting physico-chemical properties of chitosan gel. Adv Chitin Sci 5:212–215
35. Hasan S (2005) Development of materials for the removal of metal ions from radioactive and nonradioactive waste streams. PhD thesis, University of Missouri, Columbia, MO
36. Bukzem AL, Signini R, Dos Santos DM et al (2016) Optimization of carboxymethyl chitosan synthesis using responses surface methodology and desirability function. Int J Biol Macromol 85:615–624
37. Ujiang Z, Diah M. Rashid AHA, Halim AS (2011) The development, characterization, and application of water-soluble chitosan. Biotechnol Biopolymers. [Magdy Elnashar (Ed.)]. ISBN 978-953-307-179-4. Available from http:// www.intechopen.com/books/biotechnology-of-biopolymers/the-development-characterization-and application-of-water-soluble-chitosan
38. Tzaneva D, Djivoderova M, Petkova N et al (2017) Rheological properties of the cosmetic gel including carboxymethyl chitosan. J Pharm Sci Res 9(8):1383–1387. (a) Wang W, Meng Q, Li Q et al (2020) Chitosan derivatives and their application in biomedicine. Int J Mol Sci 21:487. https://doi.org/10.3390/ijms2102048
39. Lang G, Clausen T (eds) (1989) The use of chitosan in cosmetics chitin and chitosan: sources, chemistry. Biochem Phys Prop Appl 139–147 (Skjak-Braek G, Anthonsen TSP (eds) Elsevier Sciences Publishers, London, New York)
40. Wu Q-X, Lin D-Q, Yao S-J (2014) Design of chitosan and its water-soluble derivatives-based drug carriers with polyelectrolyte complexes. Mar Drugs 12:6236–6253
41. Kahya N (2019) Water-soluble chitosan derivatives and their biological activities: a review. Polym Sci 5(1:3):1–11
42. Zhao D, Yu S, Sun B et al (2018) Biomedical applications of chitosan and its derivative nanoparticles. Polymers 10:462. https://doi.org/10.3390/polym10040462
43. Chen RH, Heh RS (1999) Sun protection and skin hydration effects and physicochemical properties of sunscreen lotion containing water-soluble chitosans. J Appl Cosmetol 17:56–71
44. dos Santos KSCR, Silva HSRC, Ferreira EI, Bruns RE (2005) 32 factorial design and response surface analysis optimization of N-carboxybutyl chitosan synthesis. Carbohydr Polym 59(1):37–42 (2005)

45. Wang W (2006) A novel hydrogel crosslinked hyaluronan with glycol chitosan. J Mater Sci Mater Med 17:1259–1265
46. Chen L, Du Y, Wu H, Xiao L (2002) Relationship between molecular structure and moisture-retention ability of carboxymethyl chitin and chitosan. J Appl Polym Sci 83(6):1233–1241
47. Chaiwong N, Leelapornpisid P, Jantanasakulwong K et al (2020) Antioxidant and moisturizing properties of carboxymethyl chitosan with different molecular weights. Polymers 12:1445. https://doi.org/10.3390/polym12071445
48. Wang Q, Wang L, Wang A (2005) Hydration and moisturizing property of N-succinyl-chitosan. Riyong Huaxue Gongye 35(4):223–226
49. Muzzarelli R, Cucchiara M, Muzzarelli C (2002) N-Carboxymethyl chitosan in innovative cosmeceutical products. J Appl Cosmetol 20:201–208
50. Jimtaisong A, Saewan N (2014) Utilization of carboxymethyl chitosan in cosmetics. Int J Cosmet Sci 36:12–21
51. Mourya VK, Inamdar NN, Tiwari A (2010) Carboxymethyl chitosan and its applications. Adv Mat Lett 1(1):11–33
52. Crompton KE, Prankerd RJ, Paganin DM, Scott TF, Horne MK, Finkelstein DI, Gross KA, Forsythe JS (2005) Morphology and gelation of thermosensitive chitosan hydrogels. Biophys Chem 117(1):47–53
53. Liu T-Y, Chen S-Y, Lin Y-L, Liu D-M (2006) Synthesis and characterization of amphipathic carboxymethyl-hexanoyl Chitosan hydrogel: water-retention ability and drug encapsulation. Langmuir 22(23):9740–9745
54. Vanlerberghe G, Sebag H (1976) Cosmetic composition for the skin containing a chitosan derivative, US Patent No. 3,953,608
55. Anchisi C, Meloni MC, Maccioni AM (2006) Chitosan beads loaded with essential oils in cosmetic formulations. J Cosmet Sci 57(3):205–214
56. (a) Park CM, Kwon SS, Jeon SH, Nam GW, Kim DH, Kim YJ, Kim HK, Kang HH (2006) Cosmetic composition effective for controlling secretion of sebum comprising mixture powder of mesoporous silica with chitosan-lipase conjugate. Repub Korean Kongkae Taeho Kongbo, KR 2006078339 A 20060705. (b) Kwon SS, Jeon SH, Park CM, Shim MK, Nam GW, Yi SH, Kim DH, Chang IS, Shon JK, Kim JM (2006) Mesoporous silica complex powder containing chitosan-lipase conjugates capable of decomposing sebum without skin stimulation, and manufacturing method thereof. Repub Korean Kongkae Taeho Kongbo, KR 2006067294 A 20060620. Application: KR 2004-105553 20041214
57. Liping L, Kexin L, Huipu D, Jia L, Jie Z (2020) Study on preparation of a chitosan/vitamin C complex and its properties in cosmetics. Nat Prod Commun 15(10):1–9. https://doi.org/10.1177/1934578X20946876
58. Malik S (2001) Antiaging skin care products containing chitosan, ascorbic acid, and niacinamide. U.S. Patent Application (2004), 13 pp. US 2004047827 A1 20040311 Application: US 2002-241406 20020911. (a) Thom E, Gudmundsen O, Wadstein J (2001) The effect of a new skin ointment on skin thickness and elasticity. J Appl Cosmetol 19(2):51–57 (2001)
59. Casadidio C, Peregrina DV, Gigliobianco MR et al (2019) Chitin and chitosans: characteristics, eco-friendly processes, and applications in cosmetic science. Mar Drugs 17:369. https://doi.org/10.3390/md17060369
60. Lee S-M, Liu K-H, Liu Y-Y (2013) Chitosonic® acid as a novel cosmetic ingredient: evaluation of its antimicrobial, antioxidant and hydration activities. Materials 6:1391–1402. https://doi.org/10.3390/ma6041391
61. Bakry AM, Abbas S, Ali B et al (2016) Microencapsulation of oils: a comprehensive review of benefits, techniques, and applications. Compr Rev Food Sci Food Saf 15:143–182
62. Morganti P, Palombo M, Fabrizi G et al (2013) New insights on antiaging activity of chitin nanofibril-hyaluronan block copolymers entrapping active ingredients: in vitro and in vivo study. J Appl Cosmetol 31:1–29
63. Morganti P, Fabrizi G, Palombo M et al (2012) New chitin complexes and their antiaging activity from inside out. J Nutr Health Aging 16(3):242–245

64. Morganti P, Ciotto PD, Carezzi F et al (2014) Skin lightening efficacy of new formulations enhanced by chitin nanoparticles delivery system. Note I. J Appl Cosmetol 32:57–71
65. Morganti P, Palombo M, Tishchenko G et al (2014) Chitin-Hyaluronic nanoparticles: a multifunctional carrier to deliver antiaging active ingredients through the skin. Cosmetics 1:140–158
66. Gomaa YA, El-Khordagui LK, Boraei NA, Darwish IA (2010) Chitosan microparticles incorporating a hydrophilic sunscreen agent. Carbohyd Polym 81:234–242
67. Ikuko I, Yoneda T, Omura Y et al (2015) Protective effect of chitin urocanate nanofibers against ultraviolet radiation. Mar Drugs 13:7463–7475
68. Morganti P, Fabrizi G, Bruno C, James B (2001) A new cosmetic-carrier chitosan based. J Appl Cosmetol 19:83–88
69. Sakulwech S, Lourith N, Ruktanonchai U, Kanlayattanakul M (2018) Preparation and characterization of nanoparticles from quaternized cyclodextrin-grafted chitosan associated with hyaluronic acid for cosmetics. Asian J Pharm Sci 13:498–504
70. Rajashree S, Rose C (2018) Studies on an antiaging formulation prepared using aloe vera blended collagen and chitosan. Int J Pharm Sci Res 9(2):582–588
71. Gopal V, Prakash YG, Rose C, Rajashree S (2015) Cosmeceuticals from pharmaceutical aids. Acta Biomed Sci 2(2):68–72
72. Schneider G, Kroepke R, Kaden W, Winkler G (2001) Face cleansers containing chitosan and ethanol. Eur Pat Appl 14. (EP 1129698 A1 20010905 Application: EP 2001-103169 20010210)
73. Morganti P, Fabrizi G, Guarneri F et al (2000) Environment friendly green chemicals. J Appl Cosmetol 18(2):51–63
74. Valachová K, Šoltés L (2021) Versatile use of chitosan and hyaluronan in medicine. Molecules 26:1195. https://doi.org/10.3390/molecules26041195
75. Kim D-G, Jeong Y-Il, Choi C, Roh S-H, Kang S-K, Jang M-K, Nah J-W (2006) Retinol-encapsulated low molecular water-soluble chitosan nanoparticles. Int J Pharm 319(1–2):130–138
76. Lang G et al (1990) Process for making quaternary chitosan derivatives for cosmetic agents, US Patent, 4,921,949
77. Frade ML, de Annunzio SR, Calixto GMF et al (2018) Assessment of chitosan-based hydrogel and photodynamic inactivation against *Propionibacterium acnes*. Molecules 23:473. https://doi.org/10.3390/molecules23020473
78. Friedman AJ, Phan J, Schairer DO et al (2013) Antimicrobial and anti-inflammatory activity of chitosan-alginate nanoparticles: a targeted therapy for cutaneous pathogens. J Investig Dermatol 133:1231–1239
79. Jimtaisong A, Saewan N (2014, November/December) Use of chitosan and its derivatives in cosmetics. H&PC Today- Household and Personal Care Today, vol 9, no 6
80. Grollier JF et al (1999) Composition for the treatment of keratin fibers, based on amphoteric polymers and cationic polymers, US Patent No. 5,958,392
81. Gross P et al. (1979) Hair setting lotion containing a chitosan derivative, US Patent No. 4,134,412
82. Gross P et al (1980) Hair shampoo and conditioning lotion, US Patent No. 4,202,881
83. Morganti G, Coltelli M-B (2021) Smart and sustainable hair products based on chitin-derived compounds. Cosmetics 8:20. https://doi.org/10.3390/cosmetics8010020
84. Grollier JF et al (1983) Composition for the treatment of keratin fibers, based on amphoteric polymers and cationic polymers, US Patent No. 4,402,977
85. Sionkowska A, Lewandowska K, Planecka A (2014) Biopolymer blends as potential biomaterial and cosmetic materials. Key Eng Mater 583:95–100
86. Grabska-Zielińska S, Sionkowska A (2019) The properties of hair covered by conditioners containing collagen, chitosan and hyaluronic acid. EJMT 3(24):11–17
87. Grabska-Zielińska S, Sionkowska A (2022) Surface property modification of collagen, hyaluronic acid, and chitosan films with the neodymium laser. Polysaccharides 3:178–187. https://doi.org/10.3390/polysaccharides3010008

# References

88. Narshana M, Ravikumar P (2018) An overview of dandruff and novel formulations as a treatment strategy. Int J Pharm Sci Res 9(2):417–431
89. Morganti P, Palombo M, Cardillo A et al (2012) Anti-dandruff and anti-oily efficacy of hair formulations with a repairing and restructuring activity. The positive influence of the Zn-Chitin nanofibrils complexes. J Appl Cosmetol 30:149–159
90. Kim HS, Kwon H-K, Lee DH et al (2019) Poly ((γ-Glutamic Acid)/Chitosan hydrogel nanoparticles for effective preservation and delivery of fermented herbal extract for enlarging hair bulb and enhancing hair growth. Int J Nanomed 14:8409–8419
91. Lang G et al (1988) Cosmetic compositions based upon N-Hydroxypropyl-chitosans, new N-Hydroxylpropyl chitosans, as well as process for the production, US Patent No. 4,780,310
92. Lang G (1985) Cosmetic composition based upon chitosan derivatives, New chitosan derivatives as well as process for the production, US Patent No. 4,528,283
93. Behr A, Seidensticker T (2020) Carbohydrates from the sea-chitin, chitosan and algae, Chemistry of renewables. Springer, Berlin, Heidelberg, pp 177–189. https://doi.org/10.1007/978-3-662-61430-3_9
94. Erpaçal B, Adigüzel O, Cangül S, Acartürk M (2019) A general overview of chitosan and its use in dentistry. Int Biol Biomed J.Winter 5(1):1–11
95. Kmiec M, Pighinell L, Tedesco MF, Silva MM, Reis V (2017) Chitosan-properties and applications in dentistry. Adv Tissue Eng Regen Med 2(4):205–211
96. Irfani NF, Gunawan HA, Amir LR (2018) Effect of chitosan application on the decrease enamel demineralization process in vitro (surface damage test). IOP Conf Ser J Phys Conf Ser 1073:052006. https://doi.org/10.1088/1742-6596/1073/5/052006
97. Carvalho MMSG, Stamford TCM., Santos EP et al (2011) Chitosan as an oral antimicrobial agent. In: Mendez-Villas A (eds) Science against microbial pathogens: communicating current research and technological advances, pp 542–550
98. Costa EM, Silva S, Tavaria FK, Pintado MM (2013) Study of the effects of chitosan upon *Streptococcus mutans* adherence and biofilm formation. Anaerobe 20:27–31
99. Costa EM, Silva S, Pina C, Tavaria FK, Pintado MM (2014) Antimicrobial effect of chitosan against periodontal pathogens biofilms. SOJ Microbial Infect Dis 2(1):1–6
100. Costa EM, Silva S, Pina C, Tavaria FK, Pintado MM (2012) Evaluation and insights into chitosan antimicrobial activity against anaerobic oral pathogens. Anaerobe 18:305–309
101. Visveswaraiah PM, Prasad D, Johnson S (2014) Chitosan a novel way to intervene in enamel demineralization—an invitro study. Int J Curr Microbiol App Sci 3(11):617–627. (a) Costa EM, Silva S, Madureira AR et al (2014) A comprehensive study into the impact of a chitosan mouthwash upon oral microorganism's biofilm formation in vitro. Carbohydr Polym 101:1081–1086
102. Chen C-Y, Chung Y-C (2012) Antibacterial effect of water-soluble chitosan on representative dental pathogens *Streptococcus mutans* and *Lactobacilli brevis*. J Appl Oral Sci 20(6):620–627
103. Davoudi Z, Rabiee M, Houshmand B et al (2018) Development of chitosan/gelatin/keratin composite containing hydrocortisone sodium succinate as a buccal mucoadhesive patch to treat desquamative gingivitis. Drug Dev Ind Pharm 44(1):40–55
104. Jaiswal N, Sinha D-J, Singh U-P et al (2017) Evaluation of antibacterial efficacy of chitosan, chlorhexidine, propolis and sodium hypochlorite on *Enterococcus faecalis* biofilm: an *in vitro* study. J Clin Exp Dent 9(9):1066–1074
105. Ruan Q, Liberman D, Bapat R et al (2016) Efficacy of amelogenin-chitosan hydrogel in biomimetic repair of human enamel in pH-cycling systems. J Biomed Eng Inform 2(1):119–128
106. Savitha A, Srirekha A, Vijay R et al (2019) An in vivo comparative evaluation of antimicrobial efficacy of chitosan, chlorhexidine gluconate gel, and their combination as an intracanal medicament against *Enterococcus Faecalis* in failed endodontic cases using real-time polymerase chain reaction (qPCR). Saudi Dent J 31:360–366
107. Mukherjee J, Ruan Q, Liberman D (2016) Repairing human tooth enamel with leucine-rich amelogenin peptide-chitosan hydrogel. J Mater Res 31(5):556–563

108. Ikono R, Vibrian A, Wibowo I et al (2019) Nanochitosan antimicrobial activity against *Streptococcus mutans* and *Candida albicans* dual-species biofilms. BMC Res Notes 12:383
109. Ibrahim A, Moodley D, Uche C et al (2021) Antimicrobial and cytotoxic activity of electrosprayed chitosan nanoparticles against endodontic pathogens and Balb/c 3T3 fibroblast cells. Sci Rep 11:24487. https://doi.org/10.1038/s41598-021-04322-4
110. Elshinawy MI, Al-Madboly LA, Ghoneim WM, El-Deeb NM (2018) Synergistic effect of newly introduced root canal medicaments; ozonated olive oil and chitosan nanoparticles, against persistent endodontic pathogens. Front Microbiol 9:1371. https://doi.org/10.3389/fmicb.2018.01371
111. Wang N, Ji Y, Zhu Y et al (2020) Antibacterial effect of chitosan and its derivative on *Enterococcus faecalis* associated with endodontic infection. Exp Ther Med 19:3805–3813
112. Loyola-Rodríguez JP, Torres-Méndez F, Espinosa-Cristobal LF et al (2019) Antimicrobial activity of endodontic sealers and medications containing chitosan and silver nanoparticles against *Enterococcus faecalis*. J Appl Biomater Funct Mater 17(3):2280800019851771. https://doi.org/10.1177/2280800019851771
113. Parolia A, Kumar H, Ramamurthy S et al (2020) Effectiveness of chitosan-propolis nanoparticle against *Enterococcus faecalis* biofilms in the root canal. BMC Oral Health 20:339. https://doi.org/10.1186/s12903-020-01330-0
114. Nasr M, Diab A, Roshdy NN, Hussein A (2021) Assessment of antimicrobial efficacy of nano chitosan, chlorhexidine, chlorhexidine/nano chitosan combination versus sodium hypochlorite irrigation in patients with Necrotic Mandibular Premolars: a randomized clinical trial. Open Access Maced J Med Sci 9(D):235–242. https://doi.org/10.3889/oamjms.2021.7070
115. DaSilve L, Finer Y, Friedman S et al (2013) Biofilm formation within the interface of bovine root dentin treated with conjugated chitosan and sealer containing chitosan nanoparticles. J Endod 39(2):249–253. https://doi.org/10.1016/j.joen.2012.11.008
116. Arancibia R, Maturana C, Silva D et al (2013) Effects of chitosan particles in periodontal pathogens and gingival fibroblasts. J Dent Res 92(8):740–745
117. Pavez L, Tobar N, Chacon C et al (2018) Chitosan-triclosan particles modulate inflammatory signaling in gingival fibroblasts. J. Periodont Res 53:232–239

# Index

**A**
Abiotic factors, 307
Abrasion resistance, 323
Absolute binding energy, 238, 239
2-acetamido-2-deoxy-D-glucose units, 382
2-acetamido-2deoxy-β-D-glucopyranose, 31
Acetone, 281, 283
Acetylamino group (-NHCOCH3), 6
Acetylated amine groups, 33
Acetylcholinesterase (AChE), 281
Acetyl content of chitin and chitosan, 93
Acetyl groups, 30, 31
Acid dissociation constant, 35
Acid resistance, 163
Acid treatment, 30
Acne vulgaris, 378
Activation energy, 204
Active binding sites, 73
Activity coefficients, 68, 202
Activity of protonated chitosan, 106
Acylation, 109, 163
Acylchitins, 389
Addition/elimination type reactions in chitosan, 113
Adhesive nature of chitosan nanoparticles, 131
Admolecules, 219
Adsorbate-adsorbent interactions, 183
Adsorbate concentration, 167
Adsorption capacity, 175
Adsorption controlling parameters, 170
Adsorption energy, 182
Adsorption equilibrium, 174
Adsorption in packed beds, 207
Adsorption isotherm, 174
equilibrium adsorption, 167
Adsorption mechanism, 167
Adsorption mechanism of $^{99}$Mo onto resin, 313
Adsorption methods, 167
Adsorption of metal ions, 162
Adsorption potential, 201
Adsorption wavefront, 207
Adsorption zone, 208
Agar, 8
Agglutination of cells, 296
Agglutination of erythrocytes, 296
Agglutination potential, 296
Aggregation of colloidal particles, 162
Alcohol permeability, 373
Aldol reaction, 106
Alginate, 13
Alginate-chitosan hybrid gel beads, 164
Alkaline method, 30
Alkylation of chitosan, 109
A-chitin, 22, 79
A-chitin, β-chitin, and γ-chitin structure, 81
Amide I, 90
Amide I band, 90
Amide II, 90
Amine groups, 170
Amine (-NH$_2$) groups, 36
Amino acid, 28
2-amino-3-deoxy-β-D-glycopyronase, 31
Ammonia, 283
Amount of metal ion adsorbed per unit mass of adsorbent, 169
Amperometric enzyme sensor, 280
Amperometric glucose biosensors, 254
Amperometric response of GOx/Pt/FGS/chitosan/GCE, 259

Amylopectin, 2
Amylose, 2
Angiogenesis, 294, 296, 303
Anionic crosslinking agent, 103
Anomalous diffusion, 196
Antiaging, 386
Antiaging cosmetic products, 386
Antibacterial effectiveness of chitosan, 305
Anticancer activity of chitosan, 307
Anti-caries treatment, 392
Anticoagulant, 300
Anti-inflammatory, 300
Antimicrobial activities of chitosan, 303, 347
Antimicrobial agents, 330
Antimicrobial durability, 327
Antimicrobial effectiveness of chitosan, 303
Antimicrobial finishing of textiles, 330
Antimicrobial mechanism, 345
Antimicrobial mode of action, 329
Antimicrobial nanomaterials, 331
Antimutagenic effects, 309
Antioxidant, 384
Antioxidant activities of chitosan, 309
Antiparallel, 91
Antiparallel chain, 90
Antitumor activity of chitosan, 307
Antitumor effects, 309
Apparatus for preparing chitosan beads, 143
Aqua-complex formation, 170
Arginine, 301
Arrhenius equation, 204
Arsenic (As), 160
    As(III), 160
    As(V), 160
Ascorbate interference, 259
Ascorbic acid, 257
Ash content, 60
Atmospheric plasma treatments, 333
Atomic Absorption Spectrometer (AAS), 169
Au/Ag/Au/chitosan-graphene oxide sensor, 277
(AuNPs) doped polyelectrolyte, 259
Average Relative Error (ARE) function, 191
Avogadro's number, 181
Axial dispersion, 208
4-azidobenzoic, 104
2,2-Azino-bis-(3-ethylbenzothiazoline-6-sulfonic acid) (ABTS), 385
Azure dye (AZU), 271

# B

Bacterial endotoxin, 41
Bacterial growth inhibition, 330
Bacterial infectious disease, 381
Bacterial proliferation, 301, 303
Bacterial protease, 28
Bacterial proteinase, 28
Bactericidal activity, 82
Batch adsorption process, 168
Beta-glucan, 12
B-(1-4)-2-amino-2-deoxy-D-glucose, 382
β-(1-4-bonds, 28, 32
B-chitin, 22, 79
B-chitin, functional biopolymer, 83
B-chitin has weak intermolecular forces, 82
B-nicotinamide adenine dinucleotide (NADH), 267, 271
BET equation, 189
Bidentate, 224
Bilayer wound dressing, 334
Bile acids, 350
Binding energy, 218
Binding energy shift, 218
Binding sites, 73
Binuclear, 224
Binuclear complexes, 165
Bio-absorbability, 115
Bioactive agents, 8
Bioactive compounds, 300
Bioactive properties of chitosan, 306
Bioburden, 41
Biofilm, 390
Biogenic elicitors, 342
Biological activity, 69
Biological and structural properties, 33
Biological (enzymatic), 25
Biological Oxygen Demand (BOD), 324
Biological properties, 40, 294
Biomedical sensing technologies, 249
Biomedical sensors, 249
Biopesticide, 341
Biosensor using chitosan and Prussian blue, 254
Bipolar plate, 365
Blends, 121
Blight and harmful insects, 350
Blood coagulation, 295
Blood Coagulation Time (BCT), 296
Blood glucose metabolism, 351
Blood platelets, 293
Bond, 36
Bone cancer
    Osteosarcoma, 310

Bone matrix, 302
Bone matrix protein, 302
Bone regeneration, 302
    bone tissue engineering, 302
Bound water molecules, 117
Bragg's law, 86
Breakthrough adsorption capacity, 206
Breakthrough curve, 168
Bridge models, 211
Brunauer-Emmett-Teller (BET) method, 66
Brunauer, Emmett, and Teller isotherm
    BET isotherm, 174
Brunauer's classification of adsorption
    isotherms, 168
Buckled chain structure, 80
Burns management, 294
Burn wound, 301
Burn wound treatment, 334
1, 2, 3, 4-Butane tetracarboxylic acid
    (BTCA), 327


C

Cadmium (Cd), 158
    Cd(II)), 158
Capillary condensation, 174
Carbaryl insecticide, 281
Carbofuran, 339
Carbon nanotubes/chitosan, 259
Carbon nanotubes (CNT), 260
Carbonyl groups, 90
Carboxylate ions, 111
Carboxymethyl-based chitosan derivatives, 384
Carboxymethyl derivatives of chitosan, 382, 384
Carboxymethyl groups, 110
Carboxymethyl-hexanoyl chitosan hydrogel, 384
Cardiovascular disease, 309
Carrageenan, 13
Catalase and peroxidase enzyme activities, 341
Catalyst layer, 361
Catalytic activity, 257
Cathode depolarization, 369
Cathode polarization, 369
Cathodic response of the chitosan-biosensor, 257
Cationic network, 382
Cationic polyelectrolytes with permanent charges, 116
Cell death, 310

Cell membrane, 351
Cell performance, 366
Cellular DNA, 343
Cellular entry process, 131
Cellular functions, 381
Cellular growth of fibroblasts, 385
Cellular lysis, 306
Cellular metabolism, 300
Cellulose, 1, 5, 7
Cellulose macromolecules, 324
Cell walls of fungi, 17
Central medulla, 380
Chain conformation, 38
Chain entanglement concentration, 137
Chain repulsion and diffusion of protons, 162
Chain scission reactions, 37, 107
Chelate formation, 211
Chelating agent, 51
Chemical, 25
Chemical adsorption, 172
Chemical bonding, 119
Chemical changes, 87
Chemical crosslinked products of chitosan, 106
Chemical crosslinking, 104
Chemical crosslinking process, 119, 120
Chemical formula of chitosan derivatives, 390
Chemical formula of PNIPAm, 125
Chemical modification of chitosan, 105, 109
Chemical Oxygen Demand (COD), 162, 324
Chemical process, 26
Chemical resistance, 380
Chemical shift, 239
Chemical stability, 90
Chemical structure, 93
Chemisorption, 194
Chi-square ($\chi^2$) test, 192
Chitin, 1, 6, 8, 17
Chitin and chitosan markets, 10
Chitin and chitosan properties, 32
Chitinase, 28
Chitin constituent, 17
Chitin content in various sources, 23
Chitin content of crustacean, 24
Chitin decolorization using hypochlorite, 25
Chitin formation, 17
Chitin in fungal cell wall, 17
Chitin nanofibers, 132

Chitin nano-whiskers, 132
Chitin-protein fiber matrix, 131
Chitin synthase, 25
Chitin whiskers, 132
Chitobiose, 79
Chitosan, 9, 369
Chitosan-acrylamide (Ch-Ac) grafting, 326
Chitosan-acrylamide hybrid compound, 326
Chitosan acyl synthesis, 113
Chitosan argininamide, 386
Chitosan-arginine derivatives, 301
Chitosanase, 28
Chitosan-associated *Brassica olearaceae* L, 350
Chitosan-Based Acetone Sensor (CBAS), 283
Chitosan-based electrolyte systems, 371
Chitosan-based hair care product, 388
Chitosan-based Microporous Composite Material (MPCM) resin, 313
Chitosan-based scaffolds, 294, 304
Chitosan-based sensor by layer-by-layer (LBL) method, 261
Chitosan-based skin cream, 387
Chitosan-based transducer, 275
Chitosan-Caffeic Acid (CCA), 387
Chitosan-Carbon Nanotube (CS-CNT), 267
Chitosan-cellulose blend membrane, 274
Chitosan characterization, 51
Chitosan citrate, 327
Chitosan-coated fertilizer, 348, 352
Chitosan-coated perlite bead, 144
Chitosan-cyanuric chloride hybrid (Ch-Cy), 326
Chitosan degradation mechanisms, 37
Chitosan degradation processes, 37
Chitosan derivatives, 110, 111, 383
Chitosan derivatives for cosmetics, 390
Chitosan–ferrocene/graphene oxide/glucose oxidase, 258
Chitosan fibrils, 33
Chitosan Film Sensor (CFS), 283, 284
Chitosan glycerol, 388
Chitosan glycerol-phosphate-based hydrogel, 121
Chitosan gold (chitosan-AuNP) (AuNP), 254
Chitosan-graphite composite biosensor, 270
Chitosan hydrophilic nature, 369
Chitosan–hydroxyapatite (HA), 252
(Chitosan/kappa-carrageenan) complex (PEC), 259

Chitosan is soluble, 107
Chitosan-lipase conjugates, 385
Chitosan market, 10
Chitosan-mediated Inhibition, 344
Chitosan membranes, 251, 256
Chitosan modified gels, 121
Chitosan-montmorillonite nanocomposite, 271
Chitosan-nanocomposite based sensor, 267
Chitosan nanofibers, 131
Chitosan nanoparticle preparation mechanism, 127
Chitosan nanoparticles, 125
Chitosan nanoparticles coated-NPK fertilizers, 348
Chitosan-poly(propylene) imine (CS-PPI), 325
Chitosan/polyvinyl alcohol cladding, 281
Chitosan preparation, 21
Chitosan properties, 42
Chitosan sample preparation for the 1H NMR technique, 95
Chitosan sample shows a diffraction peak, 86
Chitosan scaffold, 252
Chitosan solubility, 58
Chitosan solubility- acids, 107
Chitosan solution viscosity
  viscosity of chitosan solution, 137
Chitosan stability, 41
Chitosan surface area, 66
Chitosan use in textiles, 323
Chitosan uses in cosmetics, 377
Chitosan-UV grafting, 326
Chitosan/Vitamin C Complex (CSVC), 385
Chitosonic® Acid, 385
  carboxymethyl hexanoyl chitosan, 385
Chloroalcohols, 33
Chloroform, 281, 283
Chlorogenic acid sensor, 262
Cholesterol, 350
Chromium (Cr), 159
  Cr(III), 159
  Cr(VI), 159
Chromophore, 71
Classification of polymer hydrogels, 119
Clausius-Clapeyron Equation, 200
Clinical crown, 380
Clinical diagnosis, 258
Clinical medicine, 384
Clot, 293
Coagulant agent, 162
Coagulating/flocculating agent, 162

Coagulating agent
  coagulation agent, 162
Cobalt hexacyanoferrate nanoparticles (CoNP), 260
Coefficient of determination ($R^2$), 192
Co(II), 275
Collagen, 291, 300, 378
Colloidal particles, 162
Colloidal structures, 125
Colon-specific drug, 300
Color strength (K/S value), 324
Combining Lewis base, 114
Commercial sources of chitin, 22
Complex conformational equilibria, 93
Composite film, 251
Composition of crustacean shell wastes, 24
Concentration distribution, 207
Condensation process, 173
Conduction of electrons, 361
Conductivity of graphene, 258
Conductometric titration, 55
CoNP-CNT-CS, 260
Contact time, 172
Contamination of agricultural products, 339
Control bleeding, 293
Controlled release of fertilizer, 346
Coordination bonding, 157
Coordination bonds, 212
Coordination complexes, 239
CO poisoning, 369
Copolymer hydrogels, 119
Copolymers, 121
Copper (Cu), 158
Core binding energies, 72
Correlation coefficients (r), 191
Cortex, 380
Cortical cells, 380
Cosmetic compositions, 388
Cosmetic creams, 378
Cosmetic formulations, 381
Cosmetic products, 377
Cotton leafworm, 345
Coulomb repulsion, 324
Covalent and non-covalent interactions, 71
Crease recovery angle, 327
Critical biological properties, 40
Critical electric field, 134
Critical environmental impact, 323
Critical pH, 381
Critical structures of the resin, 314
Critical temperatures, 120
Crop resistance, 339
Crosslinked chitosan, 105

Crosslinked chitosan and their uses, 108
Crosslinked hydrophilic polymers, 117
Crosslinked rigidity, 105
Cross-linking, 71
Crosslinking agents, 59, 104
Crosslinking density, 117
Crosslinking of chitosan, 163
Crosslinking time, 104
Crustaceans, 6, 21
Crustacean shells, 17, 26
Crystal diffraction pattern, 39
Crystal lattice, 86
Crystalline-frozen water, 66
Crystalline lamellae, 4
Crystalline nanofibrils, 132
Crystalline polymorphic forms, 79
Crystalline structure, 88, 91
Crystalline structure of chitin, 19
Crystallinity index, 86
Crystallinity of $\alpha$-chitin, 82
CS/PB/MWNT, 260
Curvularia Leaf Spot (CLS) disease, 350
Cutaneous pathogens, 378, 387
Cutaneous wound healing process, 294
Cuticle, 380
Cyanine 3 (Cy3)-labeled siRNA, 310
Cyanuric chloride, 274
Cyclic voltammetry and chronoamperometry, 255
Cysteine-based thin membrane, 380
Cytocompatibility, 115
Cytokines, 295
Cytotoxicity, 106


**D**

Dansyl, 278
Davis equation, 68, 202
Deacetylation of chitin, 30
Decolorization agent, 27
Defense enzymes, 341, 343
Defense mechanisms of plants, 341
Degradation mechanisms of chitosan, 37
Degradation of chitosan, 34, 40, 107
Degree of acetylation, 94, 95
Degree of crystallinity, 38
Degree of Deacetylation (DD), 30, 33, 55, 256, 369
Degree of Deacetylation (DD) calculation, 92
Degree of Deacetylation (DD) of chitosan, 52, 55
Degree of deacetylation estimation, 94

Degree of ionization, 57, 170
Degree of ionization of chitosan, 36
Degree of neutralization, 56
Degree of swelling of chitosan, 58
Demineralization, 26, 381
Demineralization step, 60
Demineralized shrimp shells, 29
Dendritic cells, 309
Dental caries, 381, 390
Dental enamel, 392
Dental plaque, 390, 391
Dental plaque formation, 391
Depigmentation, 26
Depolymerization, 28
Deproteinisation, 26
Deproteinization of crustacean shell wastes, 30
Deprotonation of amine groups, 211
Desorption constant, 195
Detection of biomolecules, 280
Dextran, 8
Dextrin, 11
5′,5″-dibromopyrogallolsulfonphthalein, 281
Dichloromethane (DCM), 136
Dielectric constants, 281, 371
Dietary acid, 381
Dietary fiber, 9, 11, 351
Dietary supplements, 392
Differential Scanning Calorimetry (DSC), 87
Diffraction peak of chitosan, 86
Diffuse Reflective UV- vis analysis (DRUV), 71
Diffusion parameters, 198
Digestion tract, 351
1,2-dihydroxybenzene (catechol), 264
Dimensionless heterogeneity factor, 184
Dimensionless separation factor, 179
Dimethoxymethylsilane, 264
Dimethyl adipimidate, 104
Dimethylol Dihydroxy Ethylene Urea (DMDHEU), 327
2,2-diphenyl-1-picrylhydrazyl (DPPH), 385
2,2-diphenyl-1-picrylhydrazyl free radical, 385
Direct Borohydride Fuel Cell (DBFC), 371
Direct Methanol Fuel Cell (DMFC), 357, 366, 371
Disaccharides, 111
Dissociation constant ($p$Ka), 35, 56
Disulfide, 325
Disulfide bonds, 253
DNA hybridization, 280
Dopamine, 272, 280
Dopamine and uric acid, 273
Downy mildew disease Perl millet, 340
(D–R) equation, 201
(D–R) isotherm, 182
Drug loading capacity, 130
Drug Master File (DMF), 43
Drug release capability, 130
Drug release mechanism, 130
Drug-resistant bladder cancer, 310
DSC thermograms of a) chitin, b) chitosan, 89
Dual-network hydrogel, 118
Durable Press (DP) finishing, 327
Dye exhaustion, 324
Dynamic adsorption, 205

E
Easter linkage, 114
Effective adsorption process, 167
Efficiency of DMFCs, 368
Elastic and hydrophobic retractive forces, 120
Electrical conductivity, 369
Electrical free energy, 56
Electric power, 367
Electrocatalysis of the Prussian blue, 255
Electrochemical activity, 69
Electrochemical biosensors, 264
Electrochemical changes, 253
Electrochemiluminescence (ECL), 277
Electrogenerated chemiluminescence (ECL) biosensor, 278
Electrolyte material, 357
Electron-donor groups, 163
Electronegative colloidal particles, 162
Electronic spin state, 73
Electronic transition of chitosan, 71
Electron Spin Resonance (ESR) spectroscopy, 73
Electrons transferred per mole of fuel, 363
Electron-transfer efficiency, 259
Electron transfer mediator, 253
Electron transport, 257
Electrospinning apparatus electrospinning setup, 134
Electrospinning of nanofibers, 140
Electrospinning process, 133, 134
Electrospinning setup, 134
Electrospun nanofibrous scaffolds, 303
Electrostatic attraction, 213

Index 411

Electrostatic interaction, 211
Electrostatic repulsion, 38
Electrostatic repulsive forces, 58
Eliciting property, 341
Elongation at break, 324
Elongation at breakpoint, 71
Elovich equation, 194
Enamel, 380
Enamel demineralization, 381
Enamel-like apatite crystals, 393
Endodontics, 394
Endogenic pathogens, 395
Endoplasmic Reticulum (ER) stress, 310
Endothelial cells, 303
Endothermic peaks, 88
Endotoxic content, 41
Energy conversion efficiency, 363
Energy-Dispersive Spectroscopic (EDS) X-ray microanalysis, 62
Energy Dispersive Spectroscopy (EDS), 232
Energy distribution range, 189
Energy-generating electrochemistry, 369
Engineered biochar, 166
Enthalpy change, 198
Enthalpy factor, 198
Entropy change, 198
Entropy factors, 198
Enzymatic, 32
Enzymatic biosensing of hydrogen peroxide, 261
Enzymatic deproteinization, 28
Enzymatic process, 28, 29, 109
Enzymatic treatment, 325
Enzyme activities, 340
Enzyme treatment, 333
Epidermal regeneration, 296
Epifluorescence microscopy, 265
Epithelialization, 294, 300
Equilibrium adsorption isotherm, 193
Equilibrium and kinetic studies, 163
Equilibrium batch adsorption, 168
Equilibrium isotherms, 173
Error analysis for isotherm studies, 190
Error function Marquardt's Percent Standard Deviation (MPSD), 191
Error functions, 191
Erythrocytes, 294, 295
*Escherichia coli*, 305
Essential nutrients, 346
Essential oils, 385
Esterification reactions, 109
Ethanol content in water, 266

Ethanol in alcoholic beverages, 280
Etherification reactions, 109
Ethylenediamine-Tetra-Acetic Acid (EDTA), 28
Ethylene oxide, 389
Exhaustion of reactive dyes, 325
Exothermic peaks, 88
Experimental parameters, 170
Extra Cellular Matrix (ECM), 291
Extraction efficiency of chitin, 22
Extraction of chitin, 25, 27

**F**
Face cleanser, 386
Faraday constant, 363
Fat absorption, 351
$Fe_3O_4$ incorporated chitosan film, 268
Fecal excretion, 350
Feed additive, 350
Ferredoxin, 283
Ferrocene, 255
Ferrocene-doped silica (FcDS) nanoparticles, 259
Ferrocene groups (Fc), 258
Fertilizers, 346
Fiber diameter, 137
Fiber-optic pH sensor, 274
Fibrin, 293
Fibroblast cells, 300
Fibroblast formation, 296
Fibroblasts and keratinocytes, 296
Fickian diffusion, 196
Film-forming properties, 340
Fingerprint peaks, 98
Finishing agent, 323
Fish tissues, 6
Flavin Adenine Dinucleotide (FAD), 257
Flavin Adenine Dinucleotide (FAD)/carbon nanotube/chitosan system, 259
Flocculating agent, 162
Flow field, 365
Fluorescence emission, 267
Fluorogenic probe, 275
Fluorophore, 279
Foliar uptake, 348
Food industry, 350
Food packaging, 69
Food preservation, 351
Formation of chitin, 20
Fourier-Transform Infrared (FTIR) spectroscopy, 90
Fourier Transform Infrared Spectra, 90

Fractional attainment of the metal ion, 196
Frayed end, 380
Free radicals, 307
Freundlich adsorption isotherm, 163
Freundlich equation, 181
Fructose, 1
Fuel cells, 357
Fuel cells reactions, 359
Fuel cell stack, 365
Fuel crossovers, 369
Functional biopolymer, 83
Functional biosensor using chitosan, 254
Functional food production, 350
Fungal growth, 351

**G**
Γ-chitin, 79
Γ-irradiation, 107
Gapain, 28
Gas distribution layer, 366
Gas separator, 360
Gastrointestinal (GI) tract, 350
Gaussian energy distribution, 182
Gelling rate, 106
Gelling time, 105
Gene expression profiles, 340
Gene therapy, 310
Genetically modified methods, 25
Genetic mutation, 307
Genipin, 122, 303
Gibbs free energy, 198, 363
Gingival fibroblasts, 398
Glass transient temperature, 69, 88
Glass transition temperature ($T_g$), 88
Glass transition temperature (Tg), 69
Glassy Carbon Electrodes (GCE), 254
Gluconic acid, 253
Glucosamine residues, 40
Glucosamine units, 36, 126
Glucose, 1
Glucose biosensors, 253, 254
Glucose microsensor, 255
Glucose oxidase (GOx), 253, 256
Glucose oxidase (GOx)–graphene–chitosan nanocomposites, 257
Glucose oxidase/Pt/functional graphene sheets/chitosan (GOx/Pt/FGS/chitosan), 258
Glucose oxidation, 257
1-4-β-glucosidic bonds, 79
Glutaraldehyde, 104, 126, 251
Glutaric dialdehyde, 260, 267

Glyceraldehyde, 1
Glycidol, 389
Glycidyl-trimethyl-ammonium- chloride, 389
Glycochitosan gel, 386
Glycogen, 1, 4, 7
Glycosidic bonds, 2
Glycosidic (covalent) bonds, 30
Glycosidic linkages, 1, 36, 58
Glycosidic oxygen, 30
Glycosylated metalloporphyrin, 279
Gold-Prussian Blue (Au-PB), 254
Good Manufacturing Practice (GMP), 43
Grafted and ungrafted chitin, 166
Gram-negative and gram-positive bacteria, 300
Graphene nanocomposite, 257
Graphene Oxide (GO), 373
Gravimetric method, 59, 60
Group frequency peaks, 98
Growth factor-b1(TGF-b1), 296
Guided Tissue Regeneration (GTR), 300
Gum and mucilage, 12

**H**
$H_2O_2$, 261
$H_2O_2$ sensors, 268, 269
Hair care conditioner, 389
Hair care products, 388
Hair cosmetics, 380
Hair cosmetics formulation, 388
Hair fibers, 379, 380
Hair flakes, 389
Hair gloss, 380
Hair growth-promoting product, 389
Hair setting agents, 388
Hair shaft, 379
Hair sprays, 389
Hair structure, 379
Hair surface structure, 380
Health of the cuticle, 380
Heat of adsorption, 173, 200, 204
Heat of condensation, 173
Heat of reaction, 173
Heat-storage finishes for textiles, 326
Heavy metals in process wastewaters, 158
Hemicellulose, 11
Hemostasis, 40
Hemostasis mechanism, 296
Hemostasis phase, 293
Hemostatic dressing products, 296
Hemostatic dressings, 295

Hemostatic mechanism, 296
Henderson–Hasselbalch equation, 36
Henry's law equation, 182
Heparin, 300
Herbicide formulation, 348
Hetero-bifunctional reagents, 104
Heterogeneity of the surface, 182, 204
Heterogeneity of the surface site energy, 182
Heterogeneous, 32
Heterogeneous adsorbents, 183
Heterogeneous adsorption systems, 184
Heterogeneous process of deacetylation, 31
Heterogeneous surface, 181
Heteropolysaccharides, 6
Hexafluoroacetone, 33
Hexafluoroacetone, 1,1,1,3,3,3-hexafluoro-2-propanol (HFIP), 135, 136
Hexafluoroisopropanol, 33
Hexanal, 283
High electrical conductivity, 257
High energy density, 358
Homo-bifunctional reagents, 104
Homogeneous, 32
Homogeneous fiber networks, 137
Homogeneous method, 31
Homogeneous process of deacetylation, 32
Homopolymer, 119
Homotactic sites, 189
Human blood samples, 255
Human IgG assay, 271
Humidification, 363
Hyaluronic Acid (HA), 9, 300, 381, 384
HYBRID error function, 191
Hybrid film, 281
Hybrid hydrogel, 121
Hybrid hydrogel (blends), 121
Hydrating agent, 383
Hydro active membrane, 301
Hydrocarbon solvent, 26
Hydrodynamic volume, 52
Hydrogel network structure, 118
Hydrogels, 117, 300
Hydrogel structure, 117
Hydrogen atoms, 358
Hydrogen bonds, 6, 359
Hydrogen fuel cell, 357
Hydrogen gas permeability, 371
Hydrogen Oxidation Reaction (HOR), 361
Hydrogen peroxide, 278
Hydrolysis of acetamide groups, 30
Hydrolysis of uranyl nitrate, 235

Hydrolysis product, 170
Hydrolyze, 28
Hydrolyzed chitosans, 324
Hydrolyzed polyacrylamide (HPAM), 388
Hydronium ions, 359
Hydrophilicity, 36
Hydrophilicity of chitin, 19
Hydrophilic nature, 369
Hydrophilic proton exchange sites, 357
Hydrophilic sites, 117
Hydrophilic sunscreen agent, 386
Hydrophobic, 33
Hydrophobic domains, 357
Hydrophobic interaction, 36
Hydrophobicity, 388
Hydrophobic moieties, 113
Hydroquinone, 271
Hydroxo-complexes, 170
Hydroxyapatite (HAp), 303, 380
Hydroxyl group, 6
Hygienic textiles, 333
Hypodermis, 291
Hysteresis loop, 174

**I**
Immune-modulatory, 307
Immune stimulation effect, 309
Immunoassay, 271
Immunogenicity of DNA vaccination, 310
Inductively Coupled Plasma Mass Spectrometry (ICP-MS), 169
Inflammation phase, 294
Inflammatory milieu, 296
Inhibiting bacterial growth, 330
Inhibitory index, 340
    inhibition effect, 341
Insulin, 310
Interaction forces, 173
Inter- and intra-molecular hydrogen bonds., 110
Intercalated biopolymer, 270
Intermolecular forces of β-chitin, 82
Intermolecular H-bond, 33
Intermolecular hydrogen, 36
Intermolecular hydrogen bond, 33
Intermolecular hydrogen bonding in chitosan, 107
Intermolecular interaction, 51, 222
Internal transport of protons, 357
Interpenetrating Networks (IPNs), 121
Inter-sheet hydrogen bonding, 94
Intramolecular hydrogen bonding in chitosan, 107

Intramuscular administration, 310
Intra-particle diffusion, 163, 195
Intrinsic conductivity, 372
Intrinsic properties of chitosan, 34
Intrinsic viscosity, 38, 52
Inulin, 7, 12
Ion-exchange, 183
Ion exchange capacity, 371
Ion hopping, 371
Ionic channels, 357
Ionic conductivity, 370
Ionic crosslinking, 103
Ionic crosslinking of chitosan with TPP, 130
Ionic gelation, 103, 127
Ionic molecules, 211
Ionic reaction, 324
Ionic strength, 68
Ionotropic gelation, 128
Irradiation of chitosan, 107
IR wavelength of functional groups of chito, 91
Isosteric heat, 200
Isosteric heat of adsorption, 200
Isosteric heat of gas adsorption, 184
Isotherm, 183

**J**
Jovanovic equation, 185

**K**
Katchalsky's equation, 35
Katchalsky–Spitnik equation, 56
Keratinocytes, 291
Kinetic control mechanism, 172
Kinetic energy, 200

**L**
Lab-on-a-Chip type sensors, 249
Lagergren equation, 193
Langmuir equation, 175
Langmuir isotherm, 175
Langmuir two-site model, 228
Langmuir type, 173
Layer-by-layer deposition of chitosan film, 256
Layer-by-layer (LBL), 251
LBL self-assembly technique, 252
L-cysteine., 165
Leading edge, 209
Lead (Pb), 160

Pb(II), 160
Pb(IV), 160
Length of the Unused Bed (LUB), 209
Leucine-Rich Amelogenin Peptide (LARP)-chitosan hydrogel, 392
Levamisole (LEV) assay, 279
Lewis acid-base interaction, 220
Ligand mobility, 162
Limiting adsorption capacity, 173
Limiting behavior, 184
Limiting issues related to DMFC electrodes, 369
Linear regression, 191
Lipid metabolism, 351
Lipids and pigments of crustacean shells, 26
Lipophilic and acidic substances, 351
Liposome, 309
Liquid phase fermentation, 28
*Listeria innocua*, 305
Lithium chloride (LiCl), 135
Lithium ions, 371
Liver, 350
L-lactate in human serum, 280
*Loligo* squid pens, 79
Long-range attractive forces, 201
Loss of cuticle, 380
Lower Critical Solution Temperature (LCST), 124
Low molecular mass, 25
Low-specific activity (n, $\gamma$) $^{99}$Mo, 315
L-type Langmuir isotherm, 181
LUB equilibrium method, 208
Lucifer Yellow VS dye (LYVS), 264
Lyophilized, 58
Lysinoalanine, 28
Lysozyme, 28, 40
Lysozyme adsorption, 166
Lysozyme and N-acetyl-glucosoaminidase, 296

**M**
Macromolecular structure, 69
Macromolecules, 126
Macrophages, 296
Magnetic Resonance Imaging (MRI), 260
Maillard reaction, 109
Malathion, 283
Maleic acid anhydride, 385
Malondialdehyde (MDA), 341
Marine algae, 8
Marine polysaccharides, 13
Mark–Houwink–Sakurada constant for chitosan, 53

Mark–Houwink–Sakurada equation, 52
Mass transfer zone, 207, 209
Mattifying effect, 385
Maturation phase, 294
Maxillofacial reconstruction material, 303
Maximum adsorption capacity, 171
Maximum cell voltage (E), 363
Maximum Contamination Level (MCL), 158–161
Mean free energy of sorption, 182
Mechanical properties of chitin and chitosan, 71
Mechanical properties of chitosan, 71
Mechanical properties of electrospun nanofiber, 137
Mechanism for adsorption of uranium, 239
Medical applications, 291
Medical-grade chitosan, 40
Medical isotope separation, 310
Medulla, 380
Melanin, 378
Melting temperature ($T_m$)
Membrane/electrode assembly, 360
Mercury (Hg), 161
Mercury (Hg), Hg(II), 161
Mesenchymal cell
 stem cell, 303
Metal binding capacity, 162
Methane diphosphonic molecules, 372
Methane phosphonic acid, 372
Methanol permeability, 371
Methylamine, 283
Methyltrimethoxysilane (MTOS), 255
Michaelis-Menten constant, 254
Microbial attack, 329, 381
Microbial demineralization processes, 30
Microbial factors, 344
Microcantilever, 279
Microelectromechanical (MEMS), 280
Microencapsulate, 348
Microfibrils, 6, 131
Microporous adsorbents, 174
Micro-shielding on the resin's critical surface, 314
Microwave-assisted hydrothermal reaction, 273
Microwave method, 25
Minerals, 28
Minimal Inhibitory Concentrations (MIC), 307
Minimum growth Inhibitory Concentrations (MIC), 345

Minimum Inhibitory Concentration (MIC), 329, 387
Mitochondrial damage, 310
Mitogenesis, 296
Modifiable polymers, 109
Modified sol-gel/chitosan composite, 255
Moisture absorption and retention, 384
Moisture adsorption isotherms, 70
Moisture loss, 380
Moisture retention, 381
Moisture retention ability, 383
Moisture sorption isotherm, 69
Moisture sorption onto chitosan, 69
Moisture vaporization, 66
Moisturize the skin, 378
Moisturizing effect, 383
Moisturizing properties, 384
Mold growth, 350
Molecular conformation, 58
Molecular conformation in chitosan, 36
Molecular entanglements, 118
Molecular fingerprint, 90
Molecular recognition, 249
Molecular structure of chitin, 79
Molecular structure of chitin and chitosan, 17
Molecular Weight (MW), 38, 52
Molecular weight of chitosan (Mw), 369
Molecular weight of the adsorbate, 210
Monoclinic crystal symmetrical structure, 82
Monodentate, 226
Monolayer adsorption, 172, 181
Monolayer models, 174
Monosaccharides, 1
Montmorillonite, 270
Mousse products, 389
Mouthwash and toothpaste, 392
Mucoadhesive properties, 126, 307
Mucopolysaccharides, 8, 381
Mucoproteins, 8
Mucosal membrane, 307
Mucosubstance, 8
Multibasic acids, 105
Multifunctional junctions, 118
Multifunctional monomers, 104
Multilayer adsorption, 181
Multilayer models, 189
Multiple layers of the adsorbate molecule, 174
Mycelium biomass fermentation, 165

## N

N-acetyl-D-glucosamine, 79
N-acetyl-D-glucosamine units, 30
N-acetyl glucosamine, 6, 28, 40
$N$-acetyl-glucose-amine of chitosan, 300
N-Acylation, 113
Nafion, 257
Nafion$^{TM}$, 359
Nail and hair care, 377
Nail polish, 389
N-alkylation modification, 110
N-alkylation reaction, 111
N-alkylchitosan, 110
N-alkyl derivatives, 110
Na$^+$-montmorillonite, 270
Nanofibers, 131
Natural carboxylic acids, 127
$N$-carboxymethyl chitosan, 301, 383
$N$-deacetylation, 84
Negative value of entropy, 200
Neovascularization, 302
Neurotoxic, 106
Neurotransmitters, 272
Neutron activation (n, $\gamma$) $^{99}$Mo, 312
N-hydroxy-propyl chitosan, 389
Nickel (Ni), 159
  Ni(II), 159
Nitrate ions, 271
Nitric Oxide (NO), 264
Nitric Oxide (NO) sensor, 262
Nitrogen release, 348
Nitromethane, 265
NMR spectra, 94
N, N'- di- acetylchitobiose, 28
N, N-dimethylacetamide (DMAC), 135
NO$_3^-$ ion, 268
N, O-acylated chitosan, 114
Non-equilibrium adsorption, 208
Non-immunogenic, 381
Non-saline treated plants, 348
Novel Corn Seed Coating Agent (NCSCA), 340
N-(2-Pyridylmethyl) chitosan (PMC), 276
Nuclear Magnetic Resonance (NMR) spectroscopy, 93
Nuclear medicine, 315
Nuclear Regulatory Commissions (NRC), 312
Nucleic acid hybridization, 280
Nucleophilic alkylation, 116
Nucleophilic attack, 105
Nucleophilic-substitution reactions, 114
Nutrient delivery, 348

Nutrient uptake, 340
Nylosan dye, 266

## O

Obesity and hyperlipidemia, 351
O-acylation of chitosan, 113
Ohmic polarization, 369
One-site Langmuir equation, 226
Open-Circuit Voltage (OCV), 372
Operating principle of smart gels, 123
Operating temperature, 359
Oral diseases, 377
Oral health care, 392
Oral pathogens, 391
Organic solvents, 33
Organic solvents for chitosan, 108
Organophosphorus compounds, 283
Organo-solubility of chitosan, 114
Orthorhombic cell, 80
Osteoblasts, 302
Osteoinductiveness, 115
Overall rate of adsorption, 208
Oxalate acid, 277
Oxalate ion, 278
Oxidative degradation of chitosan, 36
Oxidative stress, 307, 378
Oxidative wool pretreatment, 326
Oxoanions, 165
Oxygen Reduction Reaction (ORR), 361

## P

Papain, 28
Paracrystalline, 33
Parallel chain, 91
Parallel molecular structure, 94
Parallel structure, 85
Paraoxon, 265
Particle diffusion, 195
Particles, 162
Particle size and zeta potential, 129
Particle sizes of the chitosan nanoparticles, 128
Pathogenesis-Related (PR) proteins, 341, 343
Pathogenic microorganisms, 344
Pathogens of dental caries, 392
Pectin, 12
Pendant models, 211
Pepsin, 28
Percentage of ash, 60
Percentage of moisture, 59
Periodontal diseases, 398

# Index

Periodontal pathogens, 391, 398
Permeability, 69
Permeability characteristics, 69
Permeation enhancing effect, 307
Permselectivity, 264
Peroxidase (POD), 343
Phagocytosis process, 294
Phase transition, 59
Phase transition of chitosan, 35
PH-dependent drug release, 130
Phenolic compounds, 264, 342
Phenylalanine Ammonia-Lyase (PAL), 343
Phenyl Benzimidazole Sulfonic Acid (PBSA), 386
PH indicators, 274
PH measurement, 273
Phosphonic acid groups, 371
Phosphorus pentoxide, 115
Phosphorylated chitosan, 115, 372
Phosphorylation of chitosan, 115
Phosphotungstic Acid (PTA), 372
Photoelectron spectroscopy, 72
PH-sensitive, 300
Physical adsorption, 172
Physical and chemical cross-linking, 126
Physical and chemical properties of perlite, 142
Physical changes, 87
Physical characteristics, 307
Physical characterization, 51
Physical crosslinking method, 103
Physiological levels, 257
Physiological materials, 40
Physisorption, 183
Phytopathogen, 342
Phytosanitary agent, 339
Pigment residue, 26
Pileor Sheet of chitin chains, 80
Planar amperometric glucose microsensor, 254
Plant defense mechanisms, 342
Plant defense signaling pathways, 343
Plant disease resistance mechanisms, 340
Plant diseases, 341
Plant growth, 340
Plant-pathogen interaction, 344
Plant pathogens, 341
Plasticizer, 69
Plasticizing or swelling, 41
Platelet-Derived Growth Factor-AB (PDGF-AB), 296
Platelets, 294
Point-of-care applications, 258

Point of Zero Charge (PZC), 67, 217
Polanyi potential theory, 201
Polyaminated chitosan, 163
Polycondensation reaction, 104
Polydextrose, 12
Polydispersity, 129
Polyelectrolyte architecture, 265
Polyelectrolyte Multilayer (PEM) thin film, 265
Polyion complex, 115
Polymer blending, 125
Polymer degradation, 41
Polymer Electrolyte Membrane (PEMFC), 357, 366
Polymeric chains of $\alpha$-chitin, 82
Polymeric nanofiber, 133
Polymer jet, 134
Poly (monomethyl itaconate) (PMMI), 267
Poly(N-isopropyl acrylamide) (PNIPAm), 123
Poly perfluorosulfonic acid, 359
Polyphenol Oxidase (PPO), 343
Polypyrrole–Chitosan–Iron oxide (Ppy–CS–Fe3O4) nanocomposite films, 260
Polysaccharides, 1
Polyvinyl Alcohol (PVA), 371
Pore diffusion control, 195
Pore size distribution, 231
Porous electrodes, 362
Positive enthalpy, 200
Positive value of entropy change, 200
Potential field, 201
Potentiometric titration, 55, 67
Poultry and swine nutrition, 350
Power density, 363
Preparation of chitosan, 30, 32
Process for coating chitosan onto perlite, 139
Production of chitin, 22
Projected area of a molecule, 180
Proliferative phase, 294
Proline, 278
Properties of chitosan-based membranes, 370
Properties of nanofibers, 131
Properties of the spinning solution, 134
Propylene oxide, 389
Protease, 28
Protective antioxidants, 378
Protective film, 381
Proteinase, 28
Proteolytic enzymes, 28

Proteolytic microorganisms, 28
Protonation of free amino groups, 59
Protonation of the amine group of chitosan, 114
Proton conduction, 360
Proton conductivity, 369
Proton donor, 94
Proton Exchange Membrane (PEM), 358
Proton movement mechanism, 357
Proton transportation, 358
Prussian blue, 254
Prussian blue layer, 255
Prussian blue (PB) nanoparticles and multiwalled carbon nanotubes (MWNTs) in chitosan, 260
Prussian White (PW), 254
Pseudo-first-order equation, 193
Pseudo-first-order kinetics, 193
*Pseudomonas maltophili*, 28
Pseudo-second-order equation, 194
*Ptinus* beetle, 79
Pulmonary gene delivery, 309
Pyrene (Py) and β-cyclodextrin (β-CD) units, 265
Pyrogallol, 278
Pyrogenicity, 41
Pyrrolidonecarboxylic Acid (PCA)-chitosan, 386
PZC value of pure chitosan, 69

## Q

Quaternary ammonium groups (PRDQA), 372
Quaternary ammonium substituents, 116
Quaternary salts of chitosan, 116
Quaternization, 109
Quaternized chitosan, 116
Quaternized cyclodextrin-grafted chitosan, 386
Quenching efficiency, 265

## R

Radiation dose, 314
Radiation-induced crosslinking, 107
Radiation-induced grafting, 107
Radiation-induced reaction, 107
Radical scavenging activity, 385
Radiolysis, 107
Radiolysis of water molecules, 37
Radiopharmaceutical, 310
Raman excitation frequencies for functional groups, 98

Raman spectra, 97
Raman spectroscopy, 97
Rate-controlling steps, 163, 164
Reaction mechanism for chitosan perlite bead formation, 144
Reactive functional groups, 382, 383
Reactive hydroxyl radicals, 37
Reactive Oxygen Species (ROS), 307, 378
Ready-to-eat foods, 351
Reagentless Chemiluminescence (CL), 278
Redox reaction, 357
Reducing sugar, 30
Reducing water loss, 350
Reductive alkylation of Schiff bases, 109
Re-epithelialization, 295
Regulatory issues, 41
Relative Humidity (RH), 69
Release of $^{99m}$Tc from the column prepared from resin, 313
Remineralization, 381
Remodeling stage, 294
Renewable energy sources, 358
Repulsive electrical forces, 134
Repulsive electrical potential, 162
Repulsive forces, 136
Residual ash, 60
Reverse micellar, 127
Rheological properties, 136
Rhizobacteria, 340
Rigid crystalline domains, 382
Rigid network, 126
Role of bacteria, 381
Root canal, 392

## S

Salinity, 340
Saliva, 381
Salt-free dyeing of cotton fabric, 325
Sample preparation for the $^1$H NMR technique, 94
Scanning Electron Microscopy (SEM) analysis, 61
Scar formation, 296
Schematic diagram of a biochip system, 250
Schiff base, 126
Schiff base reaction, 105, 110
Schizophrenia medicine clozapine, 273
Seaweeds, 8
Secretion of sebum, 385
Seed-coating agent AMULET, 340
Seed-coating technology, 339
Seed germination, 339, 340

Index 419

Seed moisture, 340
Seed plasma membrane's permeability, 340
Selectivity coefficient of the chitosan sensor, 271
Self-assembly, 132
Self-crosslinked hydrogels, 121
Semipermeable film, 340
Sensing capacity for $NO_3^-$ ion, 268
Sensing layer for metal ion detection, 277
Sensitive saliva glucose sensor, 261
Shape of equilibrium isotherms, 208
Shielding concept, 314
Signal processing systems, 249
Signal-to-noise ratio, 256
Silica gel-crosslinked chitosan adsorbents, 165
Silver-selective sensor, 276
Single cell fuel cell, 358
Sips isotherm, 184
Site-specific chemical modifications, 109
Sizing and desizing agent, 323
Skin
  skin structure, 291
Skincare cosmetics, 382
Skin cleaner, 387
Skin damage, 378
Skin fibroblasts, 386
Skin health, 377
Skin hydration, 386
Skin lightening, 386
Skin lipid, 386
Skin moisturizing agent, 385
Skin photoaging process, 378
Skin remodeling, 386
Skin treatment, 377
Smart gels, 119
Smart gel synthesis, 124
Sodium alginate biopolymers, 372
Sodium Hypophosphite (SHP) monohydrate, 327
Sodium Tripolyphosphate (TPP), 127
Solid-phase fermentation, 29
Solid-state $^{15}N$ NMR, 94
Solubility of chitin and chitosan, 35
Solubility of chitosan, 33, 36, 58
Solubilization of chitosan, 36
Solvated chitosan, 114
Sonochemical coating, 333
Sorbate concentration, 172
Sorption mechanism of uranium, 230
Sorption mechanisms, 164, 211
Sorption rate, 195
Sources of chitin, 22

Soybean seeds, 340
Specific power, 363
Spinning solvent, 136
Split ends (trichoptilosis), 380
Spoilage microorganisms, 351
Spontaneous reaction, 198
Squid tendons, 25
Stability of -chitin$\alpha$, 81
*Staphylococcus aureus*, 305
Starch, 1, 2, 7
Steric configuration of OH-groups, 165
Stern-Volmer equation, 265
Stimulatory effect, 296
Stoichiometric wavefront, 209
Stomach environment, 350
Storage polysaccharides, 1, 2
Strain-stress curves of crosslinked PVA/chitosan, 138
Stress-strain behavior, 71
Structural and biological properties of chitosan, 35
Structural homogeneity, 94
Structural marine polysaccharides, 79
Structural polysaccharides, 5
Structural properties, 35
Structure of $\beta$-chitin, 82
Structure of chitosan hydrogel, 122
Sub-critical vapors, 182
Subcutaneous tissues, 291
Sugar rings of chitin, 80
Sugars and proline, 340
Sulfoalkylation of chitosan, 389
Sulfonated chitosan-Calcium Oxide (CaO), 373
Sulfonated Polyethersulfone (SPES), 372
Sulfonic acid ions, 359
Sulfosuccinic Acid (SSA), 251, 372
Sulfated chitosan, 114
Sulfated chitosan synthesis process, 114
Sum of the absolute error (EABS), 191
Sum of the Squares Error (SSE), 191
Sun Protection Factor (SPF), 386
Surface area of chitosan, 67
Surface area of the adsorbent, 181
Surface charge, 170
Surface charge analysis of chitosan, 68
Surface charge of the CFOH bead, 216, 217
Surface charge of the chitosan, 67
Surface chemistry, 170
Surface coverage, 200
Surface functional groups, 170
Surface heterogeneity, 204
Surface morphology, 51

Surface morphology of the nanoparticles, 127
Surface oxyanion interaction, 218
Surface tension, 136
Surface tension of the polymer solution, 134
Swelling behavior of chitosan, 58
Swelling effect, 41
Swelling ratio, 118
Swelling state, 120
Symbiotic relation, 340
Synthesis of mRNA and proteins, 343
Synthetic sizing agent, 323

**T**
$^{99m}$Tc/$^{99}$Mo generator, 312
Technological limitations of DMFC electrodes, 368
Teeth, 380
Temkin equation, 183
Temperature-dependent adsorption mechanism, 182
Tensile properties, 71
Tensile strength, 71, 138, 323, 388
Tetraarylporphyrins, 275
Tetracycline, 300
2,2,6,6-tetramethylpiperidine-1-oxyl (TEMPO), 132
Textile finishing processes, 325
Textiles' functional properties, 330
Textile sizing agent, 324
Theoretical electrical energy, 363
Thermal crosslinking, 103
Thermal degradation, 66
Thermal properties, 88
Thermal stability, 64
Thermal stability of chitosan, 64
Thermodynamically active functional groups, 121
Thermodynamic parameters, 205
Thermogravimetric Analysis (TGA), 64, 232
Thermoregulation, 379
Thermoreversible, 125
Thermosensitive hydrogel, 120, 121, 384
Thermo-sensitive injectable systems, 121
Thiocholine, 282
Thioglycolic acid-capped CdS quantum dots, 265
Thiolated chitosans, 271
Thiolated gold electrodes, 257
Three-dimensional scaffolds, 252

Tightly bound to hydrophilic groups, 66
Tissue engineering, 303, 304
Tissue repair, 296
Top-down techniques, 126
Total Suspended Solids (TSS), 162
Toxic fungicides, 339
Trailing edge, 210
Transepidermal Water Loss (TEWL), 386
Transmission Electron Micrograph (TEM), 64
Transmittance of chitosan materials, 59
Transmittance of chitosan nanoparticles, 59
Transport mechanism, 370
Traumatized tissue, 303
Treat burn wounds, 334
Treatment of acne vulgaris, 387
Trifluoroacetic Acid (TFA) solution, 136
Trimethylamine, 277, 278
Trypsin, 28
Tubeworms, 79
Tuna proteinase, 28
Turbidity, 162
Twisted Intramolecular Charge Transfer (TICT), 279
Two-dimensional van der Walls equation, 189
Tyrosinase, 264

**U**
Ultimate strength, 71
Ultraviolet Radiation (UVR), 378
Units cell of α-chitin, 33
Uric acid, 257
Urocanic acid, 386
UVA (long wave) photons, 378
UVB (short wave) photons, 378
UV radiation crosslinking of chitosan, 107

**V**
Vacuum, 25
Valence shell, 72
Van der Waals forces, 172, 324
Vascularization, 302
Viral infection, 344
Viscoelastic properties, 59
Viscosity-average molecular weight ($M_v$), 52
Viscosity of gel, 382
Viscosity of the chitosan, 38
Voltage drop, 363
Voltage power supply, 133
Volume phase transition, 123

# Index

## W

Water activity ($a_w$), 69
Water build-up, 366
Water management, 366
Water solubility of chitosan, 109
Water-soluble chitosan derivatives, 107, 382, 383
Water-soluble chitosans, 163
Water-soluble natural polysaccharides, 120
Water-soluble synthetic polymers, 120
Weakly surface-bound water, 66
Wearability of fabrics, 329
Weathered hair, 380
Weathering, 380
Weight loss, 66
Weight loss analysis, 66
White blood cells, 294
Wool processing, 325
Wool properties, 333
Working temperature, 368
Wound dressing, 293, 300
Wound exudates, 300
Wound healing, 293, 295
Wound healing process, 291, 293, 296
wound healing steps, 293
Wounds, 291
Wrinkle Recovery Angle (WRA), 327
Wrinkle resistance, 327
Wrinkle resistance property, 327
Wrinkling behavior of cellulose fabrics, 327

## X

X-Ray Diffraction (XRD) pattern of chitin and chitosan, 84
X-ray mapping, 232
X-ray Photoelectron Spectroscopy (XPS) analysis, 72
XRD pattern for chitosan, 39
Xylan, 7

## Y

Young's modulus, 138

## Z

$Zn^{2+}$ and $Cu^{2+}$ detection in $H_2O$, 276

Printed by Books on Demand, Germany